ISOSPIN PHYSICS IN HEAVY-ION COLLISIONS AT INTERMEDIATE ENERGIES

ISOSPIN PHYSICS IN HEAVY-ION COLLISIONS AT INTERMEDIATE ENERGIES

BAO-AN LI AND W. UDO SCHRÖDER

Nova Science Publishers, Inc.
Huntington, New York

Library of Congress Cataloging-in-Publication Data

Isospin physics in heavy-ion collisions at intermediate energies/Bao-An Li and W. Udo Schröder (editors)

p. cm.

Includes bibliographical references and index.

ISBN 1-56072-888-4

1. Isobaric spin. 2. Heavy ion collisions. I. Li, Bao-An. II. Schröder, Wolf-Udo.

QC793.3S6 I78 2001
539.7'5- -dc21

227 Main Street, Suite 100
Huntington, New York 11743
Tele. 631-424-6682 Fax 631-425-5933
e-mail: Novascience@earthlink.net
Web Site: http://www.nexusworld.com/nova

Printed in the United States of America

Contents

Preface

A new chapter is opening up for nuclear science: the exploration of phenomena associated with the neutron-proton asymmetry or charge density degrees of freedom of nuclear matter, "Isospin Physics" for short. The ultimate goal of this emerging field is to obtain a better understanding of the properties of asymmetric nuclear matter, in particular, the isospin-dependence of the nuclear equation of state. Of course, early on in the history of nuclear science, Weizsäcker and his contemporaries already realized the influence of this asymmetry on the nuclear ground-state energy. He and his followers, who introduced refined concepts of a nuclear droplet made of a binary liquid, noticed that asymmetric nuclei with unequal numbers of neutrons (N) and protons (Z) become less stable with the square of this asymmetry, more precisely with $(N\text{-}Z)^2/(N+Z)$, as compared with their $N = Z$ isobars. The scaling constant was found to be a_{sym} $23.5\ MeV$, an energy of substantial magnitude. For heavy nuclei, the Coulomb interaction shifts the maximum nuclear stability slightly to neutron-rich nuclides. This asymmetry dependence of the nuclear ground-state energy was traced back to the Fermionic nature of nucleons and the attractive interactions between unlike nucleons.

Nuclear research in the past half century has concentrated on an exploration of the structure and the interactions of stable nuclides, those found in the vicinity of the "beta-stable valley" outlined by the nuclear liquid-drop model energetics. It has been found that the nuclear isospin, I, with component, $I_3 = (N\text{-}Z)/2$, provides an ordering scheme, i.e., that it defines a symmetry for stationary states of nuclei with sufficiently small Coulomb self-interactions. In addition, studies of nuclear reactions between complex projectile and target nuclei have shown that the isospin asymmetry of the reaction partners plays an important role for the evolution of multi-nucleon exchange processes, which form the basis of nuclear transport in such reactions. Asymmetry-dependent nuclear liquid-drop energetics act as a driving force for these processes, even though isospin equilibrium seems hardly ever to be reached. Hence, isospin degrees of freedom were found important for an understanding of nuclei and nuclear interactions.

The limited range of (stable and meta-stable) nuclei available for study in the past has hampered both nuclear structure and reaction studies. Current understanding of nuclear structure suffers from ambiguities in the effective interactions that appear consistent with nuclear structure close to the beta-stable valley but do not provide unique extrapolations to unknown domains in nuclear mass and isospin. The same effective interactions affect collisions between complex nuclei, where they cause compression, dissipation, subsequent expansion, disintegration, and clusterization phenomena. Thus, the study of nuclear reactions and their dependence on reactant isospin asymmetry provides information on the properties of asymmetric nuclear matter at densities both higher and lower than that of normal nuclear matter, complementing information sought in exploration of nuclear structure. It also provides information on in-medium nucleonic interactions, e.g., nucleon-nucleon scattering and correlations, important constraints on realistic nuclear theory. Such information is essential not only for nuclear science, but also for other fields such as chemistry and material science interested in clusterization effects and for cosmology. In particular, knowledge of the equation of state of asymmetric nuclear matter is needed for understanding the structure, chemical composition and evolution of neutron stars or the dynamics of supernova explosions.

The new scientific opportunities seen for Isospin Physics have inspired the nuclear science community to expend extraordinary efforts to open up this new field. Several radioactive beam facilities are currently being upgraded and will come online shortly, providing access to a much broader range of neutron-proton asymmetries in nuclei than previously available. In addition, new and very powerful accelerator facilities and detector systems are being planned, expanding these capabilities even further.

As the nuclear science community is devising plans for the mid-term future of the field, it seems timely to take stock of the current understanding and likely prospects of this new branch of nuclear science, and to motivate new vigorous explorations. These are the tasks, which the editors attempt to accomplish with this book. Therefore, in the 19 individual chapters of this book, active workers in the field present their respective views on important aspects of nuclear interactions relevant for a better understanding of nuclear isospin degrees of freedom and their applications. Aspects covered include

- microscopic many-body theories for asymmetric nuclear matter and astrophysical applications,
- transport theories for nuclear reactions involving neutron-rich nuclei,
- phenomenological, statistical and thermal models for phase transitions in asymmetric nuclear matter,
- linear response theories for collective modes in asymmetric nuclei,
- first experimental results on reactions at intermediate energies involving strongly asymmetric nuclear systems.

One may find that these works include first experiments and their, sometimes tentative, interpretation and predictions of phenomena by theoretical models based on different combinations of sound physical principles and "educated guesses", and even outrightly speculative discussions. Needless to say, the collection of studies included in this book is incomplete, sometimes inconsistent, and does not yet project a complete and satisfactory picture of isospin-dependent nuclear phenomena. These studies do not generally provide definitive answers but certainly help pose important questions to be addressed by future exploration. It is our hope and expectation that the themes covered by individual chapters of this book will evolve in time, as new experimental data and theoretical methods become available. Hopefully, at some stage in the not-too-distant future, experimental and theoretical endeavors starting now will provide a thorough explanation of interesting isospin dependent nuclear reaction phenomena, completing the solution of the nuclear many-body problem, a problem which is fundamental to our understanding of nature. The authors of the chapters in this book are thanked for their energy and enthusiasm to get these efforts underway.

We would like to thank P.J. Siemens for a critical reading of the book manuscript and his helpful comments. The work of B.A. Li has been sponsored by the National Science Foundation Grant No. PHY-0088934 and the Arkansas Science and Technology Authority Grant No. 00-B-14, and that of W. Udo Schröder by the United States Department of Energy Grant No. DE-FG02-88ER40414.

Bao-An Li and W. Udo Schröder. December, 2000.

Contributors

R. Alfaro Department of Chemistry, Indiana University, USA
V. Baran INFN, Laboratori Nationali del Sud, Catania, Italy and NIPNE, Bucharest, Romania
Wolfgang Bauer National Superconducting Cyclotron Laboratory, and Department of Physics and Astronomy, Michigan State University, USA
Ignazio Bombaci Dipartimento di Fisica, University di Pisa and INFN Sezione di Pisa, Italy
A.Bonasera INFN, Laboratori Nationali del Sud, Catania, Italy
R.J. Charity Department of Chemistry and Physics, Washington University, USA
Lie-Wen Chen Institute of Modern Physics, Chinese Academy of Science
Ph. Chomaz Ganil (DSM-CEA/IN2P3-CNRS), Caen, France
J. Cibor Institute of Nuclear Physics, Krakow, Poland
B.M. Colonna INFN, Laboratori Nationali del Sud, Catania, Italy
Frank Daffin Department of Physics and Astronomy, Michigan State University, USA
J. Dempsey Department of Chemistry and Physics, Washington University, USA
V. Duflot Ganil (DSM-CEA/IN2P3-CNRS) and LPC Caen, France
W. A. Friedman Department of Physics, University of Wisconsin, USA
U. Fuhrmann Fachbereich Physik, University of Rostock, Germany
F. Gulminelli LPC Caen, IN2P3-CNRS, ISMRA et Université, Caen, France
S. Das Gupta Department of Physics, McGill University, Canada
K. Hagel Cyclotron Institute, Texas A&M University, USA
Gen-Ming Jin Institute of Modern Physics, Chinese Academy of Science
T. Keutgen Cyclotron Institute, Texas A&M University, USA
Che Ming Ko Department of Physics and Cyclotron Institute, Texas A&M University, USA
Bao-An Li Department of Chemistry and Physics, Arkansas State University, USA
Jian-Ye Liu Institute of Modern Physics, Chinese Academy of Science
U. Lombardo Dipartimento di Fisica, University di Catania and INFN-LNS, Catania, Italy.
W.G. Lynch National Superconducting Cyclotron Laboratory, and Department of Physics and Astronomy, Michigan State University, USA
Z. Majka Department of Physics, Jagiellonian University, Poland
N. Marie Lab. Phys. Corp., ISMRA , Caen, France
Klaus Morawetz Lab. Phys. Corp., ISMRA and Ganil, Caen, France
H. Miller Department of Physics, University of Colorado, USA
M. Murray Cyclotron Institute, Texas A&M University, USA
J.B. Natowitz Department of Chemistry and Cyclotron Institute, Texas A&M University, USA
J. Pochodzalla Max-Planck-Institut für Kernphysik, Heidelberg, Germany
S.K. Samaddar Saha Institute of Nuclear Physics, Calcutta, India
W.Udo Schröder Department of Chemistry, University of Rochester, USA
B. Serot Department of Physics, Indiana University, USA
W.Q. Shen Shanghai Institute of Nuclear Research, Chinese Academy of Science
P.J. Siemens Department of Physics, Oregon State University, USA
L.G. Sobotka Department of Chemistry and Physics, Washington University, USA
P. Staszel Department of Physics, Jagiellonian University, Poland
Jan Tõke Department of Chemistry, University of Rochester, USA
M. Di Toro Dipartimento di Fisica, University di Catania and INFN-LNS, Catania, Italy.
W. Trautmann Gesellschaft für Schwerionenforschung mbH, Darmstadt, Germany
M. B. Tsang National Superconducting Cyclotron Laboratory, and
Department of Physics and Astronomy, Michigan State University, USA
R. Wada Cyclotron Institute, Texas A&M University, USA
Rainer Walke Max-Planck-Institut for Demographic Research, Rostock, Germany
Gary D. Westfall National Superconducting Cyclotron Laboratory, and Department of Physics and Astronomy, Michigan State University, USA
S.J. Yennello Department of Chemistry and Cyclotron Institute, Texas A&M University, USA
Feng-Shou Zhang Institute of Modern Physics, Chinese Academy of Science
Zhi-Yuan Zhu Shanghai Institute of Nuclear Research, Chinese Academy of Science
Wei Zuo Institute of Modern Physics, Chinese Academy of Science

EOS of Asymmetric Nuclear Matter

U. Lombardo[1] and W. Zuo [2]

1 Introduction

The study of the equation of state (EOS) of nuclear matter in states of high charge-asymmetry has been driving a strong interest over the last years. The reason is that it spans a wide spectrum of systems, from atomic nuclei more or less far away from the stability valley up to stellar objects evolving via the supernovae state to neutron stars. This study can also promote the investigation on a rich variety of phenomena ranging between the two extreme situations of symmetric nuclear matter and pure neutron matter. In fact the EOS is expected to exhibit a deep modification along its isospin path connecting the two extreme states of nuclear matter. Some relevant issues are in order:

- very diluted state

  The system is characterized by the competition between the free Fermi-gas behaviour and a gently attractive mean-field. In this state a major role is played by long-range correlations such as pair correlations. Nuclear matter is a superfluid state of neutron-neutron (nn) and proton-proton (pp) Cooper pairs in the density range $0.01 \div 0.03 fm^{-3}$ [1]. At higher density the condensation of neutron-proton (np) pairs in a singlet-isospin state starts to compete with the two previous superfluid states. Since pair correlations are localized around the Fermi energy, a small excess of neutrons or protons blocks the levels near the Fermi surface and the superfluid state rapidly disappears [2]. The np pairing could set in heavy-ion collisions (HIC) during the expanding phase following the impact between the two ions and compete with the formation of α particles and other light fragments[3].

- low density region

  A bulk phase transition from a Fermi gas to a quantum liquid, predicted theoretically at subnuclear density, is signaled by the observed multiple emission of intermediate mass fragments in both central and peripheral HIC. In the theoretical models the coexistence region is marked by a negative compressibility (spinodal region), which is expected to shrink with increasing neutron excess because neutron matter is unstable with respect to purely nuclear force. The shrinking rate is controlled by the symmetry energy E_{sym}, i.e. the slope of the EOS $E(\rho, \beta)$ vs β^2, being $\beta = (N - Z)/A$ the asymmetry parameter. In the same context one may predict a decreasing critical temperature for systems with increasing neutron excess [4]. One predictable [5] isospin effect on the phase transition is the isospin fractionation: the liquid phase is more proton-rich than the vapour. Indeed the isospin symmetric state is energetically favourite in the liquid phase wherein the attractive nuclear interaction is more active. Equivalently, protons are driven into the liquid phase by the proton mean field which is much more attractive than the neutron one in a neutron rich system. An indication of this effect comes from the analysis of the isotopic composition of the gas phase in the multifragmentation of $Sn + Sn$ at E/A = 50 MeV [6].

- high density

  One main goal of the study of HIC is to get information on the compressibility of nuclear matter by means of the measurements of sideward flows. In asymmetric matter the isospin effect on the compressibility is directly related to symmetry energy. In some theoretical models E_{sym} is monotonically increasing with density and then the neutron-rich matter becomes stiffer than symmetric matter at the same density; some others predict that E_{sym} is decreasing above the

[1]Dipartimento di Fisica, 57 Corso Italia, I-95129 Catania, Italy and INFN-LNS, 44 Via S. Sofia, I-95123 Catania, Italy

[2]Institute of Modern Physics, Academia Sinica, Lanzhou, China

saturation density and then the neutron-rich matter becomes softer than symmetric matter. Following the terminology of the Colonna and Di Toro contribution to this book we refer to the two cases as asy-stiff and asy-soft EOS, respectively. A comparison between numerical simulations of HIC with the measured flows can give an unambiguous answer to the density dependence of E_{sym}. Lowering the beam energy the transverse flow vanishes, signalling the transition from the hydrodynamic regime driven by the NN collisions to the mean-field regime. Since this transition results from a delicate balance between mean field and NN in-medium cross section (and Coulomb repulsion between protons) we can expect a strong isospin dependence of the balance energy, which is the beam energy corresponding to vanishing flow. This fact has been already confirmed experimentally by the comparison of the balance energy measured in collisions Ni+Ni and Fe+Fe isotopes having the same atomic number but different asymmetry [7].

- very high density

  Dense nuclear matter with high isospin asymmetry could be formed in astrophysical sites during the evolution of our universe. An process, which has been intensely studied is the supernovae explosion. During the high-density phase of the presupernova core collapse, the proton concentration into the core reaches the value of about 1/3 due to the process of β-capture and, as a consequence, the compressibility of nuclear matter is reduced by about 30% [8]. The softening of the EOS is required by the dynamical simulations [9] based on prompt explosion model [10]. The remnant of a supernova explosion is a neutron star. The observed mass of neutron stars has been addressed as a constraint on the EOS of nuclear matter. Indeed the Tolman-Oppenheimer-Volkov (TOV) equation determining the structure of neutron stars is crucially depending on the compressibility of asymmetric nuclear matter. The asymmetry rate along the density profile from the surface to the inner core is determined by the density dependence of the asymmetry energy [11].

The purpose of the present chapter is to give an updated report on the status of the EOS of asymmetric nuclear matter. Section II is devoted to an overview of the microscopic approaches. In Section III we discuss the isospin effects on the EOS of nuclear matter at zero temperature, specifically the $\beta^2 - law$, the symmetry energy and the spinodal region. In Section IV we present in the context of the extended Brueckner-Hartree-Fock (EBHF) approximation some important single-particle properties such as the nucleon mean-field enhanced by the rearrangement term, the neutron and proton effective masses and mean free paths. The extension of the EOS of asymmetric nuclear matter to finite temperature is briefly discussed in Section V, with the focus mainly on the interplay between mechanical and chemical instabilities.

2 Survey of the many-body approaches

In the nonrelativistic approaches nuclear matter is assumed to be made of a collection of point-like nucleons interacting through an instantaneous two-body force. In the mean-field approximation effective interactions are employed such as Skyrme or Gogny forces which are density-dependent so as to incorporate the nuclear medium effects. It has been mainly applied to predict properties of nuclei near the stability valley and, more recently, extended to nuclei near the neutron and proton drip lines. More fundamental theories adopt the so called *realistic* interactions, namely the bare NN interactions which fit the experimental phase shifts of the nucleon-nucleon scattering.

A nonrelativistic approach is the Brueckner theory, which is based on the linked cluster expansion of the ground-state energy in terms of the G-matrix. This latter plays a role of in-medium interaction and renormalizes the short-range nuclear repulsion. The perturbation expansion of the energy per particle can be ordered according to the number of hole lines of the corresponding diagrams, which rapidly converges at low densities. In the lowest order the Brueckner theory can be interpreted as a mean-field theory usually called Brueckner-Hartree-Fock (BHF) approximation. A decisive progress has been made by calculating the three hole-line contributions. This requires to solve the Bethe-Faddeev equation, which is the Faddeev equation for the in-medium three-body problem. The

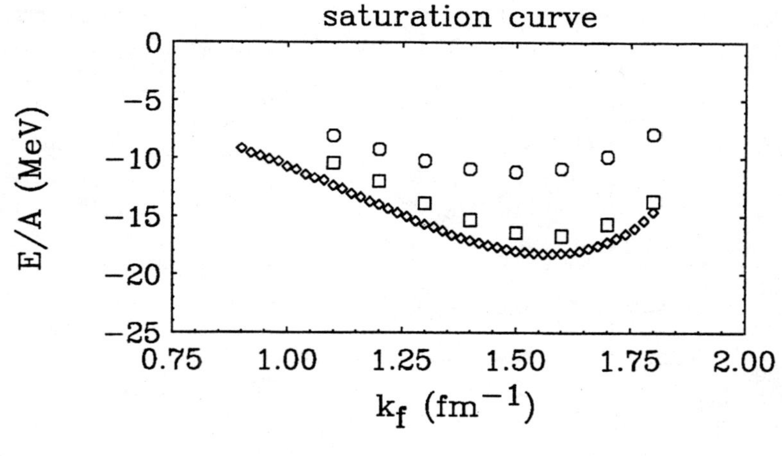

Figure 1: Convergence of the hole-line expansion in the BBG approach. A BHF calculation with the gap choice (circles) is compared with a BHF calculation with the continuous choice (diamonds), both at two hole-line level of approximation. A calculation with the continuous choice including also three hole-line diagrams (squares) shows the more rapid convergence of the continuous choice.

convergence of the hole-line expansion has been proved within the BBG context with the standard choice for the single-particle (s.p.) auxiliary potential by Day and Wiringa [12]. This problem has been revisited recently in the context of the continuous choice [13]. The results, plotted in Fig. 1, show that the hole-line expansion rapidly converges when adopting the continuous choice for the auxiliary potential. In this case, in fact, the three hole-line correction to the energy is much less than in the case of the standard choice. Anyhow the strong overlap of the results obtained within the two different approaches at the three hole-line level of approximation [13] demonstrates that the hole-line expansion is not depending on the choice of the auxiliary potential.

In the variational method the expectation value of the Hamiltonian is minimized in a subspace of Hilbert space spanned by a trial many-body wave function of the form $\Psi = F\Phi$, where Φ denotes the wavefunction of non interacting nucleons. The central quantity here is the correlation function F. Cluster expansions of the variational energy such as the hypernetted chain expansion provide an upper bound for the ground-state energy of nuclear matter. A recent variational calculation of the nuclear matter EOS with $A18$ is reported in Fig. 2 [14], where is also shown for comparison the BHF including three hole-line terms.

The Brueckner EOS including three-body correlations is comparable to the variational one: in both cases the predictions for the equilibrium (saturation) energy per particle is in good agreement with the empirical value, while the saturation density is too large, as shown in Fig. 2. Both predictions fall into the "Coester line" in the density-energy diagram: either the density is correct, but the energy is too small or the energy is correct, but the density too high. The failure of the nonrelativistic approaches to explain the nuclear saturation can be traced to the underlying model of structureless nucleons interacting through a bare two-body force. In order to remedy such failure a relativistic version of the Brueckner theory, the so called Dirac-Brueckner theory, has been devised [15]. Instead of the bare interaction it adopts a meson-exchange model of nuclear potential (OBEP potential) and the s.p. motion is described by the in-medium Dirac equation. The relativistic version of the BBG equation, the so called Thompson equation, contains a density-dependent NN interaction, arisen from the self-consistency requirement on the in-medium Dirac spinors, which allows it to reproduce simultaneously the equilibrium properties of nuclear matter and the experimental phase shifts of NN scattering (along with the deuteron properties). The EOS from this approach is depicted in Fig. 3 (DBHF).

A more fundamental approach is the relativistic mean-field (RMF) theory, where the interaction between baryons is described by the exchange of neutral scalar and vector mesons. In this case

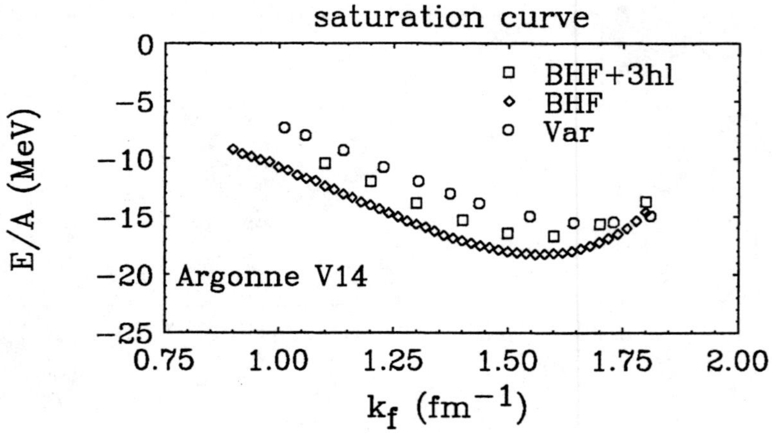

Figure 2: Non-relativistic EOS: Comparison between Brueckner and variational results

the meson parameters are adjusted to reproduce the properties of nuclear matter without passing through the NN phase shifts. In its original version [16], the two parameters of the model were adjusted to the empirical saturation point of symmetric nuclear matter. But the corresponding EOS turned out to be too stiff. The compression modulus, $K \simeq 545MeV$ at equilibrium, is indeed too large compared to the value of $210 \pm 30MeV$, extracted from the experimental monopole giant resonance in heavy nuclei [17]. The model was improved by introducing non-linear terms in the lagrangian and the corresponding parameters were fit to the properties of nuclei. The EOS for symmetric nuclear matter, plotted in Fig. 3, corresponds to the parameters taken from Ref. [18]. In this case $K \simeq 242MeV$ in better agreement with the empirical value. It is also close to the nonrelativistic prediction since the effect of the more pronounced curvature of the relativistic EOS is more or less compensated by that of its much less equilibrium density as shown in Fig. 3. In a recent approach [19], alternatively, it has been proposed to use a density-dependent baryon-nucleon coupling constants, as suggested by Dirac-Brueckner theory. However the determination of these effective parameters requires the support of a microscopic theory, the Dirac-Brueckner in that context.

As we saw earlier nonrelativistic theories with only a two-body force fail to reproduce the empirical saturation point extracted from the mass formula. Brueckner calculations of three and four hole-line contributions to the energy per particle have shown that the hole-line expansion rapidly converges, but the prediction of the saturation point is improved by no means. In this respect the Brueckner and variational predictions of symmetric nuclear matter do not differ from each other. On the other hand, the success of Dirac-Brueckner calculations seems to indicate that the mechanism of saturation is a purely relativistic effect.

This effect can be taken into account in a nonrelativistic approach by introducing a three-body force (TBF), which simulates virtual excitations of nucleons and other elementary processes, which the model of point-like nucleons cannot take into account. An extended study of TBF force from the microscopic point of view was made already a decade ago by Grange et al. [20], who proposed a TBF based on the meson-exchange model. In Fig. 4 a set of processes which are believed to give important contributions to the TBF is depicted. It contains the contribution due to the medium modification of the two-meson ($\pi\pi, \pi\rho, \rho\rho$) exchange part of the NN interaction with nucleon excitations (Δ or Roper resonance)(graph a), the contributions associated to non-linear meson-nucleon (graph b) and meson-meson coupling (graph c); last contribution describes the virtual excitation of a nucleon-antinucleon pair (graph d). The last one is supposed to give a very important relativistic effect on the EOS of nuclear matter [21]. Recently the TBF including the processes depicted in Fig. 4 has been probed in a Brueckner calculation using $AV18$ as two-body force. In order to avoid the extension of the BBG many-body approach required by a full treatment of the TBF the calculation adopts

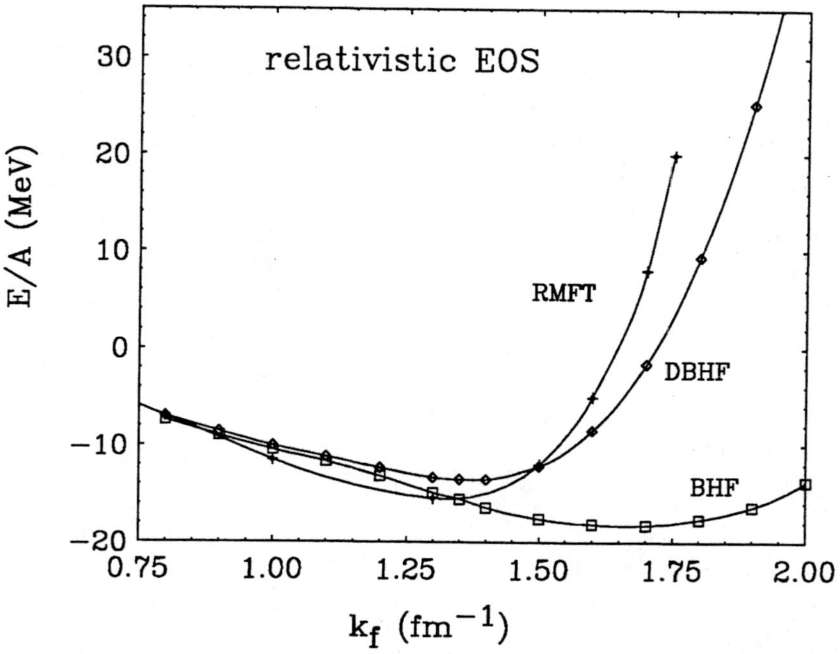

Figure 3: Relativistic EOS

the approximation devised in Ref. [20]. It is based on reducing the TBF to an effective two-body force. This goal is achieved using the procedure of averaging out the third particle by means of the correlated wave function. The effective two-body interaction is defined in r-space as follows:

$$< \vec{r}_1\vec{r}_2|V_3|\vec{r}_1'\vec{r}_2' >= \frac{1}{4}Tr\sum_n \int d\vec{r}_3 d\vec{r}_3' \phi_n^*(\vec{r}_3')(1-\eta(r_{13}'))(1-\eta(r_{23}'))$$

$$\times W_3(\vec{r}_1', \vec{r}_2'\vec{r}_3'|\vec{r}_1\vec{r}_2\vec{r}_3)\phi_n(r_3)(1-\eta(r_{13}))(1-\eta(r_{23})) \quad (1)$$

where the trace is taken with respect to spin and isospin of the nucleon 3. The function $\eta(r)$ is the average over spin and momenta in the Fermi sea of the defect function. It is self-consistently calculated along with the defect function at each step of the Brueckner iterative procedure. As shown in Fig. 5, the three-body force affects the EOS in such a way that in the low-density domain the energy per nucleon is practically unchanged, but at high density it rises up significantly. One main result for the EOS of symmetric nuclear matter is that the saturation density turns decisively improved towards the empirical value, from $0.265fm^{-3}$ to $0.19fm^{-3}$. At the same time there is no appreciable change in the saturation energy. This is a quite desirable feature of the three-body force, since the result for the saturation energy in BHF calculation, based on a purely two-body interaction, is already in good agreement with its empirical counterpart. All that makes us more confident with the general guess on the importance of the three-body force for the saturation problem of nuclear force.

No less important is the effect of TBF on the stiffness of the EOS measured by the compressibility modulus

$$K = 9\rho^2 \frac{\partial^2 E_A}{\partial \rho^2} \quad (2)$$

calculated at the saturation density. The effect of three-body force on the compressibility is to enhance its value from $250MeV$ to $265MeV$. Despite the large effect of TBF on the saturation curve,

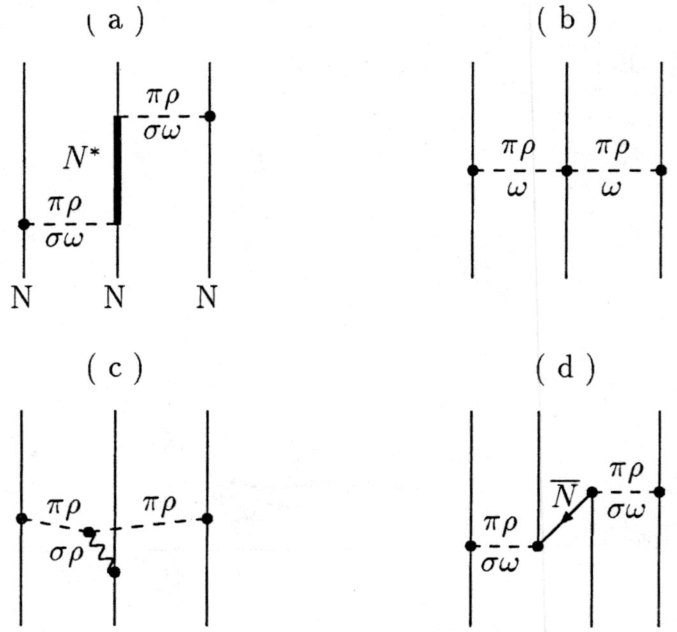

Figure 4: Typical processes contributing to the nuclear three-body force.

this change is rather small due to the fact that the more pronounced curvature of the saturation curve is balanced by the smaller value of the saturation density.

Due to the uncertainties present in a fully microscopic theory like the medium-renormalization of the meson masses and coupling constants, long ago the Urbana group proposed a phenomenological approach [22] to the TBF. The phenomenological TBF contains an attractive term $V_{ijk}^{2\pi}$ due to two-pion exchange with excitation of an intermediate Δ-resonance and a repulsive central term V_{ijk}^{R}. The two constants appearing in this TBF were fit in such a way to reproduce simultaneously the empirical saturation point and the properties of light nuclei ($^3H, ^4He$). A recent variational calculation, which adopts the $AV18$ combined with the Urbana IX TBF [14], is shown in Fig. 5. The resulting equilibrium energy of about $-12MeV$ for symmetric nuclear matter disagrees with the empirical value and no significant improvement is achieved by including relativistic boost corrections [23]. This discrepancy casts some doubt on the choice of constraining the TBF by means of the binding energy of light nuclei.

In Fig. 5 it is also plotted the EOS's obtained from a BHF calculation with two-body $AV18$ force and phenomenological Urbana TBF [24] (longdashes). Despite using two different TBF, the two BHF EOS's do not significantly differ from each other. In the calculation with the Urbana TBF the two free parameters were adjusted not only on the empirical density but also on the corresponding energy.

3 EOS of asymmetric nuclear matter

For many years the asymmetric nuclear matter was a topic of pure theoretical interest [25, 26] except for its applicability to neutron stars. Recently the possibility of reaching high neutron-proton asymmetries with RIB facilities has been stimulating a number of studies for its EOS both in non-relativistic [27, 28, 29, 32, 33] and relativistic approaches [34, 35, 36, 37, 38].

In asymmetric nuclear matter, in addition to the total density ρ, the isospin content of the system

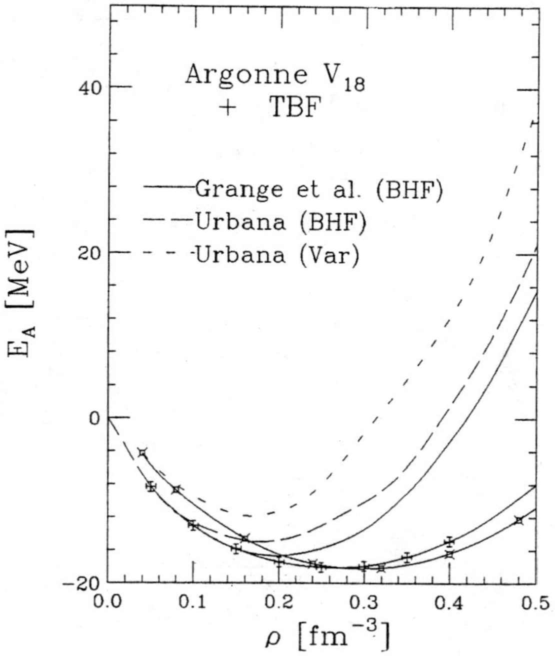

Figure 5: Non relativistic EOS with a microscopic (Grange et al.) and phenomenological (Urbana) three-body force. The two lower curves, plotted for comparison, are the variational (crosses) and BHF (stars) results with the only two-body force

must be fixed. A convenient quantity is the asymmetry parameter defined as

$$\beta = \frac{\rho_n - \rho_p}{\rho}. \tag{3}$$

where ρ_n and ρ_p are the neutron and proton densities, respectively. From now on we assume $\rho_n \geq \rho_p$. Neutrons and protons form a binary Fermi fluid and two distinct Fermi seas appear, one for protons $k_F^p = [\frac{3\pi^2}{2}(1-\beta)\rho]^{1/3}$ and one for neutrons $k_F^n = [\frac{3\pi^2}{2}(1+\beta)\rho]^{1/3}$.

3.1 *Parabolic law*

Some investigations [27, 28, 29] extrapolate the properties of asymmetric nuclear matter from the two extreme situations of symmetric nuclear matter ($\beta = 0$) and pure neutron matter ($\beta = 1$) assuming the empirical β^2 law [39] to be valid for any value of the asymmetry parameter. This assumption has been numerically verified within the BHF approximation [32] and confirmed later within the lowest order variational [33] and DBHF [37] approach. As shown in Fig. 6 the parabolic law is nicely fulfilled in a wide range of baryonic densities. The numerical results lie on a linear fit performed with only the first three values of the asymmetry parameter. The empirical parabolic law

$$E_A(\rho, \beta) = E_A(\rho, 0) + E_{sym}(\rho)\beta^2 \tag{4}$$

taken from the nuclear mass table can be thus extended up to the highest asymmetry of nuclear matter. Eq. (4) can be considered as the β^2 expansion of the binding energy per particle truncated at the lowest order. It is worth noticing that only even powers of the asymmetry parameter may occur in the expansion because Argonne V_{14}, which is the input for the calculations shown in Fig. 6, is an isospin-symmetry conserving interaction. A deviation from the β^2-law might arise at the three hole-line order of the BBG expansion. But, except at high density, the three hole-line contribution to the energy turns out to be rather small [13, 41] when adopting the continuous choice for the

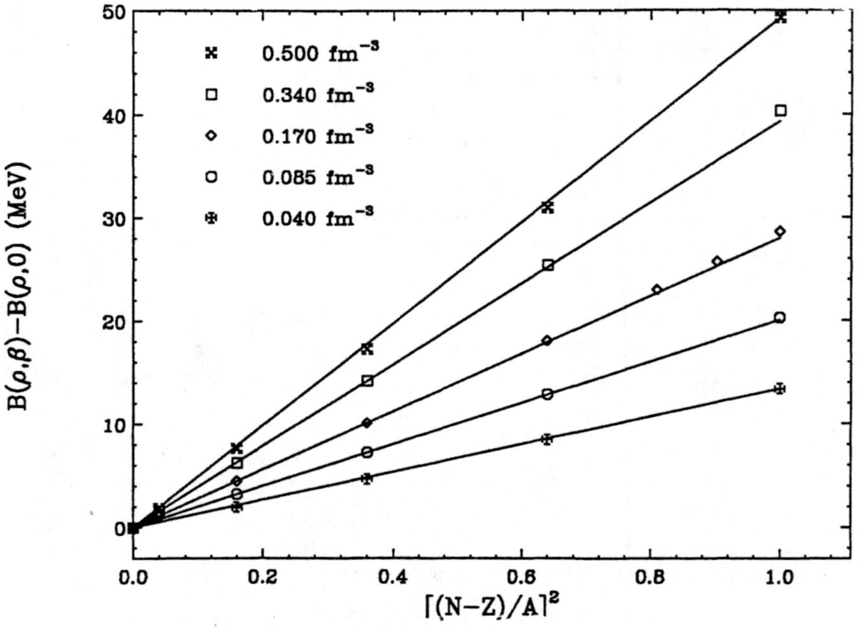

Figure 6: Binding energy in asymmetric nuclear matter. Results taken from

auxiliary potential. The β^4 term has been calculated in the DB approach. It is almost negligible even at high density [37]. To get a correction as large as 10% one should go up to $\rho = 0.7fm^{-3}$.

The quadratic dependence of the energy on the asymmetry parameter can be traced back not only to the assumed charge-symmetry of the NN interaction, but it is also related to the approximation inherent into the theory. Within a mean-field theory such as BHF, the charge symmetry is constraining the G-matrix to fulfill the approximate relations

$$< G_{pp} >_p = < G_{nn} >_n \qquad < G_{np} >_p = < G_{pn} >_n \tag{5}$$

where $< .. >_{n,p}$ is the average value of G-matrix in the (n,p) Fermi sphere. In Tab. I the isospin contributions to the energy are reported as a function of β^2 for three value of the density. The T=0 contribution is dominated by the SD channel, where the tensor force is mostly concentrated. It saturates above $\rho = 0.17fm^{-3}$ in connection to the vanishing of the correspondent phase shift at $E_{lab} \simeq 300MeV$.

The enhancement of the energy per particle vs β is mainly due to the sd-channel suppression. The non relativistic predictions of the EOS of pure neutron matter are depicted in Fig. 7. For the case of BHF approach the EOS corresponds to the two hole-line approximation. In fact the three hole-line contribution turns out to be quite small[41] (due to the small values taken by the wound parameter in neutron matter, which are close to 0.1 [45]).

The variational calculation [14, 33], also reported in Fig. 7, substantially deviates from the Brueckner one above the saturation density. This discrepancy is a puzzling aspect of the comparison between variational and Brueckner approaches. In fact in the case of symmetric nuclear matter we found that the agreement is satisfactory taking into account that the variational prediction is an upper estimate of the energy. Considering the dominant role of sd channel in asymmetric nuclear matter, as previously discussed, one may guess that this discrepancy is to be traced to a different treatment of the tensor force in the two approaches.

Table 1: Breakdown of the energy per particle (in MeV) into isospin components.

ρ	β	E/A	E_p	$E_p(T=0)$	$E_p(T=1)$	$E_p(SD)$
0.04	0.0	-7.476	-16.256	-9.641	-6.615	-9.921
0.04	0.2	-6.931	-15.913	-9.211	-6.702	-9.601
0.04	0.4	-5.528	-15.092	-8.266	-6.837	-8.507
0.04	0.6	-2.787	-13.348	-6.078	-7.270	-6.465
0.04	0.8	1.031	-10.961	-3.497	-7.464	-3.613
0.04	1.0	5.918	-8.019	0.0	-8.019	0.0
0.17	0.0	-15.816	-38.851	-19.013	-19.838	-19.306
0.17	0.2	-14.726	-38.274	-18.285	-19.991	-18.577
0.17	0.4	-11.357	-36.453	-16.255	-20.198	-16.319
0.17	0.6	-5.749	-33.460	-12.293	-21.167	-12.544
0.17	0.8	2.215	-29.250	-6.989	-22.261	-7.156
0.17	1.0	12.830	-23.736	0.0	-23.736	0.0
0.4	0	-13.402	-54.153	-19.855	-34.298	-19.992
0.4	0.2	-11.641	-53.299	-19.068	-34.231	-19.214
0.4	0.4	-6.577	-50.972	-16.84	-34.132	-16.986
0.4	0.6	1.998	-47.023	-12.999	-34.022	-13.130
0.4	0.8	14.207	-41.457	-7.457	-34.000	-7.529
0.4	1.0	30.480	-34.208	0.0	-34.208	0.0

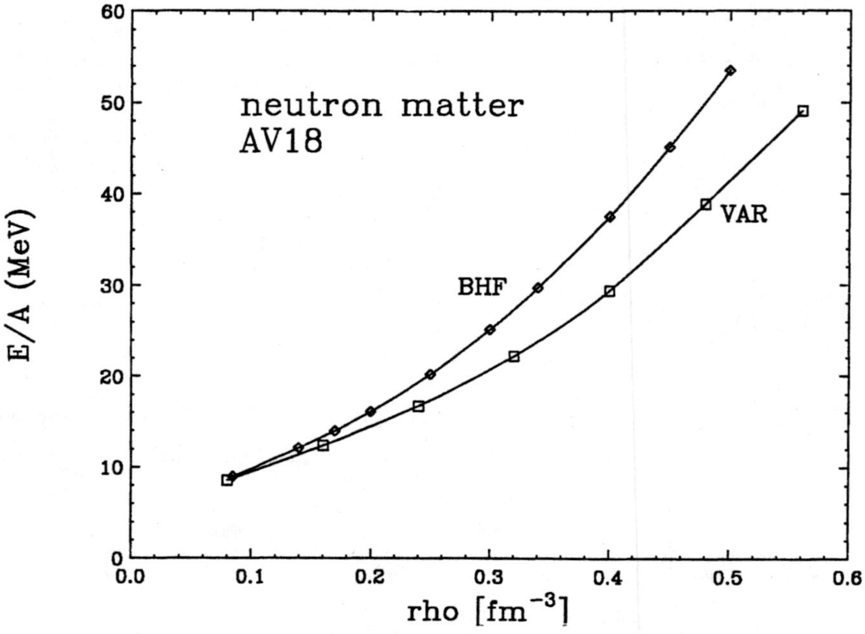

Figure 7: EOS of pure neutron matter matter.

3.2 Symmetry Energy

The symmetry energy E_{sym} measures the variation rate of the energy per particle versus isospin asymmetry ranging from symmetric nuclear matter to pure neutron matter. Its properties, including the density dependence, are of great interest for several issues of nuclear physics and nuclear astrophysics.

The symmetry energy is defined as

$$E_{sym}(\rho) = \frac{1}{2}\left[\frac{\partial^2 E_A(\rho,\beta)}{\partial\beta^2}\right]_{\beta=0}. \qquad (6)$$

In Fig. 8 the different contributions to the symmetry energy, plotted as a function of density are reported from a BHF calculation [40]. The kinetic contribution monotonically increases as $\rho^{\frac{2}{3}}$ according to the free Fermi-gas model. In the figure (right panels), the isoscalar and isovector contributions of potential part are plotted separately.

The density dependence of the symmetry energy is dominated at high density by the kinetic contribution, where the opposite behavior vs density of the potential contributions from $T = 0$ and $T = 1$ channels results in a very flat density dependence of the symmetry potential. As already found in the previous paper [32], the most important contribution to the $T = 0$ component is due to the deuteron ${}^3S_1$-${}^3D_1$ coupled channels of the interaction, which exhibits a maximum at $\rho \simeq 0.3 fm^{-3}$. This peak can be traced back to the behavior of the corresponding phase shifts of nucleon-nucleon scattering in terms of the energy: the attractive ${}^3S_1$ channel dominates at low energy whereas the repulsive ${}^3D_1$ dominates at high energy. Two terms compensate each other at the energy $E_{lab} \simeq 4E_F \simeq 200 MeV$, where E_F is the Fermi energy corresponding to $\rho \simeq 0.3 fm^{-3}$.

The BHF prediction of the symmetry energy at the empirical saturation density ($\rho = 0.17 fm^{-3}$) of about 28 MeV is in quite good agreement with the value extracted from the mass table. As far as the β^2-law is satisfied the symmetry energy can be calculated by a good accuracy as difference between the binding energy of pure neutron matter and that of symmetric nuclear matter:

$$E_{sym}(\rho) = E_A(\rho, 1) - E_A(\rho, 0), \qquad (7)$$

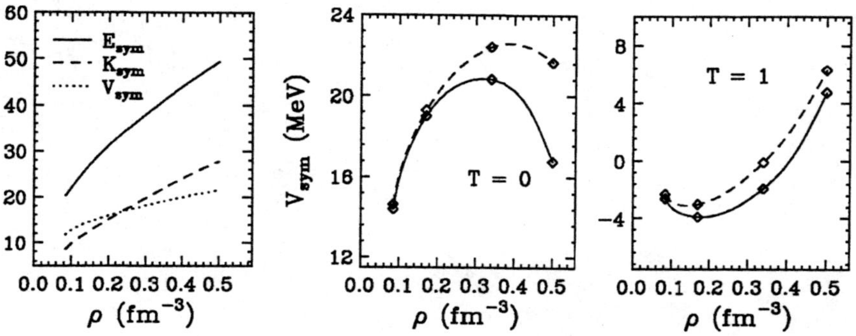

Figure 8: Breakdown of the symmetry energy. Left panel: total symmetry energy E_{sym}, kinetic part K_{sym} and total potential part V_{sym}. Middle and right panels: scalar and isovector components of V_{sym} from Argonne V_{14} (solid lines) and separable Paris potential (dashed lines).

But one would refrain from applying it at very high density where deviations from the parabolic law could be sizeable.

In Fig. 9 (left panel) the results of the Brueckner calculation are plotted together with the variational calculation performed with the same Argonne V_{14} potential [29]. The two predictions without three-body force exhibit a systematic disagreement which increases with the density and reaches about 30% at $\rho = 0.5 fm^{-3}$.The same discrepancy has also been observed in our previous calculations for asymmetric nuclear matter [32] and may have been anticipated from the calculation for neutron matter showed in Fig. 7. This discrepancy has been a puzzle since long time as it does not occur in the case of symmetric nuclear matter, as discussed previously, and also the BHF prediction turns out to be in excellent agreement with a calculation done within the lowest-order constrained variational approach [33], which is less sophisticated than that from hypernetted chain approach. This calculation, shown in Fig. 9 for comparison, is also obtained from Argonne V_{14}. More details on these controversial aspects are given in Ref. [40].

The most serious aspect of this disagreement is that in the variational approach with or without three-body force [29] (but also in the HF approximation for a class of Skyrme interactions such as SkM^* [30]), the curve $E_{sym}(\rho)$ at high density bends down and $V_{sym}(\rho)$ becomes negative (asysoft EOS) when adopting the Urbana UV14 interaction plus a three-body force. At the contrary the Brueckner prediction exhibits a monotonic increase (asystiff EOS) independently of whether the three-body force is included or not. This is displayed in Fig. 9 (right panel). At variance with the variational approach, the phenomenological TBF [24] turns out to be in good agreement with the microscopic one as far as it is worked out in the framework of Brueckner approach. We also found a good agreement with the DBHF prediction. This achievement give more support to the guess that the relativistic effects can be incorporated in a non relativistic approach via a TBF.

In the relativistic mean-field theory the same monotonic increase is easily understood in terms of the ρ meson exchange, which leads to a repulsive symmetry potential at all densities [37]. However a quite monotonic increase is also obtained, without including any ρ meson, merely as an effect of the exchange term [42].

3.3 Spinodal region

The saturation density and the compression modulus can be fit by the simple polynomial expressions:

$$\rho_0(\beta) = 0.253\,(1 - 0.6\beta^2 - 0.886\beta^4) \qquad (8)$$

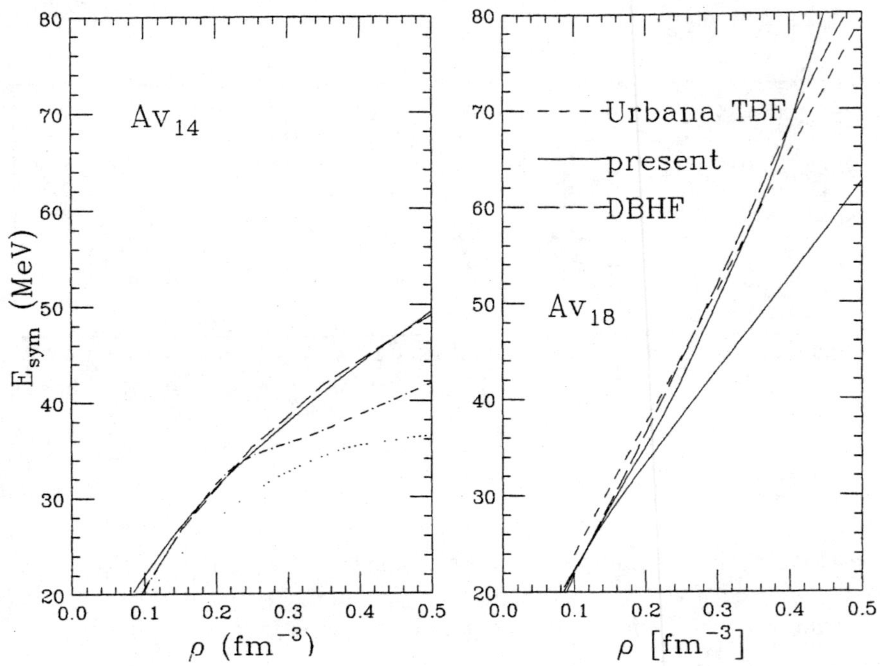

Figure 9: Left panel: Different predictions for the symmetry energy from calculations using Argonne V_{14}. The BHF result is shown as a solid line whereas the other three curves correspond to variational hypernetted chain expansion without (dotted) and with (dotdashed) TBF -wff- and lowest-order constrained approximation -mod-, respectively. Right panel: TBF effect on the symmetry energy and comparison with the relativistic DBHF prediction. The solid curve is the BHF result including the microscopic TBF described in sect.II (upper) and without TBF (lower), the short dashed curve is the BHF result with phenomenological Urbana TBF, the long dashed curve is the DBHF result.

and

$$K_0(\beta) \;=\; 187.6\,(1 - 1.766\beta^2 - 0.392\beta^4) \tag{9}$$

The two quantities follow the parabolic law only at low asymmetry. At higher asymmetry the next power is responsible for a rapid decrease of both quantities. It is instructive to derive analytical expressions connecting these quantities to the symmetry energy. Expanding $E_A(\rho,\beta)$ around the minimum, from Eq. (4) one gets [43]

$$\rho_0(\beta) \;=\; \rho_0(0) - \frac{9\rho_0^2(0)}{K_0(0)}\left(\frac{dE_{sym}(\rho)}{d\rho}\right)_0\ \beta^2 \tag{10}$$

where $\rho_0(\beta)$ denotes the equilibrium density at the asymmetry β and $K_0(0)$ the corresponding compressibility. Assuming $E_{sym}(\rho)$ to be a smooth function of ρ, which has been proved to be a reasonable approximation, a simple expression for the compressibility can be derived at the order β^2

$$K_0(\beta) \;=\; K_0(0) - 18\rho_0(0)\left(\frac{dE_{sym}(\rho)}{d\rho}\right)_0\ \beta^2 \tag{11}$$

The two quantities crucially depend on the slope of the symmetry energy at the equilibrium density. In the range where ρ_0 falls E_{sym} is a monotonically increasing function and then both $\rho_0(\beta)$ and $K_0(\beta)$ linearly decrease with β^2. An important consequence of such a behaviour is that the spinodal region shrinks as increasing asymmetry. This effect should be observed in processes where a liquid-gas phase transition occurs such as intermediate energy HIC (see for instance the contribution by M.

Colonna and M. Di Toro to this book). A simple estimate of the critical asymmetry parameter β_0, for which the spinodal region disappears and the liquid-gas phase cannot be supported any longer, is obtained from Eq. (11)

$$\beta^2 = \frac{K_0(0)}{18E_{sym}^0} \tag{12}$$

assuming E_{sym} do vary linearly around the saturation density. For $K_0(0) \simeq 200MeV$ and $E_{sym} \simeq 30MeV$ it results $\beta_0 \simeq 0.6$. So the isospin suppression of the liquid-gas phase could provide a new source of information for the nuclear compressibility.

4 Extended BHF approach

The most extensive microscopic investigation of the single-particle properties in nuclear matter has been performed in the framework of Brueckner theory [51, 45, 40, 52]. Despite the drawback of missing saturation density the BHF without three-body force is a reasonable approximation in the density range typical of nuclei. But the pure BHF approximation corresponding to the lowest order in the hole-line expansion of the self-energy (or mass operator) is not an adequate approximation to describe single-particle properties [46]. This can be seen from the point of view of the Landau definition of quasi-particle energy as variational derivative of the energy functional [47] with respect to the occupation numbers. According to that, one derives from the BHF energy functional

$$\varepsilon_k = \varepsilon_k^{BHF} + \sum_{k_1 k_2} < k_1 k_2 | \frac{\delta G}{\delta n_k} | k_1 k_2 >_A \tag{13}$$

where the first term ε_k^{BHF} is the pure BHF single-particle energy, represented by the first diagram of the hole-line expansion for the self-energy in Fig. 10. The second term, which is contributed by higher order diagrams depicted also in Fig. 10, is the correction to the BHF mean field introduced by the ground-state particle-hole excitations. It corresponds to the rearragement contribution appearing in the HF formalism due to the density dependence of the Skyrme force. The treatment of the single-particle potential going beyond the BHF approximation will be referred to as extended BHF (EBHF) approximation. A detailed description of the self-energy in the EBHF scheme is given in the Appendix. As we will see later the EBHF correction is responsible for the deviation of the self-energy from the linear behavior as a function of the isospin asymmetry.

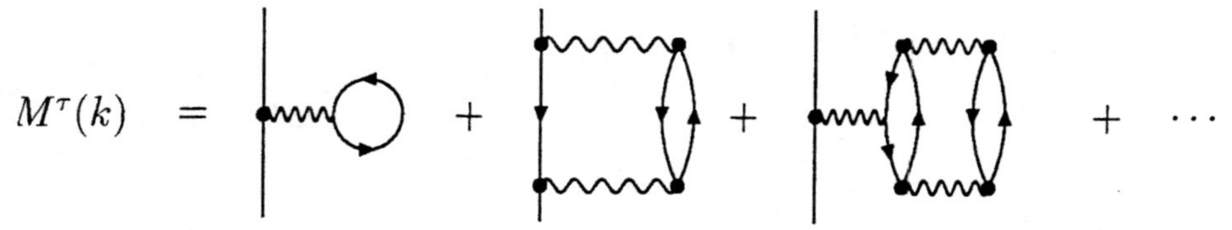

Figure 10: Hole–line expansion of the mass–operator (nucleon self–energy)

4.1 Fermi Energy and Hugenholtz-Van Hove theorem

The justification of the EBHF approximation basically relies on the Landau definition of quasi-particle energy . But an additional support could be supplied by the study of the Hugenholtz-Van

Hove (HVH) theorem [48].

Strictly speaking, the HVH theorem concerns only nuclear matter at the saturation point ($P = 0$): It states that the energy per nucleon must be exactly equal to the Fermi energy in symmetric nuclear matter. In the case of asymmetric nuclear matter (two-component system) at zero temperature, the HVH theorem can be generalized via the thermodynamic relation

$$\frac{E(\rho,\beta)}{A} + \frac{P(\rho,\beta)}{\rho} = Y^p E_F^p(\rho,\beta) + Y^n E_F^n(\rho,\beta) \tag{14}$$

where the pressure $P(\rho,\beta) = \rho^2(\partial E(\rho,\beta)/\partial\rho)$, and $Y^p = \rho_p/\rho$ and $Y^n = \rho_n/\rho$ are the proton and neutron fractions, respectively. The Fermi energy is calculated from the quasi-particle energy spectrum at Fermi surface according to Eq. (A2) of the Appendix. From the data reported in Tab. II one can appreciate to what extent the HVH theorem is fulfilled by the EBHF approximation. In the forth column the left hand side of Eq. (14) is calculated for several asymmetries (density fixed at $\rho = 0.17 fm^{-3}$). One would notice that, despite the total density is fixed at the empirical saturation value, the mechanical stability does not occur at $\beta = 0$ but is somehow shifted to higher asymmetry, because, as previously seen, in the Brueckner theory with two-body force the saturation density is not $\rho = 0.17 fm^{-3}$. The last three columns provide different approximations for the right hand side of Eq. (15). The pure BHF approximation by itself is far from fulfilling the HVH theorem. Including the unrenormalized g.s. correlations (indicated by BHF+M_2 in the table), where the Fermi energy is calculated from Eq. A4), provides some improvement which however is not sufficient to restore the HVH theorem. One needs to include both the rearrangement and the renormalized contributions (Eq. (A15)) in order to reach a satisfactory agreement within less than 10% (as shown in the last column of Tab. II). This result is in keeping with the uncertainty in the calculation of the pressure because the binding energy curve is rather flat as a function of density.

Table 2: As a function of the asymmetry parameter (first column) are reported (in MeV) the physical quantities involved in the Hugenholtz-Van Hove theorem: pressure over ρ (2nd column), energy per nucleon (3rd column) and 'weighted' chemical potentials of asymmetric nuclear matter in different approximations, as discussed in the text. $Y^p = Z/A$ and $Y^n = N/A$ are the proton and neutron fractions, respectively. The total density is $\rho = 0.17 fm^{-3}$.

β	P/ρ	E/A	$P/\rho + E/A$	$Y^p E_F^p + Y^n E_F^n$ BHF	$Y^p E_F^p + Y^n E_F^n$ BHF+M_2	$Y^p E_F^p + Y^n E_F^n$ EBHF
0.0	−5.02	−15.92	−20.94	−34.27	−28.50	−19.28
0.2	−4.40	−14.73	−19.13	−32.29	−26.43	−17.35
0.4	−3.27	−11.36	−14.63	−26.28	−20.74	−12.42
0.6	0.08	−5.75	−5.67	−16.56	−11.44	−4.41
0.8	3.76	2.24	6.00	−2.40	2.07	7.42
1.0	8.91	12.83	21.74	16.28	19.67	22.39

4.2 Neutron and proton mean fields

For asymmetric nuclear matter the neutron mass operator M^n is different from the proton mass operator M^p. In Fig. 11 the on-shell values of the real part $V_1^\tau(k)$ (upper panels) and imaginary part $W_1^\tau(k)$ (lower panels) of M_1^τ are reported for different values of the asymmetry parameter β at fixed $\rho = 0.17 fm^{-3}$. As already noticed in Ref. [32], the isospin dependence of the neutron and proton mean fields is determined to a large extent by the attractive $T = 0$ SD channel, which contains most of the tensor component of the nuclear force. Accordingly, the proton and neutron mean fields, the former becoming more attractive and the latter less attractive as increasing the asymmetry parameter, fulfill a linear law in β. This result supports from a microscopic point of view the validity of the so called Lane potential [53].

It is worth noticing that a crossing point occurs for both V_1^p and V_1^n, where the isospin effect on neutron and proton mean fields vs β is inverted. It can be understood in terms of the interplay between the pp (nn) and pn (np) effective interactions, as it can be easily seen from the approximation

$$V_1^p(k) \simeq \frac{1}{2}(1-\beta)\rho < G^{pp} > + \frac{1}{2}(1+\beta)\rho < G^{pn} > \tag{15}$$

$$V_1^n(k) \simeq \frac{1}{2}(1-\beta)\rho < G^{np} > + \frac{1}{2}(1+\beta)\rho < G^{nn} > \tag{16}$$

where $< G^{\tau\tau'} >$ is an average of G-matrix in the Fermi sea and independent of the asymmetry parameter. This approximation is suggested by the almost linear dependence of V_1^n and V_1^p on β and, in fact, is numerically fulfilled with a good accuracy (see also Fig. 11). The crossing point in momentum space is determined by the occurrence of $< G^{pp} >=< G^{pn} >$ for V_1^p and $< G^{np} >=< G^{nn} >$ for V_1^n at a certain value of momentum which does not depend upon β. A signature of the inversion of the isospin effect at the crossing point could be found in those collective observables measured in HIC which are sensitive to the momentum dependence of the mean field. In particular, one experimental evidence of this crossing could be the disappearance of the isospin dependence of the transverse flows well above the balance energy ($\simeq 150 AMeV$) [44].

The imaginary part W_1^τ of the mass operator M_1^τ arises from the virtual collisions of a single nucleon with a neutron or a proton of the background, promoting it to a particle state. W_1^τ is vanishing below the Fermi momentum k_F^τ due to the Pauli blocking. It is worth noticing that reducing the proton Fermi momentum implies a less Pauli blocking for protons. This means that high asymmetric nuclear matter is less transparent to the proton propagation.

At high density, say $\rho = 0.5 fm^{-1}$, the proton mean field becomes less sensitive to β as shown in Fig. 12. This is contrary to the case of lower densities (up to $0.34 fm^{-3}$ in our calculations) where the proton mean field is more and more attractive for high asymmetry. There could be two reasons for that. At relative low density, the SD attractive component is dominant. First, increasing β means increasing the neutron excess, and, as a result, the proton feels more SD tensor attraction from the neutron background. However, at relative high density, say $\rho = 0.5 fm^{-3}$, the high l channels start to compete and to play an increasingly important role. Second, as far as the density increases, the SD force becomes less and less attractive. The second order terms of the on-shell mass operator M_2^τ are plotted in Fig. 13. The real part V_2^τ (upper panels) gives the contribution to the mean field due to the coupling of the single-particle motion with the ground state particle-hole excitations. As it is well known, V_2^τ is repulsive and reduces to a large extent the pure BHF mean field V_1^τ which is too attractive compared with the phenomenological optical potential [46, 40]. The imaginary part W_2^τ plays a role complementary to W_1^τ: it describes the virtual collisions of a single nucleon of the background with an excited neutron or proton, making it to decay into a hole state. W_2^τ is vanishing above the Fermi momentum k_F^τ. In order to focus on only the isospin dependence, we plot in Fig. 14 the mass operator as a function of the β at $k = 0 fm^{-1}$ except for the imaginary part of M_1 for which a value of k above the Fermi momentum has to be taken. The first-order contribution has the linear behavior for both the real part and the imaginary part as expected from phase-space arguments. The slope of $|W_1^{pn}|$ is more pronounced than that of $|W_1^{np}|$ since the neutron particle-hole excitations coupled to a proton in a particle state are more favoured than the proton particle-hole excitations coupled to a neutron.

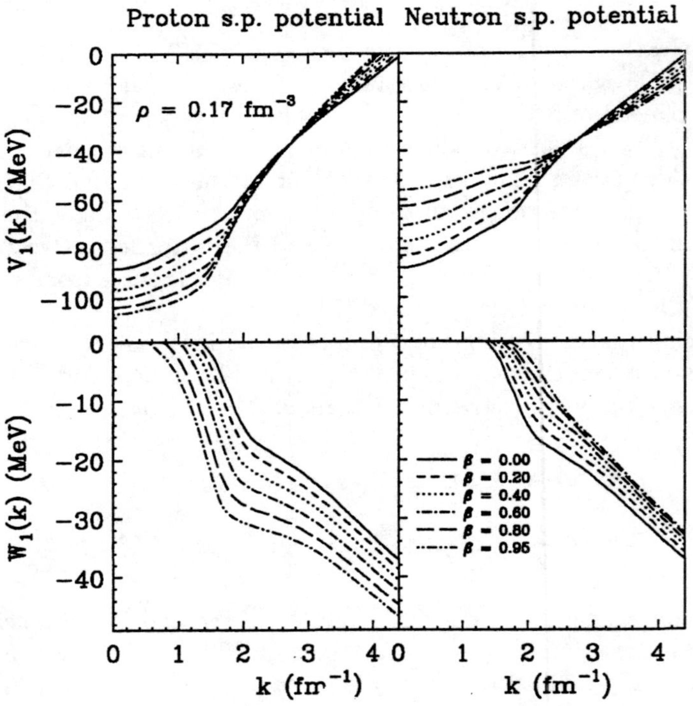

Figure 11: On-shell BHF self-energy at the empirical saturation density.

The isospin dependence of the second-order contribution M_2^τ is affected by the coupling between the nucleon hole states and particle-hole excitations (see Eq. (A4)), which yields a non-linear variation of the mixed components M_2^{pn} and M_2^{np} vs β. The non-linearity is much more sizeable for V_2^{pn} and W_2^{np} that can be easily explained as a phase-space effect as well, *i.e.* of the interplay between the neutron and proton phase-spaces as increasing neutron excess. From the s.p. potentials one may extract the proton and neutron Fermi energies in different approximations. These are reported in Tab. III for three densities. A remarkable property is that the shift between proton and neutron Fermi energies is almost insensitive to the approximation in which the two quantities are calculated despite they are very different in magnitude going from one approximation to another one. This fact can have important consequences for the study of the neutron star crust, because its isospin composition is strongly related to that shift.

4.3 Effective mass

The effective mass incorporates the non-local part of the mean field which makes the local part less attractive for a nucleon travelling with momentum $k > 0$. It is defined as

$$\frac{m_\tau^*(k)}{m} = \frac{k}{m}\left(\frac{dE^\tau(k)}{dk}\right)^{-1} \tag{17}$$

The momentum dependence of m^* is characterized by the wide bump inside the Fermi sphere due to the high probability amplitude for particle-hole excitations near the Fermi surface [50]. The effect of correlations is a flattening of the slope of the mean field around the Fermi energy, which implies an enhancement of the effective mass at k_F with respect to BHF value [45, 52]. This result is shown in Fig. 15, where an increase from 0.8 to 0.92 is observed for symmetric nuclear matter at the saturation density. In the same figure the isospin dependence of the neutron (upper curve) and

Table 3: Proton and neutron Fermi energies calculated in different approximations and compared with the symmetry energy. The results corresponding to four values of the asymmetry parameter β for each of the three densities are reported.

ρ	β	ϵ_F^p	ϵ_F^n	$\Delta\epsilon_F$	ϵ_F^p	ϵ_F^n	$\Delta\epsilon_F$	ϵ_F^p	ϵ_F^n	$\Delta\epsilon_F$	$4\beta E_{sym}$
(fm^{-3})		BHF			BHF+$M_2^{(R)}$			EBHF			
0.085	0.2	−33.99	−18.05	15.94	−30.44	−14.18	16.26	−23.00	−6.92	16.08	16.22
	0.4	−42.98	−10.77	32.20	−39.75	−7.01	32.74	−32.60	−0.16	32.44	32.45
	0.6	−51.71	−3.31	48.40	−49.26	0.15	49.41	−43.24	5.67	48.91	48.67
	0.8	−61.32	3.68	65.00	−59.63	6.58	66.21	−54.85	10.49	65.34	64.90
0.170	0.2	−45.94	−23.19	22.75	−40.45	−17.08	23.37	−31.01	−8.24	22.77	23.00
	0.4	−58.08	−12.65	45.43	−53.46	−6.72	46.74	−44.30	1.25	45.55	46.00
	0.6	−71.73	−2.77	68.96	−68.10	2.72	70.82	−59.60	9.39	68.99	69.00
	0.8	−86.22	6.91	93.13	−83.99	11.63	95.62	−76.42	16.73	93.15	92.00
0.340	0.2	−47.53	−16.11	31.42	−38.89	−5.75	33.14	−24.04	7.12	31.16	32.30
	0.4	−64.84	−1.79	63.05	−57.48	8.88	66.36	−42.31	20.22	62.53	64.59
	0.6	−82.75	12.21	94.96	−77.17	22.85	900.02	−62.44	32.35	94.79	96.89
	0.8	−103.15	25.48	128.63	−99.87	35.52	135.39	−86.32	42.95	129.27	129.18

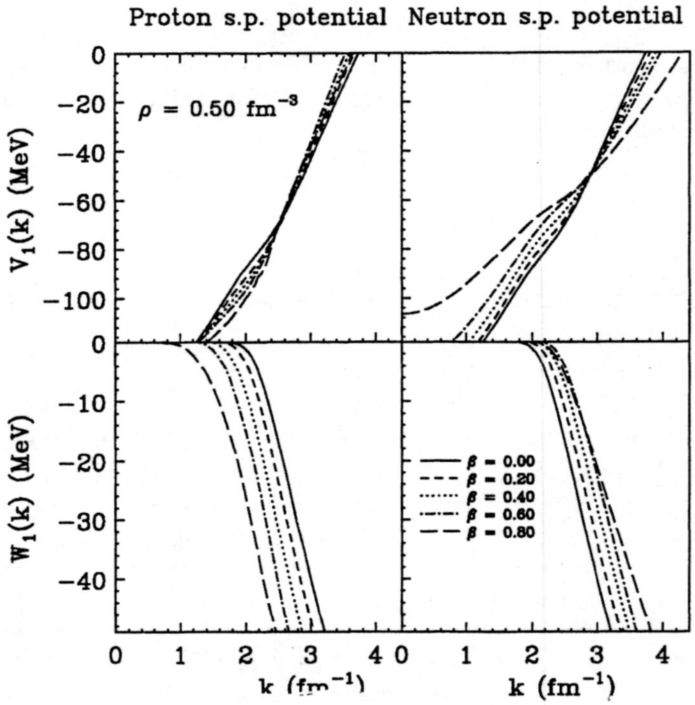

Figure 12: On-shell BHF self-energy at high density

proton (lower curve) effective masses is also reported. In both the BHF and EBHF calculations, m_n^* increases and m_p^* decreases as increasing β. Compared to the BHF approximation, the corrections of EBHF shift m_n^* and m_p^* to higher values, a feature which can be traced to the depletions of the proton and neutron Fermi surfaces due to the g.s. correlations. The value of m_p^* calculated from EBHF approaches its BHF value as increasing β since the correlations become smaller.

4.4 Mean free path

Information on the in-medium cross section or, equivalently, on the mean free path of a nucleon travelling inside a nuclear medium can be obtained from the transparency of a nucleus measured in $(e, e'p)$ reactions [54] and, in general, from nucleon-induced reactions at low energy [55]. The underlying assumption is that the behavior of a nucleon located at the position $\vec{r}$ in a nucleus is the same as a nucleon in nuclear matter at density $\rho(\vec{r})$. Such an assumption is the well known local-density approximation (LDA) [57]. The mean free path is intimately related to the imaginary part of the optical potential or, equivalently, to the imaginary part of the mean field. The latter comes from the collisions of a single nucleon with the background of neutrons and protons: a nucleon with momentum $k \geq k_F$ can collide with a neutron or proton of its Fermi sea and promote it to a particle state, or a nucleon with momentum $k \leq k_F$ interacting with an excited neutron or proton can make it decay into a hole state. The first process is related the imaginary part of M_1, the second one to the imaginary part of M_2, both of which have been plotted in the lower panels of Figs. 11,12 and 13. But, in the case of asymmetric matter the collisions between like and unlike nucleons yield contributions to the mean field which are very different.

The mean free path λ_τ is given by

$$\lambda_\tau(E) = \frac{\hbar^2 k(E)}{2\tilde{m}_\tau} \frac{1}{|\mathrm{Im} M_\tau(k(E), E)|},$$

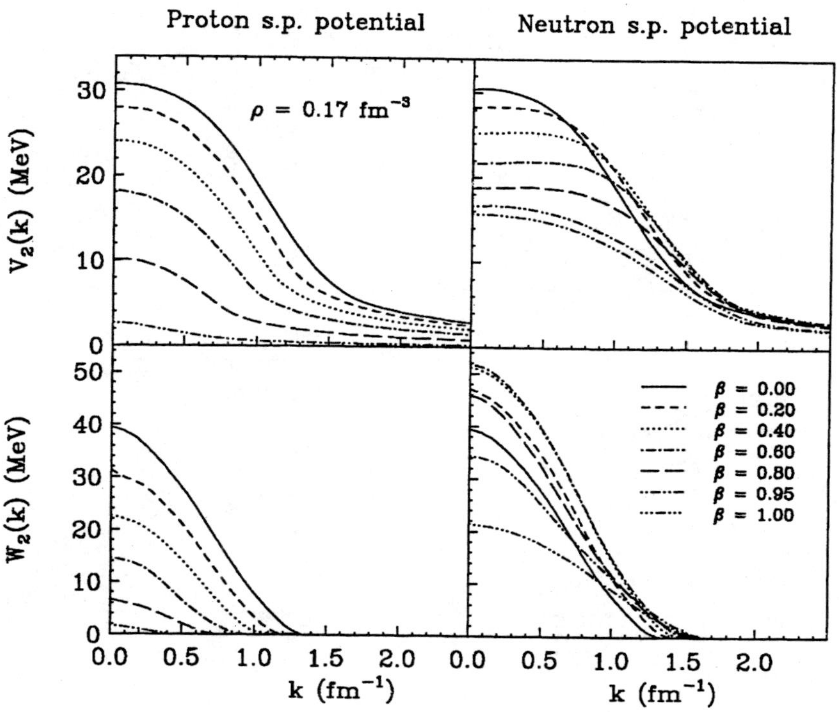

Figure 13: On-shell correlation self-energy

where $\tilde{m}_\tau$ is the so called k-mass, and E is the single-particle energy [56]. In Fig. 16 the proton (upper panels) and neutron (lower panels) mean free paths calculated within the EBHF approximation are shown for three values of the total density. In each panel the values of λ_τ for several asymmetries are plotted as a function of single-particle energy. The most relevant effect of the isospin-asymmetry is the increasing deviation from the symmetric values (solid lines), upward for λ_n and downward λ_p, as increasing asymmetry. The non-vanishing values of neutron and proton mean free paths below their respective Fermi energies are effect of g.s. correlations which prevents a full occupancy of the Fermi spheres. Comparing with the BHF calculation it turns out that the correlation effects tend to rise the asymptotic value of the mean free path from about $3fm$ up to about $4fm$ at the saturation density of symmetric nuclear matter [52].

In Fig. 17 it is shown how the isospin dependence of the inverse λ develops as increasing nuclear matter densities at a fixed value of the energy. Except for very small asymmetries the shift of λ_p and λ_n is not symmetric with respect to their common value at $\beta = 0$. At any density the slope of the neutron inverse λ is less than the proton one. This effect can be traced to the reduction, as increasing neutron excess, of the proton particle-hole excitations contributing to the neutron optical potential. Moreover, the EBHF λ_n seems to reach the asymptotic value of pure neutron matter much faster than in the uncorrelated case. The most striking effect of the isospin-asymmetry is the sizeable reduction the proton mean free path at high asymmetry. Accordingly, nuclear surface would result more transparent to neutrons than protons in nucleon-induced reactions on nuclei near the neutron drip-line. This effect would be more pronounced at higher density as shown in Fig. 17.

5 EOS at finite temperature

The EOS of nuclear matter at finite temperature exhibits several points of interest. One is the study of liquid-gas phase transition which is signaled by the production of intermediate mass fragments in

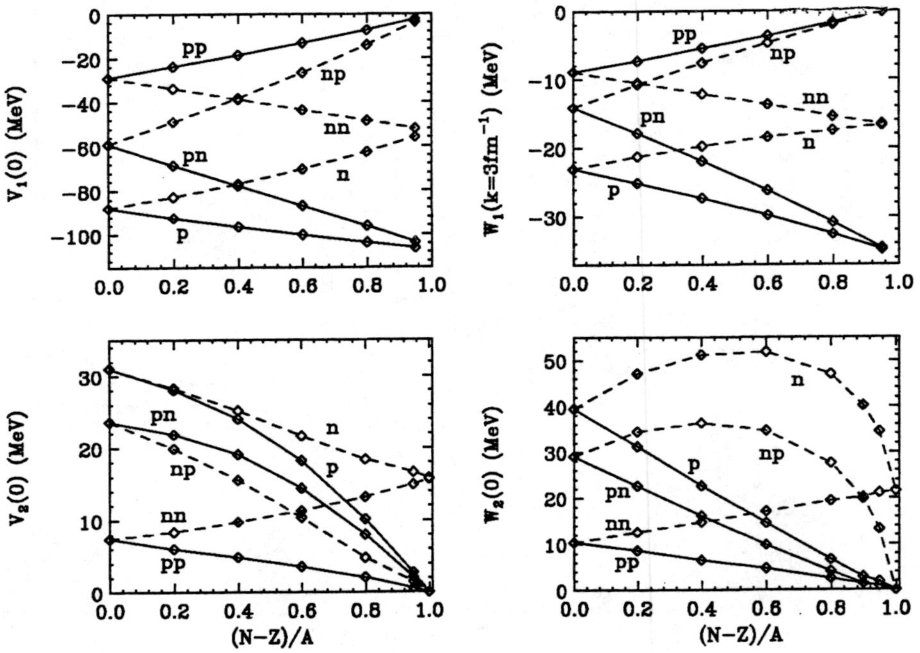

Figure 14: Self-energy vs asymmetry parameter at $\rho = 0.17 fm^{-3}$

HIC. The isospin effect on the phase transition has not received enough attention, because only small charge-asymmetries have been so far involved in such collisions. The new RIB facilities will soon offer a chance to study the phenomenology coming from exotic beams with high charge-asymmetry.

In the study of astrophysical compact objects the importance of the isospin degree of freedom has been recognized since long time. One may just mention two important fields where the EOS of hot asymmetric nuclear matter has been invoked. In the prompt explosion model of a II-type supernova the β-capture process drives the star to the last stage of the collapse where the proton fraction is about 0.33. Since the future evolution of the star is intimately related to its stiffness, an accurate knowledge of the compression modulus is demanded. For such a hot and asymmetric system indeed the compression modulus may have reached a very low value (see Tab. IV, below). Moreover the cooling mechanism of a protoneutron star via direct URCA processes requires to know with high accuracy the symmetry energy in hot nuclear matter. In fact the proton fraction is very sensitive to the symmetry energy [1].

Only few microscopic calculations of the nuclear EOS have been performed at finite temperature for symmetric nuclear matter and much less for asymmetric nuclear matter (for an exhaustive list of references see Ref. [65]).

The most appealing feature which is common to RMF and BHF approaches based on bare interaction as well as the HF using phenomenological effective interaction is the close similarity of the predicted nuclear matter EOS to the classical Van der Waals equation. This can be simply explained in terms of the typical feature of the nuclear force, which contains a short-range repulsive and a long-range attractive components as the intermolecular force. The straightforward consequence is the possible onset of a liquid-gas phase transition in correspondence to the region of mechanical instability.

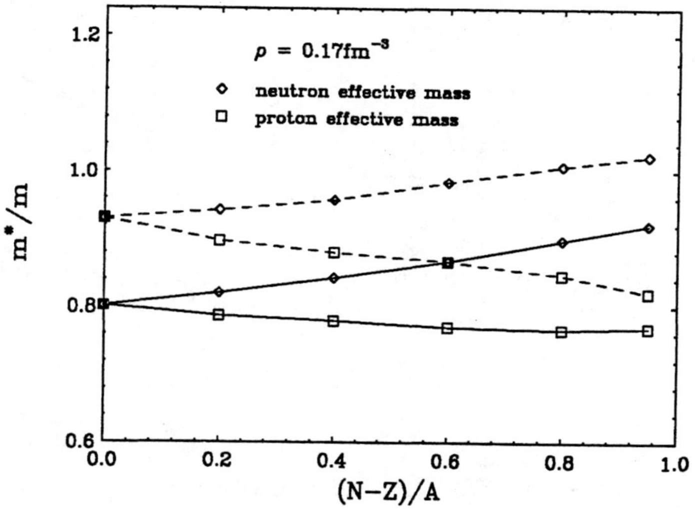

Figure 15: Effective mass.

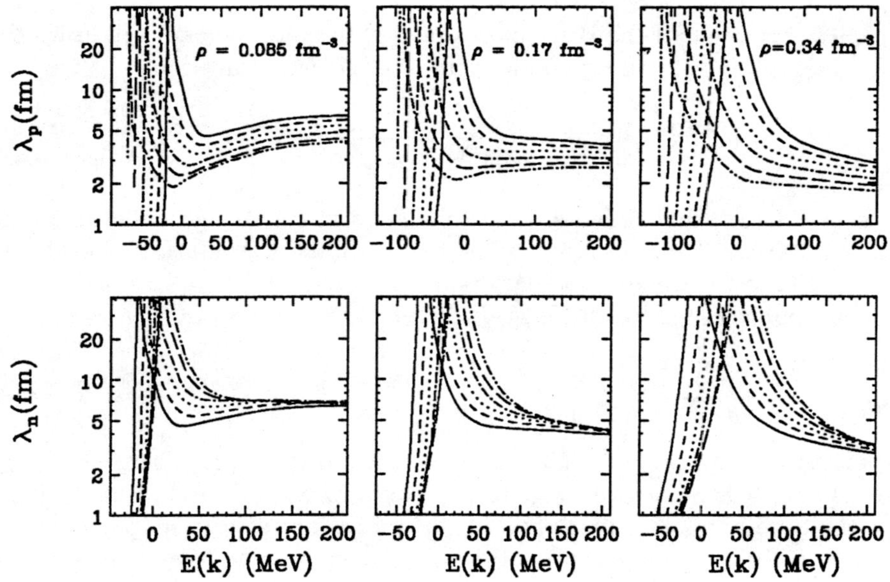

Figure 16: Mean free path.

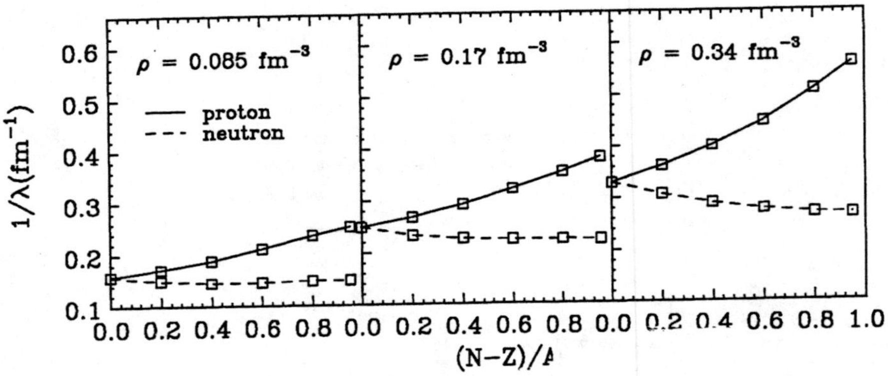

Figure 17: Mean free path vs asymmetry.

5.1 EOS for hot asymmetric nuclear matter from BHF

The finite temperature BBG formalism has been developed years ago by Bloch and De Dominicis [58]. For the present purpose one can follow the procedure leading to the finite temperature extension of the nonrelativistic mean-field approximation with phenomenological Skyrme-like forces. In this approximation the energy is given by

$$E = \sum_k \frac{\hbar^2 k^2}{2m} n_k(T) + \frac{1}{2}\sum_k \sum_{k'} < kk'|G|kk' >_A n_k(T) n_{k'}(T) \tag{18}$$

This is the simplest extension of the BHF approximation to finite temperature in which the occupation numbers, instead of being just step-functions, are the Fermi functions

$$n_{\vec{k}}^{(p,n)}(T) = \frac{1}{1 + \exp\left[(e_k^{(p,n)} - \mu^{(p,n)})/k_B T\right]} \tag{19}$$

The G-matrix is calculated from the Bethe-Goldstone equation suitably generalized at finite temperature,i.e. replacing the cold Pauli operator by the thermal Pauli operator $Q^{k,k'} = (1 - n_k(T))(1 - n_{k'}(T))$ and, accordingly, the thermal auxiliary potential.

The entropy is approximated by the Fermi-gas limit

$$S = -\sum_k [n_k(T) \lg n_k(T) + (1 - n_k(T)) \lg(1 - n_k(T))] \tag{20}$$

In hot nuclear matter the role played by the energy is replaced by the free energy $F_A = E_A - T \cdot S_A$. A BHF calculation of F_A is reported in Fig. 18 for several values of T and β. The minimum of F_A corresponds to the equilibrium density at finite temperature, i.e. to vanishing pressure. Increasing temperature, the equilibrium point disappears at a given asymmetry. As soon as asymmetry is high enough ($\beta = 0.8$ in the figure) the stability is lost at any temperature. Roughly speaking, in this approximation the temperature mainly affects the kinetic energy. But its slope in terms of β^2 is insensitive to variation of T as shown by the almost parallel lines in Fig. 19 (left panel). Thus the symmetry energy is almost insensitive to even large variations of T, as shown in the same figure (right panel). A systematic BHF calculation of proton and neutron mean field (without correlation effects) is reported in Fig. 20. Rising temperature from 0 to 10 MeV both neutron and proton mean fields do not change remarkably. The only effect is to smooth out the typical wiggles around the

Table 4: Compression modulus vs β and T from FPLyon interaction. Data taken from Ref. [62]

β	T = 0	T = 4	T = 6
0.0	218.7	198.9	172.8
0.2	204.32	175.5	159.3
0.4	166.73	147.6	124.2
0.6	104.4	79.2	45.0
0.8	17.23		

Table 5: Critical temperature. Data taken from Ref. [62]

$\beta =$	0.0	0.2	0.4	0.6	0.8
FPLyon	15.	14.5	12.8	10.2	4.0
SkIII	20.	17.5	15.9	13.0	7.0

Fermi momentum. As a consequence the enhancement of the effective mass is strongly reduced when heating the system [59, 60]. Otherwise the isospin effect on the mean fields turns unchanged.

Other important properties of hot asymmetric nuclear matter, including criticality, have not yet systematically investigated in other contexts but the mean-field approach with phenomenological forces. These calculations miss the temperature dependence of microscopic effective interaction, which, in the context of G-matrix, is contained in the Pauli blocking term and in the dispersive effect of the mean field. However, despite the parameters of a Skyrme-like interaction fitting the thermal G-matrix are strongly T-dependent [61], the EOS does not change appreciably due to a large cancellation. The most systematic study of hot asymmetric nuclear matter in the non relativistic limit has been performed in the framework of the thermal Hartree-Fock approximation with phenomenological interactions [65, 62]. In Fig. 21 a typical EOS expressed as pressure vs total density is plotted for $\beta = 0.33$ and temperature from $T = 0$ up to $T = T_c$, where T_c is the critical temperature for the liquid-gas phase transition.

The most relevant isospin effect is that the EOS becomes softer when increasing the neutron excess as the system becomes less stable. In Tab. IV the compression modulus is reported as a function of the asymmetry parameter for three values of temperature. As a combined effect of asymmetry and temperature one may conclude that a very rich-neutron system cannot sustain high temperatures.

5.2 Liquid-gas phase transition

The EOS plotted in Fig. 21 has a behavior typical of the Van der Waals EOS at finite temperature. The density range for which the pressure shows a negative slope delimits the region of mechanical instability. In equilibrium processes it corresponds to a liquid-gas phase transition for a nuclear system. It has been supposed to be responsible of the multifragmentation events in HIC.

In Tab. V the critical temperature T_c vs β is reported. It has been obtained in a HF calculation [62] using the FPLyon interaction [63]. This result is comparable with the prediction obtained in the isospin lattice gas model[64] and also in the RMF theory [4].

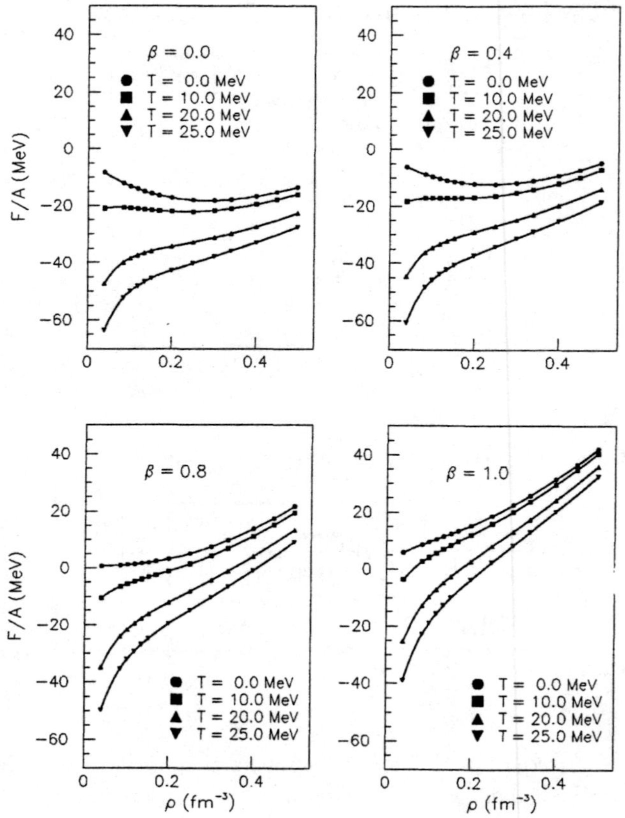

Figure 18: Free energy of asymmetric nuclear matter in the BHF approximation.

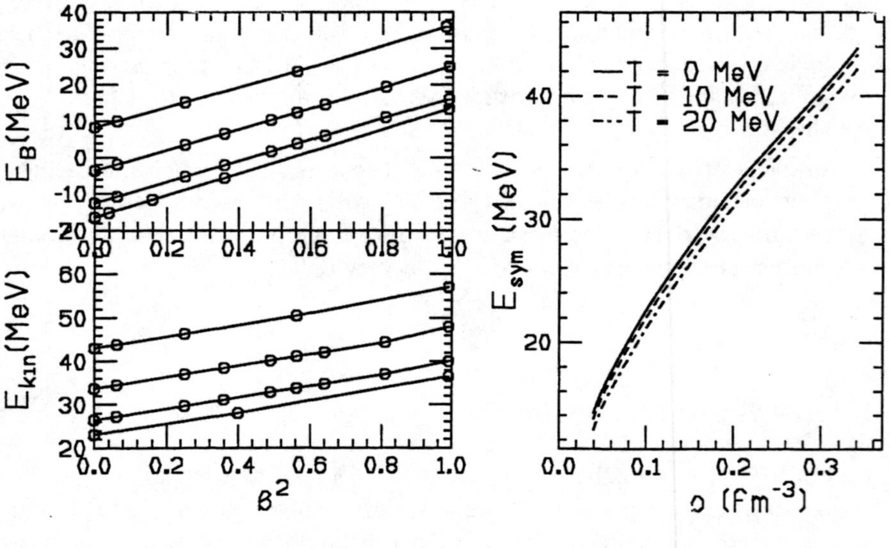

Figure 19: Symmetry energy at finite temperature (right panel)in the BHF approximation. Total and kinetic energy are plotted for T=0,5,10,20 MeV from the bottom (left panel).

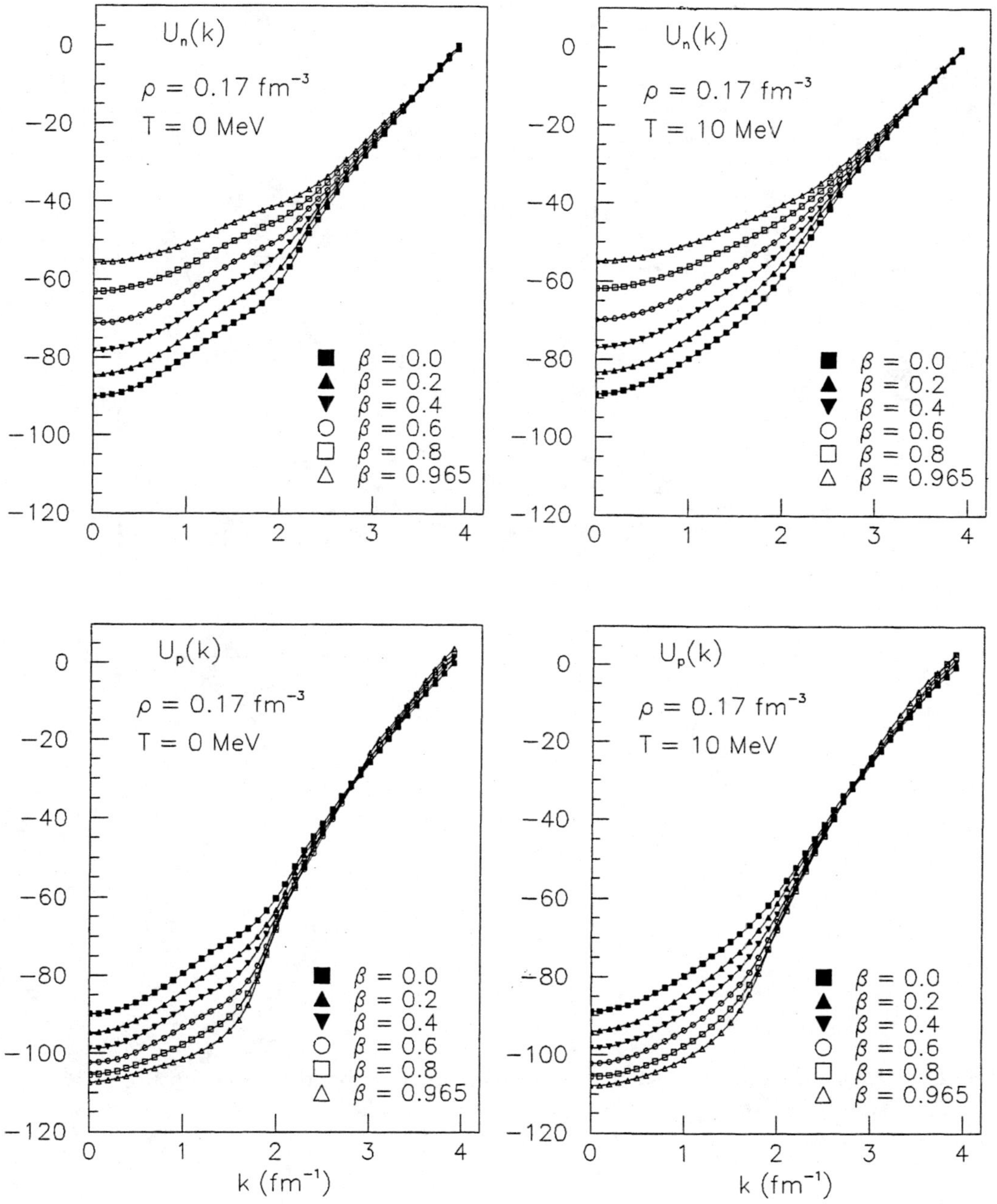

Figure 20: Proton and neutron mean field for hot nuclear matter.

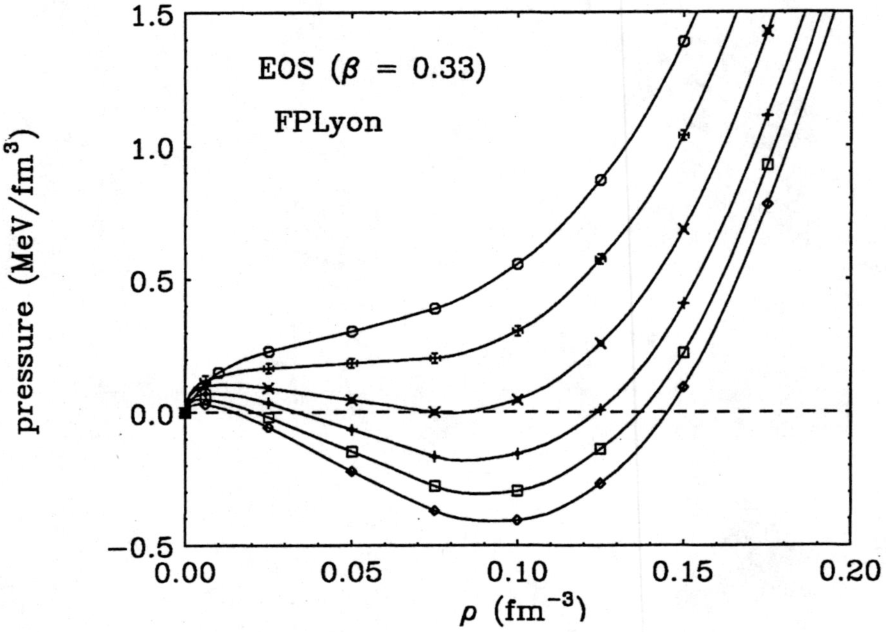

Figure 21: EOS for asymmetric nuclear matter. Isotherms correspond to temperature T = 4, 6, 8, 10.6, 13.5, 16 MeV, respectively, from below. Data taken from Ref. [62]

5.3 Chemical instability

Besides the mechanical instability, asymmetric nuclear matter could support a chemical instability related with the binary nature of the system. Recently the so called isospin fractionation observed in HIC [6] has been interpreted as a signature of such a chemical instability. On the theoretical side only very few studies on the interplay between chemical and mechanical instability have been reported in the literature [4, 65].

The conditions for these instabilities to occur in the case of neutron-rich nuclear matter are given, respectively, by the following inequalities

$$(\frac{\partial p}{\partial \rho})_{T,\beta} \leq 0 \quad ; \quad (\frac{\partial \mu_n}{\partial \beta})_{T,p} \leq 0, \tag{21}$$

where p is the pressure, μ_n the neutron chemical potential.

The domains of the two instabilities are plotted in Fig. 22 for a couple of temperatures. Increasing asymmetry and/or temperature the domain of mechanical instability gradually shrinks whereas that of chemical instability expands. This happens on the side of the liquid phase as well as on the side of the gaseous phase (notice the logarithmic scale in Fig. 22).

One may take care of the chemical instability by means of a Maxwell construction on the isobars of the EOS. As a result one gets the binodal coexistence line in the plane pressure-asymmetry which allocates the isospin content of either phase. This is plotted in Fig. 23. A detailed discussion of its physical meaning is given Ref. [4]. Here we show for comparison the predictions derived from two different interactions in order to illustrate the influence of the symmetry energy. The SkIII force [66] (solid line) gives an isospin-soft symmetry energy whereas the Lyon force [67] (dashed line) gives an isospin-stiff symmetry energy[68]. The latter is in agreement with the microscopic Brueckner calculations as discussed before. The most remarkable difference between the two predictions is that the critical pressure for SkIII is almost twice as large as the one from Lyon force.

The isospin fractionation can be explained in terms of simple energetic arguments provided the system evolves through equilibrium states, Let us consider in fact the interacting Fermi system of N neutrons and Z protons at zero temperature (extension to $T > 0$ is straightforward). Suppose the system drops into the spinodal region and splits into a liquid phase with density ρ_L and a gaseous phase with density ρ_G in the $\beta = (N - Z)/A = const$ curve. In the parabolic approximation the total energy is given by

$$E(\rho,\beta) = E_0(\rho_L) + \beta_L^2 E_{sym}(\rho_L) + E_0(\rho_G) + \beta_G^2 E_{sym}(\rho_G), \tag{22}$$

where E_0 is the energy of homogeneous symmetric nuclear matter and E_{sym} is the symmetry energy. The isospin distribution of the two phases accommodates itself in order that the energy of the whole system is the lowest with the constraint of particle and charge conservation. This provides a simple relation of the asymmetry ratio to the symmetry energy in the form

$$\frac{\beta_G}{\beta_L} = \frac{E_{sym}(\rho_L)}{E_{sym}(\rho_G)}. \tag{23}$$

Since the symmetry energy is a monotonically increasing function of the density, at least in the relevant density domains, β_G is always larger than β_L which means that the gaseous phase is more neutron-rich than the initial system as a whole, and, at contrary, the liquid phase is less neutron-rich. It is worth to notice that the asymmetry ratio does not depend on the initial asymmetry.

The interplay between the two instabilities could be also the origin of the chemical effects discussed by Colonna and Di Toro in their contribution to this book.

6 Conclusions

In this chapter the EOS of asymmetric nuclear matter has been presented and discussed from various points of view, from relativistic and nonrelativistic mean-field theories to the variational, Brueckner

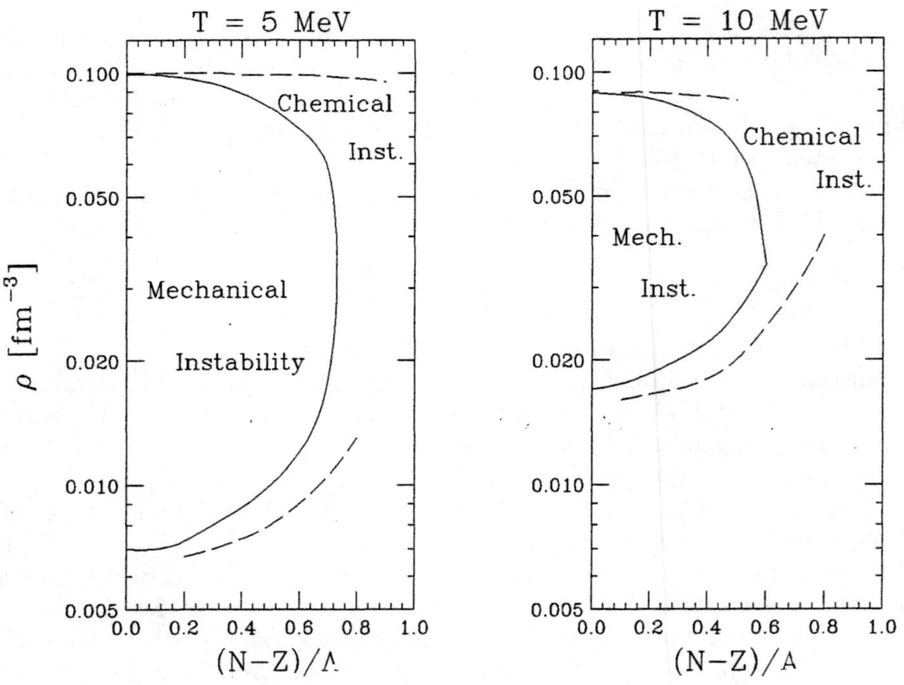

Figure 22: Regions of mechanical and chemical instability for a couple of temperatures.

and Dirac-Brueckner approaches. Since our interest was focussed mainly on the isospin effects we ranged our discussion from symmetric nuclear matter to pure neutron matter.

An achievement common to many calculations is that the β^2-law introduced in the empirical mass formula to describe neutron or proton-rich nuclei, can be extended, in the case of nuclear matter, up to high asymmetry and density. The β^2-law justifies the major role played by the symmetry energy in several fields of nuclear physics and astrophysics. At present properties of E_{sym}, including the density dependence, suffer of the controversial predictions except for its value at the saturation point. But the knowledge of its magnitude sometimes is of crucial importance as in the case of the neutron-star cooling, to settle whether or not the efficient mechanism of URCA processes can be operative. In HIC with exotic beams regions of large charge asymmetry could be formed whose properties are expected to be very sensitive to the EOS of asymmetric nuclear matter and hopefully will allow us to decide whether nuclear matter is asy-stiff or asy-soft as to the density dependence of the symmetry energy. This point is of extreme importance in the study of supernovae explosion, since the further evolution from the stage of asymmetric nuclear matter ($\beta = 1/3$) is depending on the stiffness of its EOS.

The isospin suppression of the compression modulus is established without any doubt. As a consequence the spinodal region is shrinking with increasing asymmetry. Signatures of this effect should be found in multifragmentation events again in HIC physics.

The parabolic law is also responsible for the linear isospin dependence of the proton and neutron potentials (Lane potential), effective masses and inverse mean free paths. Combining the isospin effects of both mean field and mean free path one may predict the behavior of collective flows in HIC with high charge-asymmetry. Suitable for this analysis is the balance energy for being the borderline between mean-field and collisional regime.

Isospin non linear effects on the single-particle properties are predicted in the context of EBHF approximation as due to ground state correlations. Their influence becomes relevant at high asymmetry so that possible signatures have to be expected in very high asymmetry regions such as in

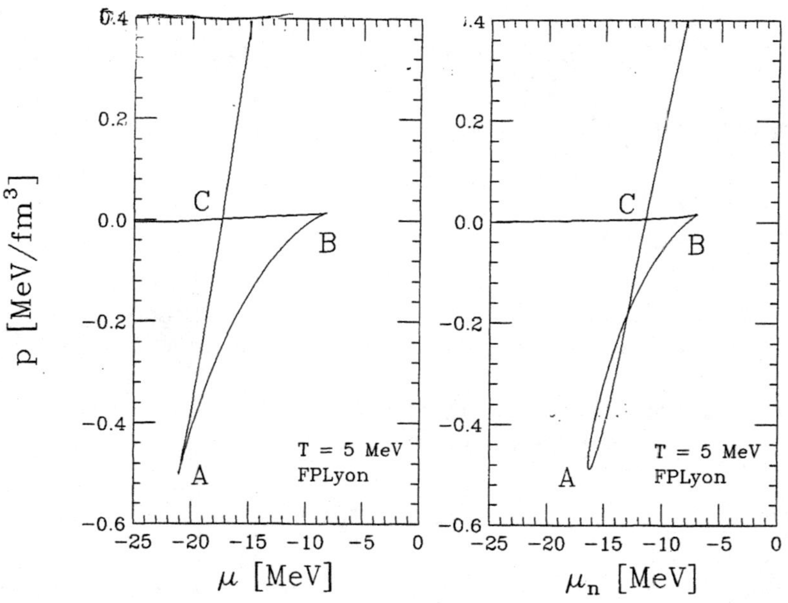

Figure 23: Binodal curve for the two interactions of the previous figure.

the neck region or in skin overlapping region of HIC. The neutron and proton chemical potential are remarkably affected by the contributions beyond the BHF approximation. This could have a far reaching consequences again for the physics of neutron star crust. In fact, the proton chemical potential in asymmetric nuclear matter is a very important ingredient in locating the inner boundary of the neutron star crust. But the difference between the neutron and proton chemical potentials and, as a consequence, the proton fraction are almost unchanged in EBHF with respect to BHF approximation.

In the wider context of thermodynamics of asymmetric nuclear matter one may investigate the simultaneous influence of temperature and isospin. In particular, sticking to mean field approximation with phenomenological forces, an important outcome is that the symmetry energy seems to be insensitive to heating the system. Finally, an important effect of isospin is the appearance of chemical instability in the EOS of asymmetric nuclear matter. The interplay between mechanical and chemical instability can be investigated in the disassembling phase in HIC. From this competition we could expect from a neutron-rich compound system evidence for a more symmetric liquid phase and more asymmetric vapour phase.

The final goal is, on one side, that the isospin physics could provide important constraints on the EOS of asymmetric nuclear matter and the theoretical investigation can open the route to explore far away the instability regions of nuclear matter, on the other side.

Acknowledgments. Discussions with Prof. M. Di Toro and Dr M. Colonna are gratefully acknowledged.

Appendix

Mass operator and quasi–particle energy

The mass operator is a complex quantity defined as

$$M^{\tau}(k,\omega) = V^{\tau}(k,\omega) + iW^{\tau}(k,\omega), \tag{1}$$

Its real part can be identified with the potential energy felt by a neutron ($\tau = n$) or a proton ($\tau = p$) with momentum $\vec{k}$ and energy ω (hereafter we will write out explicitly the isospin index τ). In the spirit of the BBG theory, the mass operator $M^{\tau}(k,\omega)$ can be expanded in a perturbation series according to the number of hole lines [49]. The first terms of this expansion are shown as Goldstone diagrams in Fig. 10.

In analogy with the case of symmetric nuclear matter, the neutron and proton quasi–particle energies $E^{\tau}(k)$ are the solutions of the energy–momentum relation

$$E^{\tau}(k) = \frac{\hbar^2 k^2}{2m} + V^{\tau}(k, E^{\tau}(k)), \tag{2}$$

i.e. $E^{\tau}(k)$ is obtained from the on–shell values of the real part of the mass operator.

To the lowest order in the hole-line expansion the mass operator is given by (the first diagram in Fig. 10)

$$M_1^{\tau}(k,\omega) = \sum_{\tau'} \sum_{\vec{k}'\sigma'} n^{\tau'}(k') \langle kk'|G^{\tau\tau'}(\omega + \epsilon^{\tau'}(k'))|kk'\rangle_A \equiv \sum_{\tau'} M_1^{\tau\tau'}(k,\omega). \tag{3}$$

In this approximation the quasi–particle energy $E_1^{\tau}(k)$ coincides with the BHF s.p. energy ε_k^{BHF}, which plays also the role of auxiliary potential in the selfconsistent procedure of solving the BHF equations.

The next contribution to the perturbative expansion of the mass operator is given by the so called *rearrangement* term $M_2^{\tau}(k,\omega)$ [50]. This contribution arises from the occupation probability dependence of the G–matrix due to Pauli operator and, in finite nuclei, accounts partly for the difference of the absolute value of the single particle energy and the removal energy. The associated Goldstone diagram is the second one shown in Fig. 10. M_2^{τ} is a second-order diagram in the G-matrix and accounts for particle–hole excitations in nuclear matter. Its expression, extended to asymmetric nuclear matter, reads

$$\begin{aligned} M_2^{\tau}(k,\omega) &= \frac{1}{2} \sum_{\tau'} \sum_{\vec{k}'\sigma'} (1 - n^{\tau'}(k')) \sum_{k_1 k_2} n^{\tau}(k_1) n^{\tau'}(k_2) \frac{|\langle kk'|G^{\tau\tau'}(\epsilon^{\tau}(k_1) + \epsilon^{\tau'}(k_2))|k_1 k_2\rangle_A|^2}{\omega + \epsilon^{\tau'}(k') - \epsilon^{\tau}(k_1) - \epsilon^{\tau'}(k_2) - i\eta} \\ &\equiv \sum_{\tau'} M_2^{\tau\tau'}(k,\omega) \end{aligned} \tag{4}$$

where $\epsilon^{\tau}(k)$ is again the s.p. spectrum in BHF approximation. In this approximation for the mass operator (*i.e.*, $M^{\tau}(k,\omega) \simeq M_1^{\tau}(k,\omega) + M_2^{\tau}(k,\omega)$) the quasi–particle energy, Eq. (2), is given by the approximate relation

$$\begin{aligned} E_2^{\tau}(k) &= E_1^{\tau}(k) + Z_2^{\tau}(k) V_2^{\tau}(k, E_1^{\tau}(k)) \\ &= \frac{\hbar^2 k^2}{2m} + V_1^{\tau}(k, E_1^{\tau}(k)) \end{aligned} \tag{5}$$

where

$$Z_2^{\tau}(k) = \left\{ 1 - \frac{\partial}{\partial\omega} \Big(V_1^{\tau}(k,\omega) + V_2^{\tau}(k,\omega) \Big) \right\}^{-1}_{\omega = E_1^{\tau}(k)} \tag{6}$$

is an approximation of the *quasi-particle strength* for asymmetric nuclear matter, which is defined by

$$Z^\tau(k) = \left\{ 1 - \frac{\partial}{\partial\omega}\left(V^\tau(k,\omega) \right) \right\}^{-1}_{\omega=E^\tau(k)} \tag{7}$$

Due to many-body correlations the two Fermi seas are partially depleted, and the correlated momentum distributions $\tilde{n}^\tau(k)$ differ from the uncorrelated ones $n^\tau(k) = \theta(k - k_F^\tau)$. To account for this physical effect, one considers the contribution $M_3^\tau(k,\omega)$ (last diagram in Fig. 10) given by [50, 51]

$$M_3^\tau(k,\omega) = -\sum_{\tau'}\sum_{\vec{h}'\sigma'} \kappa_2^{\tau'}(h')\langle kh'|G^{\tau\tau'}(\omega + \epsilon^{\tau'}(h'))|kh'\rangle_A \tag{8}$$

where the letter h refers to "hole" state with momentum smaller than k_F^τ, and

$$\kappa_2^{\tau'}(h') = -\left[\frac{\partial}{\partial\omega}M_1^{\tau'}(h',\omega)\right]_{\omega=\epsilon^{\tau'}(h')} \tag{9}$$

is at the lowest-order the depletion of neutron (proton) Fermi sea[50, 51], *i.e.* $\kappa_2^{\tau'}(h')$ is the probability that a neutron (proton) hole-state ($|\vec{h}'| \le k_F^{\tau'}$) is empty. This correction to the mass operator can also be attributed to the probability dependence of the energy denominator in the Bethe-Goldstone equation. Let us consider now the sum

$$\widetilde{M}_1^\tau(k,\omega) \equiv M_1^\tau(k,\omega) + M_3^\tau(k,\omega) = \sum_{\tau'}\sum_{\vec{h}'\sigma'}\left[1 - \kappa_2^{\tau'}(h')\right]\langle kh'|G^{\tau\tau'}(\omega + \epsilon^{\tau'}(h'))|kh'\rangle_A \tag{10}$$

$\tilde{n}_2^{\tau'}(h') = \left[1 - \kappa_2^{\tau'}(h')\right]$ is the second order approximation for the correlated momentum distribution and $\widetilde{M}_1^\tau(k,\omega)$ is the so called *renormalized* BHF approximation for the off–shell mass operator (compare to Eq. (3)).

An accurate approximation consists in using the average value of the depletion, which is

$$\kappa^{\tau'} = \kappa_2^{\tau'}(h' = 0.75k_F^{\tau'}) \tag{11}$$

Then Eqs. (8) and (10) yield

$$M_3^\tau(k,\omega) \approx -\sum_{\tau'}\kappa^{\tau'}M_1^{\tau\tau'}(k,\omega) \tag{12}$$

$$\widetilde{M}_1^\tau(k,\omega) \approx \sum_{\tau'}\left[1 - \kappa^{\tau'}\right]M_1^{\tau\tau'}(k,\omega) \tag{13}$$

From the similar considerations, a renormalization correction should also be brought from the four hole-line terms to the second–order contribution to M_2^τ in order to take into account the fact that the hole-state k_1 in Eq. (4) is partially empty (see also ref.[51] for symmetric nuclear matter). Along the same line of the previous correction one gets the *renormalized* M_2, which is approximately given by

$$\widetilde{M}_2^\tau(k,\omega) = \sum_{\tau'}\left[1 - \kappa^{\tau'}\right]M_2^{\tau\tau'}(k,\omega) \tag{14}$$

The renormalized contributions can also be traced to the functional dependence of the G-matrix on the quasi-particle occupation numbers within the Landau theory of Fermi liquids. It can be shown, in fact, that taking the functional derivative of the binding energy (at two hole-line level) includes also the terms of third and fourth order in the self-energy, the effect of which has been just discussed. Taking into account all the corrections discussed above, from Eq. (2) one can get, to the first approximation, the following expression for the quasi-particle energy [51],

$$E^\tau(k) \simeq \frac{\hbar^2k^2}{2m} + V_1^\tau(k,\epsilon_k^\tau) + Z^\tau(k)\sum_{\tau'}\left[-\kappa^{\tau'}V_1^{\tau\tau'}(k,\epsilon_k^\tau) + (1-\kappa^{\tau'})V_2^{\tau\tau'}(k,\epsilon_k^\tau)\right] \tag{15}$$

where $\epsilon_k^\tau = \hbar^2k^2/(2m) + V_1^\tau(k, \epsilon_k^\tau)$ is simply the BHF Fermi energy and the quasi-particle strength is calculated from Eq.(7) in the new approximation. The approximation in which the binding energy is truncated at the lowest hole-line order (two hole-line) and the single-particle energy is renormalized according to the Landau definition of quasi-particle energy, as discussed above, is called extended BHF (EBHF) approximation [51, 45, 52].

References

[1] for a recent review see: U. Lombardo, in *Nuclear Methods and the Nuclear Equation of State*, M. Baldo Ed. (World Scientific, Singapore, 1999), Chapt.9, p.458.

[2] A.Sedrakian, T.Alm, U.Lombardo, Phys. Rev. **C55** (1997) R582 ; U. Lombardo and A. Sedrakian, Phys. Rev. Lett. **84** (2000) 602.

[3] M. Baldo, U. Lombardo and P. Schuck, Phys. Rev. **C52** (1995) 975.

[4] H.Mueller and B.D.Serot, Phys.Rev. **C52** (1995) 2072.

[5] V.Baran, A.Larionov, M.Colonna and M.Di Toro, Nucl.Phys. **A632** (1998) 28.

[6] H.S.Xu et al. "Isospin fractionation in nuclear multifragmentation",MSUCL-1137, October 1999.

[7] G. Westfall, Nucl. Phys. **A630** (1998) 27c.

[8] I. Bombaci, T.T.S. Kuo and U. Lombardo, Phys. Reports. **242** (1994) 165 .

[9] E. Baron, J. Cooperstein and S. Kahana. Phys. Rev. Lett. **55** (1985) 126; Nucl. Phys. **A440** (1985) 744.

[10] S. Kahana, Ann. Rev. Nucl. Part. Sci. **39** (1989) 231.

[11] Baldo, J. Cugnon, A. Lejeune and U. Lombardo, Nucl. Phys. **A536** (1991) 349.

[12] B.D. Day and R.B. Wiringa, Phys. Rev. **C32** (1985) 1057.

[13] H. Q. Song, M. Baldo, G. Giansiracusa and U. Lombardo, Phys. Rev. Lett. **B81** (1998) 1584.

[14] A. Akmal and V.R. Pandharipande, Phys. Rev. C56 (1997) 2261¿

[15] for a recent review see: R. Brockmann and R. Machleidt, in *Nuclear Methods and the Nuclear Equation of State*, M. Baldo Ed. (World Scientific, Singapore, 1999), Chapt.2, p.121.

[16] J. D. Walecka, Ann. Phys. **83** (1974) 491.

[17] J. P. Blaizot, Phys. Reports **65** (1980) 171.

[18] B. D. Serot and J. D. Walecka, Int. Journ. of Modern Physics **E6** (1997) 515.

[19] S. Typel and H.H. Wolter, Nucl. Phys. **A656** (1999) 331.

[20] P. Grangé, A. Lejeune, M. Martzolff and J.-F. Mathiot, Phys. Rev. **C40** (1989) 1040.

[21] G.E. Brown, W. Weise, G. Baym and J. Speth, Comm. Nucl. Part. Phys.

[22] J. Carlson, V.R. Pandharipande and R.B. Wiringa, Nucl.Phys. **A401** (1983) 59.

[23] A. Akmal, V.R. Pandharipande and D. G. Ravenhall, Phys. Rev. **C58** (1998) 1804.

[24] M. Baldo, I. Bombaci and G. F. Burgio, Astron. Astrophys. 328 (1997) 274.

[25] K. A. Brueckner, S. A. Coon and J. Dabrowski, Phys. Rev. **168** (1968) 1184.

[26] P. J. Siemens, Nucl. Phys. **A141** (1970) 225.

[27] I. E. Lagaris and V.R. Pandharipande, Nucl. Phys. **A369** (1981) 470.

[28] J. Cugnon, P. Deneye and A. Lejeune, Z. Phys. **A328** (1987) 409.

[29] R. B. Wiringa V. Fiks, and A. Fabrocini, Phys. Rev. **C38** (1988) 1010.

[30] H. Krivine, J. Treiner and O. Bohigas, Nucl. Phys. **A336** (1980) 155.

[31] A. Lejeune, U. Lombardo and W. Zuo, Nuclear matter EOS with a three-body force, to appear in Phys. Lett. B.

[32] I. Bombaci and U. Lombardo, Phys. Rev. C44 (1991) 1892.

[33] G. H. Bordbar and M. Modarres, Phys. Rev. **C57** (1998) 714.

[34] B. ter Haar and R. Malfliet, Phys. Rev. Lett. **59** (1987) 1652.

[35] L. Engvik, M. Jorth-Jensen, E. Osnes, G. Bao, E. Oestgaard, Phys. Rev. Lett. **73** (1994) 2650.

[36] H. Huber, F. Weber and M. K. Weigel, Phys. Rev. **C51**(1995) 1790.

[37] C. H. Lee, T. T. S. Kuo, G. Q. Li and G. E. Brown, Phys. Rev. **C57** (1998) 3488

[38] F. de Jong and H. Lenske, Phys. Rev. **C57** (1998) 3099.

[39] W. D. Myers, At. Data and Nucl. Data Tables bf 17 (1976) 411.

[40] W. Zuo, I. Bombaci and U. Lombardo, Phys. Rev. **C60** (1999) 024605.

[41] M. Baldo, G. Giansiracusa and U. Lombardo and H. Q. Song, Phys. Letters **B473** (2000) 1.

[42] V. Greco, M. Colonna, M. Di Toro, G. Fabbri and F. Matera, *"Effective field models of nuclear matter: the role of Fock terms"* Proceedings XXXVIII International Meeting In Nuclear Physics, Bormio 2000.

[43] M. Di Toro et al., Prog. Part. Nucl. Phys. **42** (1999) 125-136.

[44] M. Di Toro, private communication.

[45] W. Zuo,G. Giansiracusa, U. Lombardo, N. Sandulescu and H.-J. Schulze, Phys. Letters **B421** (1998) 1.

[46] P. Grangé, J. Cugnon and A. Lejeune, Nucl. Phys. **A473** (1987) 365.

[47] G. E. Brown, Rev. Mod. Phys. **43** (1971 1.

[48] N.M. Hugenholtz and L. Van Hove, Physica (Amsterdam) **24** (1958) 363.

[49] J. Hüfner and C. Mahaux, Ann. Phys. (N.Y.) **73** (1972) 525.

[50] J.P. Jeukenne, A. Lejeune and C. Mahaux, Phys. Rep. **25C** (1976) 83.

[51] M. Baldo, M. Bombaci, G. Giansiracusa, U. Lombardo, C. Mahaux and R. Sartor, Phys. Rev. **C41** (1990) 1748.

[52] W. Zuo, U. Lombardo and H.J. Schulze, Phys. Lett. **B432** (1998) 241.

[53] A.M. Lane, Nucl. Phys. **35** (1962) 676.

[54] V. R. Pandharipande and S. C. Pieper, Phys. Rev. C45 (1992) 791.

[55] M. Avrigeanu, A. Harangozo, V. Avrigeanu and A. N. Antonov, Phys. Rev. **C54** (1996) 2538, **C56** (1997) 1633.

[56] C. Mahaux, Phys. Rev. **C28**, (1983) 1848.

[57] P. Ring and P. Schuck, "The Nuclear Many-Body Problem", Springer-Verlag, New York, 1980.

[58] C. Boch and C. De Dominicis, Nucl. Phys. **7** (1958) 459; Nucl. Phys **10** (1959) 509.

[59] A. Lejeune, P. Grangé, M. Martzolff and J. Cugnon, Nucl. Phys. **A453** (1986) 189.

[60] M.Baldo,I.Bombaci,L.S.Ferreira,G.Giansiracusa and U.Lombardo, Phys.Lett. **B215** (1988) 19.

[61] J.Cugnon, A.Lejeune, M.Baldo, U.Lombardo, Nucl.Phys.**A492** (1989) 1.

[62] D.G. Catalano,*Equation of state for nuclear matter with Skyrme interactions*, Thesis, 1999.

[63] J. Meyer *Forces effectives aux frontières de la stabilité* XI Ecole Joliot-Curie de Physique Nucléaire. Maubuisson,1993.

[64] T.T.S. Kuo, S. Ray, J. Shamanna and R.K. Su, Int. J. Mod. Phys. **E5** (1996) 303.

[65] Bao-An Li, C.M.Ko and W.Bauer, Int.J.Mod.Phys. **E7** (1998) 147-229.

[66] J. Meyer, *Forces effectives aux frontières de la stabilité*, IX Ecole Joliot-Curie de Physique Nucléaire, Maubuisson, 1993.

[67] D. Catalano, G. Giansiracusa and U. Lombardo, CRIS2000 Conference *Nucl. Phys. A* (in press).

[68] M. Colonna, M. Di Toro, G. Fabbri and S. Maccarone, Phys. Rev. **C57** (1998) 1410; M. Colonna, M. Di Toro, A. Guarnera, M. Zielinska-Pfabe, H.H. Wolter, Nucl. Phys. **A642** (1998) 449.

Equation of State for Dense Isospin Asymmetric Nuclear Matter for Astrophysical Applications Equation of State for Isospin-Asymmetric Nuclear Matter and Neutron Star Properties

I. Bombaci *

1. Introduction

The theoretical understanding of heavy ion collision experiments, and the study of the structure and evolution of astrophysical compact objects, demand a detailed knowledge of the equation of state (EOS) of nuclear matter (NM) under extreme physical conditions well far from those that can be found in ordinary nuclei. For example, matter formed in the high–dense stage of a pre–supernova core collapse [1,2] is highly asymmetric, having a neutron–proton asymmetry $\beta = (N - Z)/A \sim 0.33$. The density in the pre–supernova core, during the maximum compression stage, can reach $\sim$ 4–6 times the normal saturation density ($n_0 = 0.16$ fm^{-3}) of symmetric NM, and the central temperature has been estimated [3] to be around 30–70 MeV (assuming an entropy per nucleon S $= 1-2$ k$_{Boltz}$).

Neutron stars sample a wide density range of the EOS of dense matter. From ~ 10 g/cm^3 in the surface to several times NM saturation density ($\rho_0 \sim 2.8\times10^{14}$ g/cm^3) in the center of the star. According to our present knowledge, the cross section of a neutron star should look like the one schematically shown in Fig. 1. The details of the internal constitution of the neutron star are sensitive to the particular high density EOS employed in the calculations. However, in general, a neutron star has a characteristic layered structure. The existence of different layers is a consequence of the onset of different regimes in the EOS. Very schematically these layers can be identified as follows:

(1) **surface** ($\rho_{\rm Fe} = 7.9$ g/cm$^3 < \rho < 10^6$ g/cm^3). Matter in the neutron star's surface is composed by a lattice of ^{56}Fe nuclei (the endpoint of thermonuclear burning). As the density increases, more and more electrons detach from their respective nuclei, and at $\rho \sim 10^4$g/cm^3, one has complete ionization.
(2) **outer crust** (10^6 g/cm$^3 < \rho < \rho_n^{drip}$). The outer crust is a solid layer consisting of a Coulomb lattice of heavy nuclei ($A > 56$) in β–equilibrium with a degenerate relativistic electron gas. Going deeper into the crust (to higher density regions) nuclei in the lattice become more and more neutron rich, because the electron capture processes $e^- + p \to n + \nu_e$, lower the total energy of the system.* When the density reaches the value $\rho_n^{drip} \simeq$

*Dipartimento di Fisica Universitá di Pisa, and INFN Sezione di Pisa, Via Buonarroti 2, 56127 - Pisa, Italy

*The sequence of equilibrium nuclides mainly depends on the nuclear binding energy B(A,Z), which can be calculated using a semiempirical mass formula. The crucial point is that in the case of the neutron star's crust, one has to extrapolate the semiempirical mass formula far from the region in which their parameters

4.3 10^{11} g/cm^3 (neutron drip density), nuclei are so neutron rich that neutron states in the continuum begin to be filled.
(3) **inner crust** ($\rho_n^{drip} < \rho < \rho_{NM}$). Above the neutron drip density, the crust consist of a lattice of neutron rich nuclei embedded in a ultra–relativistic electron gas and in a neutron gas. Due to the nuclear pairing force, neutrons forms Cooper pairs and are expected to be in a $^1S_0$ superfluid state.
(4) **nuclear matter core** ($\rho_{NM} < \rho < \rho_{exo}$). When the density reaches a value ρ_{NM} =(1.8 — 2.2) 10^{14} g/cm^3, nuclei merge together and a phase transition to NM takes place [†] In this region the star consist of asymmetric nuclear matter in β–equilibrium with e^-, e^+, μ^- and μ^+. The composition of matter is mainly ruled by the density dependence of the nuclear symmetry energy. Neutron–neutron pairs form in a $^3P_0$ superfluid state and proton–proton pairs are in a $^1S_0$ superconducting state.
(5) **exotic core** ($\rho > \rho_{exo}$). Here by definition, we call exotic core of a neutron star the region above the density (ρ_{exo}) for which non nucleonic degrees of freedom become important. The existence, the properties, and the exact composition of the exotic core region are strongly dependent on the high density EOS. The possible appearance of such an exotic core, has enormous consequences for the neutron star and black hole formation mechanism [9,10].

Unfortunately large uncertainties are still present in the theoretical treatment of this ultra–dense regime [11]. Therefore, in the present work, we consider a more conventional picture assuming the NS core is composed only by an uncharged mixture of neutrons, protons, electrons and muons in equilibrium with respect to the weak interaction (β–stable matter).

The study of asymmetric nuclear matter represents also the first step for a microscopic theory of the structure of nuclei far from the valley of beta stability. This "Terra Incognita" is going to be explored in the near future thanks to a new generation of experimental facilities with high intensity radioactive nuclear ion beams [12,13]. Moreover, dynamical simulations of collisions between neutron–rich nuclei show that the main reaction mechanisms including fragmentation are quite sensitive to the density dependence of the nuclear symmetry energy [14,15,16]. Such calculations mainly make use of phenomenological Skyrme–like forces where the symmetry energy at high density can also be in strong disagreement with the one extracted from some microscopic predictions.

2. Realistic models for the nuclear EOS

Since the origin of nuclear physics, two different methods have been developed to attach the nuclear many–body problem. One is what is usually called the *microscopic approach* and the other is the *phenomenological approach.*

---

have been determined from available nuclear data (valley of beta stability).

†The phase transition to NM is expected to happen [4] through an intermediate step, in which one has the transition from spherical blobs of nuclear matter (neutron rich nuclei) embedded in the neutron and electron gas, to a phase in which balloons filled by neutron and electron gases are surrounded by NM ("swiss cheese" phase), since in this way the sum of surface and Coulomb energy can be reduced. However, further theoretical studies [5,6,7,8] have shown that the transition to nuclear matter could take place via the formation of non–spherical shapes of nuclear matter generally called *nuclear pasta.* With increasing density, the NM spherical blobs, first turn in rodlike structures ("spaghetti"), and then to platelike ones ("lasagna"). For higher densities one has the complementary nuclear shapes, i.e., the so-called "antispaghetti", and "swiss cheese".

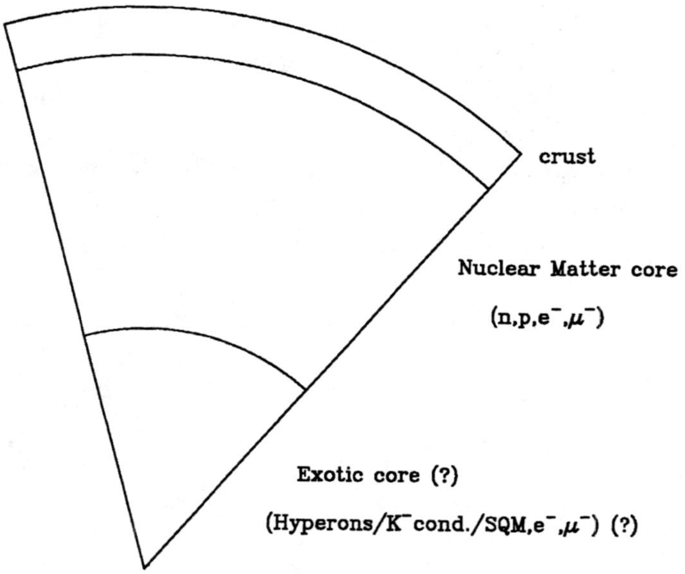

Figure 1: Schematic cross section of a neutron star as described in the text. The thicknesses of the various layers are not drawn in scale.

In a microscopic approach the input is the two–body nucleon–nucleon (NN) interaction, described by so-called *realistic* interactions like the Argonne, Bonn, Nijmegen, Paris, Urbana potentials (for a discussion on modern realistic interactions see ref.s [17,18] and references therein quoted). The theoretical basis to construct these realistic NN potentials is the meson–exchange theory of nuclear forces. In this scheme, nucleons, nucleon resonances (*e.g.* $\Delta(1232)$), and mesons such as π, ρ and ω, are incorporated in a potential representation. The various parameters in the potential are then adjusted to reproduce the experimental data for the two–body problem (deuteron properties and NN phase shifts). Then one has to solve the complicated many–body problem to get the EOS. Various methods to solve the nuclear many–body problem have been developed. In the present work, we will briefly describe the Brueckner–Bethe–Goldstone (BBG) theory. We will not discuss alternative many-body approaches, like the variational method [19,20,21,22,23], or the relativistic extension of the BBG theory, usually referred to as the Dirac–Brueckner theory [24,25,26,27]. We refer the interested reader to a few recent works [22,28,29,30,31], where the EOS of asymmetric NM is calculated within these many–body approaches.

In a phenomenological approach the input is a density-dependent effective NN interaction. The parameters appearing in this phenomenological interaction are adjusted to reproduce various properties of nuclei (*e.g.* binding energies, radii) and the empirical saturation properties of NM. The most popular of this kind of interaction is the Skyrme interaction [32,33]. The EOS can then be deduced from the effective interaction in a very simple and straightforward manner, and it is possible to derive analytical expressions for many thermodynamical quantities. The phenomenological approach is obviously less fundamental than the microscopic one, but nevertheless it has a certain number of advantages. First of all the simplicity of calculations to obtain the EOS compared to the complicated many-body calculations required in the microscopic approach. Moreover, the phenomenological EOS can be easily extrapolated far from the region in which the effective interaction parameters have been tuned (properties of nuclei and saturation properties of NM). However, as we will see in the following of this chapter, the behaviour of dense asymmetric NM is strongly related to quantity like the nuclear symmetry energy (in particular its density dependence) and to the density–, momentum–, and asymmetry–dependence of single particle properties like the neutron and proton mean field (effective masses, chemical potentials, etc.). All these quantities are very poorly constrained by experimental data. Therefore, a crude extrapolation to extreme conditions of high density and neutron–proton asymmetry is not to be fully trusted, or at least can be quite arbitrary. In fact, one can introduce phenomenological models of the EOS which equally reproduce the saturation properties of symmetric NM and finite nuclei, but which are very different in the high density and high asymmetry region. Microscopic determinations of the EOS can be very useful to clarify this ambiguity.

In any case, each model for the EOS of nuclear matter must satisfy some fundamental requisites arising from experimental data and from basic physical laws. In the present work, we call "realistic" an equation of state of NM which satisfies the following basic requirements:

(i) It must reproduce the empirical saturation point for symmetric NM

$$n_0 = 0.17 \pm 0.01 \text{ fm}^{-3}, \qquad E_0/A = -16 \pm 1 \text{ MeV}; \tag{1}$$

(ii) it must give a symmetry energy at the saturation point compatible with nuclear phenomenology

$$E_{sym}(n_0) = 30 \pm 2 \text{ MeV}, \tag{2}$$

and "well behaved" at high densities;

(iii) for symmetric NM the incompressibility at saturation must be compatible with the values extracted from phenomenology [34,35]

$$K_0 = 220 \pm 30 \text{ MeV}; \tag{3}$$

(iv) the speed of sound in NM must not exceed the speed of light in vacuum (*causality condition*), at least up to the relevant densities

$$s/c = \left(\frac{dP}{d\varepsilon}\right)^{1/2} \leq 1. \tag{4}$$

In the following, first we will present some microscopic calculations of the EOS of asymmetric NM in the framework of the non-relativistic Brueckner–Bethe–Goldstone (BBG) theory. We will not deal with the saturation problem, but we will mainly focus our attention on the analysis of several physical quantities (*e.g.* density dependence of the symmetry energy, proton and neutron single particle properties, *etc.*). Then we will introduce a phenomenological model for the EOS of (isospin) asymmetric NM, using the previous results from microscopic calculations as a guideline to constrain our phenomenological EOS in the "low density" ($n \sim n_0$) region, and in addition to the constraints given by the saturation properties of NM and finite nuclei.

3. Microscopic EOS: the Brueckner–Bethe–Goldstone theory

Asymmetric nuclear matter is a two-component system consisting of Z protons and N neutrons, interacting via the strong nuclear forces (the Coulomb interaction between protons is not considered by definition) within a volume V, in the thermodynamic limit:

$$A = N + Z \to \infty, \qquad V \to \infty, \qquad n \equiv \frac{A}{V} \to \text{finite}. \tag{5}$$

The neutron and proton number densities are related to the total number density $n = n_p + n_n$, and to the isospin-asymmetry parameter (or simply the *asymmetry parameter*)

$$\beta = \frac{n_n - n_p}{n} \tag{6}$$

via the equations:

$$n_n = \frac{1}{2}(1+\beta)\,n, \qquad n_p = \frac{1}{2}(1-\beta)\,n. \tag{7}$$

The Fermi momentum $k_F^{[p]}$ of protons and $k_F^{[n]}$ of neutrons is related to the corresponding number density n_τ as follows:

$$n_\tau = \frac{1}{3\pi^2}(k_F^{[\tau]})^3 \tag{8}$$

(with $\tau = p$, or $\tau = n$) and to the total number density n:

$$k_F^{[\tau]}(n,\beta) = \left[\frac{3\pi^2}{2}(1 \pm \beta)n\right]^{1/3} \tag{9}$$

where the upper sign holds for neutrons and the lower for protons.

The extension of the Brueckner–Bethe–Goldstone (BBG) theory to finite temperature T is based on a linked cluster expansion of the grand canonical potential Ω, in analogy to the usual BBG linked cluster expansion of the energy per particle of symmetric NM at zero temperature.

The basic ingredient of the BBG approach is the reaction matrix G. In the case of asymmetric NM, the reaction matrix depends on the isospin components of the two colliding nucleons. Therefore we have different reaction matrices for proton-proton scattering, neutron-neutron scattering, and neutron-proton scattering. These satisfy the generalized Bethe-Goldstone equations:

$$G_{\tau\tau'}(n,\beta,T;\omega) = v + v\sum_{k_a k_b} \frac{|k_a k_b\rangle Q_{\tau\tau'} \langle k_a k_b|}{\omega - e_\tau(k_a) - e_{\tau'}(k_b)} G_{\tau\tau'}(n,\beta,T;\omega), \tag{10}$$

where v is the bare nucleon-nucleon (NN) interaction, ω is the starting energy. Here, we consider NM with no spin excess (*spin-symmetric* NM), therefore in eq.(10) and in the following equations we drop the spin indices to simplify the formalism. $Q_{\tau\tau'}$ is the finite temperature Pauli operator

$$\begin{aligned} Q_{\tau\tau'} &\equiv Q_{\tau\tau'}(k_a,k_b;n,\beta,T) = \\ &= \left[1 - f_\tau(k_a,T,k_F^{[\tau]}(n,\beta))\right]\left[1 - f_{\tau'}(k_b,T,k_F^{[\tau']}(n,\beta))\right] \end{aligned} \tag{11}$$

which can be expressed in terms of the Fermi distribution

$$f_\tau(k,T,k_F^{[\tau]}(n,\beta) = \Big[1 + exp\Big(\frac{e_\tau(k) - \mu_\tau}{T}\Big)\Big]^{-1} \tag{12}$$

where $\mu_\tau = \mu_\tau(n,\beta,T)$ is the chemical potential for the species τ, and

$$e_\tau(k) = e_\tau(k;n,\beta,T) = \frac{\hbar^2}{2m}k^2 + U_\tau(k;n,\beta,T) \tag{13}$$

is the single particle energy. [‡] The Brueckner–Hartree–Fock (BHF) approximation for the single particle (s.p.) potential U_τ using the *continuous choice* is

$$\begin{aligned} U_\tau(k;n,\beta,T) &= \sum_{\tau'} U_{\tau\tau'}(k;n,\beta,T) \\ &= \sum_{\tau'}\sum_{k'} f_{\tau'}(k';T,k_F^{[\tau']})\langle kk'|G_{\tau\tau'}(n,\beta,T;e_\tau(k)+e_{\tau'}(k'))|kk'\rangle_A \end{aligned} \tag{14}$$

where we use the notation $|kk'\rangle_A \equiv |kk' - \delta_{\tau\tau'}k'k\rangle$ to indicate antisymmetrization of the matrix element for identical particles pair ($\tau = \tau'$). In this generalized BBG approach equations (10)–(14) have to be solved selfconsistently.

In the BHF approximation the total energy is:

$$E(n,\beta,T) = E^{kin} + D_2 \ , \tag{15}$$

where

$$E^{kin}(n,\beta,T) = \sum_\tau\sum_k f_\tau(k;n,\beta,T)\ \frac{\hbar^2}{2m}k^2 \ , \tag{16}$$

is the total kinetic energy, and

$$D_2(n,\beta,T) = \frac{1}{2}\sum_\tau\sum_k f_\tau(k;n,\beta,T)\ U_\tau(k;n,\beta,T) \equiv \tag{17}$$

$$\frac{1}{2}\sum_{\tau\tau'}\sum_{kk'} f_\tau(k;n,\beta,T)\ f_{\tau'}(k';n,\beta,T)\ \langle kk'|G_{\tau\tau'}(n,\beta,T;e_\tau(k)+e_{\tau'}(k'))|kk'\rangle_A \tag{18}$$

is the BHF total potential energy. It has been recently shown [36] that the three hole-line contributions (D_3) to the BBG expansion are very small as compared to D_2, both in the case of symmetric NM and pure neutron matter, at zero temperature, when the continuous choice for s.p. potential is used.

The total entropy S can be evaluated in the approximation of a non-interacting Fermi gas of quasi-particle in the mean field $U_\tau(k;n,\beta,T)$, and is given by

$$S(n,\beta,T) = -\sum_\tau\sum_k \Big\{f_\tau(k)\ln f_\tau(k) + [1 - f_\tau(k)]\ln[1 - f_\tau(k)]\Big\} \ . \tag{19}$$

The free energy can be calculated using the standard thermodynamic relation

$$F = E - TS \ . \tag{20}$$

[‡] We assume the neutron and proton rest masses equal to their average value m.

In this scheme, the only input quantity we need is the bare NN interaction v in the generalized Bethe–Goldstone equation (10). In this respect the BBG approach can be considered as a microscopic one. The Brueckner G-matrix accounts for the complicated many-body effects on the in medium NN interaction. It provides the starting point for a microscopic determination of the effective nuclear interaction to be used in shell-model calculations of nuclear properties. This is the philosophy behind the well known Kuo–Brown effective interaction [37] and successive developments based on the so-called *folded diagrams method* [38,39].

Various properties of cold asymmetric NM have been calculated, within the BBG approach, by Bombaci and Lombardo [40] (BL), and in ref. [3] for supernova matter (*i.e* $\beta \sim 0.33$, and constant entropy per nucleon $\sim 1\ k_{Boltz}$), using the separable form of the Paris potential [41]. Also, the EOS at finite temperature for symmetric nuclear and pure neutron matter has been recently calculated by Baldo and Ferreira [42] for the Argonne v_{14} NN interaction.

3.1. Symmetry energy

Under the assumption of charge symmetry of nuclear interactions ($v_{nn} = v_{pp}$), the energy per particle $B(n,\beta) \equiv E(n,\beta)/A$ of asymmetric NM at zero temperature can be written as

$$B(n,\beta) = B(n,0) + S_2(n)\beta^2 + S_4(n)\beta^4 + S_6(n)\beta^6 + ... \qquad (21)$$

The nuclear symmetry energy is defined by

$$E_{sym}(n) \equiv \frac{1}{2}\frac{\partial^2 B}{\partial \beta^2}\Big|_{\beta=0} \qquad (22)$$

and is equal to the lower term in β in the expansion (21). If the higher order terms, S_4, S_6,..., in the expansion (21) are zero, for all densities, the energy per particle of asymmetric nuclear matter can be written as

$$B(n,\beta) = B(n,0) + E_{sym}(n)\beta^2, \qquad (23)$$

and the symmetry energy can then be expressed in term of the difference of the energy per particle between symmetric and neutron matter:

$$E_{sym}(n) = B(n,1) - B(n,0)\ . \qquad (24)$$

The parabolic form (23) has been assumed by many authors to evaluate the energy per particle of asymmetric matter by interpolation between that of symmetric and neutron matter. The validity of equation (23) has been checked microscopically by BL [40] in the framework of the BBG theory, and by Bordbar and Modarres [43] with the lowest order constrained variational method. In Fig. 2, we report the results (symbols) of ref. [40], for the difference $B(n,\beta) - B(n,0)$, as calculated selfconsistently according to equations (15)-(17). For comparison, in the same figure, we show the quantity $E_{sym}(n)\beta^2$ (straight lines), with E_{sym} calculated according to its definition (22) and using the three values $\beta = 0.0$, 0.2, 0.4 for the asymmetry parameter. As we can see the parabolic law (23) is fulfilled very well for all the range of densities considered in Fig. 2. In Fig. 3, we show the symmetry energy versus the number density n. The circles (curve Paris(BL)) represent the results of BL for the

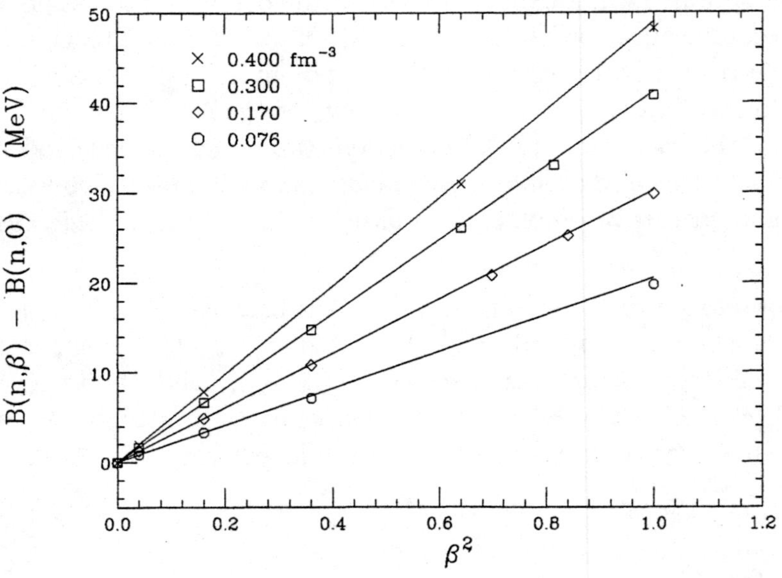

Figure 2: Total binding energies, at four different densities, as obtained from the self-consistent BHF calculations (symbols). These results are compared with the parabolic fits (straight lines) obtained from the first three values of β (0.0, 0.2, 0.4). The slope of each line gives the corresponding symmetry energy.

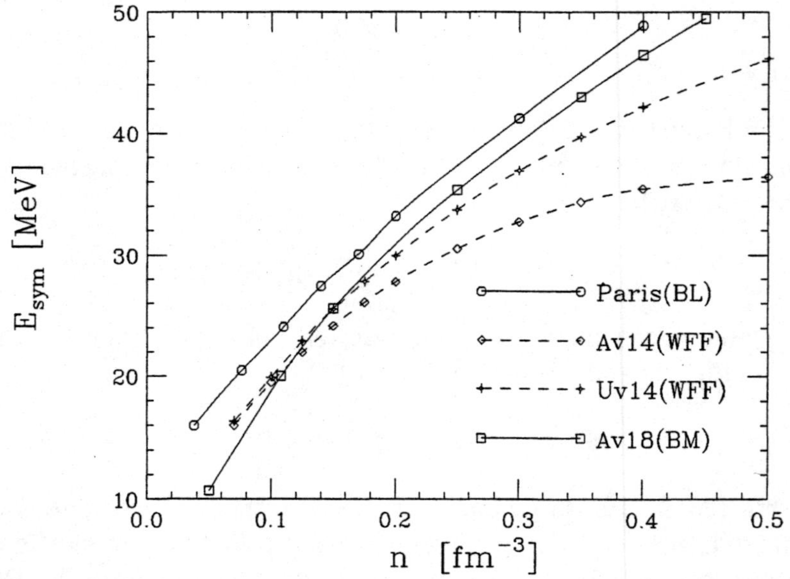

Figure 3: Symmetry energy versus number density. Microscopic calculations from ref. [40] (curve Paris(BL)), and ref. [43] (curve Av18(BM)). Curves labeled Av14(WFF) and Uv14(WFF) refer to the variational calculations of ref. [22] with the Argonne and Urbana v_{14} potentials (with only two-body forces hamiltonian). The results by WFF make use of the parabolic approximation (23).

Paris potential, obtained using eq. (22). The calculated value at $n = 0.17$ fm^{-3} is 30 MeV, which is in good agreement with the value 28–32 MeV, extracted from the semiempirical mass formula [34]. The squares (curve Av18(BM)) depict the results of ref. [43] with the Argonne v_{18} potential. In the same figure, we also show E_{sym} as calculated by Wiringa Ficks and Fabrocini [22] (WFF) in the framework of a *variational* approach, in the case of Argonne v_{14} (Av_{14}) and Urbana v_{14} (Uv_{14}) NN potentials. The WFF symmetry energy is calculated from eq. (24), *i.e.* assuming *a priori* the validity of the parabolic approximation.

3.2. Single particle potential

The self-consistent BHF neutron and proton single particle (s.p.) potential energy (or simply the s.p. *potential*) is reported in the left panel of Fig. 4 as a function of the s.p. momentum, for different values of the asymmetry parameter β and for the density $n = 0.17$ fm^{-3}. The proton mean field U_p become more attactive, while the neutron mean field U_n become more repulsive going from symmetric ($\beta = 0$) to neutron ($\beta = 1$) matter. The same qualitative behaviour has been found by the authors of ref. [40] for all the other values of the density ($n < 0.4\ fm^{-3}$) considered in their work. In the right panel of the same figure, the single particle potentials, for $k = 0$ and $n = 0.17\ fm^{-3}$, are plotted as a function of the neutron–proton asymmetry β together with their component $U_{\tau\tau'}$ defined according to eq. (14). U_{pp} and U_{np} are the components of the mean field felt, respectively, by a proton and a neutron due to the proton background, which, as expected, vanish at $\beta = 1$. U_{nn} and U_{pn} are the same mean field components generated by the neutron background.

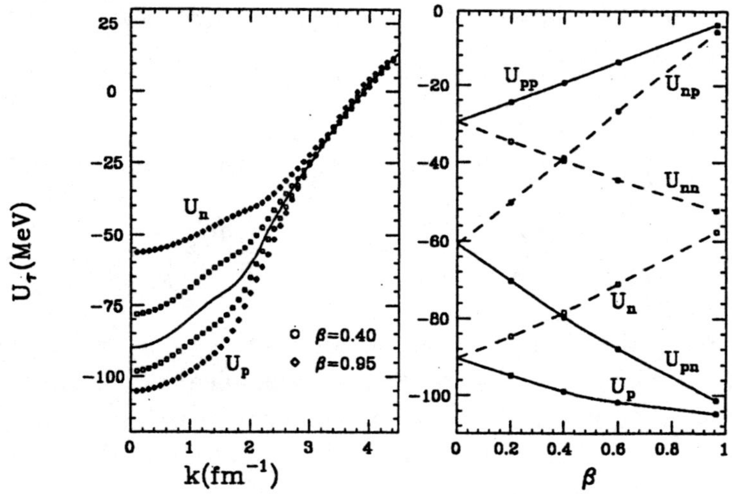

Figure 4: Microscopic single particle potential. Left panel: U_n and U_p versus k, for different values of asymmetry parameter. The continuous line gives the sp. potential in symmetric NM. Right panel: the s.p. potentials, at $k = 0$, are plotted as a function of β together with their component $U_{\tau\tau'}$ defined according to eq. (14). The number density is taken equal to $0.17\ fm^{-3}$. Adapted from Bombaci & Lombardo ref. [40].

The β–dependence of U_n and U_p is almost linear and nearly symmetric with respect to their common value at $\beta = 0$ for small β. This support from a microscopic point of view the validity of the so–called Lane potential [44]. The effect of the neutron–proton asymmetry can be clarified by splitting the proton and neutron mean field into their component, in such a way to single out their phase space dependence. Using eq. (14) the BHF s.p. potential can be written:

$$U_n = U_{nn} + U_{np} = u_{nn}n_n + u_{np}n_p = \frac{1}{2}n(u_{nn} + u_{np}) + \frac{1}{2}n(u_{nn} - u_{np})\beta \qquad (25)$$

$$U_p = U_{pp} + U_{pn} = u_{pp}n_p + u_{pn}n_n = \frac{1}{2}n(u_{pp} + u_{pn}) - \frac{1}{2}n(u_{pp} - u_{pn})\beta \qquad (26)$$

where u_{nn} is the average value of G_{nn} in the neutron Fermi sphere ($k' \leq k_F^{[n]}$), u_{np} is the average value of G_{np} in the proton Fermi sphere ($k' \leq k_F^{[p]}$), and u_{pp}, u_{pn} have similar definitions.

Suppose now,

$$u_{nn} = u_{pp}, \qquad u_{np} = u_{pn} \qquad (27)$$

and introduce the quantities [§] : $\Sigma \equiv u_{nn} + u_{np} = u_{pp} + u_{pn}$, and $\Delta \equiv u_{nn} - u_{np} = u_{pp} - u_{pn}$. Then eq.s (25),(26), can be written as

$$U_n = \frac{1}{2}n\Sigma + \frac{1}{2}n\Delta\,\beta \qquad (28)$$

$$U_p = \frac{1}{2}n\Sigma - \frac{1}{2}n\Delta\,\beta \qquad (29)$$

The results summarized in Fig. 4, are in agreement with the previous analysis (eq.s (28),(29)). In particular, at $n = 0.17\ fm^{-3}$ and $k = 0$, and in the case of "low asymmetry" ($\beta \leq 0.2$), we found that u_{nn}, u_{pp} differ by less than 4% and u_{np}, u_{pn} differ by less than 2%. We verified that conditions (27), and consequently eq.s (28),(29), are less and less fulfilled going to higher neutron–proton asymmetries or to higher densities.

3.3. Effective mass

The neutron and proton effective masses in nuclear medium are defined according to

$$\frac{m^*_\tau}{m} = \left\{\frac{m}{\hbar^2 k}\frac{de_\tau}{dk}\right\}^{-1}_{k_F^{[\tau]}} = \left\{1 + \frac{m}{\hbar^2 k}\frac{dU_\tau}{dk}\right\}^{-1}_{k_F^{[\tau]}} \qquad (30)$$

where e_τ is the single particle energy. The calculated m^*_n and m^*_p are plotted in Fig. 5 as a function of β for $n = 0.17\ fm^{-3}$. The resulting scissors-like dependence of $m^*_\tau(\beta)$ can be interpreted in terms of the same analysis performed for the s.p. potential.

Using eq.s (28),(29) the effective mass can be written as

$$\frac{m^*_\tau}{m} = \left\{\left(1 + \frac{m}{2\hbar^2 k}n\frac{d\Sigma}{dk}\right) \pm \left(\frac{m}{2\hbar^2 k}n\frac{d\Delta}{dk}\,\beta\right)\right\}^{-1}_{k_F^{[\tau]}} \qquad (31)$$

where the upper sign refers to neutrons ($\tau = n$) and the lower sign to protons ($\tau = p$). For $\beta \ll 1$, and using the Taylor expansion eq. (31) reads

$$\frac{m^*_\tau}{m} = \frac{1}{a_\tau}\left\{1 \mp \frac{1}{a_\tau}\frac{m}{2\hbar^2}n\left(\frac{1}{k}\frac{d\Delta}{dk}\right)_{k_F^{[\tau]}}\beta + \ldots\ldots\right\} \qquad (32)$$

[§] Clearly, even under the assumptions (27), $u_{pn}, u_{np} \neq u_{nn}, u_{pp}$. In fact, u_{pn} and u_{np} receive contributions from both singlet ($T = 0$) and triplet ($T = 1$) isospin channels, whereas u_{nn} and u_{pp} only from triplet isospin channels.

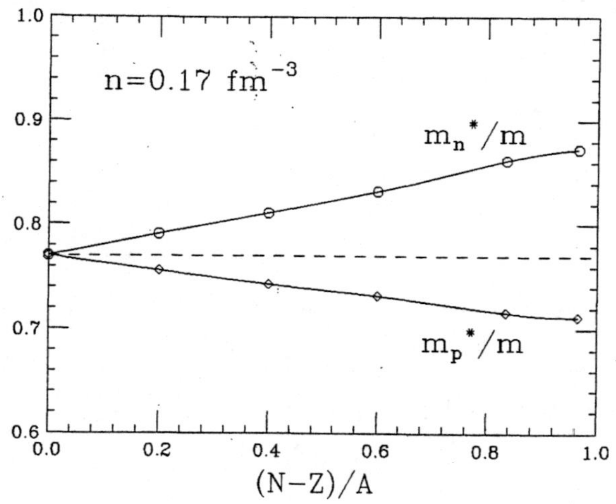

Figure 5: Neutron and proton effective masses (eq. (30)) for different values of the asymmetry parameter β. Adapted from ref. [40].

with

$$a_\tau(\beta) \equiv \left(1 + \frac{m}{2\hbar^2 k} n \frac{d\Sigma}{dk}\right)_{k_F^{[\tau]}} \tag{33}$$

After a Taylor expansion of $a_\tau(\beta)$ and $(\frac{1}{k}\frac{d\Delta}{dk})_{k_F^{[\tau]}}$, retaining only linear terms in β, one has:

$$m_\tau^*(n) \simeq m_0^*(n)\left\{1 \pm b(n)\ \beta\right\} \tag{34}$$

$m_0^*(n)$ being the nucleon effective mass in symmetric NM, and $b(n)$ a constant which depends on the derivatives of Σ and Δ with respect to k calculated at $k_F = (3\pi^2/2)^{1/3} n^{1/3}$.

3.4. Microscopic EOS with three-body forces

Non–relativistic calculations, based on purely two–body interactions, fail to reproduce the correct saturation point for symmetric nuclear matter [45]. This well known deficiency is commonly corrected introducing three-body forces (TBF). Unfortunately, it seems not possible to reproduce the experimental binding energies of light nuclei and the correct saturation point accurately with one simple set of TBF (see ref. [22]). Relevant progress has been made in the theory of nucleon TBF, but a complete theory is not yet available. A realistic model for nuclear TBF has been introduced by the Urbana group [46]. The Urbana model consists of an attractive term $V_{ijk}^{2\pi}$ due to two–pion exchange with excitation of an intermediate Δ-resonance, and a repulsive phenomenological central term V_{ijk}^R

$$V_{ijk} = V_{ijk}^{2\pi} + V_{ijk}^R \tag{35}$$

The two–pion exchange contribution is a cyclic sum over the nucleon indices i, j, k of products of anticommutator $\{,\}$ and commutator $[,]$ terms

$$V_{ijk}^{2\pi} = A \sum_{cyc} \Big(\{X_{ij}, X_{jk}\}\{\tau_i \cdot \tau_j, \tau_j \cdot \tau_k\} \tag{36}$$

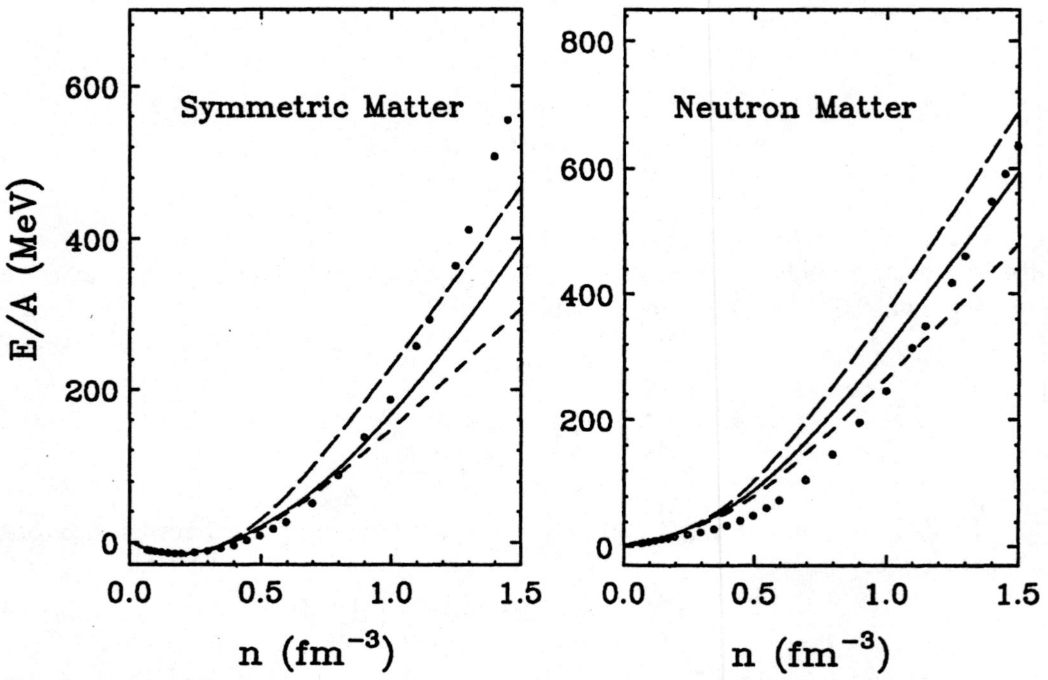

Figure 6: The energy per baryon E/A is plotted vs. the number density n for symmetric matter (panel (a)) and for neutron matter (panel (b)). Several EOS's are shown, *i.e.* non-relativistic Brueckner calculations with three-body forces (solid line, Paris potential; short dashed line, AV14 potential) and a relativistic Dirac-Brueckner one (long dashes, DBHF). For comparison a variational calculation (with AV14+UVII interaction) is also reported (open circles, WFF).

Table 1. Saturation properties for the microscopic equations of state of ref. [47]. The number density n is given fm^{-3}. The saturation energy E_0/A, the incompressibility K_0, and the symmetry energy $E_{sym}(n_0)$ are given in MeV.

EOS	*interaction*	n_0	E_0/A	K_0	$E_{sym}(n_0)$
BBB1	Av14+TBF	0.178	-16.46	253	34.6
BBB2	Paris+TBF	0.176	-16.01	281	34.9

$$+ \quad \frac{1}{4}[X_{ij}, X_{jk}][\tau_i \cdot \tau_j, \tau_j \cdot \tau_k]\Big), \tag{37}$$

where

$$X_{ij} = Y(r_{ij})\sigma_i \cdot \sigma_j + T(r_{ij})S_{ij} \tag{38}$$

is the one–pion exchange operator, σ and τ are the Pauli spin and isospin operators, and $S_{ij} = 3[(\sigma_i \cdot r_{ij})(\sigma_j \cdot r_{ij})/r_{ij}^2 - \sigma_i\sigma_j]$ is the tensor operator. $Y(r)$ and $T(r)$ are the Yukawa and tensor functions, respectively, associated to the one–pion exchange, as in the two–body potential.

The repulsive part is taken as

$$V_{ijk}^R = U \sum_{cyc} T^2(r_{ij})T^2(r_{jk}) \tag{39}$$

The constants A and U in the previous equations can be adjusted to reproduce observed nuclear properties. Within this scheme the Urbana group [46] found $A = -0.0333$ and $U = 0.0038$ by fitting properties of light nuclei (^{3}H, ^{4}He), the so-called Urbana VII parameterization (UVII).

Recently Baldo, Bombaci and Burgio [47] (hereafter BBB) introduced the same Urbana three-nucleon model within the BHF approach. To incorporate the TBF in the Brueckner scheme, the authors of ref. BBB followed the method of Lejeune *et al.* [48]. The TBF is reduced to an effective two-body force by averaging on the position of the third particle, assuming that the probability of having two particles at a given distance is reduced according to the two-body correlation function. The resulting effective two-body force is of course density dependent. The parameters A and U are then adjusted in order to reproduce closely the correct saturation point of symmetric NM, since for neutron stars studies this is an essential requirement, and there is no reason to believe that TBF be the same as in light nuclei. The values used by BBB are $A = -0.0329$ and $U = 0.00361$ both for the Paris and the AV14 potentials.

The corresponding EOS obtained using the Paris (solid line) and the AV14 potentials (short dashed line) are depicted in Fig. 6 for symmetric (panel (a)) and neutron matter (panel (b)). The saturation properties for these EOS models are given in Tab. 1.

Notice that the value for the nuclear incompressibility is very close to the recent phenomenological estimate of Myers [34]. In the same figure, we show the EOS obtained within the variational many–body approach by WFF when the UVII parameterization [46], is added to the AV14 two–body force (open circles).

In Fig. 6 we plot also the EOS from a recent Dirac-Brueckner calculation [27] (DBHF) with the Bonn–A two–body force (long dashed line). In the low density region ($n < 0.4\ fm^{-3}$), both BHF equations of state with TBF and DBHF equations of state are very similar, whereas at higher density the DBHF is stiffer. The discrepancy between the non-relativistic and relativistic calculation of the EOS can be easily understood by noticing that the DBHF treatment is equivalent [49] to introduce in the non-relativistic BHF the three-body force corresponding to the excitation of a nucleon-antinucleon pair, the so-called Z-diagram [50]. The latter is repulsive at all densities. In BHF treatment, on the contrary, both attractive and repulsive three-body forces are introduced, and therefore a softer EOS can be expected.

4. The Skyrme interaction and EOS

An alternative approach to the use of microscopic nuclear effective interactions *a la* Kuo–Brown [37,38], makes use of purely phenomenological interactions. The most popular of these are the Skyrme interactions [32]. Here we consider a Skyrme interaction of the following form: ¶

$$\begin{aligned} v_{ij} &= t_0(1+x_0P_\sigma)\delta(\mathbf{r}) + \frac{1}{2}t_1\left[\mathbf{k}'^2\delta(\mathbf{r}) + \delta(\mathbf{r})\mathbf{k}^2\right] + t_2\mathbf{k}'\cdot\delta(\mathbf{r})\mathbf{k} \\ &+ \frac{1}{6}t_3(1+x_3P_\sigma)\left[n(\mathbf{R})\right]^\gamma\delta(\mathbf{r}) + iW_0\sigma\cdot\left[\mathbf{k}'\times\delta(\mathbf{r})\mathbf{k}\right] \end{aligned} \tag{40}$$

with $\mathbf{r}=\mathbf{r}_i-\mathbf{r}_j$, $\mathbf{R}=(\mathbf{r}_i+\mathbf{r}_j)/2$, $\mathbf{k}=\frac{1}{2i}(\nabla_i+\nabla_j)$, $\mathbf{k}'=-\frac{1}{2i}(\nabla_i+\nabla_j)$, with the convention that $\mathbf{k}'$ acts on the wave function to its left. P_σ is the spin exchange operator

$$P_\sigma = \frac{1}{2}(1+\sigma_i\cdot\sigma_j), \qquad\qquad \sigma=\sigma_i+\sigma_j, \tag{41}$$

n denotes the nucleon number density, and $\delta(\mathbf{r})$ is the Dirac δ-function. The coefficients t_0, t_1, t_2, t_3, x_0, x_3 and W_0 are phenomenological parameters. The Skyrme interaction eq.(40) contains a zero-range part, a momentum dependent part which accounts for non-local contributions to the interaction (t_1 and t_2 terms), a density dependent term and a spin-orbit term. Since the work of Vautherin and Brink [33] Skyrme forces have been widely and successfully used to calculate various nuclear properties within the Hartree–Fock method.

Consider now *spin-symmetric* (but isospin-asymmetric) NM. Then the energy density $\varepsilon = E/A$ (including the rest mass contribution) obtained from the interaction (40) is: ‖

$$\begin{aligned} \varepsilon(n_n,n_p) &= \varepsilon_n^{kin} + \varepsilon_p^{kin} + \frac{1}{2}t_0\left[\left(1+\frac{1}{2}x_0\right)n^2 - \left(\frac{1}{2}+x_0\right)(n_n^2+n_p^2)\right] + \\ &+ \frac{1}{12}t_3\left[\left(1+\frac{1}{2}x_3\right)n^2 - \left(\frac{1}{2}+x_3\right)(n_n^2+n_p^2)\right]n^\gamma \\ &+ \frac{1}{4}(t_1+t_2)(\tau_n+\tau_p)n + \frac{1}{8}(t_2-t_1)(\tau_n n_n+\tau_p n_p) + nmc^2 \end{aligned} \tag{42}$$

¶ With respect to the "standard" form of the Skyrme interaction, here we do not consider momentum dependent spin-exchange forces. In other words, using the usual notation for the phenomenological parameters, we take $x_1 = x_2 = 0$.

‖ All contributions to the energy density involving spatial derivatives are equal to zero in NM due to the translational invariance. Moreover, the spin-orbit term in the Skyrme interaction is not relevant for our discussion, since in the present work we consider spin-symmetric NM.

with

$$\varepsilon_i^{kin} = \frac{\hbar^2}{2m}\tau_i, \qquad \tau_i = \frac{3}{5}(3\pi^2)^{2/3} n_i^{5/3}, \qquad i = n, p \tag{43}$$

The energy per nucleon $B \equiv E/A = \varepsilon/n - mc^2$, is then:

$$\begin{aligned} B(n,\beta) &= B^{kin}(n,\beta) + \frac{1}{4}t_0\left[\frac{3}{2} - \left(\frac{1}{2}+x_0\right)\beta^2\right] n + \frac{1}{24}t_3\left[\frac{3}{2} - \left(\frac{1}{2}+x_3\right)\beta^2\right] n^{\gamma+1} \\ &+ \frac{1}{8}(t_1+t_2)\frac{3}{5}\left(\frac{3\pi^2}{2}\right)^{2/3}\left\{(1+\beta)^{5/3}+(1-\beta)^{5/3}\right\} n^{5/3} \\ &+ \frac{1}{32}(t_2-t_1)\frac{3}{5}\left(\frac{3\pi^2}{2}\right)^{2/3}\left\{(1+\beta)^{8/3}+(1-\beta)^{8/3}\right\} n^{5/3} \end{aligned} \tag{44}$$

where

$$B^{kin}(n,\beta) = \frac{1}{2}\left\{(1+\beta)^{5/3}+(1-\beta)^{5/3}\right\} B_0^{kin}(n)\ , \tag{45}$$

$$B_0^{kin}(n) = \frac{3}{5}E_F(n) = \frac{3}{5}\frac{\hbar^2}{2m}\left(\frac{3\pi^2}{2}n\right)^{2/3}, \tag{46}$$

B_0^{kin} being the kinetic energy per particle of symmetric NM.

4.1. Symmetry energy

As we will discuss in section 6, the density dependence of the nuclear symmetry energy plays a crucial role in determining the proton concentration in dense β-stable matter. The latter quantity is relevant to model the thermal evolution of neutron stars [51]. To distinguish between phase space effects and interaction effects, we write the energy per particle of asymmetric NM in the following way

$$B(n,\beta) = B^{kin}(n,\beta) + B^{pot}(n,\beta)\ , \tag{47}$$

where $B^{kin}(n,\beta)$ is the kinetic energy per particle of asymmetric NM given by eq.s (45) and (46)

In the parabolic approximation (23) the potential energy per particle can be written

$$B^{pot}(n,\beta) = V_0(n) + E_{sym}^{pot}(n)\beta^2\ , \tag{48}$$

where V_0 is the potential energy per particle of symmetric nuclear matter and E_{sym}^{pot} is the potential symmetry energy. The kinetic contribution to the symmetry energy is

$$E_{sym}^{kin} = \frac{1}{3}E_F(n). \tag{49}$$

In the case of the Skyrme interaction (40), adopted in the present work, E_{sym}^{pot} is given by

$$E_{sym}^{pot} = -\frac{1}{4}t_0\left(\frac{1}{2}+x_0\right) n + \frac{1}{6}t_2\left(\frac{3\pi^2}{2}\right)^{2/3} n^{5/3} - \frac{1}{24}t_3\left(\frac{1}{2}+x_3\right) n^{\gamma+1} \tag{50}$$

The next term in the β^{2n}-expansion (21) for the energy per nucleon of asymmetric NM is:

$$S_4 = S_4^{kin} + S_4^{pot} = \frac{1}{81}E_F(n) + \frac{1}{648}\left(\frac{3\pi^2}{2}\right)^{2/3} n^{5/3}(3t_1+t_2). \tag{51}$$

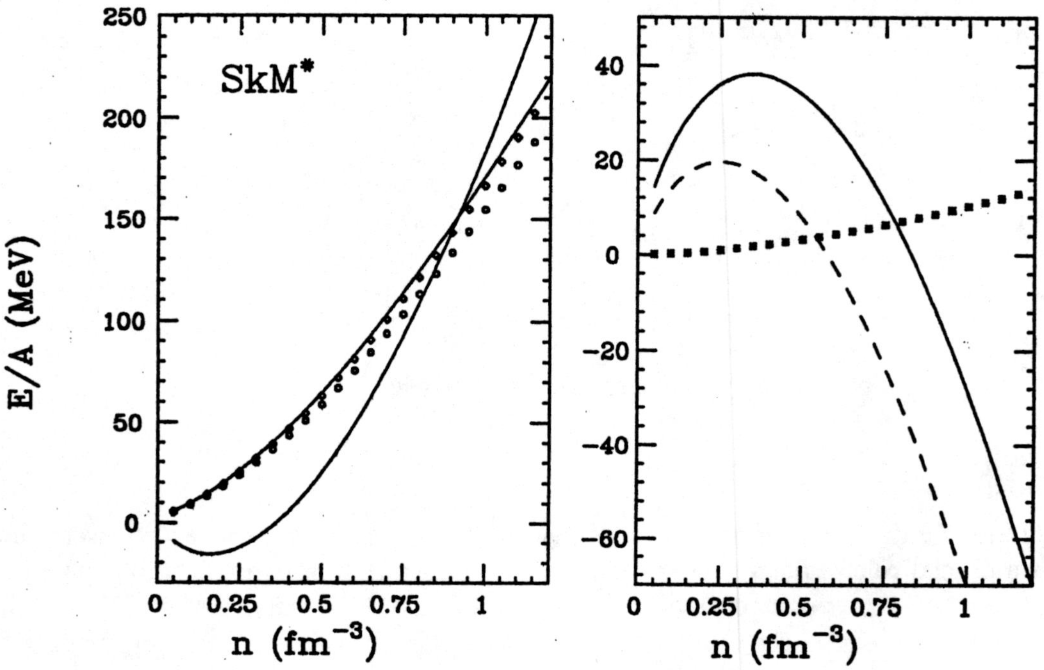

Figure 7: Left panel: Energy per nucleon for symmetric NM (lower continuous curve) and neutron matter (upper continuous line) obtained using eq.(44). The circles give the energy per nucleon for neutron matter $B(n,1)$ using the parabolic approximation (23), while the diamonds give $B(n,1) = B(n,0) + E_{sym}(n) + S_4(n)$. Right panel: The nuclear symmetry energy E_{sym} (continuous line), the potential symmetry energy E^{pot}_{sym} (dashed line) and the β^4-potential energy contribution S^{pot}_4 (star symbols) are plotted as a function of the number density. Results in both panels are relative to the SkM* interaction.

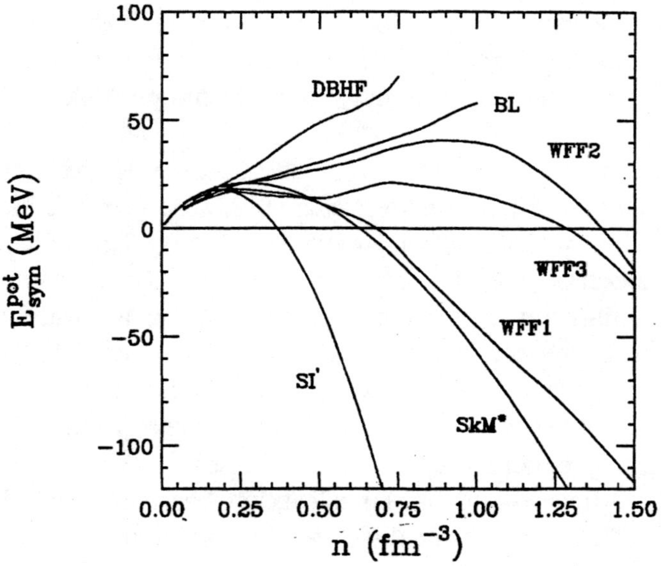

Figure 8: Potential part of the nuclear symmetry energy. The curve labeled by BL refers to the calculations of ref. [40], extended here to higher density than those reported in ref. [40]. Curves SkM* and SI' are relative to two Skyrme interactions, curve DBHF to the Dirac–Brueckner calculations from ref. [27]. Finally, curves WFF refer to the variational calculations of ref. [22] for different interactions.

In the left panel of Fig. 7, we plot the energy per nucleon $B(n,0)$ for symmetric NM (lower curve) and for neutron matter (upper curve), as obtained using eq.(44) and for the SkM* interaction. The circles, in the same figure, refer to the energy per nucleon $B(n,1)$ for neutron matter using the parabolic approximation (23), while the diamonds give $B(n,1) = B(n,0) + E_{sym}(n) + S_4(n)$. Clearly, in the case of standard Skyrme interactions, the parabolic approximation (23) breaks down at high density. In addition, in the high density regime (n larger than $\sim$ 0.92 fm^{-3} for the SkM* interaction) neutron matter has a lower energy per particle with respect to symmetric NM. With a few exception (see *e.g.* ref. [52]), this is a common feature of most of the standard Skyrme interactions.

For the SkM* interaction, in the right panel of Fig. 7, we show the total symmetry energy E_{sym} (continuous line) and its potential energy contribution E^{pot}_{sym} (dashed line). As we see, $E^{pot}_{sym}(n)$ becomes negative as the density increases, and its decrease with density, can not be compensated by the positive contribution of the kinetic symmetry energy E^{kin}_{sym}. As a results the total symmetry energy becomes negative for densities larger than about 0.83 fm^{-3} (for n larger than about 0.45 fm^{-3} in the case of the SI$'$ interaction). Finally, the star symbols in the same figure, depict the potential energy contribution S^{pot}_4 to the β^4-term.

Next, in Fig. 8, we compare E^{pot}_{sym}, for Skyrme interactions (SI$'$ and SkM*), with the results from microscopic calculations. All models are in good agreement in the low density region, and reproduce very well the empirical value at saturation density E_{sym} =28–32 MeV, extracted from nuclear data systematics. At high densities, however, both phenomenological and microscopic EOSs are in strong conflict. Especially puzzling is the different density dependence of E^{pot}_{sym} predicted by microscopic models. In particular, the disagreement between Brueckner type calculations, both non–relativistic (BBG) [40,28] and relativistic (Dirac–Brueckner) [27,28], and variational type calculations [22]. The origin of this discrepancy and its relations with the many body technique used in the calculations need to be clarified.

In the variational type calculations (ref. [22], and similar works), the variational expectation value E_{var} of the Hamiltonian is calculated in a diagrammatic cluster expansion (FHNC-SOC) [20]. The variational trial wave function used in Ref. [22] does not contain the correlations which arise from L^2, $L^2(\sigma_i \cdot \sigma_j)$ and $(L \cdot S)^2$ terms of the nucleon-nucleon potential. Finally, spin-orbit correlations are not treated accurately (see the discussion in ref. [22]). All these NN correlations are included in a selfconsistent way in the BHF approach. All the approximations mentioned above could give large uncertainties in the calculated expectation value of the energy in the high density region.

A very promising developing field, which could shed some new light on this problem, is the physics with radioactive (exotic) nuclear ion beams. It has been recently shown (see [14,15] and the other chapters in the present book) that collective transverse flows in heavy ion collisions at medium energies (E/A = 40 – 120 MeV) are strongly sensitive to the density dependence of the nuclear symmetry energy. All present data seem to indicate [14,15] the need for a strong density dependence of E^{pot}_{sym}.

4.2. Single particle energy

According to the Landau definition, the single particle energy can be obtained taking the functional derivative of the energy density with respect to the distribution function f_τ

$$e_\tau = \frac{\partial \varepsilon}{\partial f_\tau} \tag{52}$$

In the particular case of the Skyrme interaction, e_τ is given by

$$e_\tau(k;n,\beta) = \frac{\hbar^2}{2m}k^2 + U_\tau(k;n,\beta) = \frac{\hbar^2}{2m}k^2 + U'_\tau(k;n,\beta) + \tilde{U}_\tau(n,\beta) \tag{53}$$

with

$$U'_\tau(k;n,\beta) = \frac{1}{16}\Big[(3t_1 + 5t_2) \pm (t_2 - t_1)\beta\Big]\; n\; k^2 \tag{54}$$

$$\begin{aligned} \tilde{U}_\tau(n,\beta) &= \frac{3}{4}t_0\; n + \frac{1}{8}\Big(1 + \frac{1}{2}\gamma\Big)t_3\; n^{\gamma+1} - \frac{1}{24}\gamma\Big(\frac{1}{2} + x_3\Big)t_3\; n^{\gamma+1}\beta^2 \\ &\mp \frac{1}{2}\Big[t_0\Big(\frac{1}{2} + x_0\Big)n + \frac{1}{6}t_3\Big(\frac{1}{2} + x_3\Big)n^{\gamma+1}\Big]\beta \\ &+ \frac{3}{40}\Big(\frac{3\pi^2}{2}\Big)^{2/3}\Big[\frac{1}{2}(t_1 + 3t_2)(1 \pm \beta)^{5/3} + (t_1 + t_2)(1 \mp \beta)^{5/3}\Big]\; n^{5/3} \end{aligned} \tag{55}$$

here and in the following formulae which refer to a nucleon with a non specified isospin third component, the upper sign refers to neutrons ($\tau = n$) and the lower sign to protons ($\tau = p$). Notice that in the case of the Skyrme interaction, the momentum dependence of the s.p. energy is quadratic, and can be fully incorporated in the nucleon effective mass, which is equal to

$$\frac{m^*_\tau}{m} = \frac{1}{(1+\alpha_\tau(\beta)\,n)} = \frac{1}{(1+\alpha_0 n) \mp \alpha_1 n\beta} \tag{56}$$

with $\alpha_\tau(\beta) = \alpha_0 \mp \alpha_1\beta$ and

$$\alpha_0 \equiv \frac{2m}{\hbar^2}\frac{1}{16}(3t_1+5t_2), \qquad\qquad \alpha_1 \equiv \frac{2m}{\hbar^2}\frac{1}{16}(t_1-t_2) \tag{57}$$

Particularly in symmetric nuclear matter

$$\frac{m^*_0}{m} = \frac{1}{1+\alpha_0 n}. \tag{58}$$

When β and n fulfil the condition $x \equiv \alpha_1 n\beta/(1+\alpha_0 n) \ll 1$, we can use Taylor expansion for eq. (56), then to the lower order in x we obtain ,

$$\frac{m^*_\tau}{m} \simeq \frac{m^*_0}{m}\left[1 \pm \left(\frac{m^*_0}{m}\alpha_1 n\right)\beta\right] \tag{59}$$

which is linear in the asymmetry parameter β in agreement with the results from microscopic calculations (compare to eq. (34)) and results in Fig. 5). Since for all Skyrme parameterizations $t_1 > t_2$, when the condition $x \ll 1$ is fulfilled, we find a dependence of the effective mass on the isospin-asymmetry parameter β in agreement with the microscopic BHF calculations reported in Fig. 5.

The s.p. energy can be rewritten in terms of the effective mass as

$$e_\tau(k;n,\beta) = \frac{\hbar^2}{2m^*_\tau}k^2 + \tilde{U}_\tau(n,\beta) \tag{60}$$

To make a more transparent comparison with the microscopic calculations, it is instructive to consider the Taylor expansion of the s.p. energy around $\beta = 0$. Up to linear contributions in β this expansion gives:

$$e_\tau(k;n,\beta) \simeq \frac{\hbar^2}{2m^*_\tau(n,\beta)}k^2 + U_0(n) \pm U_{sym}(n)\beta \tag{61}$$

with

$$U_0(n,\beta) = \frac{3}{4}t_0\, n + \frac{1}{8}\left(1+\frac{1}{2}\gamma\right)t_3\, n^{\gamma+1} + \frac{3}{80}\left(\frac{3\pi^2}{2}\right)^{2/3}(t_1+3t_2)n^{5/3}, \tag{62}$$

$$U_{sym}(n) = \left\{\frac{1}{16}\left(\frac{3\pi^2}{2}\right)^{2/3}(t_2-t_1)n^{5/3} - \frac{1}{2}t_0\left(\frac{1}{2}+x_0\right)n - \frac{1}{12}t_3\left(\frac{1}{2}+x_3\right)n^{\gamma+1}\right\}, \tag{63}$$

or in terms of the s.p. energy $e_0(k;n)$ in symmetric NM,

$$e_\tau(k;n,\beta) \simeq e_0(k;n) \mp \frac{1}{16}(t_1-t_2)n\,k^2\,\beta \pm U_{sym}(n)\,\beta \tag{64}$$

The previous equation, clearly shows that for $k \neq 0$ (and for moderate densities) the isospin dependence associated with the term U_{sym}, is balanced by the "effective mass term" $(t_2-t_1)nk^2/16$. For example, in the case of the SkM* interaction, for $n = 0.16$ fm^{-3} and $k = 3$ fm^{-1}, $U_{sym} = 38.56$ MeV, and $(t_2-t_1)nk^2/16 = -49.05$ MeV.

5. A generalized Skyrme–like EOS

5.1. Energy density

We now introduce a phenomenological EOS of hot asymmetric NM [53] (see also ref. [54,11]) in a general and flexible form suitable both for astrophysical and heavy ion collisions applications. Our EOS can be deduced from effective nuclear interaction in analogy with a Skyrme EOS.

We start with the energy density, which is given by

$$\varepsilon(n_n, n_p, T) = \varepsilon^{kin} + \varepsilon^A + \varepsilon^B + \varepsilon^C + n\, mc^2 \,, \tag{65}$$

where $n\, mc^2$ is the rest mass energy density (we neglect the neutron–proton mass difference),

$$\varepsilon^{kin}(n_n, n_p, T) = \frac{2}{(2\pi)^3}\int d^3k f_n(k)\frac{\hbar^2}{2m}k^2 + \frac{2}{(2\pi)^3}\int d^3k f_p(k)\frac{\hbar^2}{2m}k^2 \tag{66}$$

is the kinetic energy density, f_n and f_p are the Fermi distribution functions,

$$\varepsilon^A \equiv \frac{2}{3}\frac{A}{n_0}\left[\left(1+\frac{1}{2}x_0\right)n^2 - \left(\frac{1}{2}+x_0\right)\left(n_n^2+n_p^2\right)\right], \tag{67}$$

$$\varepsilon^B \equiv \frac{4}{3}\frac{1}{n_0^\sigma}B\frac{\left[\left(1+\frac{1}{2}x_3\right)n^2 - \left(\frac{1}{2}+x_3\right)\left(n_n^2+n_p^2\right)\right]n^{\sigma-1}}{1+\frac{4}{3}\frac{1}{n_0^{\sigma-1}}B'\left[\left(1+\frac{1}{2}x_3\right)n^2 - \left(\frac{1}{2}+x_3\right)\left(n_n^2+n_p^2\right)\right]n^{\sigma-3}}\,, \tag{68}$$

$$\varepsilon^C(n_n, n_p, T) \equiv \sum_{i=1}^{N_{range}} \varepsilon^{C_i} \equiv$$

$$\sum_{i=1}^{N_{range}}\left\{\frac{4}{5}\frac{1}{n_0}(C_i+2z_i)n\left[\frac{2}{(2\pi)^3}\int d^3k f_n(k)g(k,\Lambda_i) + \frac{2}{(2\pi)^3}\int d^3k f_p(k)g(k,\Lambda_i)\right]\right.$$

$$\left.+\frac{2}{5}\frac{1}{n_0}(C_i-8z_i)\left[n_n\frac{2}{(2\pi)^3}\int d^3k f_n(k)g(k,\Lambda_i) + n_p\frac{2}{(2\pi)^3}\int d^3k f_p(k)g(k,\Lambda_i)\right]\right\} \tag{69}$$

are the contributions due to nuclear interaction. The two terms ε^A and ε^B arise from local contact nuclear interaction which led to power density contributions like in the standard Skyrme EOS. They are assumed to be temperature independent. The term ε^C describes the effects of finite range interactions according to the choice of the function $g(k)$ as we now discuss.

(1) BGBD EOS:

$$N_{range} = 1\,, \qquad B' = 0\,, \qquad g(k) = \left[1 + \frac{k^2}{\Lambda_1^2}\right]^{-1} \tag{70}$$

where the parameter Λ_1^{-1} has the dimension of a length and is associated with a finite range term in the effective nuclear force. With this choice for the momentum dependent function g, if we set the parameter $B' = 0$, and in the particular case of zero temperature and symmetric NM one recovers the Gale–Bertsch–Das Gupta [55] (GBD) EOS, which has been widely used in heavy ion collision numerical simulations:

$$\begin{aligned}\varepsilon_{GBD} &= \frac{3}{5}\frac{\hbar^2}{2m}\left(\frac{3\pi^2}{2}\right)^{2/3} n^{5/3} + \frac{1}{2}\frac{A}{n_0}n^2 + \frac{B''}{\sigma+1}\frac{n^{\sigma+1}}{n_0^\sigma} \\ &+ \frac{2}{\pi^2}C_1\Lambda_1^3 u\left[\frac{u^{1/3}}{\Lambda_1/k_{F0}} - \arctan\frac{u^{1/3}}{\Lambda_1/k_{F0}}\right], \qquad (71)\end{aligned}$$

where $B'' \equiv (\sigma+1)B$, $u \equiv n/n_0$, and k_{F0} is the Fermi momentum of symmetric NM corresponding to the saturation density n_0. Therefore with the choice (70) our generalized Skyrme–like EOS (65)–(69) gives a generalization of the GBD EOS to the case of asymmetric NM and finite temperature. In the following we will refer to it as the BGBD EOS.

(2) Skyrme EOS:

$$N_{range} = 1\,, \qquad B' = 0\,, \qquad g(k) = k^2 \qquad (72)$$

In this case one has from eqs. (65)–(69) the energy density for a Skyrme EOS. This is more clearly seen if we redefine the parameters A, B, C_1, z_1, σ in eqs. (65)–(69) using the usual notations for the Skyrme EOS

$$\begin{aligned}t_0 &\equiv \frac{4}{3}\frac{1}{n_0}A\,, \qquad t_3 \equiv 16\frac{1}{n_0^\sigma}\frac{B}{\sigma+1}\,, \qquad \gamma \equiv \sigma - 1 \\ t_1 &\equiv 16\frac{1}{n_0}z_1\,, \qquad t_2 \equiv \frac{16}{5}\frac{1}{n_0}(C_1 - 3z_1). \qquad (73)\end{aligned}$$

(3) BPAL EOS:

$$N_{range} = 2\,, \qquad g(k,\Lambda_i) = \left[1 + \frac{k^2}{\Lambda_i^2}\right]^{-1}. \qquad (74)$$

In the BPAL model we introduce two finite range terms: one corresponding to a long–range attraction and the other to a short–range repulsion. In the particular case of cold (T=0) symmetric NM one recovers the EOS of Prakash Ainsworth and Lattimer [56] (PAL).

(4) SL EOS:

$$N_{range} = 2\,, \qquad g(k) = 1 - \frac{k^2}{\Lambda_1^2} \qquad (75)$$

in the present case the finite range interactions are approximated by effective local interactions by retaining only the quadratic momentum dependence.

The parameters A, B, B', C_1 (C_2) and σ are adjusted to reproduce the empirical saturation properties of cold symmetric NM. B' is introduced to maintain causality in the high density region of the EOS. With an appropriate choice of these parameters it is also possible to have different values for the nuclear incompressibility K_0. Numerical values for the BGBD, BPAL and SL EOSs are given in Tab. 2. The additional parameters x_0, x_3, z_1 (z_2) give the possibility to model different density dependence of the nuclear symmetry energy, constrained to the empirical value at saturation density. The numerical values reported in Tab. 4 have been fixed taking $E_{sym}(n_0) = 30\ MeV$ and assuming three different forms for the density dependence of the potential part E_{sym}^{pot} of the symmetry energy, in which it varies approximatively as $\sqrt{u}$, u and $2u^2/(1+u)$.

5.2. Pressure

From thermodynamics we know that the pressure is related to the internal energy E through

$$P = -\frac{\partial E}{\partial V}\Big|_{S,A} . \tag{76}$$

With a change of variable:

$$P = n^2 \frac{\partial E/A}{\partial n}\Big|_{S,A} = n_0 u^2 \frac{\partial E/A}{\partial u}\Big|_{S,A} = n \frac{\partial \varepsilon}{\partial n}\Big|_{S,A} - \varepsilon \tag{77}$$

For the BGBD EOS, in the particular case of zero temperature, from eqs. (65)–(69) and (70) one gets:

$$\begin{aligned}
P(u,\beta,0) &= \frac{1}{5}E_F(n_0) + \Big\{(1+\beta)^{5/3} + (1-\beta)^{5/3}\Big\}\, n_0 u^{\frac{5}{3}} \\
&+ \frac{1}{3}A\Big\{\frac{3}{2} - \Big(\frac{1}{2}+x_0\Big)\beta^2\Big\}\, n_0 u^2 \\
&+ \frac{2}{3}\sigma B\Big\{\frac{3}{2} - \Big(\frac{1}{2}+x_3\Big)\beta^2\Big\}\, n_0 u^{\sigma+1} \\
&+ \frac{1}{10}\Big(4C_1 + 8z_1\Big)\lambda_1^3\Big\{\frac{(1+\beta)^{1/3} + (1-\beta)^{1/3}}{\lambda_1} \\
&- \frac{\lambda_1(1+\beta)^{1/3}}{\lambda_1^2 + (1+\beta)^{2/3}u^{2/3}} - \frac{\lambda_1(1-\beta)^{1/3}}{\lambda_1^2 + (1-\beta)^{2/3}u^{2/3}}\Big\}\, n_0 u^{\frac{4}{3}} \\
&+ \frac{1}{10}\Big(C_1 - 8z_1\Big)\lambda_1^3\Big\{\frac{(1+\beta)^{4/3} + (1-\beta)^{4/3}}{\lambda_1} \\
&- (1+\beta)\frac{\lambda_1(1+\beta)^{1/3}}{\lambda_1^2 + (1+\beta)^{2/3}u^{2/3}} \\
&- (1-\beta)\frac{\lambda_1(1-\beta)^{1/3}}{\lambda_1^2 + (1-\beta)^{2/3}u^{2/3}}\Big\}\, n_0 u^{\frac{4}{3}}
\end{aligned} \tag{78}$$

where $\lambda_1 \equiv \Lambda_1/k_{F0}$, and $E_F(n_0)$ is the Fermi energy of symmetric NM at saturation.

In Fig. 9 we plot the isothermal EOS in the case of the BGBD EOS for two different values of the asymmetry parameter β. Our phenomenological EOS show the typical Van der Waals behaviour, in a good qualitative agreement with the results of microscopic calculations [21,57,42]. The critical temperature for the liquid–gas phase transition, we obtain in the case of BGBD EOS, is $T_c(\beta{=}0) = 14.7$ MeV for symmetric nuclear matter, to be

Table 2. Parameters for the phenomenological EOS. Parameters A, B, B', C_1, (C_2), and σ in eq.s (65)–(69) have been determined by fitting the equilibrium properties of symmetric NM for some input values of the compression modulus K_0. We have employed: $n_0 = 0.16\ fm^{-3}$, $E_0/A = -16\ MeV$, $m^*(n_0) = 0.7\ m$, $U(n_0, k=0) = -76.34\ MeV$. Finite range parameters have been taken $\Lambda_1 = 1.5k_{F0}$ and $\Lambda_2 = 3k_{F0}$. Parameters x_0, x_3, z_1, (z_2), are fixed to give different density dependence of nuclear symmetry energy. In all cases, the symmetry energy at the nuclear saturation density is taken to be 30 MeV. The notation BPALn_1n_2 and SLn_1n_2 is used to denote different EOSs; n_1 refers to different values of K_0, and $n_2 = 1$, 2 and 3 indicates, respectively, a $\sqrt{u}$, u and $2u^2/(1+u)$ dependence of the potential symmetry energy on the density. Parameters A, B, B', C_1, C_2, z_1, z_2, and K_0 are given in MeV; all other entries in the table are dimensionless numbers.

EOS	K_0	A	B	B'	σ	C_1	C_2
BGBD	215	−144.00	203.300	0.0	7/6	−75.00	–
BPAL1n	120	75.940	−30.880	0.0	0.498	−83.84	23
BPAL2n	180	440.940	−213.410	0.0	0.927	−83.84	23
BPAL3n	240	−46.650	39.450	0.3	1.663	−83.84	23
SL1n	120	3.706	−31.155	0.0	0.453	−41.28	23
SL2n	180	159.470	−109.040	0.0	0.844	−41.28	23
SL3n	240	−204.010	72.704	0.3	1.235	−41.28	23

EOS	K_0	x_0	x_3	z_1	z_2
BGBD	215	0.000	−0.400	−3.420	–
BPAL11		−0.689	0.577	−14.000	16.690
BPAL12	120	−1.361	−0.244	−13.910	16.690
BPAL13		−1.903	−1.056	−1.830	5.090
BPAL21		0.086	0.561	−18.400	46.270
BPAL22	180	−0.410	−0.105	−9.380	24.050
BPAL23		−1.256	−1.358	−11.670	−10.900
BPAL31		0.376	0.246	−12.230	−2.980
BPAL32	240	0.927	−0.227	−11.510	8.380
BPAL33		1.654	−1.112	3.810	13.160
SL12	120	−3.548	−0.500	−13.355	2.789
SL22	180	−0.410	−0.105	9.380	−4.421
SL32	240	−0.442	−0.500	−13.387	2.917

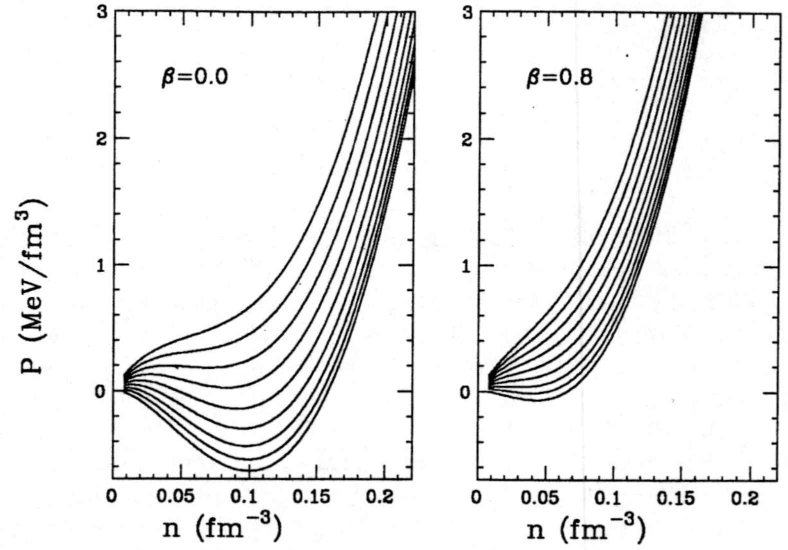

Figure 9: BGBD EOS. Isotherms of pressure versus nucleon number density for two different values of the asymmetry parameter β. Isothermic curves refer to temperatures T = 0, 4, 6, 8, 10, 12, 14, 16, 18 MeV, from the lower to the upper curve.

compared with $T_c(\beta{=}0)\sim$ 20 MeV from microscopic calculations [57,42]. As expected [3], T_c decreases going from symmetric to pure neutron matter. Similar results have been obtained for other members of our generalized Skyrme–like EOSs introduced by eq.s (65)–(69), like *e.g.* the BPAL or the SL EOSs.

5.3. Single Particle energy

In the particular case of the BGBD EOS, e_τ is given by

$$\begin{aligned}
e_\tau(k;u,\beta,T) &\equiv \frac{\hbar^2}{2m}k^2 + U_\tau(k;u,\beta,T) \\
&= \frac{\hbar^2}{2m}k^2 + \frac{1}{5}\Big[5C_1 \pm (C_1 - 8z_1)\beta\Big]\; u\; g(k) + \\
&+ \; Au + B''u^\sigma + \Big[-\frac{2}{3}(\sigma-1)\frac{B''}{\sigma+1}\Big(\frac{1}{2}+x_3\Big)u^\sigma\Big]\beta^2 + \\
&\pm \; \Big[-\frac{2}{3}A\Big(\frac{1}{2}+x_0\Big)u - \frac{4}{3}\frac{B''}{\sigma+1}\Big(\frac{1}{2}+x_3\Big)u^\sigma\Big]\beta + \\
&+ \; \frac{4}{5}\frac{1}{n_0}\bigg\{\frac{1}{2}(3C_1 - 4z_1)\frac{2}{(2\pi)^3}\int d^3k\; f_\tau(k;u,\beta,T)\; g(k,\Lambda_1) \\
&+ \; (C_1 + 2z_1)\frac{2}{(2\pi)^3}\int d^3k\; f_{\tau'}(k;u,\beta,T)\; g(k,\Lambda_1)\bigg\} \qquad (79)
\end{aligned}$$

To compare with the microscopic results (see Fig. 4) discussed before, in Fig. 10 we show the s.p. spectrum $U_\tau(k;u,\beta)$, at zero temperature, in the case of the BGBD EOS. In particular, the left panel shows the momentum dependence of U_τ at normal saturation density n_0 and for various values of the neutron–proton asymmetry β. The central panel shows the density dependence of U_n and U_p, at $k = 0$, for different asymmetries. Finally,

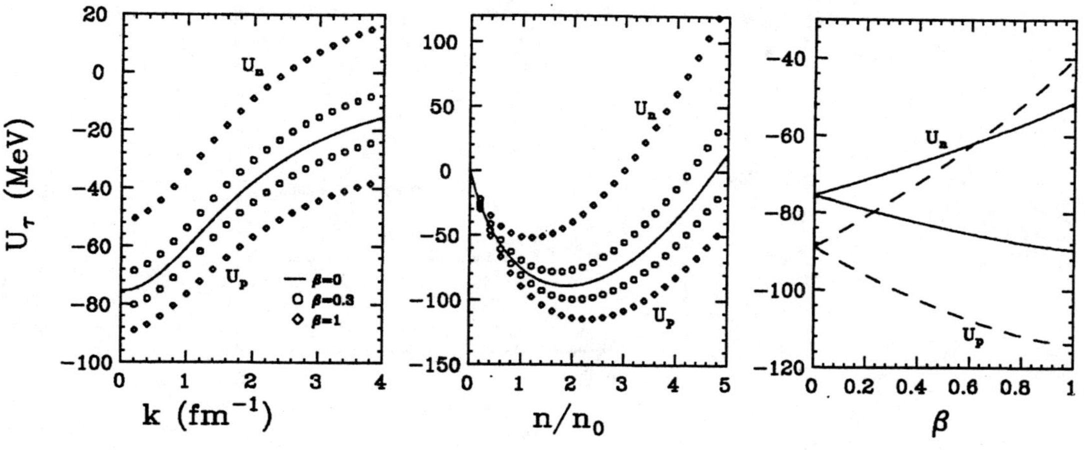

Figure 10: Single particle potential for neutrons (U_n)and protons (U_p) in asymmetric NM at zero temperature in the case of the BGBD EOS. Left panel: Momentum dependence of U_τ, at normal saturation density n_0, for three values of the asymmetry parameter ($\beta = 0$, 0.3, 1.0). Central panel: Dependence on the number density (in unit of n_0), at $k = 0$, for β = 0, 0.3, 1.0 (symbols as in the left panel). Right panel: Neutron and proton s.p. potentials versus β for $k = 0$, at saturation density n_0 (continuous lines) and $n = 2n_0$ (dashed lines).

in the right panel U_n and U_p are plotted as a function of β, for $n = n_0$ (continuous lines) and $n = 2n_0$ (dashed lines) for a fixed ($k = 0$) value of the single particle momentum.

5.4. Effective mass

For the BGBD model, m^*_τ at zero temperature, is given by:

$$\frac{m^*_\tau}{m} = \left\{ 1 + \frac{-\frac{2m}{\hbar^2}\frac{1}{\Lambda_1^2}\frac{1}{5}\left[5C_1 \pm \left(C_1 - 8z_1\right)\beta\right]u}{\left[1 + \left(\frac{k_{F0}}{\Lambda_1}\right)^2\left(1 \pm \beta\right)^{2/3} u^{2/3}\right]^2} \right\}^{-1} \tag{80}$$

The nucleon effective mass, within this model, decreases from 1 (at $n = 0$), reaches a minimum value (at different densities for proton and neutron) and then increases going again to one as the density goes to infinity. Notice that in the case of the Skyrme interaction, the effective masses are monotonically decreasing function of the density. This dissimilarity is a consequence of the different momentum dependence (choice of the function $g(k)$ in eq.s (65)–(69)) assumed in the two models.

The effective masses are plotted in Fig. 11 as a function of the number density for charge neutral asymmetric NM in β-equilibrium with electrons and muons, in the case of the phenomenological BGBD model.

5.5. Symmetry energy with BGBD and BPAL models

In view of the large uncertainty on the functional form of E^{pot}_{sym} (see Fig.8), and having in mind its role on the structure and evolution of astrophysical compact objects, in our

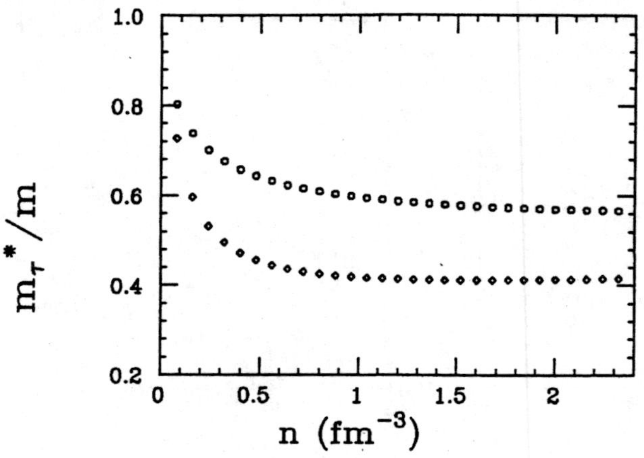

Figure 11: The proton (diamonds) and neutron (circles) effective masses as a function of the number density for β-stable nuclear matter, for the BGBD equation of state.

new phenomenological EOS, we choose to have the freedom to model different density dependence of E^{pot}_{sym}. This is possible with an appropriate choice for the values of the parameters x_0, x_3, z_1, (z_2). For the BPAL and SL models in Tab. 2, we choose one of the three possibilities

$$E^{pot}_{sym} \sim \sqrt{u}, \quad u, \quad \frac{2u^2}{1+u}. \tag{81}$$

The BGBD parameterization, in Tab. 2, corresponds to $E^{pot}_{sym} \sim u \equiv n/n_0$.

5.6. *Speed of sound*

As we said, a basic feature of any "realistic" EOS, regardless the details of the interactions among matter constituents, is the causality condition (4), where the energy density ε includes the rest mass energy density.

Causality is violated by many phenomenological EOSs, which have been used in hydrodynamical simulations of heavy ion collisions and supernova explosions or in the study of neutron stars structure. To highlight this point, we plot in Fig. 12a the ratio s/c for two typical Skyrme EOSs (SI$'$ and SkM*). Dashed lines refer to symmetric NM and continuous lines to pure neutron matter. The open dots on the latter curves represent the central density of the maximum mass (M_{max}) configuration for stars with a pure neutron matter core. As we see, in the case of SI$'$ EOS, the central region of stars with M$\sim$M$_{max}$, becomes superluminar. The speed of sound in charge neutral asymmetric NM in β–equilibrium with electrons and muons, is plotted in Fig. 12b for some of our models for the EOS. All BPAL models fulfil causality up to the maximum central density (open dot on each curve) found in the center of the maximum mass neutron star. In the same figure, we plot s/c for a few microscopic EOSs: the upper dashed line refers to the BBB2 EOS (Paris+TBF) [47], the lower dashed line refers to the BBB1 EOS (Av14+TBF), and the curve labeled WFF to the results of ref. [22] with the (Av14+UVII) interaction.

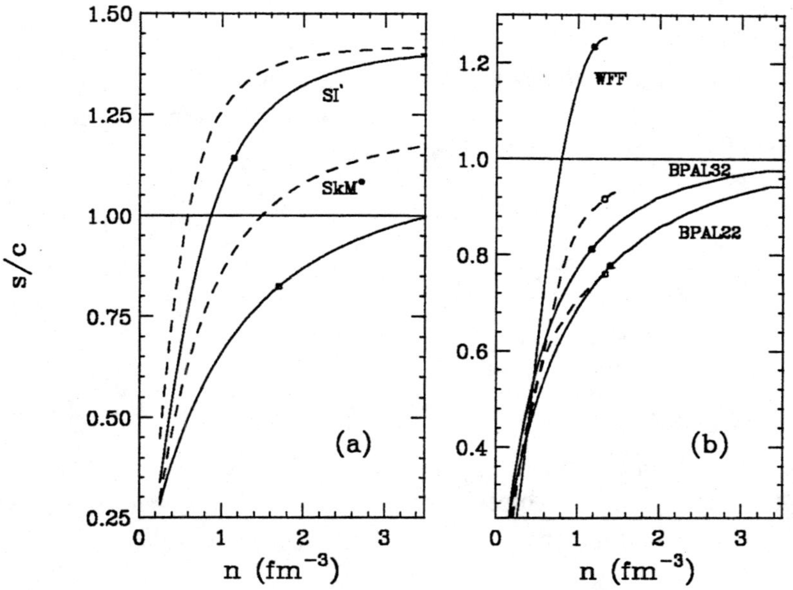

Figure 12: Speed of sound. *(a)* symmetric (dashed line) and pure neutron (continuous line) matter. *(b)* β-stable matter. Curve WFF refers to the calculations of ref. [22] with AV14+UVII interaction. The two dashed lines refer to the BBB2 EOS (upper line) and BBB1 EOS (lower line). The open dot represents the speed of sound at the center of the maximum mass neutron star.

The comparison of the results discussed in the previous pages, shows a fair good general agreement with the results obtained within the microscopic approach. Therefore, our EOS is able to reproduce bulk and s.p. properties of asymmetric NM as predicted by microscopic many body theories.

6. EOS for β–stable nuclear matter

The composition of the neutron star material is determined by the requirement of charge neutrality and equilibrium with respect to the weak interactions (β–stable matter).

To begin with we assume the constituents of the neutron star are neutrons, protons and electrons. The equilibrium condition for the weak reactions

$$p + e^- \rightarrow n + \nu_e \tag{82}$$

$$n \rightarrow p + e^- + \bar{\nu}_e \tag{83}$$

can be written in the following way:

$$\mu_e = \mu_n - \mu_p \equiv \hat{\mu} \tag{84}$$

which is valid for the neutrino–free neutron star (μ_{ν_e}=$\mu_{\bar{\nu}_e} = 0$). Therefore, in the case of untrapped neutrinos, the leptonic chemical potential is determined once the difference $\hat{\mu} \equiv \mu_n - \mu_p$ is known. The latter quantity can be expressed as:

$$\hat{\mu} = -\frac{\partial E/A}{\partial x}\bigg|_n = 2\frac{\partial E/A}{\partial \beta}\bigg|_n \tag{85}$$

where E/A is the energy per nucleon of asymmetric NM, $x = Z/A = n_p/n$ is the proton fraction ($\beta = 1 - 2x$), and the partial derivative are taken for constant nucleon number density n. In eq. (85) we neglect the neutron–proton mass difference. The chemical equilibrium condition eq. (84) makes clear the interplay of the nuclear and weak interactions in determining the composition (the value of the proton concentration in the present case) of the neutron star material.

Local charge neutrality means that the total electric charge in a "small" volume inside the star should vanish. In terms of the number density of charged particles, for our specific case, we have

$$n_p = n_e \tag{86}$$

The electron number density can be expressed as a function of its chemical potential. In the ultrarelativistic limit for non–interacting electrons one has:

$$n_e = \frac{1}{3\pi^2}\frac{1}{(\hbar c)^3}\mu_e^3 \tag{87}$$

Using eq.s (87) and (84), the charge neutrality condition can be written

$$3\pi^2(\hbar c)^3 n x - \hat{\mu}^3 = 0 \tag{88}$$

of the type $F(n, x) = 0$, which defines in an implicit way the proton fraction at equilibrium and can be solved numerically for each given and fixed value of the nucleon number density n. From eq. (88) we see that the exact composition of matter in charge neutral chemical equilibrium is ruled not only by the weak interaction but also by the strong interaction through $\hat{\mu}$. In the parabolic approximation (23) for the energy per particle of asymmetric NM, one has

$$\hat{\mu} = 4E_{sym}(n)(1-2x) \tag{89}$$

and the proton fraction is the solution of the equation

$$3\pi^2(\hbar c)^3 nx - \Big[4E_{sym}(n)(1-2x)\Big]^3 = 0 \tag{90}$$

This equation, clearly shows the crucial role of the nuclear symmetry energy in the determination of the proton abundance inside a neutron star.

Solving eq. (90) at saturation density and assuming $E_{sym}(n_0) = 30$ MeV, one has for the equilibrium proton fraction $x_{eq}(n_0) \simeq 3.76 \times 10^{-2}$.

Consider now the case the constituents of the neutron star are neutrons, protons, electrons, and muons. The threshold density for the appearance of muons is when the electron chemical potential is greater than the muon rest mass:

$$\mu_e > m_\mu c^2 \sim 105.6 \text{ MeV}, \tag{91}$$

at this point it is energetically favorable for the electrons to convert to muons via the weak process

$$e^- \to \mu^- + \nu_e + \bar{\nu}_\mu \tag{92}$$

The following processes are also possible:

$$p + \mu^- \to n + \nu_\mu \tag{93}$$

$$n \to p + \mu^- + \bar{\nu}_\mu \tag{94}$$

For neutrino–free matter ($\mu_{\nu_e}{=}\mu_{\nu_\mu}{=}\mu_{\bar{\nu}_e}{=}\mu_{\bar{\nu}_\mu} = 0$) chemical equilibrium with respect to the weak processes gives:

$$\mu_e = \hat{\mu}, \qquad\qquad \mu_\mu = \mu_e, \tag{95}$$

The charge neutrality condition reads:

$$n_p = n_e + n_\mu \tag{96}$$

The muon number density can be expressed as a function of the muon chemical potential μ_μ (the Heaviside step function is denoted by $\theta(x)$):

$$n_\mu = \frac{1}{(\hbar c)^3}\frac{1}{3\pi^2}[\mu_\mu^2 - (m_\mu c^2)^2]^{3/2}\theta(\mu_e - m_\mu c^2)\,. \tag{97}$$

Then, the charge neutrality condition can be written:

$$3\pi^2(\hbar c)^3 nx - \hat{\mu}^3 - [\hat{\mu}^2 - (m_\mu c^2)^2]^{3/2}\theta(\hat{\mu} - m_\mu c^2) = 0 \tag{98}$$

of the type $F(n,x) = 0$, which can be solved numerically for each fixed value of n.

The value of the proton fraction is an important ingredient to model the thermal evolution of a neutron star. In fact, if the proton fraction in the core of the star, is above a critical value x^{Urca}, the so-called direct Urca processes (82), (83) can occur [51]. If they occur, neutrino emission and neutron star cooling rates are increased by a large factor compared to the standard cooling scenario [51]. The critical proton fraction has been estimated

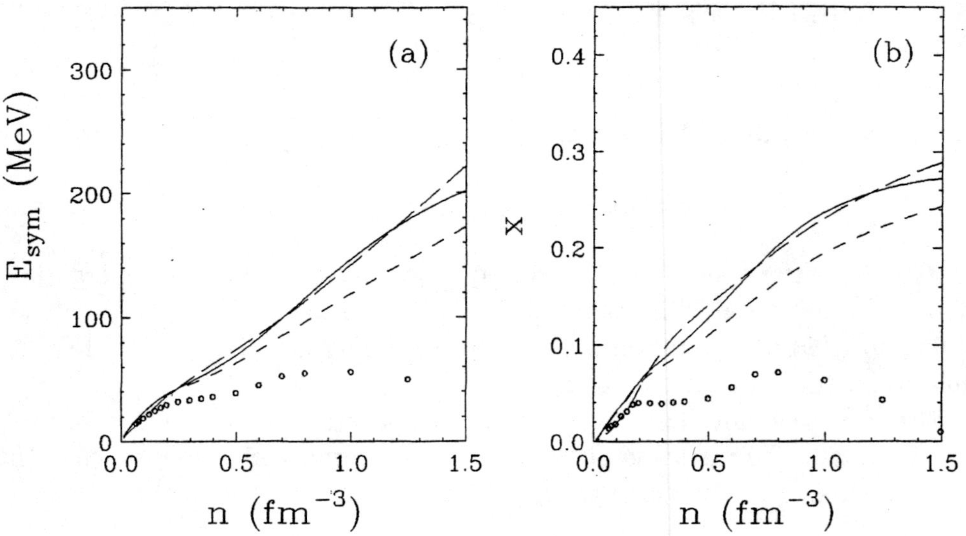

Figure 13:
The symmetry energy and the proton fraction are plotted vs. number density respectively in panel (a) and (b). Results are relative to different models for the EOS, *i.e.* non-relativistic Brueckner calculations with three-body forces (solid line, Paris potential; short dashed line, AV14 potential) and a relativistic Dirac-Brueckner one (long dashes, DBHF). For comparison a variational calculation (with AV14+UVII interaction) is also reported (open circles, WFF).

to be in the range x^{Urca} = 11–15%. However, models for the thermal history of a neutron star depend on a large number of different physical effects [58] (*e.g.* nuclear superfluidity, presence of hyperons, medium effects on the modified Urca processes, presence of light elements in the upper layers of the neutron star, *etc.*). The effect of the neutron star magnetic field on the cooling has been recently investigated by Baiko and Yakovlev [59]. These authors showed that the magnetic field can switch on the direct Urca processes below the threshold proton fraction x^{Urca} obtained in the zero magnetic field case. This multitude of potentially influential physical effects, clearly demonstrates that the present interpretation of the observational data on neutron star temperatures is too strongly model dependent to extract a direct information on the proton fraction in the star interior. The values of E_{sym} for the different EOS's are reported in Fig. 13 (panel (a)), together with the corresponding proton fraction (panel (b)). The long dashed line represents the relativistic DBHF calculations [27], whereas the non–relativistic Brueckner calculations with the AV14 + TBF and the Paris + TBF models are respectively represented by a short dashed and a solid line. We notice that in both relativistic and non–relativistic Brueckner–type calculations, the proton fraction x can exceed the "critical" value x^{Urca}. This is at variance with the WFF variational calculation (open circles), which predicts a low absolute value both for the symmetry energy and the proton fraction with a slight bend over. For the BBB1 model (AV14+TBF) the authors of ref. [47] find x^{Urca} = 13.9%, which correspond to a critical density $n^{Urca} = 0.65\ fm^{-3}$, whereas the BBB2 model (Paris+TBF) predicts the same critical proton fraction at density $n^{Urca} = 0.54\ fm^{-3}$. Therefore, within this model, neutron stars with a central density higher than n^{Urca} develop inner cores in which direct Urca processes are allowed. In Fig. 14, we show the energy per baryon of asymmetric MN

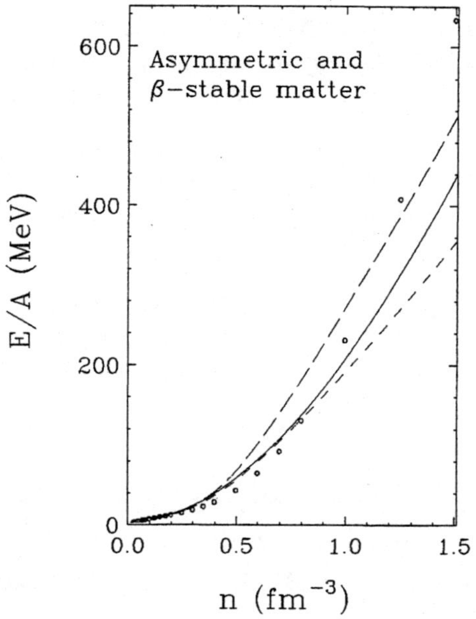

Figure 14: The energy per baryon E/A of asymmetric nuclear matter in β-equilibrium (with electrons and muons) is plotted vs. the number density n. Different curves are denoted as in the previous figure.

in β-equilibrium, for the models under consideration. We notice that the AV14 and the Paris two-nucleon interactions, implemented by the same three-body force, produce similar EOS's up to densities $n \simeq 0.8\ fm^{-3}$, whereas the high density region is quite different.

From the energy per baryon of asymmetric NM in β-equilibrium, one can calculate the nuclear contribution P_{nucl} to the total pressure of stellar matter using the thermodynamical relation

$$P_{nucl}(n) = n^2 \frac{d\,(E/A)}{dn}\bigg|_A \tag{99}$$

Then the total pressure and total mass density ρ are given by

$$P = P_{nucl} + P_{lep} \tag{100}$$

$$\rho = \frac{1}{c^2}\Big(\varepsilon_{nucl} + \varepsilon_{lep}\Big) = \frac{1}{c^2}\Big(n\ E/A + n\ m_N + \varepsilon_{lep}\Big) \tag{101}$$

P_{lep} and ε_{lep} being the leptonic contributions to the total pressure and energy density, m_N the nucleon mass and c the speed of light in the vacuum.

7. The structure equations for Neutron stars

7.1. Tolman–Oppenheimer–Volkoff equations

The hydrostatic equilibrium configurations for a self–gravitating mass distribution are the solutions of the Einstein's field equations of general theory of relativity. Hydrostatic equilibrium means an exact balance between the gravitational force and the pressure force acting upon each mass element.

Consider an isolated, non–rotating, spherically symmetric distribution of mass in hydrostatic equilibrium. Assume there is no magnetic field. Under these assumptions the Einstein's field equations take the form

$$\frac{dP(r)}{dr} = -G\frac{m(r)\rho(r)}{r^2}\left\{\Big[1+\frac{P(r)}{c^2\rho(r)}\Big]\Big[1+\frac{4\pi r^3 P(r)}{c^2 m(r)}\Big]\Big[1-\frac{2Gm(r)}{c^2 r}\Big]^{-1}\right\} \tag{102}$$

$$\frac{dm(r)}{dr} = 4\pi r^2 \rho(r) \tag{103}$$

where $G = 1.327128 \times 10^{11}\ \mathrm{km}^3/(M_\odot s^2)$ is the universal constant of gravitation, c the speed of light, $P(r)$ and $\rho(r)$ are the pressure and the mass density respectively, $m(r)$ is the gravitational mass enclosed within a "radius" r [determined by (proper surface area) $= 4\pi r^2$)].

Equations (102) and (103) are called the Tolman–Oppenheimer–Volkoff (TOV) equations [60,61]. The TOV eq.s are first–order differential equations, which have to be integrated with the following boundary conditions:

$$m(0) = 0, \qquad P(R) = P_{surf}. \tag{104}$$

The first condition means that the density $\rho(r)$ and the pressure $P(r)$ are finite at the center of the star; the second specifies the pressure at the star surface. Usually one takes $P_{surf} = 0$, or $P_{surf} = P(\rho_{\mathrm{Fe}})$, with $\rho_{\mathrm{Fe}} = 7.86\ \mathrm{g/cm}^3$ the density of solid ^{56}Fe. The second boundary condition is the definition of the surface of the star, which specifies the *radius* R of the neutron star through the surface area $4\pi R^2$.

The TOV eq.s are integrated for a given value of the central density ρ_c (or equivalently of the central pressure $P_c = P(\rho_c)$), and the solution

$$m(r;\rho_c), \qquad P(r;\rho_c) \tag{105}$$

depends parametrically on ρ_c.

7.2. Role of the EOS

The TOV equations can be solved when the relation $P(r) = P(\rho(r))$ between the pressure P and the mass density ρ is known at each point. This means that the structure of the star depends upon the EOS of the stellar material.

It is reasonable to assume matter in the neutron star to be a *perfect fluid* (no shear stresses) in a *catalyzed* configuration [62]. By *catalyzed matter*, we mean matter at baryon density n, in which the baryon abundances ($x_i = n_i/n$) are such as to absolutely minimize the total energy per baryon ($c^2\rho/n$). Such matter is in equilibrium with respect to the strong and weak interactions ("chemical" equilibrium). For a multicomponent system in chemical equilibrium, the abundances of its constituents are fixed uniquely by two thermodynamic variables [62], *e.g.*, by the baryon number density n and the total entropy per baryon s. The EOS can then be expressed as:

$$\rho = \rho(n,s) = \rho_0 + \frac{1}{c^2}\varepsilon' \equiv \frac{\varepsilon}{c^2} \tag{106}$$

$$P = P(n,s) = n\left(\frac{\partial \varepsilon}{\partial n}\right)_s - \varepsilon \tag{107}$$

where $\rho = \varepsilon/c^2$ is the total mass density, inclusive of rest–mass density ρ_0. The internal energy density ε' includes the kinetic plus the potential energy density due to the interaction of a particle of the *i-th* component with all the other particles in the system (this potential energy density does not include the gravitational potential energy density).

In addition, one must know how the entropy per baryon depends on the baryon number density, then one gets a one-parameter EOS to be used to solve the TOV equations:

$$\rho = \rho(n, s(n)), \qquad P = P(n, s(n)). \tag{108}$$

A newly born neutron star cools, within a few hours, to temperatures in the range of 10^7–10^9 K. These temperatures are very small when compared to nuclear matter energy scale (Fermi energies). Typical values for the Fermi energy in the core of a neutron star are of the order of tens to hundreds MeV (as the density increases going towards the center) .** Then we assume $s(n) = 0$ (or equivalently $T(n) = 0$) within the whole star. We will refer to such a state of matter as *cold catalyzed matter.*

The exact constituents of matter inside a neutron star and the nature of the interactions between these particles (particularly strong interactions) are poorly known. Therefore neutron stars can be viewed as a "natural laboratory" to investigate the properties of high dense matter. From astrophysical observational data on neutron stars properties (*e.g.* measured masses, period of rotation, X–ray emission,...) one can hope to constrain the high density EOS. On the other hand it is possible (as we will discuss briefly later on) to establish, based on general laws of physics, some constraints on the EOS and therefore on some properties of neutron stars.

7.3. Masses and binding energies

The total mass

$$M \equiv M_G = m(R) = \int_0^R 4\pi r^2 \rho(r) dr \tag{109}$$

is the *gravitational mass* of the neutron star.

The volume of a spherical layer of the star in the Schwarzschild metric (proper volume) is

$$dV = 4\pi\ e^{\lambda(r)/2}\ r^2 dr = 4\pi\left[1 - \frac{2Gm(r)}{c^2 r}\right]^{-1/2} r^2 dr\ . \tag{110}$$

If we call $n(r)$ the *baryon number density*, the *total number of baryons* N_B in the star is

$$N_B = \int n\ dV = \int_0^R \frac{4\pi r^2 n(r) dr}{\left[1 - \frac{2Gm(r)}{c^2 r}\right]^{1/2}}\ , \tag{111}$$

** $k_B T = 1$ MeV corresponds to $T \simeq 1.134\ 10^{10}$ K.

and the *baryonic mass* of the star is

$$M_B = m_u N_B, \tag{112}$$

where m_u is an average baryonic mass unit. M_B is the rest mass of N_B baryons (dispersed at infinity) which form the neutron star. The *proper mass* of the star is given by

$$M_P = \int \rho \, dV = \int_0^R \frac{4\pi r^2 \rho(r) dr}{\left[1 - \frac{2Gm(r)}{c^2 r}\right]^{1/2}} \,. \tag{113}$$

M_P is equal to the sum of the mass elements on the whole volume of the star, it includes the contributions of rest mass and internal energy (kinetic and interactions (other than gravitation)) of the constituents of the star [††] Since $e^{\lambda(r)/2} \geq 1$, it follows that $M_P \geq M_G$. The difference

$$E_G = (M_G - M_P)c^2 = c^2 \int_0^R 4\pi r^2 dr \rho(r) \left[1 - e^{\lambda(r)/2}\right] \leq 0 \tag{114}$$

is the *gravitational energy* of the star. Its opposite $B_G = -E_G$ is called the *gravitational binding energy.* It is the gravitational energy released moving the infinitesimal mass elements ρdV from infinity to form the star. In the Newtonian limit, from eq.(114), one recovers the well known relation for the (Newtonian) gravitational energy

$$E_G^{Newt} = -G \int_0^R 4\pi r^2 dr \frac{m(r)\rho(r)}{r} \,. \tag{115}$$

The *internal energy* is

$$E_I = (M_P - M_B)c^2 = \int_0^R \varepsilon'(r) dV \tag{116}$$

and $B_I = -E_I$ is the *internal binding energy.*

The *total binding energy* is given by

$$B = B_G + B_I = (M_B - M_G)c^2, \tag{117}$$

it is the total energy released during the formation of a static neutron star configuration from a rarefied gas of N_B baryons. The stability of a such configuration requires $B > 0$.

Therefore, the gravitational mass of the neutron star, represents the total energy ($E = M_G c^2$) of the star, including both the rest mass energy $M_B c^2$ of its constituents dispersed at infinity, and the mass–energy contribution coming from the microscopic motion and the interactions (including gravitation) between the star's constituents:

$$M_G = M_B + \frac{1}{c^2} E_I + \frac{1}{c^2} E_G = M_P + \frac{1}{c^2} E_G \tag{118}$$

In fig. 15, we show the gravitational, baryonic and proper masses as a function of ρ_c (left panel) and the associated binding energies (right panel) in the case of the BPAL22 EOS. Notice the opposite sign for B_I (< 0) and B_G (> 0). These results make clear that neutron stars are bound by gravity, not by the strong (nuclear) interaction.

[††] This particular role of the gravitational interaction which does not enter in the expression of the energy density (EOS), and which contribute to the gravitational mass (total energy) of the star through the general relativistic field equations is due to the long–range nature of the gravitational interaction.

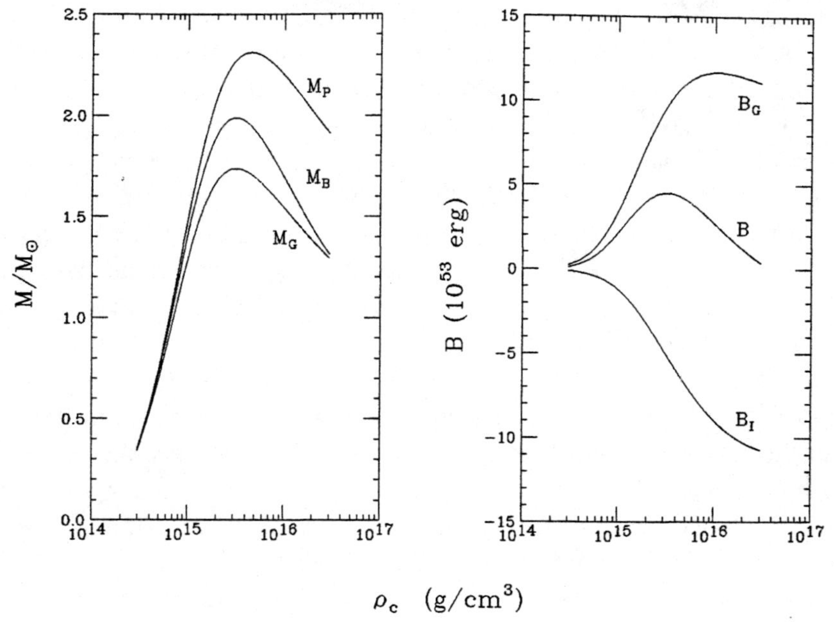

Figure 15: Masses (left panel) and binding energies (right panel) of a neutron star versus central density ρ_c assuming a generic EOS. Results are relative to the BPAL22 EOS. Configurations on the decreasing branch of the function $M_G(\rho_c)$ are unstable.

7.4. The Oppenheimer–Volkoff mass

As we said the solution (105) of the TOV equations depends parametrically on the central density, therefore the total gravitational mass (109) is a function of the central density ρ_c:

$$M_G = M_G(\rho_c) \; . \tag{119}$$

The main feature of the curve $M_G(\rho_c)$ is the following: *there is a maximum value of the gravitational mass of a neutron star that a given EOS can support.* This value M_{OV} of the mass is also referred to as the *Oppenheimer–Volkoff mass.* M_{OV} is the analogous of the Chandrasekhar mass (in Newtonian equilibrium) which is the maximum mass of a white dwarf. In the case of a white dwarf the EOS is dominated by the contribution of the relativistic electron gas (see *e.g.* ref. [63]), while for a neutron star the EOS is mainly ruled by the nature of the strong interactions. The existence of a finite value for the maximum mass of a neutron star implies that configurations with M_G greater than M_{OV} will collapse to a black hole.

Having the static equilibrium configurations (solutions of the TOV eq.s), one has to settle the question of their stability with respect to small perturbations. This problem is discussed in details in many textbooks of general relativity (see *e.g.* ref. [62,64,65]). The usual analysis of the stability problem is based on the assumption that time dependent stellar configurations, which undergo small radial perturbations, could be described by the EOS of a perfect fluid in "chemical" equilibrium (catalyzed matter). Under this hypothesis, it has been shown [62,64,65] that stable configurations can be obtained only when $dM_G/d\rho_c > 0$. This condition is necessary but is not sufficient for stability. Therefore, configurations on the decreasing branch of the function $M_G(\rho_c)$ (see Fig. 15) are unstable. It can be shown that configurations on the $M_G(\rho_c)$ curve on the left of the Oppenheimer–Volkoff mass are stable. This branch of stable stellar configurations (neutron stars) terminates (going towards lower densities) in the point $\rho_{c,min}$ where $dM_G/d\rho_c = 0$. This point sets the minimum value $M_{min}(\rho_{c,min})$ for the gravitational mass of a (stable) neutron star.

More generally, a perturbation of a static equilibrium configuration implies (local) time dependent perturbations of matter density. This in turn drives matter out of chemical equilibrium. In this case, the relevant EOS is set by the following time scales: τ_{chem}, the reaction relaxation time after which the composition of matter reaches its equilibrium state ‡‡, and the dynamical time scale τ_{dyn} of the perturbation. The stability of equilibrium configurations near M_{OV}, under this circumstance, has been recently investigated by Gourgoulhon, Haensel and Gondek [66]. They found that due to the slowness of weak interaction processes, the region of stellar configurations stable with respect to infinitesimal radial perturbations extend up to central densities larger than about 10% the central density ρ_{OV} of the Oppenheimer–Volkoff mass configuration. Moreover, using a full hydrodynamical analysis, which included effects of non-equilibrium weak interaction processes, these authors [66] showed that for finite amplitude perturbations, stellar configurations with central densities close to ρ_{OV} (which were stable with respect to infinitesimal radial perturbations) become unstable. In this sense, configurations in the vicinity of M_{OV} are metastable. The main outcome of the analysis of Gourgoulhon *et al.* [66] is that the actual maximum mass does not coincide with the Oppenheimer–Volkoff mass. This circumstance could be relevant when comparing measured neutron star masses with the values allowed by theoretical calculations.

In the present work, unless otherwise specified, we assume $M_{max} \equiv M_{OV}$, according to the usual analysis of the stability problem.

7.5. Lower bound on M_{max}

Since 1967, the year in which the first radio pulsar was discovered [67] and it was interpreted as a rotating neutron star [68,69], the masses of many neutron stars have been measured with high accuracy, and the results are shown in Fig. 16. Binary stellar systems in which one (or both) component is a compact object represent the most reliable way to measure the mass of the compact object [70,71,72].

The calculated maximum mass, for a given EOS, should be larger than the values of all the measured neutron star masses. If this is not the case, then the EOS used to calculate M_{max} has to be rejected *.

One of the most accurate measured masses is that of the pulsar PSR1913+16, which is a component of a binary neutron star system. The mass of PSR1913+16 is [73,74]

$$M_{PSR1913+16} = (1.4411 \pm 0.0007)\ M_\odot\ . \tag{120}$$

Based on the measured neutron star masses, summarized in Fig. 16, in the following we will take the mass of PSR1913+16 as an observational lower bound for the maximum mass of a neutron star

$$M_{max} \geq M^{lower} = 1.44\ M_\odot\ . \tag{121}$$

The measured neutron star masses provide us a constraint on the overall stiffness of the high density EOS, in fact each model of the EOS should be able to have stable configurations

‡‡In the case of equilibrium configurations near M_{OV}, τ_{chem} is set by weak interactions.

*The effect of rotation on the maximum mass is briefly discussed below

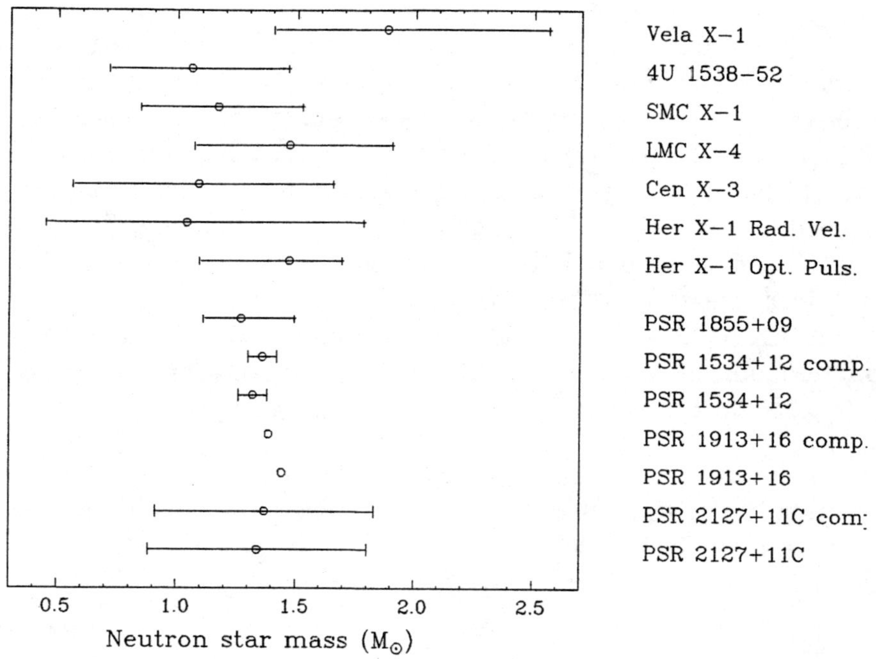

Figure 16: The currently known masses of neutron stars. Shown are both the masses for the X-ray pulsars and for the radio pulsars. Adapted from reference [71]

which can accommodate the observed neutron star masses. For example, the ideal neutron Fermi gas EOS is too "*soft*" to explain the observed neutron star masses; in fact, for this EOS $M_{max} = 0.71\ M_{\odot}$. The latter result can also be interpreted as a clear evidence coming from astrophysical observations that the nuclear interaction should be repulsive at high density or, in other words, at short nucleon–nucleon distance.

Of course, new precise measures of neutron star masses (*e.g.* for Vela X-1, or for a new discovered neutron star in a binary system) could push M^{lower} to a new value, higher than the present 1.44 $M_{\odot}$ which is consistent with existing data of Fig. 16. This will give more stringent constraints on the *stiffness* of dense matter EOS.

When comparing the neutron star masses with the theoretical predictions for M_{max}, one should consider the effects of rotation. Equilibrium sequences for rapidly rotating compact stars in general relativity have been recently constructed [75,76] in the case of realistic equations of state. These calculations [75,76] showed that rotation increases the gravitational maximum mass in a sizeable way ($\sim$ 20%). However, this large increase occurs only for rotation frequencies much higher than all observed pulsar frequencies. A more factual evaluation consist in assuming, as a limiting case, $\Omega = 4028$ rad/s. This is the rotation frequency corresponding to the most rapidly rotating pulsar presently known (PSR1937+21). Then, for a non–rotating star with $M_G = 1.400 M_{\odot}$, the corresponding (*i.e.* with the same baryonic mass) rotating star has [76] $M_G = 1.460 M_{\odot}$ ($M_G = 1.454 M_{\odot}$) using the BBB1 (BBB2) model [47] for the EOS of dense matter. This would mean an increase of 4% in M_G.

7.6. Upper bound on M_{max}

An unambiguous identification of a compact object as a neutron star or as a black hole, depends on the value of M_{max} given by nature. On the other hand the theoretical determination of M_{max} is very sensitive to the details of the poorly known high density EOS. However, it is possible to establish an upper bound on the maximum mass of a non-rotating neutron star which comes out from basic physical principles and minimal assumptions, and does not depend on the details of the high density EOS. Originally this idea dates back to the paper by Oppenheimer and Volkoff [61], who argued that an upper bound on M_{max} could be obtained using the stiffest possible EOS consistent with fundamental physical principles. Following this idea, a number of investigators [77,78,79,80,81,82] have derived various upper bounds on M_{max}. Basically all the derivations for an optimal upper bound on M_{max} are based on the following set (or subset) of assumptions:

(a) General relativity is the correct theory of gravitation.
(b) The matter is a perfect fluid described by a one parameter EOS which relates the pressure P to the mass density ρ.
(c) The density is non-negative (gravity is attractive)

$$\rho \geq 0. \tag{122}$$

(d) The EOS satisfies the so-called microscopic stability condition (le Chatelier's principle)

$$dP/d\rho \geq 0. \tag{123}$$

(e) The EOS is known below some fiducial density ρ^*.
(f) The EOS satisfies the causality condition (s is the speed of sound in the stellar medium).

$$s = \left[dP/d\rho\right]^{1/2} \leq c \tag{124}$$

Under the set of assumptions (a)–(f) it has been shown [78,80,82] that: if the fiducial density ρ^* is lower than a few times the normal saturation density of NM (lets say 10^{14} g/cm^3 $\leq \rho^* \leq 10^{15}$ g/cm^3) then

- The upper bound M^{upper} is independent on the details of the EOS below ρ^* (which we denote as EOS*).
- M^{upper} scales with ρ^* as [82]

$$M^{upper} = 6.8\left(\frac{10^{14}\ \mathrm{g/cm^3}}{\rho^*}\right)^{1/2} M_\odot \tag{125}$$

A careful treatment along these lines was first given by Rhoades and Ruffini [78]. These authors made use of a variational method to search for an extremum of the neutron star mass with respect to variations of the EOS above the fiducial density. In their paper [78] Rhoades and Ruffini chose $\rho^* = 4.6 \times 10^{14}$ g/cm^3, and used the Harrison–Wheeler EOS (see ref. [62]) below ρ^*. They found $M^{upper} = 3.2\ M_\odot$. Although Rhoades and Ruffini were aware of the fact that M^{upper} depends on the fiducial density, the value $M^{upper} = 3.2\ M_\odot$ had become so popular that it is often incorrectly quoted as an absolute optimal upper bound on M_{max}.

The sharp decrease of M^{upper} with ρ^* emphasize the great utility of even modest improvements in our knowledge of the EOS of dense matter. For example, if we know the EOS up to the normal saturation density of NM, *i.e.* $\rho^* = \rho_0 = 2.8 \times 10^{14}$ g/cm^3, then $M^{upper} = 4.06\ M_\odot$. For $\rho^* = 2\rho_0$, eq.(125) gives $M^{upper} = 2.87\ M_\odot$.

If the fiducial density is pushed to sufficiently high values ($\rho^* >> 10^{15}\ g/cm^3$), then eq.(125) will no longer be valid and M^{upper} will tend to a limiting value which is equal to the maximum mass $M_{max}(EOS^*)$ set by the EOS below ρ^*. In particular for $\rho_c^{max} \leq \rho^*$ one has $M^{upper} = M_{max}(EOS^*)$, ρ_c^{max} being the central density of the maximum mass configuration.

7.7. The EOS for the neutron star's crust

Compared to the large uncertainty which affects our knowledge of the EOS at supranuclear density, the properties of matter at subnuclear density are known with a satisfactory confidence. Bulk properties of a neutron star with $M \sim M_{max}$, slightly depend on the EOS at subnuclear density. However, the crust plays a crucial role for a number of observational properties of neutron stars: for example, thermal evolution [83], x-ray bursts, pulsars glitches [84,85]

There have been many studies on the physics of matter at subnuclear density, and we refer the interested reader to those papers [5,6,7,8]. In the calculations presented below, to describe the surface and the crust of the star we used the following equations of state: the Feynman-Metropolis-Teller [86] EOS for the surface; the Baym-Pethick-Sutherland [87] EOS for the outer crust; and the Negele and Vautherin [88] EOS for the inner crust. Then we joined these equations of state with the high-density EOSs employed to describe the core of the neutron star.

7.8. Results

In Fig. 17, we show the calculated gravitational mass as a function of the radius (panel a) and the central number density (panel b) for different microscopic equations of state. The properties of the maximum mass configuration, for each EOS, are summarized in Tab. 3. Bulk neutron stars properties (masses, radii, central densities) calculated with these modern realistic EOSs are relatively close to each other. In particular, the calculated masses are consistent with the measured masses of X–ray pulsars and radio pulsars [71] (see fig 16).

A large difference between the BBB and the WFF models concerns the threshold properties for the onset of direct Urca processes. In fact, as summarized in Tab. 4, direct Urca processes are possible according to the BBB models while they are not in the WFF model. This difference reflects the discrepancy already discussed for the symmetry energy in the two models.

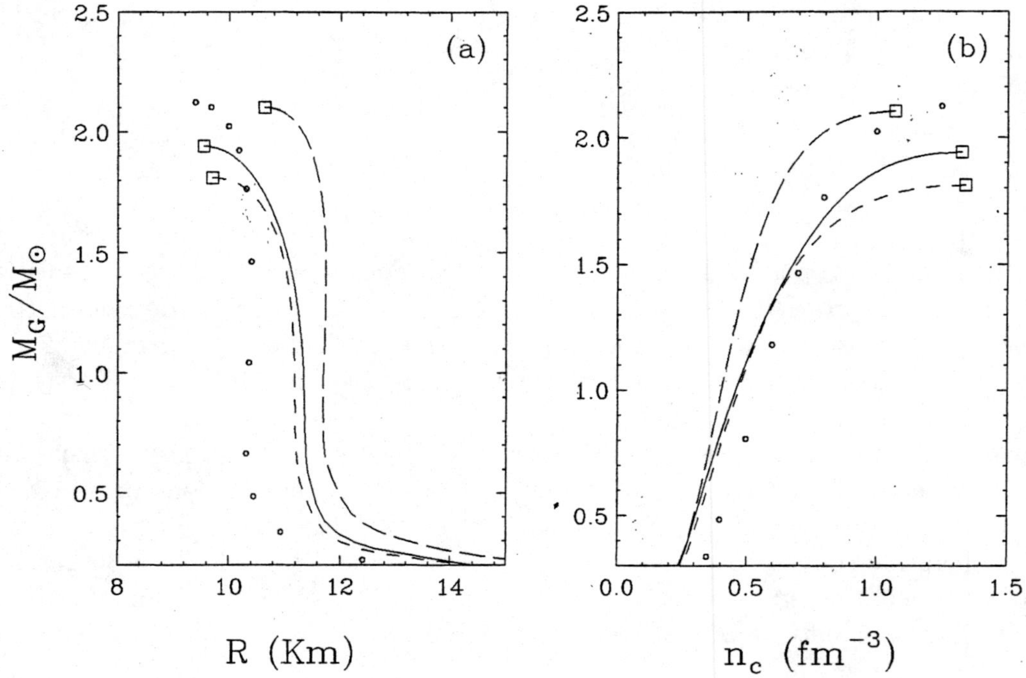

Figure 17: Gravitational mass versus radius (panel a), and versus central number density (panel b), for different microscopic equations of state. BBB1 EOS (short dashed line), BBB2 EOS (solid line), DBHF EOS (long dashes) WFF EOS (with AV14+UVII interaction) (open circles).

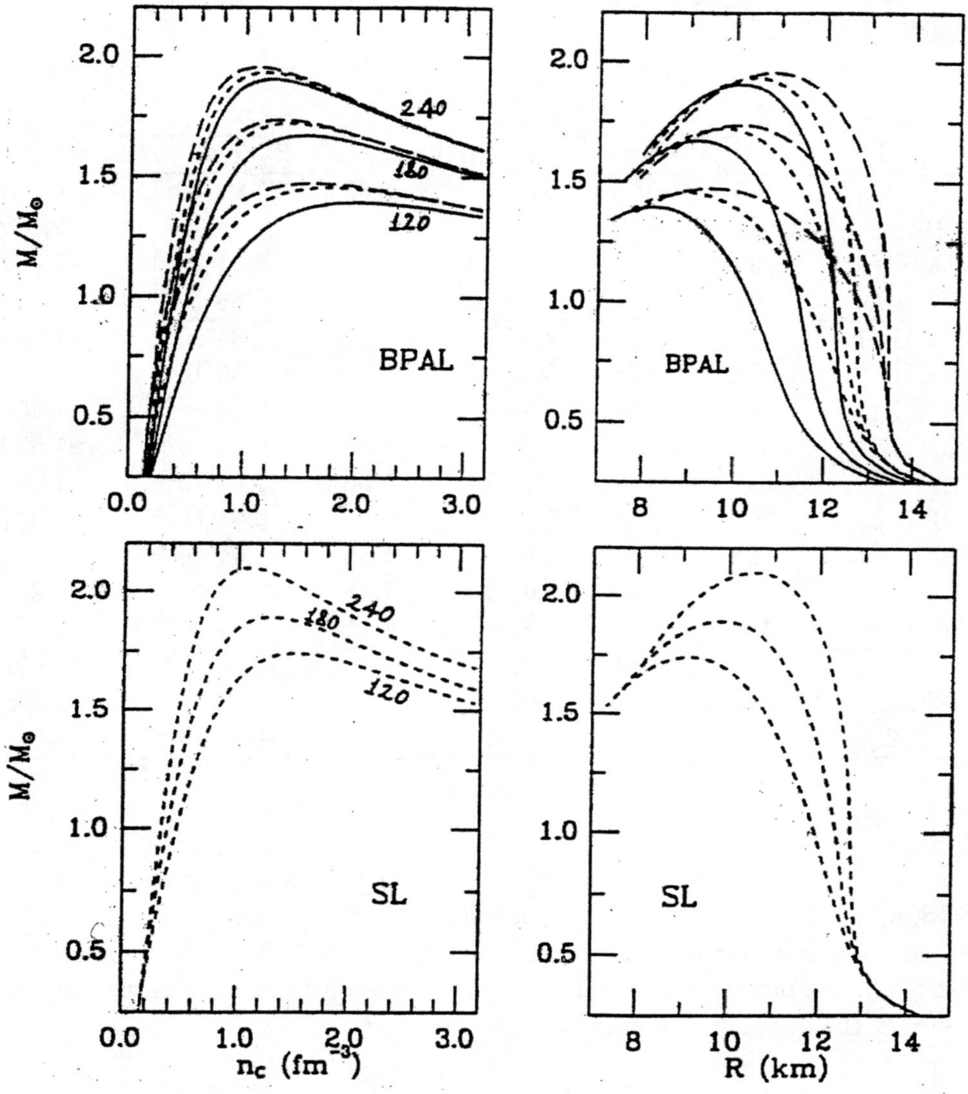

Figure 18: Gravitational mass versus central number density (left panels) and versus radius (right panels) for some of the phenomenological equation of state discussed in the text. Continuous, short-dashed, and long-dashed curves refer to E_{sym}^{pot} proportional (approximatively) to a $\sqrt{u}$, u and $2u^2/(1+u)$ respectively. Curves are labeled with the value of the nuclear incompressibility K_0 exprèssed in MeV.

Table 3. Properties of the maximum mass configuration obtained from different equations of state. M_G is the gravitational (maximum) mass, R is the corresponding radius, ρ_c the central density, n_c the central number density ($n_0 = 0.16\ fm^{-3}$), P_c the central pressure, M_B the baryonic mass.

EOS	$M_G/M_\odot$	R(km)	$\rho_c(g/cm^3)$	n_c/n_0	$P_c(dyne/cm^2)$	$M_B/M_\odot$
BBB1	1.790	9.66	$3.08\ 10^{15}$	8.53	$1.02\ 10^{36}$	2.084
BBB2	1.917	9.49	$3.19\ 10^{15}$	8.45	$1.44\ 10^{36}$	2.262
WFF	2.130	9.40	$3.00\ 10^{15}$	7.81	$2.01\ 10^{36}$	2.632
DBHF	2.100	10.60	$2.40\ 10^{15}$	6.69	$0.88\ 10^{36}$	2.510
BPAL11	1.403	8.32	$4.56\ 10^{15}$	12.91	$1.34\ 10^{36}$	1.586
BPAL12	1.465	9.04	$3.93\ 10^{15}$	10.99	$1.06\ 10^{36}$	1.646
BPAL13	1.484	9.59	$3.56\ 10^{15}$	10.36	$0.89\ 10^{36}$	1.655
BPAL21	1.684	9.29	$3.51\ 10^{15}$	9.78	$1.20\ 10^{36}$	1.938
BPAL22	1.735	9.83	$3.17\ 10^{15}$	9.00	$1.02\ 10^{36}$	1.987
BPAL23	1.750	10.22	$2.99\ 10^{15}$	8.51	$0.91\ 10^{36}$	1.992
BPAL31	1.919	10.22	$2.83\ 10^{15}$	7.85	$1.04\ 10^{36}$	2.234
BPAL32	1.947	10.54	$2.67\ 10^{15}$	7.58	$0.94\ 10^{36}$	2.261
BPAL33	1.969	10.94	$2.52\ 10^{15}$	7.11	$0.84\ 10^{36}$	2.272
SL12	1.753	9.33	$3.47\ 10^{15}$	9.74	$1.36\ 10^{36}$	2.032
SL22	1.904	9.94	$3.03\ 10^{15}$	8.25	$1.23\ 10^{36}$	2.218
SL32	2.113	10.67	$2.55\ 10^{15}$	7.10	$1.11\ 10^{36}$	2.491

To explore how the structure of a neutron star depends on the properties of the underlying EOS, we used our generalized Skyrme-like EOS. This study, in particular, enables one to investigate the role on the neutron star structure of quantities like *e.g.* the nuclear incompressibility, the momentum dependence of the effective nuclear interaction, the density dependence of the symmetry energy.

Fig. 18 shows the gravitational mass versus central density (left panels) and versus radius (right panels) for some of the BPAL and SL EOSs with different incompressibility K_0 and density dependence of E^{pot}_{sym}. The properties of the maximum mass configurations are reported in Tab. 3. From results in Fig. 18 and Tab. 3, we deduce the following quite general features for the maximum mass configuration in our EOS model: *a "stiff" EOS gives a maximum mass neutron star with a higher mass, a larger radius, and a lower central density with respect to the corresponding quantities of the maximum mass star obtained with a "soft" EOS.* The previous sentence should not be considered as a rule, the main reason being in the use of terms like "stiff"' or "soft" to describe a property of the EOS. In particular, the *stiffness* (or *softness*) is a general property of the EOS which describes how the pressure in matter changes with density. It can not be fully characterized, *e.g.*, by the value of the nuclear incompressibility, which refers to a property around n_0. In our model the *stiffness* of the EOS depends on a combined effect of the value of K_0, the density dependence of E^{pot}_{sym}, the momentum dependence of the NN interaction (choice of $g(k)$ in eq. (65)–(69). In a phenomenological EOS, like our generalized Skyrme-type EOS, all the previous physical quantities can be varied independently with a certain freedom.

Table 4. Threshold properties for the onset of direct Urca processes. x^{Urca}, $n^{Urca}(fm^{-3})$, and $M^{Urca} \equiv M_G(n^{Urca})$ are respectively the threshold proton fraction, number density and gravitational mass. M_{max} is the gravitational maximum mass and x_c the corresponding central proton fraction. Finally, $x_c(1.4\ M_\odot)$ is the central proton fraction of the star with 1.4 $M_\odot$ gravitational mass. Masses are expressed in unit of the mass of the sun.

EOS	x^{Urca}	n^{Urca}	M^{Urca}	M_{max}	x_c	$x_c(1.4\ M_\odot)$
BBB1	0.139	0.65	1.40	1.790	0.230	0.139
BBB2	0.139	0.54	1.24	1.917	0.265	0.165
WFF	no	no	no	2.130	0.045	0.066

Table 5. Properties of neutron stars with $M_G = 1.40\ M_\odot$. The central density is denoted by ρ_c. R is the (total) radius, R_{core} the radius of the nuclear matter core, ΔR_{inner} and ΔR_{outer} the thickness of the inner and outer crust respectively, $\Delta R = \Delta R_{inner} + \Delta R_{outer}$, all reported in km.

EOS	$\rho_c(g/cm^3)$	R	R_{core}	ΔR_{inner}	ΔR_{outer}	ΔR
BPAL12	2.5×10^{15}	9.98	8.56	1.15	0.27	1.42
BPAL22	1.2×10^{15}	11.81	9.63	1.75	0.43	2.18
BPAL32	0.9×10^{15}	12.60	10.06	2.05	0.49	2.54
BPAL21	1.5×10^{15}	10.91	9.38	1.19	0.34	1.53
BPAL23	1.1×10^{15}	12.48	9.67	2.33	0.48	2.81

Table 6. Crustal properties of neutron stars with different gravitational masses. Results are relative to neutron stars with a core of pure neutron matter.

EOS	$M_G/M_\odot$	$\rho_c(\mathrm{g/cm^3})$	R	ΔR_{inner}	ΔR_{outer}	ΔR
BPAL22	1.40	$0.9\ 10^{15}$	12.78	2.25	0.50	2.75
BPAL22	0.41	$0.3\ 10^{15}$	14.00	7.93	2.53	10.46

To investigate how the thickness of the neutron star crust depends on the high density EOS † in Tab. 5 we report the (total) radius, the nuclear matter core radius R_{core}, and the thickness of the inner and outer crust for different BPAL models for the EOS, and for $M_G = 1.40\ M_\odot$. In general, the stiffer the EOS, the more the extent of the crust (for a given fixed value of M_G). It is interesting to compare in Tab. 5 the results for the BPAL21, BPAL22, and BPAL23 models (which differ in the density dependence of the symmetry energy): as we see, the thickness of the crust is strongly affected by the symmetry energy. A detailed study on the crustal properties of neutron stars can be found in ref. [89], and some interesting implication for the dynamical models for neutron star glitches and postglitch behaviour is discussed in ref. [85]. In general, for a given EOS, neutron stars with a large mass ($M_G \sim M_{max}$) have thin crust, while they have a thick crust when M_G is small in comparison to M_{max}. As an example, we report in Tab. 6, the thickness of the crust in the case of neutron stars with a core of pure neutron matter, for $M_G = 1.40\ M_\odot$ and $M_G = 0.41\ M_\odot$.

8. *

References

[1] H.A. Bethe, G.E. Brown, J. Applegate, and J.M. Lattimer, *Nucl. Phys.* A **324**, 487 (1979).

[2] J.M. Lattimer, and F.D. Swesty, *Nucl. Phys.* A **535**, 331 (1991).

[3] I. Bombaci, T.T.S. Kuo, and U. Lombardo, *Phys. Rep.* **242**, 165 (1994).

[4] G. Baym, H.A. Bethe, and C.J. Pethick, *Nucl. Phys.* A **175**, 225 (1971).

[5] D.G. Ravenhall, C.J. Pethick, and J.R. Wilson, *Phys. Rev. Lett.* **50**, 2066 (1983).

[6] M. Hashimoto, H. Seki, and M. Yamada, *Prog. Theor. Phys.* **71**, 320 (1984).

[7] C.P. Lorenz, D.G. Ravenhall, and C.J. Pethick, *Phys. Rev. Lett.* **70**, 379 (1993).

[8] K. Oyamatsu, *Nucl. Phys.* A **561**, 431 (1993).

[9] G.E. Brown, and H. A. Bethe, *Astroph. Journ.* **423**, 659 (1994).

[10] I. Bombaci, *Astron. Astrophys.* **305**, 871 (1996).

[11] M. Prakash, I. Bombaci, M. Prakash, P.J. Ellis, R. Knorren and J.M. Lattimer, *Phys. Rep.* **280**, 1 (1997).

[12] R.N. Boyd, *Mod. Phys. E Suppl.* **3**, 294 (1984).

[13] H. Geissel, G. Münzenberg, and K. Riisanger, *Ann. Rev. Nucl. Part. Sci.* **45**, 163 (1995).

[14] B.-A. Li, C.M. Ko, and W. Bauer, *Int. J. Mod. Phys. E* **7**, 147 (1998).

[15] M. Di Toro, *et al.*, *Prog. Part. Nucl. Phys.* **42**, 125 (1999).

† *i.e.* we consider different equations of state for the NM core, with the EOS at subnuclear density fixed to the set of EOS referred to in subsect. 7.7

[16] M. Colonna, M. Di Toro, G. Fabbri, and S. Maccarone, *Phys. Rev.* C **57**, 1410 (1998).

[17] R. Machleidt, *Adv. Nucl. Phys.* **19**, 189 (1989).

[18] G.Q. Li, and R. Machleidt, *Phys. Rev.* C **58**, 1393 (1998).

[19] S. Fantoni, and S. Rosati, *Nuovo Cimento A* **25**, 593 (1975); *Phys. Lett.* B **84**, 23 (1979).

[20] V.R. Pandharipande, and R. B. Wiringa, *Rev. Mod. Phys.* **51**, 821 (1979).

[21] B. Friedman and, V.R. Pandharipande, *Nucl. Phys.* A **361**, 502 (1981).

[22] R.B. Wiringa, V. Fiks and A. Fabrocini, *Phys. Rev.* C **38**, 1010 (1988), (WFF).

[23] A. Akmal, and V.R. Pandharipande, *Phys. Rev.* C **56**, 2261 (1997).

[24] M.R. Anastasio, L.S. Celenza, W.S. Pong, and C.M. Shakin, *Phys. Rep.* **100**, 327 (1983).

[25] R. Brockmann, and R. Machleidt, *Phys. Rev.* C **42**, 1965 (1990).

[26] B. ter Haar, and R. Malfliet, *Phys. Rep.* **149**, 207 (1987).

[27] G.Q. Li, R. Machleidt and R. Brockmann, *Phys. Rev.* C **45**, 2782 (1992).

[28] L. Engvik, E. Osnes, M. Hjorth-Jensen, G. Bao, and Ostgaard, *Astroph. Journ.* **469**, 794 (1996).

[29] C.-H. Lee, T.T.S. Kuo, G.Q. Li, and G.E. Brown, *Phys. Rev.* C **57**, 3488 (1998).

[30] H. Huber, F. Weber, and M.K. Weigel, *Phys. Rev.* C **57**, 3484 (1998).

[31] F. de Jong, and H. Lenske, *Phys. Rev.* C **57**, 3099 (1998).

[32] T.H.R. Skyrme, *Phylos. Mag.* **1**, 1043 (1956); *Nucl. Phys.* A **9**, 615 (1959).

[33] D. Vautherin, and D.M. Brink, *Phys. Rev.* C **5**, 626 (1972).

[34] W.D. Myers and W.J. Swiatecky, *Nucl. Phys.* A **601**, 141 (1996).

[35] J.P. Blaizot, *Phys. Rep.* **64**, 171 (1980).

[36] H.Q. Song, M. Baldo, G. Giansiracusa, and U. Lombardo, *Phys. Lett.* B **411**, 237 (1997); *Phys. Rev. Lett.* **81**, 1584 (1998).

[37] T.T.S. Kuo, and G.E. Brown, *Nucl. Phys.* A **85**, 40 (1966)

[38] T.T.S.Kuo, and E.Osnes, *Folded-Diagram Theory of the Effective Interaction in Atomic Nuclei*, Springer Lecture Notes in Physics, vol.364 (Springer, Berlin, 1990).

[39] M. Hjorth-Jensen, T.T.S.Kuo, and E.Osnes, *Phys. Rep.* **261**, 125 (1995)

[40] I. Bombaci, and U. Lombardo, *Phys. Rev.* C **44**, 1892 (1991), (BL).

[41] J. Haidenbauer, and W. Plessas, *Phys. Rev.* C **30**, 1822 (1984);*Phys. Rev.* C **32**, 1424 (1985).

[42] M.Baldo, and L.S. Ferreira, *Phys. Rev.* C **59**, 682 (1999).

[43] G.H. Bordbar, and M. Modarres, *Phys. Rev.* C **57**, 714 (1998).

[44] A. M. Lane, *Nucl. Phys.* **35**, 676 (1962).

[45] F. Coester, S. Cohen, B.D. Day, and C.M. Vincent, *Phys. Rev.* C **1**, 769 (1970).

[46] J. Carlson, V.R. Pandharipande and R.B. Wiringa, *Nucl. Phys.* A **401**, 59 (1983); R. Schiavilla, V.R. Pandharipande and R.B. Wiringa, *Nucl. Phys.* A **449**, 219 (1986).

[47] M. Baldo, I. Bombaci and G.F. Burgio, *Astron. Astrophys.* **328**, 274 (1997), (BBB)

[48] A. Lejeune, P. Grangé, M. Martzolff and J. Cugnon, *Nucl. Phys.* A **453**, 189 (1986).

[49] M. Baldo, G. Giansiracusa, U. Lombardo, I. Bombaci and L.S. Ferreira, *Nucl. Phys.* A **583**, 599 (1995).

[50] G.E. Brown, W. Weise, G. Baym and J. Speth, *Comm. Nucl. Part. Phys.* **17**, 39 (1987).

[51] J. Lattimer, C. Pethick, M. Prakash and P. Haensel, *Phys. Rev. Lett.* **66**, 2701 (1991).

[52] E. Chabanat, P. Bonche, P. Haensel, J. Meyer, and R. Schaeffer, *Nucl. Phys.* A **627**, 710 (1997); *Nucl. Phys.* A **635**, 231 (1998)

[53] I. Bombaci, in *Perspectives on Theoretical Nuclear Physics*, Proceedings of the conference *"Problems in Theoretical Nuclear Physics"*, October 1995, Cortona (Italy). Ed.s I. Bombaci et al., ETS Pisa.

[54] I. Bombaci, M. Prakash, M. Prakash, P.J. Ellis, J.M. Lattimer, and G.E. Brown, *Nucl. Phys.* A **583**, 623 (1995).

[55] C. Gale, G. Bertsch, and S. Das Gupta, *Phys. Rev.* C **35**, 1666 (1987).

[56] M. Prakash, T.L. Ainsworth, and J.M. Lattimer, *Phys. Rev. Lett.* **61**, 2518 (1988).

[57] M. Baldo, I. Bombaci, L.S. Ferreira, G. Giansiracusa and U. Lombardo, *Phys. Lett.* B **215**, 19 (1988).

[58] D. Page, *Thermal evolution of isolated neutron stars.* astro-ph/9706259.

[59] D.A. Baiko, and D.G. Yakolev, *Astron. Astrophys.* **342**, 192 (1999).

[60] R.C. Tolman, *Proc. Nat. Acad. Sci. USA* **20**, 3 (1934).

[61] J. Oppenheimer and G. Volkoff, *Phys. Rev.* **55**, 374 (1939).

[62] B.K. Harrison, K.S. Thorne, M. Wakano, and J.A. Wheeler, *Gravitation Theory and Gravitational Collapse*, University of Chicago Press, Chicago, 1965.

[63] S. Chandrasekhar, *An introduction to the study of stellar structure*, (University of Chicago Press, Chicago, 1939).

[64] S.L. Shapiro, and S.A. Teukolsky, *Black Holes, White Dwarfs, and Neutron Stars* (John Wiley, New York, 1983).

[65] Ya.B. Zeldovich and I.D. Novikov, *Relativistic Astrophysics: Stars and Relativity* (The University of Chicago Press, Chicago, 1971).

[66] E. Gourgoulhon, P. Haensel and D. Gondek, *Astron. Astrophys.* **294**, 747 (1995).

[67] A. Hewish, S.J. Bell, J.D.H. Pikington, P.F. Scott, and R.A. Collins, *Nature* **217**, 709 (1968).

[68] F. Pacini, *Nature* **216**, 567 (1967).

[69] T. Gold, *Nature* **218**, 731 (1968).

[70] R.N. Manchester, and J.H. Taylor, *Pulsars*, (Freeman, San Francisco, 1977).

[71] M.H. van Kerkwijk, J. van Paradijs, and E.J. Zuiderwijk, *Astron. Astrophys.* **303**, 497 (1995).

[72] S.E. Thorsett, and Deepto Chakrabarty, *Neutron Star Mass Measurements. I. Radio Pulsar.* astro-ph/9803260. Astrophysical Journal (submitted).

[73] J.H. Taylor, and J.M. Weisberg, *Astroph. Journ.* **345**, 434 (1989).

[74] J.H. Taylor, in *Pulsar as Physical Laboratories*, Ed.s: Blanford *et al.*, The Royal Society, Oxford University Press (1993).

[75] G.B. Cook, S.L. Shapiro, and S.A. Teukolsky, *Astroph. Journ.* **424**, 823 (1994).

[76] B. Datta, A.V. Thampan, and I. Bombaci, *Astron. Astrophys.* **334**, 943 (1998).

[77] M. Nauemberg, and G. Chapline, *Astroph. Journ.* **179**, 277 (1973).

[78] C. Rhoades, and R. Ruffini, *Phys. Rev. Lett.* **32**, 324 (1974).

[79] D.J. Hegyi, T.-S. H. Lee, and J.M. Cohen, *Astroph. Journ.* **201**, 462 (1975).

[80] D.M. Chitre, and J.B. Hartle, *Astroph. Journ.* **207**, 592 (1976).

[81] J.B. Hartle, and A.G. Sabbadini, *Astroph. Journ.* **213**, 831 (1977).

[82] J.B. Hartle, *Phys. Rep.* **46**, 201 (1978).

[83] K.A. Van Riper, B. Link and R.I. Epstein, *Astroph. Journ.* **448**, 294 (1995).

[84] B. Link, R.I. Epstein, and K.A. Van Riper, Nature **359**, 616 (1992).

[85] B. Datta, M.A. Alpar, *Astron. Astrophys.* **275**, 210 (1993).

[86] R. Feynman, F. Metropolis and E. Teller, *Phys. Rev.* **75**, 1561 (1949).

[87] G. Baym, C. Pethick and D. Sutherland, *Astroph. Journ.* **170**, 299 (1971).

[88] J. W. Negele and D. Vautherin, *Nucl. Phys.* A **207**, 298 (1973).

[89] B. Datta, A.V. Thampan, and D. Bhattacharya, *Jour. Astrophys. Astr.* **16**, 375 (1995).

Phase Transitions in Warm, Isospin-Asymmetric Nuclear Matter

Horst Müller * Brian D. Serot †

1 Introduction

The determination of the properties of nuclear matter as functions of density, temperature, and the ratio of protons to neutrons is a fundamental problem in nuclear physics. To achieve this goal, one must study not only the ground and excited states of normal nuclei, but also highly excited nuclei created in nucleus–nucleus collisions and nuclei far from stability. Indeed, with the advent of radioactive beam facilities [1], the creation of transient states of nuclear matter with appreciable isospin asymmetry became feasible. In this work we consider the properties of equilibrium nuclear matter at finite temperature and arbitrary proton fraction, and we describe some new qualitative features [2] that may be relevant for the energetic collisions of heavy ions.

There have been numerous theoretical studies of the dynamics of medium-energy heavy-ion collisions [3–8]. Some of these are based on equilibrium thermodynamics and focus on the nuclear matter phase diagram [9–13]. The basic feature is the liquid–gas phase transition at low densities and moderate temperatures, and how the nuclear system evolves through various phase-separation boundaries (*binodals*) and instability boundaries (*spinodals*) [9,14,15]. Other analyses concentrate on the nonequilibrium evolution of the system by describing the early stages of the collision through cascade simulations [16, 17], for example, or the late stages of the collision using models of fragmentation or nucleation [18–20]. Although the equilibrium analysis oversimplifies this problem and is applicable only to certain aspects of the evolution, it is useful for providing an orientation to the complex dynamics by giving concrete descriptions of the phase-separation process, which may be difficult to characterize in more microscopic formulations. Here we will follow the thermodynamic approach and focus on the qualitatively new features that arise when a system with two conserved charges (*i.e.*, baryon number and electric charge) undergoes a liquid–gas phase transition. These features have been discussed earlier by Barranco and Buchler [21], who use a simple phenomenological equation of state with an interaction energy that is independent of temperature, and by Glendenning [22], who focuses on zero-temperature matter and neutron stars.

The main ingredient in the present analysis is the nuclear equation of state, and ours is based on a relativistic mean-field model involving the interaction of Dirac nucleons with scalar and vector mesons [23–25]. Recent developments in the application of these models to the structure of nuclei show that they can provide an excellent description of bulk nuclear

*Department of Physics, University of Colorado, Boulder, Colorado 80309

†small Physics Department and Nuclear Theory Center,Indiana University, Bloomington, Indiana 47405

properties throughout the periodic table, provided that nonlinear scalar and vector self-interactions are included (for a recent review, see [25]). Thus we have a way to calibrate our equation of state at zero temperature and normal nuclear densities, and then extrapolate into the warm, dilute regime appropriate for the phase transition. We note that generally an extrapolation from normal nuclei, particularly into regimes of high temperature and density, is problematic [26] because one is sensitive to model parameters that cannot be calibrated by studying systems at normal nuclear densities. However, in the regimes we are interested in, the influence of these new parameters can be expected to be negligible.

We emphasize that although we present quantitative results based on an accurate nuclear equation of state, our focus is on the new qualitative features that arise in liquid–gas phase transitions in systems with more than one conserved charge. Indeed, even the use of a relativistic mean-field theory is not essential for our discussion; any equation of state that is accurately calibrated to nuclei should produce similar results. We believe that our simplified hydrodynamic, thermodynamic, and mean-field description of nucleus–nucleus collisions is the most transparent way to do this, and we expect that signatures of these new features will survive in more sophisticated calculations. In particular, our primary result is that the liquid–gas phase transition in asymmetric nuclear matter is of *second order* [2], in contrast to the familiar first-order (van der Waals) transition that occurs in one-component systems (and in symmetric nuclear matter). If such a phase transition actually occurs in medium-energy heavy-ion collisions, then generically it should be smoother than a first-order phase transition. To be more precise, even if the assumption of thermodynamic equilibrium is totally justified, and even if the finite-size effects on the phase separation are totally negligible, we show that the generic liquid–gas phase transition *must be continuous*; for example, it occurs over a temperature range of several MeV for matter that is roughly 40% protons. Thus observable signals of this transition should be more continuous than previously expected from conventional analyses, which find a discontinuous (first-order) phase transition in the thermodynamic limit [18].

Furthermore, we find that in asymmetric matter, the relevant spinodal that signals instability to infinitesimal fluctuations arises from variations in the proton concentration at fixed pressure. This is in contrast to the usual scenario (and the one relevant for symmetric matter), in which spinodal decomposition occurs due to fluctuations in the baryon density at points of mechanical instability. Thus microscopic models that attempt to describe this decomposition using nucleation or fragmentation must allow for different proton concentrations in the liquid and gas phases.

2 Phase Transitions in Multicomponent Systems

We consider a system characterized by a hamiltonian $\hat{H}$ and a set of n mutually commuting charges $\hat{Q}_i$. Here a conserved charge does not necessarily imply an independent particle species; it includes any conserved quantity resulting from an underlying symmetry of the system, for example, electric charge, baryon number, total angular momentum, *etc.* Moreover, particle numbers may change during a process (by decay, for example), and it is actually the conserved *charges* that are relevant for the thermodynamic analysis.

The equilibrium state of the system enclosed in a volume V is completely described by the thermodynamic potential [27]

$$\Omega(T, V, \mu_i) = -\frac{1}{\beta} \ln \operatorname{Tr} \exp\Big(-\beta(\hat{H} - \sum_i \mu_i \hat{Q}_i)\Big) , \tag{1}$$

where β is the inverse temperature. Thus the average charges and the pressure can be obtained from

$$Q_j(T,V,\mu_i) = -\left(\frac{\partial\Omega}{\partial\mu_j}\right)_{T,V,\{\mu_i,i\neq j\}}, \tag{2}$$

$$p(T,V,\mu_i) = -\left(\frac{\partial\Omega}{\partial V}\right)_{T,\{\mu_i\}}. \tag{3}$$

In some regions of T, some values of the μ_i may lead to systems with the same p but different density, and this allows for the possibility of phase transitions. To be more specific, let us first perform a Legendre transformation and consider the Helmholtz free energy F. We assume a system with specified charges Q_i in a large volume V, so that surface effects can be neglected. We then have

$$\Omega(T,V,\mu_i) = -Vp(T,\mu_i), \tag{4}$$

$$Q_i \equiv V\rho_i(T,\mu_i), \tag{5}$$

and correspondingly,

$$F(T,V,Q_i) \equiv V\mathcal{F}(T,\rho_i) = \Omega(T,V,\mu_i) + V\sum_i \mu_i\rho_i(T,\mu_i), \tag{6}$$

with

$$\mu_j = \left(\frac{\partial\mathcal{F}(T,\rho_i)}{\partial\rho_j}\right)_{T,\{\rho_i,i\neq j\}}. \tag{7}$$

The system will be stable against separation into two phases if the free energy of a single phase is lower than the free energy in all two-phase configurations. This requirement can be formulated as

$$\mathcal{F}(T,\rho_i) < (1-\lambda)\mathcal{F}(T,\rho_i') + \lambda\mathcal{F}(T,\rho_i''), \tag{8}$$

with

$$\rho_i = (1-\lambda)\rho_i' + \lambda\rho_i'', \qquad 0<\lambda<1, \tag{9}$$

where the two phases are denoted by a prime and a double-prime. In formal terms, stability implies that the free energy density is a *convex* function of the densities ρ_i [28]. Convexity implies that stability against separation into two phases also guarantees stability against separation into an arbitrary number of phases (which we denote by superscripts α, β, ..., ω). That is, by Jensen's inequality on convex functions, Eqs. (8) and (9) are equivalent to

$$\mathcal{F}(T,\rho_i) < \sum_{\{\alpha\}} \lambda_\alpha\mathcal{F}(T,\rho_i^\alpha), \tag{10}$$

with

$$\rho_i = \sum_{\{\alpha\}}\lambda_\alpha\rho_i^\alpha \quad \text{and} \quad \sum_{\{\alpha\}}\lambda_\alpha = 1, \tag{11}$$

where the sums are over all phases $\{\alpha\}$. The parameters $\lambda^\alpha = V^\alpha/V$ specify the volume fraction occupied by each phase. The second equation (9) or (11) ensures that the overall charges are conserved:

$$V\rho_i = \sum_{\{\alpha\}} V^\alpha \rho_i^\alpha \quad \text{and} \quad V = \sum_{\{\alpha\}} V^\alpha . \tag{12}$$

Equations (8) and (10) are *global* criteria for the stability of the one-phase system. If these inequalities are satisfied, then it is necessarily true that the symmetric matrix

$$\mathcal{F}_{ij} \equiv \left(\frac{\partial^2 \mathcal{F}}{\partial\rho_i \partial\rho_j} \right)_T$$

is positive [28]. In contrast, whenever Eq. (8) is violated, a system with more than one phase is energetically favorable. The phase coexistence is governed by the Gibbs' conditions

$$\mu_i^\alpha(T,\rho_i^\alpha) = \mu_i^\beta(T,\rho_i^\beta) = \cdots = \mu_i^\omega(T,\rho_i^\omega) , \quad i = 1,\ldots,n , \tag{13}$$

$$p^\alpha(T,\rho_i^\alpha) = p^\beta(T,\rho_i^\beta) = \cdots = p^\omega(T,\rho_i^\omega) , \tag{14}$$

where the temperature is the same in all phases. The *local* positivity conditions on $\mathcal{F}_{ij}$ give rise to a set of n inequalities that divide the parameter space $\{T, \rho_i\}$ into stable and unstable regions. It is important to realize, however, that under conditions of phase separation, Eq. (8) may still be valid *locally* (that is, for $\rho_i', \rho_i'' \approx \rho_i$), but it may nevertheless be possible to find significantly different densities ρ_i' and ρ_i'' that violate this condition. This leads to the existence of metastable states.

These ideas are illustrated for a simple one-component system in Fig. 1. Here the free energy density at fixed temperature is shown as a function of the density. At point D, we evidently have

$$\frac{\partial^2 \mathcal{F}}{\partial\rho^2} > 0 ,$$

and the system is stable. At points A and B, $\mathcal{F}$ is still locally convex, but these points share a common tangent (chemical potential). Thus, although the stability criteria remain valid locally around A and B, the existence of the common tangent implies that these two phases with different densities can coexist. The values of T and ρ at the points A and B lie on the binodal surface in the $\{T, \rho\}$ plane, which is a one-dimensional surface in this simple example. At point C, the free energy is concave, indicating that this configuration is unstable against phase separation. For points between A and B where the second derivative of $\mathcal{F}$ remains positive, the system is metastable and can exist temporarily in a single phase, allowing for supercooling or superheating. However, there is an inflection point between A and C, and also one between B and C, where the second derivative of $\mathcal{F}$ vanishes; these points lie on the spinodal surface, which delimits the onset of instability.

According to Gibbs' phase rule [29], at most

$$K_{\text{max}} = n + 2$$

phases can coexist in equilibrium in a system with n conserved charges. Each single phase is characterized by the set $(T, \rho_i^\alpha, V^\alpha)$ of $n+2$ variables, so that if there are K phases, the total number of variables is $K(n+1)+1$. (All phases have the same temperature.) We assume that the n total charges Q_i and the total volume V are also specified, which

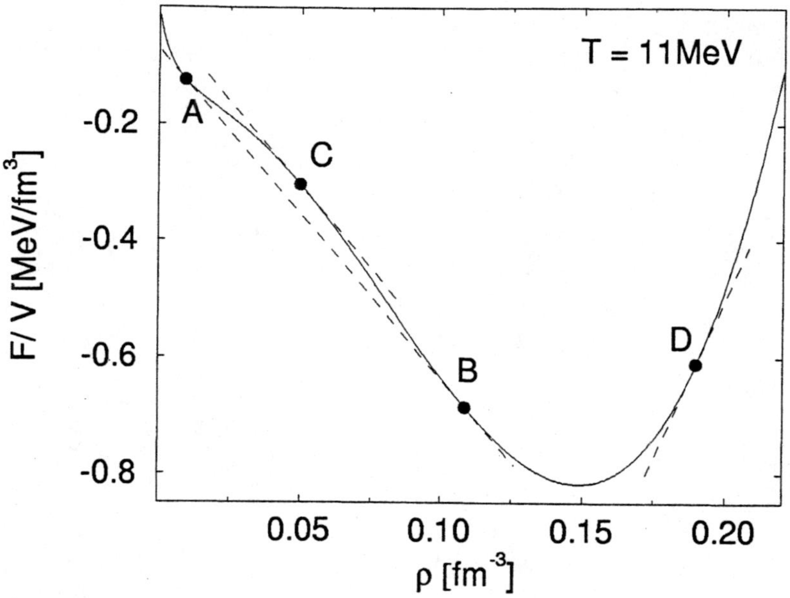

Figure 1: Free-energy density as a function of density for a one-component system at fixed temperature. For clarity, we show $\mathcal{F} - e_0\rho \equiv \mathcal{F} + (15.75\,\mathrm{MeV})\rho$, since the variations in $\mathcal{F}$ are quite small. The additional term linear in the density has no effect on the stability conditions or on Gibbs' criteria. Taken from Ref. [2].

brings the total number of variables to $K(n+1)+n+2$. These variables are restricted by the Gibbs' conditions (13) and (14), which provide $(K-1)(n+1)$ constraints, and by the conservation laws (11) on the charges (n constraints) and (12) on the total volume (1 constraint). We therefore end up with a set of $n+2$ independent variables, which we can take to be (T, Q_i, V) or (T, ρ_i, V). This result implies that the number of degrees of freedom in a system with strictly conserved charges is independent of the number of phases [30,31].

It is important to emphasize, however, an important difference between a multicomponent system and one containing a single conserved charge. Although the ratios of the total charges Q_i/Q_j remain fixed once the system has been prepared, *the ratios can be different in different phases.* Moreover, since the independent variables T, Q_i, and V determine the energetically stable state of the system, it is impossible to impose additional constraints. For example, the common (vapor) pressures and chemical potentials vary when the proportions λ^α of the phases change. These results will be illustrated explicitly for the case of asymmetric nuclear matter in Secs. 3 and 4.

A necessary condition for the coexistence of more than two phases is that any pair of these phases must be in equilibrium. In addition, as mentioned earlier, a system that is stable against separation into two phases is also stable against separation into multiple phases. Let us therefore focus on two such phases for a moment, which we will denote with a prime and a double-prime. The two sets of densities $\{\rho_i', \rho_i''\}$ that satisfy Eqs. (13) and (14) form a surface in the parameter space $\{T, \rho_i\}$; this is the phase separation boundary, or binodal. For n conserved charges and two coexisting phases, Gibbs' phase rule implies that the binodal is an n-dimensional surface. One can also show [2] that this surface encloses all

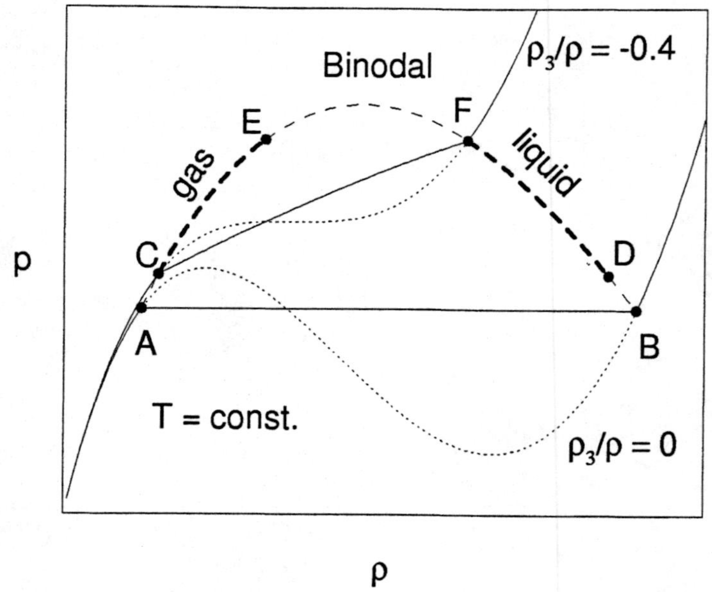

Figure 2: The Maxwell construction for symmetric and asymmetric systems. The construction for symmetric matter ($\rho_3/\rho = 0$) is indicated by the segment AB and is the same as for a one-component system. The construction for asymmetric matter with $\rho_3/\rho = -0.4$ is indicated by the segment CF and shows the qualitatively new behavior allowed in a two-component system. The asymmetry is held constant throughout the phase separation. The (dashed) binodal line is obtained from similar isotherms at other values of the asymmetry. Taken from Ref. [2].

points satisfying

$$\rho_i = (1-\lambda)\rho_i' + \lambda\rho_i'' , \qquad 0 < \lambda < 1 , \tag{15}$$

that lead to a single (unstable) configuration with a higher value for the free energy.

One feature of the binodal surface is that it may contain *critical points.* At the critical points, if they exist, the two phases can no longer be distinguished by their densities. Therefore the critical points form a line that divides the binodal surface into different regions describing either a high density (liquid) or a low density (gas) phase. Finally, we note that more than two phases can coexist if and only if each pair of phases form a binodal, and if all these binodals have a common region of intersection [29].

The binodal surface determines the stability boundaries of the system, but it remains to show how the system behaves inside, *i.e.*, how to interpolate within the metastable and unstable regions using a Maxwell construction. To explain this in more detail, we consider an isothermal compression in a situation where the system can separate into two phases.

We begin with the familiar case of a one-component system, as illustrated by symmetric nuclear matter in Fig. 2. Suppose that during the compression, the system encounters the binodal at some point A in $\{T, \rho\}$ space. At this point, the whole volume is occupied by a phase with density ρ^A, and a second phase with density ρ^B is about to emerge in an infinitesimally small volume. The two phases at A and B are connected by the Gibbs' conditions, so that they have equal temperatures, pressures, and chemical potentials. In a one-component system, B is the point at which the system leaves the two-phase region;

on the (p, ρ) diagram of Fig. 2, A and B are connected by a horizontal line, the well-known Maxwell construction. In a multicomponent system, however, as depicted in the upper portion of Fig. 2, the ratios of the charges in the emerging phase at D are generally different from those in the original preparation at C, thus violating the conservation laws. The system must therefore evolve instead through configurations that maintain the ratios of the total charges (the curve CF), and it leaves the instability region at the point F, which lies together with C on the line of constant ratios Q_i/Q_j. At this point, the original phase is present in infinitesimal quantities with densities $\{\rho_i^E\}$, while the newly created phase has evolved to point F. The configuration at F is consistent with the conservation laws, and in general, the pressure and chemical potentials in the coexisting phases have changed throughout the transition.

To determine the nature of the system between these extreme values, we must solve

$$\rho_i = (1-\lambda)\rho_i' + \lambda\rho_i'' \tag{16}$$

for given values of ρ_i, with ρ_i' and ρ_i'' lying on the binodal surface. It is important to realize that Eqs. (16) are indeed a set of n equations in n unknowns, since the ρ_i are specified, and among the $2n+1$ variables ρ_i', ρ_i'', and λ, $n+1$ can be eliminated by virtue of the Gibbs' conditions (13) and (14). Moreover, these equations yield solutions with qualitatively different characteristics. If the solutions yield all values of λ in the interval $[0,1]$, so that

$$\begin{aligned} \rho_i' &= \rho_i^C \qquad \text{for} \qquad \lambda = 0\ , \\ \rho_i'' &= \rho_i^F \qquad \text{for} \qquad \lambda = 1\ , \end{aligned}$$

then the system has undergone a phase transition. However, anticipating the subsequent discussion, there are also solutions with $0 \le \lambda \le \lambda_{\max} < 1$. In this case the system becomes unstable to phase separation, but undergoes a *retrograde condensation*: after occupying a maximal volume fraction $\lambda_{\max}$, the new phase begins to disappear, and the system leaves the instability region in the original phase. In either situation, Eq. (16) provides the desired Maxwell construction that determines the free energy in the transition region according to

$$\mathcal{F}(T,\rho_i) = (1-\lambda)\mathcal{F}(T,\rho_i') + \lambda\mathcal{F}(T,\rho_i'')\ .$$

Densities related to other extensive quantities can be computed accordingly.

We close this section by specializing the general formalism to asymmetric nuclear matter, a system of interacting neutrons, protons, and mesons. Such a system is characterized by two conserved charges: the total number of baryons

$$B = N_p + N_n \equiv V\rho \tag{17}$$

and the total charge, or equivalently, the third component of isospin

$$I_3 = \frac{N_p - N_n}{2} \equiv \frac{1}{2}V\rho_3\ . \tag{18}$$

Thus we have

$$\mathcal{F}(T,\rho,\rho_3) = -p(T,\mu,\mu_3) + \mu\rho + \mu_3\rho_3\ . \tag{19}$$

The stability condition (8) implies the following set of inequalities on the convex free energy density:

$$\frac{\partial^2 \mathcal{F}}{\partial \rho^2} > 0 , \tag{20}$$

$$\frac{\partial^2 \mathcal{F}}{\partial \rho^2}\frac{\partial^2 \mathcal{F}}{\partial \rho_3^2} > \left(\frac{\partial^2 \mathcal{F}}{\partial \rho_3 \partial \rho}\right)^2 . \tag{21}$$

To make the physical content of these conditions more transparent, it is convenient to introduce the proton fraction y defined by

$$y \equiv \frac{N_p}{N_p + N_n} = \frac{\rho + \rho_3}{2\rho} \tag{22}$$

and to rewrite the free-energy density as

$$\mathcal{F}(T, \rho, \rho_3) = \mathcal{F}(T, \rho, y) . \tag{23}$$

It is straightforward to show that the conditions (20) and (21) are equivalent to

$$\rho \left(\frac{\partial p}{\partial \rho}\right)_{T,y} = \rho^2 \left(\frac{\partial^2 \mathcal{F}}{\partial \rho^2}\right)_{T,y} > 0 , \tag{24}$$

$$\left(\frac{\partial \mu_p}{\partial y}\right)_{T,p} > 0 \quad \text{or} \quad \left(\frac{\partial \mu_n}{\partial y}\right)_{T,p} < 0 , \tag{25}$$

where we have introduced chemical potentials for protons and neutrons defined by

$$\mu_p \equiv \mu + \mu_3 \quad , \quad \mu_n \equiv \mu - \mu_3 . \tag{26}$$

The first inequality is the familiar requirement that the isothermal compressibility is positive, that is, the system is mechanically stable. The second condition reflects the special character of the binary system. It expresses "diffusive stability", which guarantees that energy is required to change the concentration in a stable system while holding the remaining variables (pressure and temperature) fixed [30].

To describe the nuclear equation of state, we use a relativistic mean-field model containing nucleons, neutral scalar and vector fields, and the isovector ρ meson field [23–25]. The neutral meson fields are self-interacting, including terms through fourth order in the fields, as proposed by Bodmer and Price [32, 33]. The desired nuclear symmetry energy is achieved using the simplest possible coupling of the ρ meson to the nucleon, and there are no ρ self-interactions. Although it is known that this way of generating the symmetry energy is a simplification of more complex mechanisms (for example, one-pion-exchange in a Hartree–Fock calculation produces significant contributions [34]), and that the dependence of the energy on the isovector density ($\rho_3 \equiv \rho_\mathrm{p} - \rho_\mathrm{n}$) may be more complicated when one is far from symmetric matter, the qualitative features studied here arise because the symmetry energy is repulsive for *all* values of ρ_3, and this repulsion grows monotonically as $|\rho_3|$ increases. Thus the present simple description is adequate for our purposes, and we expect that other models of the symmetry energy with these features will produce similar results.

The model is calibrated to reproduce empirical nuclear matter properties, which are the equilibrium density and binding energy (ρ, e_0), the nucleon effective mass at equilibrium

Table 1: Nuclear Matter Properties

Equilibrium Properties:

k_F	ρ	M^*/M	e_0	K_V^{-1}	a_4
$1.30\,\mathrm{fm}^{-1}$	$0.1484\,\mathrm{fm}^{-3}$	0.60	$-15.75\,\mathrm{MeV}$	$250\,\mathrm{MeV}$	$35\,\mathrm{MeV}$

Critical Values:

T_c	ρ_c	M_c^*/M	p_c
$14.40\,\mathrm{MeV}$	$0.04661\,\mathrm{fm}^{-3}$	0.8543	$0.2010\,\mathrm{MeV/fm}^3$

(M^*), the compression modulus (K_V^{-1}), and the bulk symmetry energy (a_4). Experience has shown that an accurate reproduction of these five properties leads to realistic results when the calculations are extended to finite nuclei [35,36]. The nuclear matter properties, as well as the properties at the critical point in symmetric matter, are given in Table 1, below. For more details and the explicit equations that determine all the thermodynamic quantities, we refer the reader to Ref. [2].

3 Application to Asymmetric Nuclear Matter

Phase transitions in binary systems are more complex than in one-component systems. In the case of nuclear matter, as discussed in Sec. 2, the global instability boundary forms a two-dimensional surface in (T, p, y) space, enclosing the region where either mechanical instability [Eq. (24)] or diffusive instability [Eq. (25)] occurs.

The full complexity of the binary system is indicated in Fig. 3, where chemical potential isobars for neutrons and protons are shown as a function of y at fixed temperature. Above a certain critical pressure p_c, the matter is stable, but for $p < p_c$, the second condition (25) is violated, and the system becomes chemically unstable. The critical isobar p_c is determined by the existence of an inflection point:

$$\left(\frac{\partial \mu_p}{\partial y}\right)_{T,p} = \left(\frac{\partial^2 \mu_p}{\partial y^2}\right)_{T,p} = 0\,. \tag{27}$$

This isobar marks the upper boundary of instability with respect to the pressure and defines a critical point (p_c, y_c) for a given temperature [30].

The Gibbs' conditions (13) and (14) for phase equilibrium demand equal pressure and chemical potentials for two phases with different concentrations. Thus the two desired solutions form the edges of a rectangle and can be found by means of the geometrical construction indicated in Fig. 3 [21,37]. The collection of all such pairs $y_1(T,p)$ and $y_2(T,p)$ form the binodal surface. For a given temperature, the two-phase region is limited from below by the pressure at equal concentrations p_{eq}, which corresponds to symmetric nuclear matter ($y = 0.5$). Correspondingly, the rectangle in Fig. 3 shrinks to a point at this particular pressure, and the system becomes stable again for $p < p_{eq}$. We did not find any cases of three-phase coexistence, which would require two rectangles with a common side.

The phase-separation boundary, or binodal surface, obtained from the preceding geometrical constructions is indicated in Fig. 4. (The shape of this surface has been previously described as a "filet mignon" [21].) Several slices at constant T are indicated, and one

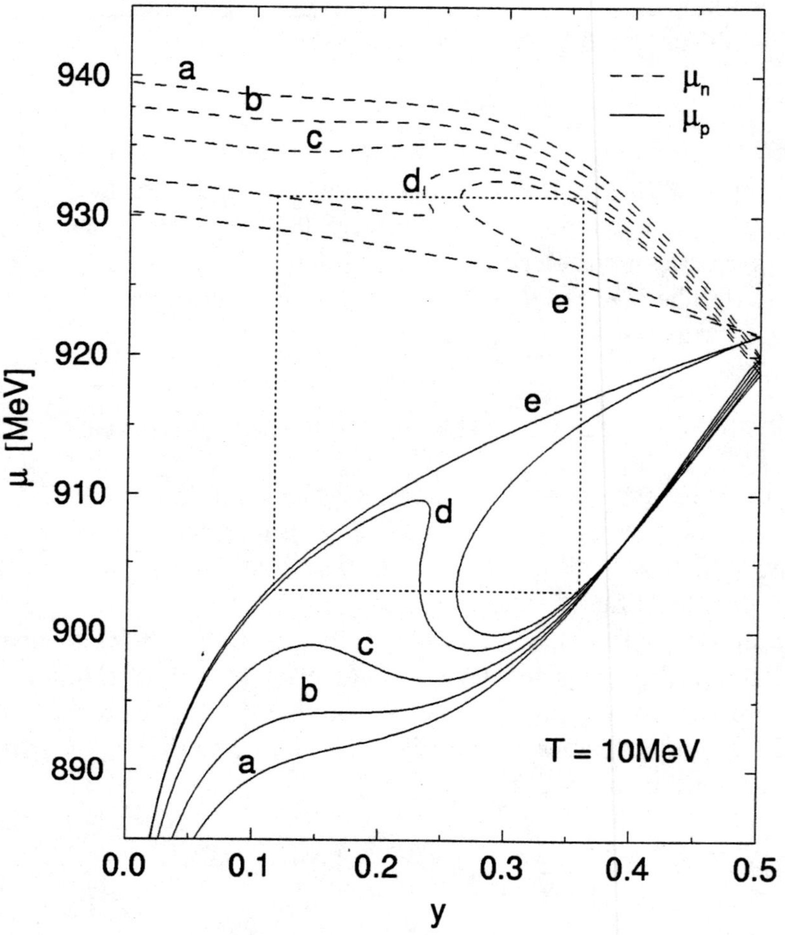

Figure 3: Chemical potential isobars at fixed temperature as a function of y. The curves labeled a through e have pressures $p = 0.25$, 0.198, 0.15, 0.10, and 0.075 MeV/fm^3, respectively. The curves labeled b are at the critical pressure $p_c = 0.198\,\mathrm{MeV/fm^3}$. Indicated by the dotted line is the geometrical construction which is used to obtain the proton fractions and chemical potentials in the two coexisting phases. Taken from Ref. [2].

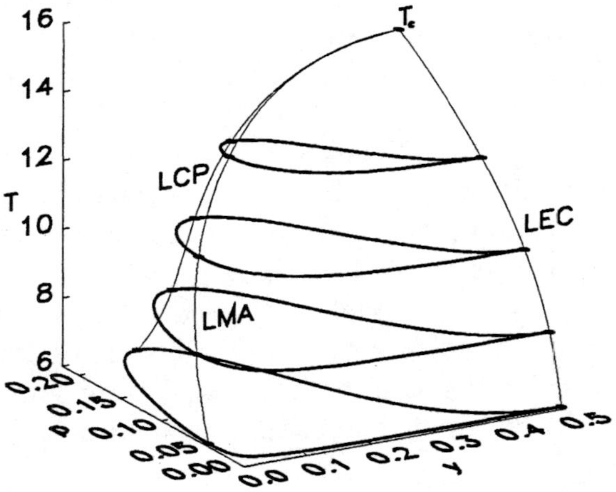

Figure 4: The binodal surface indicating the two-dimensional phase-coexistence boundary is shown in (p, T, y) space. The critical temperature $T_c(y = 0.5)$, the line of equal concentrations (LEC), the line of critical points (LCP), and the line of maximal asymmetry (LMA) are indicated. T is in MeV, p is in MeV/fm^3, and y is dimensionless. Taken from Ref. [2].

observes that the enclosed area in these sections decreases with increasing temperature until it vanishes at the critical point T_c of symmetric nuclear matter. This is also the point at which the lines LCP and LMA meet at $y = 0.5$. The line of critical points (LCP) is determined by the solutions of Eq. (27), and it can be parametrized uniquely by just one of the coordinates (T, p, y). The LCP begins at the critical point of symmetric nuclear matter $(y = 0.5, T = T_c)$ and ends at $y = y_0 = 0.057$ at $T = 0$. Note also that for a given temperature, the critical point determines the maximum pressure in the two phase region.

In addition to the LCP, we have also indicated the points on the binodal surface with the maximal asymmetry (LMA), or minimal proton fraction $y_{\rm min}$, at each temperature.[1] Any system with $y < y_{\rm min}$ is external to the two-phase region and is therefore stable. Moreover, since $y_{\rm min}$ is a monotonically increasing function of the temperature, the LMA determines the maximum temperature $T_{\rm max}$ of phase separation at any given y [21]. The LMA also begins at the critical point of symmetric nuclear matter and descends to a small, positive value of $y_{\rm min}$ at $T = 0$. This implies that pure neutron matter is stable at all temperatures in this model.

Configurations that separate into two phases each having equal numbers of neutrons and protons form the line of equal concentration (LEC). In these cases, the binodal section at any T degenerates into a point, at which the relation [29,31]

$$\left(\frac{\partial p}{\partial y}\right)_T = 0 \tag{28}$$

is satisfied. (This is an example of the Gibbs–Konowalow rule [31].) The LEC coincides with the projection $y = 0.5$ in our model, which implies that only symmetric nuclear matter

[1]As with all results in this section, our analysis is symmetric with respect to protons and neutrons, so that only the physically useful regime $0 \le y \le 0.5$ need be considered.

will separate into phases with equal concentration [21,37]. Since symmetric matter behaves as a one-component system,[2] one observes how the phase separation simplifies in this case: in an isothermal compression, the system evolves until it encounters the binodal (which is just the LEC) and then remains there until the transition is complete. In contrast, for $y \neq 0.5$, the system encounters a two-dimensional section of the binodal surface. Since the energy contains a term proportional to ρ_3^2, it is energetically favorable for asymmetric matter to separate into a liquid phase that is less asymmetric and a gas phase that is more asymmetric, rather than into two phases with equal concentration. This leads to more complex phase separations, as we discuss shortly.

For a given temperature, the binodal section is divided into two branches by the critical point and the point of equal concentration. One branch describes the system in a high-density (liquid) phase, while the other branch describes the low-density (gas) phase. These two branches contain the beginning and ending configurations of the phase transition.

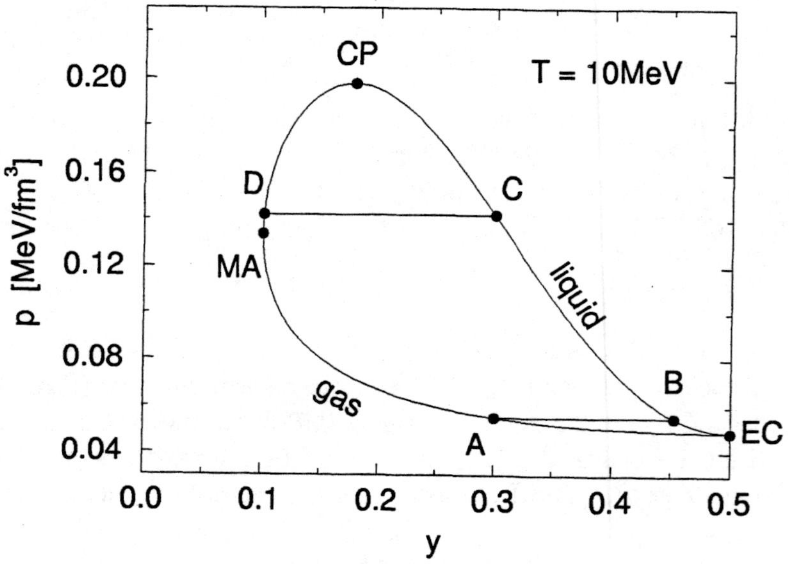

Figure 5: Binodal section at $T = 10\,\mathrm{MeV}$. The points A through D denote phases participating in a normal phase transition. The critical point (CP) and the points of equal concentration (EC) and maximal asymmetry (MA) are also indicated. Taken from Ref. [2].

We now return to the behavior of the matter under isothermal compression, to illustrate the different phase-separation scenarios. Consider the situation in Fig. 5, which shows a section through the binodal surface at $T = 10\,\mathrm{MeV}$. The critical point CP, the point of maximal asymmetry MA, and the point of equal concentration EC are indicated, and the validity of Eq. (28) is apparent. Assume that the system is initially prepared in the low-density (gas) phase with proton fraction $y = 0.3$. During the compression, the two-phase region is encountered at the point A, and now a (liquid) phase with a higher density begins to emerge. The geometrical construction described above determines the density and the proton fraction y_B of this new phase, which occurs at the point labeled B. As the system is compressed, the *total* proton fraction y remains fixed, as dictated by the conservation laws,

[2]Strictly speaking, symmetric nuclear matter is an *azeotrope*. During the liquid–gas phase separation, all equilibrium configurations have two phases with the same composition ($y = 0.5$). This type of phase equilibrium is called "indifferent equilibrium" [31]. Since the liquid–gas phase equilibrium in a one-component system is also indifferent, symmetric nuclear matter behaves as a one-component system.

but the gas phase evolves from A to D, while the liquid phase evolves from B to C. At the point C, the system leaves the region of instability. The original (gas) phase is about to disappear, and it exists in an infinitesimal volume with a density and proton fraction y_D corresponding to the point D.

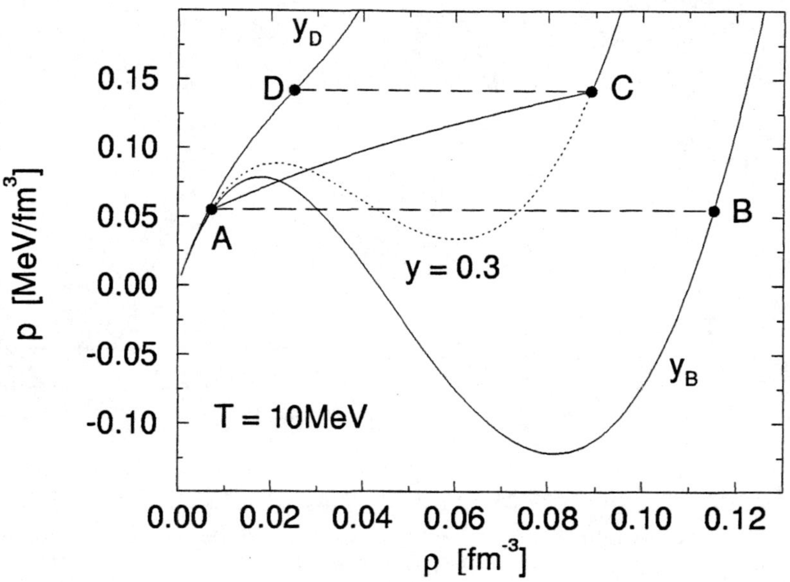

Figure 6: Isotherms for a normal phase transition at $T = 10\,\mathrm{MeV}$ and initial condition $y = 0.3$. The Maxwell construction produces the curve AC. Note that $y_D \neq 0.3 \neq y_B$. Taken from Ref. [2].

To evolve the system between configurations A and C, we must solve the equations

$$\rho = (1-\lambda)\rho' + \lambda\rho'' , \quad (29)$$

$$\rho_3 = (1-\lambda)\rho_3' + \lambda\rho_3'' , \quad (30)$$

according to (16), for densities that lie on the binodal surface, and for given values of ρ and $\rho_3 = (2y-1)\rho$. The result is the generalized Maxwell construction in the binary system. The corresponding isotherms are drawn in Fig. 6. The dotted line between A and C is the unphysical course of the pressure at the fixed total proton fraction, and the nearly straight line connecting A and C is the interpolation due to the Maxwell construction, which corresponds to the stable configuration at each intermediate density. The volume fraction λ starts with $\lambda = 0$ at A and runs through the whole interval $[0, 1]$, ending with $\lambda = 1$ at C. Since the points A and C lie on different branches of the binodal surface (as defined above), the matter has undergone a phase transition from a gas to a liquid phase.

Interestingly enough, the geometry of the binodal surface offers a second possibility. In the previous example, there is a transition between the two branches of the binodal surface because the value of y in the original phase is larger than y_c. For $y < y_c$, however, the system enters and leaves the two-phase region on the same branch, so the system remains in the same phase [2]. This *retrograde condensation* is unique to the binary system and does not occur in one-component systems.

To close this section, we consider the order of the liquid–gas phase transition. First-order phase transitions in single-component systems are characterized by discontinuities in certain physical quantities. This occurs because quantities like the density and the entropy

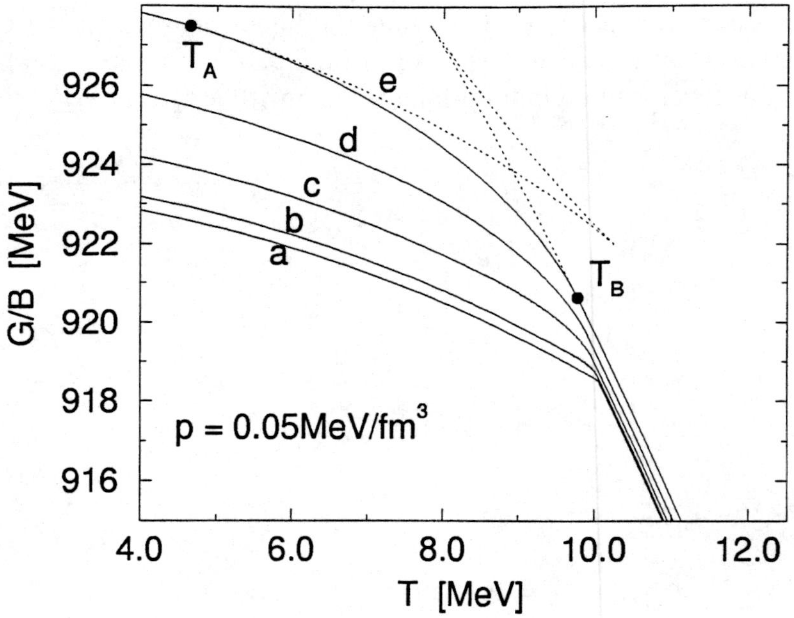

Figure 7: Specific free enthalpy at constant pressure as a function of temperature and proton fraction. The proton fractions are $y = 0.5, 0.45, 0.4, 0.35, 0.3$ for curves a through e, respectively. The phase separation is illustrated explicitly for curve e. The solid curve represents the stable mixed-phase configuration, and the dotted extensions denote metastable liquid and gas systems; T_A and T_B are the endpoints of the coexistence region. The matter is unstable along the remaining dotted curve. Taken from Ref. [2].

are different in the two distinct phases, but they remain constant throughout the phase transition. This behavior does not occur in the binary system, because the constraints of charge conservation and Gibbs' criteria force the density (and pressure) in each individual phase to change throughout the transition. We might therefore expect the transition in the binary system to be "smoother". To make this point more precise, we consider the Gibbs free energy (or free enthalpy) per nucleon

$$\frac{G(T,p,N_p,N_n)}{N_p+N_n} \equiv g(T,p,y) = y\mu_p + (1-y)\mu_n \ , \tag{31}$$

which allows us to discuss isobaric processes as a function of temperature.

This quantity is shown for several different asymmetries in Fig. 7. As discussed in Sec. 2, the Gibbs free energy must be computed according to

$$g(T,p,y) = (1-\bar{\lambda})g(T,p,y') + \bar{\lambda}g(T,p,y'') \tag{32}$$

in the transition region, where $\bar{\lambda}$ now specifies the number of particles in each phase:

$$\bar{\lambda} \equiv \frac{N_p'' + N_n''}{N_p + N_n} \ , \tag{33}$$

and where y' and y'' denote the different proton fractions. For matter with $N_p \neq N_n$ (curve e, for example), the phase transition begins at a temperature T_A and ends at $T_B > T_A$, leading to a completely smooth free enthalpy curve. In contrast, if one heats the symmetric system through the transition point (curve a), it remains at constant temperature until the transition is completed, producing a striking kink in the free enthalpy curve. The free enthalpy for various values of the proton fraction is shown in Fig. 7 to illustrate how the kink in g develops as the system becomes more symmetric.

It is also of interest to examine the entropy per nucleon, which is defined by

$$s(T,p,y) = \frac{S}{B} = -\left(\frac{\partial g(T,p,y)}{\partial T}\right)_{p,y} . \tag{34}$$

Figures 8a and 8b show the entropy as a function of temperature for several different isobars in symmetric and asymmetric matter, respectively. The kink in g for the symmetric system gives rise to a discontinuity. In contrast, the entropy in an asymmetric system is *continuous*, but there are now kinks at the end points of the phase transition. By using $dQ = Tds$, we can calculate the amount of heat that has to be transferred to the system during the course of the transition. For the symmetric system, this yields the well-known expression for the latent heat:

$$Q_L = T(s_A - s_B) ,$$

which means that the all of the energy is used to convert the liquid into vapor.

The behavior of asymmetric matter is quite different. The temperature changes during the transition and so does the entropy, so that the integration becomes nontrivial:

$$Q = \int_{s_A}^{s_B} T(s)ds ,$$

and thus some fraction of the energy is used just to heat the system, as is evident from Fig. 8b. The concept of latent heat is therefore not strictly applicable to the transition in the binary system. A well-known example of this behavior is the distillation of alcoholic beverages; the concentration of alcohol in the liquid changes throughout the distillation, which produces a change in the boiling point.

Finally, the distinct behavior of the binary system has an even more dramatic impact on the heat capacity per nucleon:

$$c_p = \frac{C_p}{N_p + N_n} = T\left(\frac{\partial s}{\partial T}\right)_{p,y} . \tag{35}$$

In symmetric matter, the discontinuity in the entropy produces an undefined heat capacity at the transition temperature. In contrast, in the general case, there are finite discontinuities in the heat capacity at the endpoints of the transition region, with finite values of c_p in between. Thus, according to Ehrenfest's definition of phase transitions [38], the *first-order* phase transition in symmetric matter becomes a *second-order* transition in the asymmetric case. In particular, in the present model, the phase transition occurs over $\approx 1.5\,\text{MeV}$ for matter that is 40% protons and over $\approx 5\,\text{MeV}$ for matter that is 30% protons. The latter concentration might be obtained in energetic collisions with radioactive ion beams.

4 Heavy-ion Collisions

We consider the energetic collision of two heavy ions within our simple thermodynamic, hydrodynamic, and mean-field picture, and concentrate on the new features that arise as a function of proton fraction. Although our discussion of the theory will encompass the entire

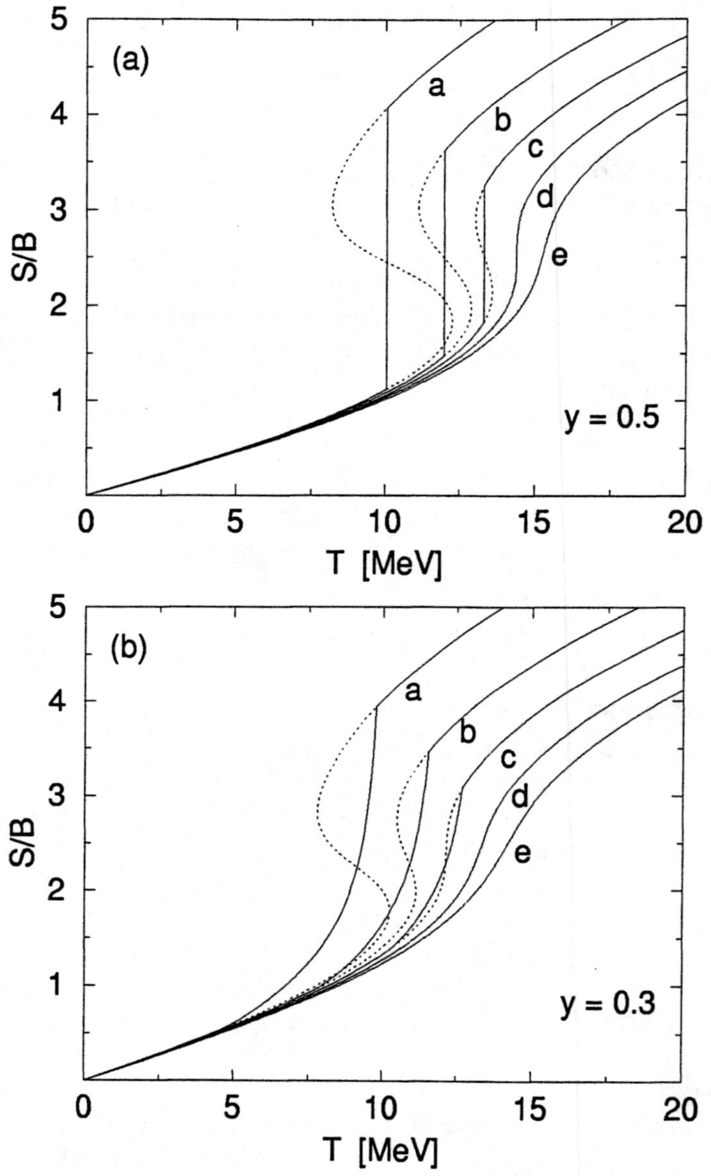

Figure 8: Specific entropy as a function of temperature for different isobars. (a) Symmetric nuclear matter with $y = 0.5$. The pressures on the curves labeled a through e are $p = 0.05, 0.10, 0.15, 0.20, 0.25\,\mathrm{MeV/fm^3}$, respectively. Curve d is at the critical pressure $p = p_c = 0.20\,\mathrm{MeV/fm^3}$. (b) Asymmetric nuclear matter with $y = 0.3$. The curves are labeled as in (a), and the critical pressure (curve d) is still $p_c = 0.20\,\mathrm{MeV/fm^3}$ (to two significant figures). Taken from Ref. [2].

regime of $y < 0.5$, to obtain reasonable estimates for the empirical size of the new effects, we will restrict consideration to $0.3 \leq y \leq 0.5$, where the lower value might be obtainable with radioactive ion beams. We shall assume that the combined system is compressed and heated and ultimately reaches equilibrium at some finite temperature, density, and pressure. As is well known, the question of whether the system actually reaches such an equilibrium state is a difficult one, which we will not attempt to answer here. We simply assume that such a state arises and follow the subsequent expansion and cooling of the nuclear matter.

The properties of the system can be deduced from Figs. 9 through 11, where the pressure is shown as a function of the baryon density for various proton fractions, temperatures, and specific entropies. Also indicated are the coexistence curves or binodals, which are determined by the Gibbs' criteria discussed earlier. The spinodals that determine the boundaries of the unstable region now generally come in three varieties, one arising from a mechanical instability, one arising from a thermal instability, and one arising from a diffusive (or chemical) instability. The conditions that determine the different spinodals will be discussed shortly. The region between the binodal and the most extensive spinodal contains metastable states, which correspond either to superheated liquid or supercooled (supersaturated) vapor.

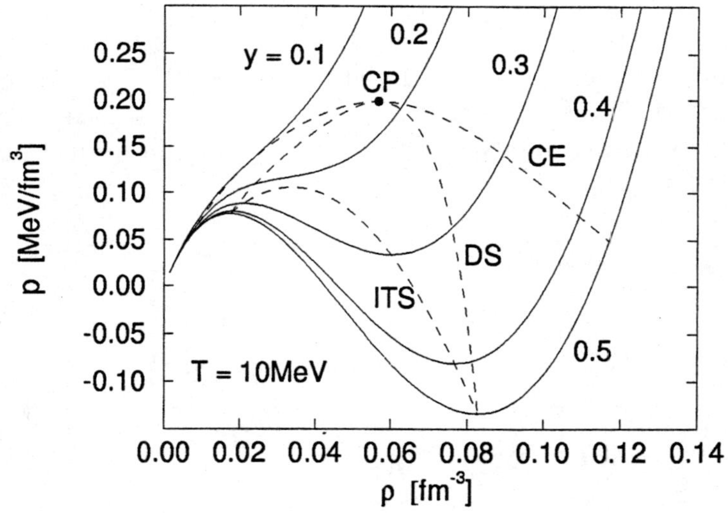

Figure 9: Pressure as a function of density at fixed temperature for various y. The critical point (CP), coexistence curve (CE), diffusive spinodal (DS), and isothermal spinodal (ITS) are indicated. The dashed curves are discussed in the text. Taken from Ref. [2].

The metastable states play a key role in understanding the phase transition. In stable configurations, Eq. (8) is satisfied, so that

$$\Delta\mathcal{F}_b \equiv (1-\lambda)\mathcal{F}(T,\rho',y') + \lambda\mathcal{F}(T,\rho'',y'') - \mathcal{F}(T,\rho,y) > 0 \tag{36}$$

holds for all densities that obey the conservation laws

$$\begin{aligned} \rho &= (1-\lambda)\rho' + \lambda\rho'' , \\ y\rho &= (1-\lambda)y'\rho' + \lambda y''\rho'' . \end{aligned} \tag{37}$$

In the metastable region, two phases can be found that produce $\Delta\mathcal{F}_b < 0$, but the energy barrier $\Delta\mathcal{F}_b$ remains positive for small variations in ρ and y about the single-phase values.

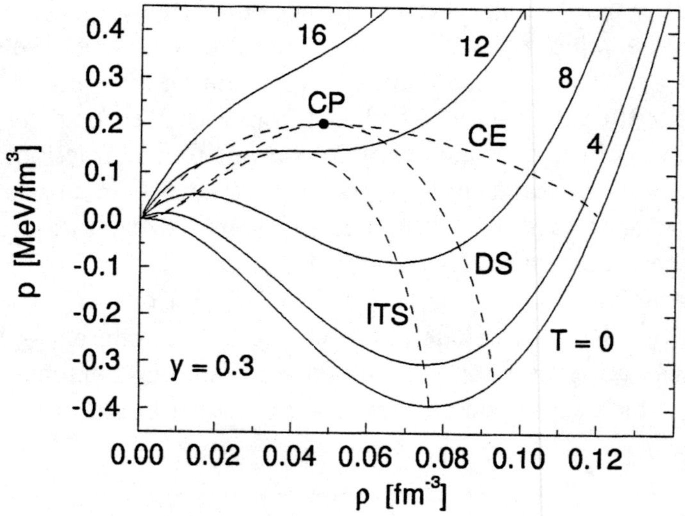

Figure 10: Pressure as a function of density at fixed $y = 0.3$. The solid curves are labeled by the temperature (in MeV). The critical point (CP), coexistence curve (CE), diffusive spinodal (DS), and isothermal spinodal (ITS) are indicated. The dashed curves are discussed in the text. Taken from Ref. [2].

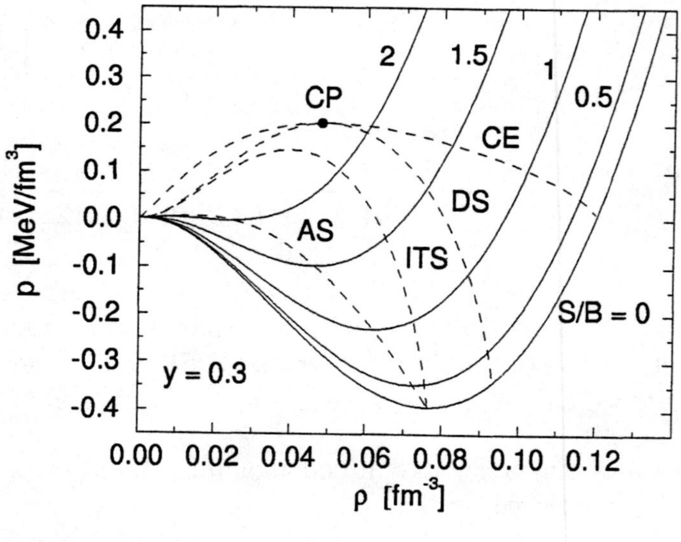

Figure 11: Pressure as a function of density at fixed $y = 0.3$. The solid curves are labeled by the specific entropy. The critical point (CP), coexistence curve (CE), diffusive spinodal (DS), isothermal spinodal (ITS), and adiabatic spinodal (AS) are indicated. The dashed curves are discussed in the text. (From Ref. [2].)

Thus the free-energy density is *locally* convex, and the energy barrier prevents phase separation due to infinitesimal fluctuations. Eventually, $\Delta\mathcal{F}_b$ becomes negative as the system enters the labile region, signaling the instability even to infinitesimal fluctuations. We postpone a discussion of the various mechanisms that can produce the phase separation and concentrate here on the behavior of the free energy in the region of the spinodals.

As demonstrated in Ref. [2], $\Delta\mathcal{F}_b > 0$ requires

$$\left(\frac{\partial p}{\partial \rho}\right)_{T,y} > 0 \tag{38}$$

and

$$\left(\frac{\partial \mu_p}{\partial y}\right)_{T,p}\left(\frac{\partial p}{\partial \rho}\right)_{T,y} > 0\,. \tag{39}$$

Therefore, as long as the conditions (38) and (39) are valid, the system is stable against small flutuations.

The spinodals are defined by the densities at which these inequalities become invalid. As is clear from Fig. 9, there are generally *four* such densities of interest in an isothermal expansion, corresponding to the intersections of the isotherm with the spinodals. Each of these four densities marks a change in the sign of the derivative $(\partial\mu_p/\partial y)_{T,p}$. At the largest density, this derivative changes sign from positive to negative, indicating the onset of the diffusive instability; nevertheless, the compressibility remains positive, so that only the second inequality (39) is violated. This also implies that the free-energy density now has a saddle point, being stable against density fluctuations at fixed y, but unstable against fluctuations in concentration at fixed pressure. Through continued expansion, one next encounters the mechanical instability, at which point $(\partial p/\partial\rho)_{T,y}$ becomes negative, but $(\partial\mu_p/\partial y)_{T,p}$ now becomes positive again, so that *both* conditions (38) and (39) are violated. As the density decreases further, one eventually reaches densities where the derivatives change sign in reverse order, producing the remaining two intersections of the spinodals with the isotherm.

The conclusion from this analysis is that the spinodal structure is qualitatively different in the asymmetric system than in the symmetric one. There are now *two* types of fluctuations: one corresponding to changes in ρ (isoscalar) and one corresponding to changes in y (isovector). Due to the form of the nuclear symmetry energy, the diffusive spinodal (DS) encloses more of the configuration space than does the isothermal spinodal (ITS). The DS passes through the critical point and includes all densities where there is also a mechanical instability, which occur in the region bounded by the ITS. *Therefore the diffusive instability defines the relevant spinodal for the asymmetric system.* The mechanical instabilities are restricted to isotherms with $y > y_*$, where y_* is defined by the inflection point

$$\left(\frac{\partial p}{\partial \rho}\right)_{T,y_*} = \left(\frac{\partial^2 p}{\partial \rho^2}\right)_{T,y_*} = 0\,.$$

As expected, both spinodals coincide at $y = 0.5$, where only mechanical instability is possible. Note also that there are isotherms that pass through the metastable region and never intersect either spinodal. These isotherms, along which retrograde condensation is possible, never become labile and allow for the system to evolve completely through the metastable region, if the process is carried out carefully enough (or fast enough).

So far, we have considered only isothermal processes, but in fact, the equilibrium evolution of the system will depend strongly on the variables that actually remain fixed during the process. As we have noted, it is not clear that *any* thermodynamic variables can be assigned to realistic situations encountered in warm, expanding nuclei. Nevertheless, intranuclear cascade calculations [10, 16] suggest that this expansion will be isentropic. To discuss isentropic (adiabatic) processes, we must generalize our formalism slightly. The appropriate state function to describe trajectories at constant specific entropy $s = S/B$ is the energy/baryon, $E/B = e(s, v, y)$, where $v = 1/\rho$. The corresponding stability condition can be formulated as:

$$e(s, v, y) < (1 - \bar{\lambda})e(s', v', y') + \bar{\lambda}e(s'', v'', y'') \ , \tag{40}$$

with

$$s = (1 - \bar{\lambda})s' + \bar{\lambda}s'' \ , \quad v = (1 - \bar{\lambda})v' + \bar{\lambda}v'' \ , \quad y = (1 - \bar{\lambda})y' + \bar{\lambda}y'' \ , \tag{41}$$

where $\bar{\lambda}$ [see Eq. (33)] determines the fraction of particles in each phase.

As in the case of the free-energy density, the global criterion Eq. (40) can be reformulated in terms of local conditions, namely,

$$\left(\frac{\partial p}{\partial \rho}\right)_{s,y} > 0 \ , \tag{42}$$

$$\frac{1}{c_p}\left(\frac{\partial p}{\partial \rho}\right)_{s,y} = \frac{1}{c_v}\left(\frac{\partial p}{\partial \rho}\right)_{T,y} > 0 \ , \tag{43}$$

$$\frac{1}{c_p}\left(\frac{\partial p}{\partial \rho}\right)_{s,y}\left(\frac{\partial \mu_p}{\partial y}\right)_{T,p} = \frac{1}{c_v}\left(\frac{\partial p}{\partial \rho}\right)_{T,y}\left(\frac{\partial \mu_p}{\partial y}\right)_{T,p} > 0 \ , \tag{44}$$

where c_p is defined in Eq. (35). The first two inequalities are relevant for both symmetric (one-component) and asymmetric (two-component) systems, and the third embodies the diffusive or chemical stability criterion that arises for asymmetric matter. Thus there will generally be three spinodals, as indicated in Fig. 11. The equivalence of the two ways of writing the stability criteria, Eq. (43) and Eq. (44), implies that the stability boundaries are *independent* of the actual process. Moreover, since in our model (see also Ref. [21]), $c_v = T(\partial s/\partial T)_{v,y}$ is always positive, the conditions (43) and (44) reduce to those discussed earlier in Eqs. (38) and (39). Thus, phase separations in isothermal processes in symmetric matter occur only due to mechanical instabilities (density fluctuations), as determined by the ITS. In asymmetric systems, the final stability condition is also relevant, leading to diffusive instabilities and the appearance of the DS as well.

In contrast, c_p undergoes several sign changes in an adiabatic expansion, so that all three criteria (42)–(44) must be considered. One sign change accompanies a sign change in the isothermal compressibility [or $(\partial p/\partial \rho)_{T,y}$], and one accompanies a corresponding change in the adiabatic compressibility [or $(\partial p/\partial \rho)_{s,y}$].[3] The resulting situation is depicted in Fig. 11. The outermost spinodal (DS) defines where Eq. (44) is violated, which occurs due to the diffusive instability, since both c_p and the compressibilities are all positive here. The middle spinodal is determined by condition (43), which is violated when c_p becomes negative; This curve is *identical* to the ITS, since c_v and $(\partial p/\partial \rho)_{s,y}$ are still positive, so $(\partial p/\partial \rho)_{T,y}$ must

[3] Note that the usual definition of the isothermal compressibility, $\kappa_T \equiv -(\partial V/\partial p)_T/V$ can be rewritten as $\rho(\partial p/\partial \rho)_T = 1/\kappa_T$ [39]. Analogous expressions hold at fixed entropy.

also become negative. In other words, the ITS encloses the region of mechanical instability in an isothermal expansion and *also* encloses the region of thermal instability in an adiabatic expansion. The mechanism for this thermal instability is interesting, because c_p changes sign by passing through infinity rather than zero. Thus a small thermal fluctuation *at constant pressure* near the ITS creates a density fluctuation, but leaves the temperature unchanged (since c_p is very large); since the isothermal compressibility becomes negative, the system is unstable to the induced density fluctuation.

Finally, the innermost curve is the adiabatic spinodal (AS), which determines when Eq. (42) is violated, signaling the onset of mechanical instability in the adiabatic process. Here c_p becomes positive again, since both compressibilities are negative here. Note also that, as discussed earlier, $(\partial \mu_p/\partial y)_{T,p}$ changes sign from positive to negative at the DS and then changes back again at the ITS, so that once any stability criterion (42)–(44) is violated, it remains invalid as one proceeds deeper into the unstable region.

Just as we found in the case of isothermal processes, the DS separates the metastable and labile regions in isentropic processes. Configurations that indicate sign changes of the compressibilities and of c_p are included by this boundary. We also observe that even in symmetric nuclear matter, the ITS determines the boundary of the labile region in *both* isothermal and isentropic processes, which has often been overlooked in previous studies [9, 10, 13, 14, 40, 41]. Although this result apparently implies that the AS is irrelevant, this conclusion is premature, since the actual mechanism for spinodal decomposition depends on the relative rate of thermal and mechanical fluctuations as the warm matter expands through the spinodal region.

Armed with this understanding of the spinodal structure, we turn now to the evolution of the warm matter. The relation between the evolution and the thermodynamic variables can be studied using Fig. 12, which shows the properties of the matter for this equation of state in the T, ρ plane for two different values of y. Equipotential surfaces and adiabats are also indicated.

The solid curves in the lower-left corners of Figs. 12a,b determine the phase behavior. The outer curve labeled CE is the coexistence curve, which is simply a section through the binodal surface at fixed $y = 0.3, 0.5$. Inside this curve, the stable configuration is a mixture of liquid and gas. Also shown is the diffusive spinodal (DS), which is determined by the condition $(\partial \mu_p/\partial y)_{T,p} = 0$, as discussed earlier. Between the spinodal and the CE, the system can exist in metastable superheated or supersaturated states.

There are several qualitative differences between the curves for asymmetric matter and for symmetric matter. First, the most extensive spinodal is defined by the diffusive instability rather than by thermal or mechanical instabilities. (We note that mechanical instability is often assumed to be the relevant one even in asymmetric systems.) Second, as is evident from Fig. 4, the critical temperature $T_c(y)$ is not unique, but varies with the concentration along the LCP. Finally, the maximum temperature of phase separation does not occur at the critical point, but rather along the LMA; it is apparent from Fig. 12a that this occurs at a lower density than the density at the critical point. We observe, however, that although these differences exist in principle, for physically accessible systems, the temperature differences are small. For example, in the present model, the critical temperature changes from $T_c(0.5) = 14.4\,\text{MeV}$ to $T_c(0.3) = 13.1\,\text{MeV}$, and $T_{\mathbf{max}}(0.3) - T_c(0.3) \approx 0.2\,\text{MeV}$. Since statistical fluctuations near the critical point are expected to prevent the determination of the critical temperature with an uncertainty smaller than 1 or 2 MeV [18], it is unlikely that these small temperature differences can be directly observed. Nevertheless, a trend in the observable signals characteristic of the critical point as a function of increasing asymmetry

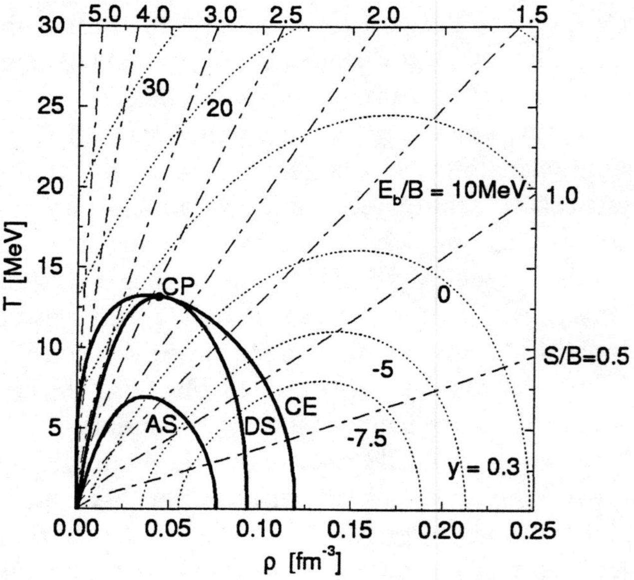

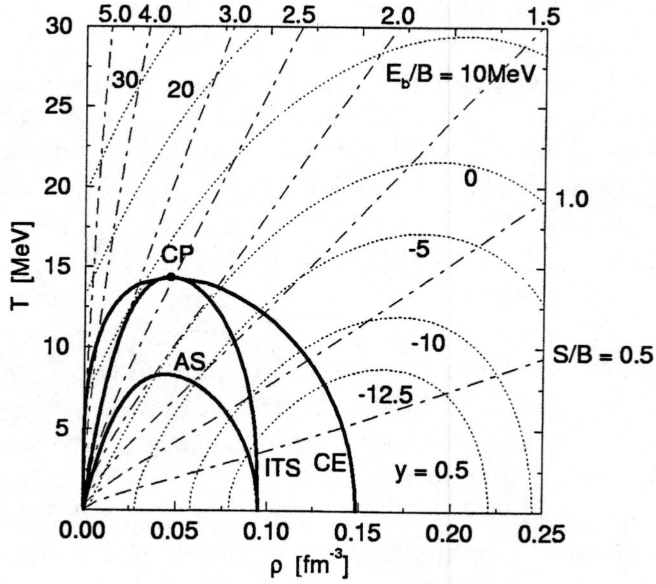

Figure 12: Properties of nuclear matter as functions of temperature and density. The dotted curves are contours of equal energy/baryon (in MeV), and adiabats are shown as dot-dashed curves. The solid lines denote the coexistence curve (CE), the diffusive spinodal (DS), the isothermal spinodal (ITS), and the adiabatic spinodal (AS). Part (a) shows results for asymmetric matter with $y = 0.3$ and part (b), for symmetric matter ($y = 0.5$). Taken from Ref. [2].

may be detectable.

The evolution of the system is a complicated process that involves hydrodynamic flow and expansion, together with dissipative effects and nonequilibrium processes like nucleation and fragmentation. As the system expands, internal energy is transformed into collective motion, which manifests itself as local flow velocity. If the expansion is highly damped, the energy of motion is rapidly transformed back into internal energy; the resulting expansion is therefore slow and proceeds along an equipotential surface. The expansion continues until $p = 0$, after which the warm nuclear fluid remains at rest and evaporates particles until it is cool. In contrast, if the motion is undamped, the expansion carries the system past hydrostatic equilibrium, where it begins to slow down as the energy of motion is returned to internal energy. Since there is no dissipation, the expansion is isentropic, and the motion is bounded by the equipotential surface of the initial hot configuration.

As noted earlier, intranuclear cascade calculations [10, 16] suggest that the expansion will be isentropic. The system will expand adiabatically into the coexistence region and become either superheated (for small specific entropy) or supersaturated (for large specific entropy). If the expansion is slow enough to allow for nucleation, bubbles of gas form or the system separates into droplets and vapor. If the nucleation is relatively slow, which is more likely, and the expansion halts before a spinodal is reached, the direction of motion is reversed and the system vibrates, ultimately evaporating neutrons to cool down. In contrast, if the system crosses a spinodal, fragmentation occurs; the relevant mechanism for the decomposition (that is, which spinodal is the relevant one) depends on the relative rates for isovector, isoscalar, and thermal fluctuations.

5 Summary

We studied a relativistic mean-field model of nuclear matter with arbitrary proton fraction at finite temperature. We analyzed the liquid–gas phase transition in a system with two conserved charges (baryon number and isospin) using the stability conditions on the free energy, the conservation laws, and Gibbs' criteria for phase equilibrium. For a binary system with two phases, the coexistence surface (binodal) is two-dimensional. The Maxwell construction through the phase-separation region was discussed, and it was shown that the stable configuration can be determined uniquely at every density. Moreover, because of the greater dimensionality of the binodal surface, the liquid–gas phase transition is continuous (second order by Ehrenfest's definition), rather than discontinuous (first order), as in familiar one-component systems. Using a mean-field equation of state calibrated to the properties of nuclear matter and finite nuclei, various phase-separation scenarios were considered. The model was then applied to the liquid–gas phase transition that may occur in the warm, dilute matter produced in energetic heavy-ion collisions. In asymmetric matter, instabilities that produce a liquid–gas phase separation arise from fluctuations in the proton concentration (chemical instability), rather than from fluctuations in the baryon density (mechanical instability).

With our simple picture that focuses on equilibrium states, it is impossible to say anything definitive about the dynamics of the phase separation. This is especially true since it is unlikely to occur through a sequence of equilibrium configurations and will involve instead superheating, supercooling, or spinodal decomposition. Nevertheless, the dynamics of the phase separation is intimately linked to the character of the phase transition; since the phase transition in asymmetric matter is not of the usual van der Waals type, there could be observable consequences in the signals used to detect the phase separation, such as the distribution of emitted mass fragments. In particular, unlike the van der Waals case, the pressure, temperature, density, and concentration of the gas and liquid phases can change throughout the phase separation process. Moreover, since the spinodal decomposition occurs generally through a diffusive instability, any microscopic model of the fragmentation process [18, 42–44] must allow the gas and liquid phases to have different concentrations.

References

[1] J. M. D' Auria, Nucl. Instrum. Meth. **B99**, 330 (1995).

[2] H. Müller and B. D. Serot, Phys. Rev. C **52**, 2072 (1995).

[3] S. Nagamiya, J. Randrup, and T. J. M. Symons, Ann. Rev. Nucl. Part. Sci. **34**, 155 (1984).

[4] D. H. Boal, Adv. Nucl. Phys. **15**, 85 (1985).

[5] J. M. Eisenberg and W. Greiner, *Nuclear Models* (North-Holland, Amsterdam, 1987), Sec. 17.1.

[6] *Nuclear Matter and Heavy-Ion Collisions*, Proceedings of the 1989 Les Houches Winter School, M. Soyeur, H. Flocard, B. Tamain, and M. Porneuf, eds. (Plenum, New York, 1989).

[7] W. Bauer, Ann. Rev. Nucl. Part. Sci. **42**, 77 (1992).

[8] L. P. Csernai, *Introduction to Relativistic Heavy Ion Collisions* (Wiley, New York, 1994).

[9] G. Bertsch and P. J. Siemens, Phys. Lett. **126B**, 9 (1983).

[10] D. H. Boal, Nucl. Phys. **A447**, 479c (1985).

[11] B. M. Waldhauser, J. Theis, J. A. Maruhn, H. Stöcker, and W. Greiner, Phys. Rev. C **36**, 1019 (1987).

[12] J. Diaz Alonso, J. M. Ibañez, and H. Sivak, Phys. Rev. C **39**, 671 (1989).

[13] V. R. Pandharipande and D. G. Ravenhall, in *Nuclear Matter and Heavy-Ion Collisions*, Proceedings of the 1989 Les Houches Winter School, M. Soyeur, H. Flocard, B. Tamain, and M. Porneuf, eds. (Plenum, New York, 1989), p. 103.

[14] P. J. Siemens, Nature **305**, 410 (1983).

[15] D. H. Boal, Ann. Rev. Nucl. Part. Sci. **37**, 1 (1987).

[16] G. Bertsch and J. Cugnon, Phys. Rev. C **24**, 2514 (1981).

[17] D. H. Boal and A. L. Goodman, Phys. Rev. C **33**, 1690 (1986).

[18] A. L. Goodman, J. I. Kapusta, and A. Z. Mekjian, Phys. Rev. C **30**, 851 (1984).

[19] L. P. Csernai and J. I. Kapusta, Phys. Rep. **131**, 223 (1986).

[20] W. G. Lynch, Ann. Rev. Nucl. Part. Sci. **37**, 493 (1987).

[21] M. Barranco and J. R. Buchler, Phys. Rev. C **22**, 1729 (1980).

[22] N. K. Glendenning, Phys. Rev. D **46**, 1274 (1992).

[23] B. D. Serot and J. D. Walecka, Adv. Nucl. Phys. **16**, 1 (1986).

[24] B. D. Serot, Rep. Prog. Phys. **55**, 1855 (1992).

[25] B. D. Serot and J. D. Walecka, Int. J. Mod. Phys. E **6** (1997) 515.

[26] H. Müller and B. D. Serot, Nucl. Phys. **A606**, 508 (1996).

[27] A. L. Fetter and J. D. Walecka, *Quantum Theory of Many-Particle Systems* (McGraw–Hill, New York, 1971), Sec. 4.

[28] A. W. Roberts and D. E. Varberg, *Convex Functions* (Academic Press, New York, 1973).

[29] L. D. Landau and E. M. Lifshitz, *Statistical Physics* (Pergamon Press, Oxford, 1969).

[30] R. Haase, *Thermodynamik der Mischphasen* (Springer, Berlin, 1956).

[31] R. Haase and H. Schönert, Int'l. Encyclopedia of Physical Chemistry and Chemical Physics, topic 13: *Mixtures, Solutions, Chemical, and Phase Equilibria*, vol. 1: *Solid–Liquid Equilibria* (Pergamon Press, NY, 1969).

[32] A. R. Bodmer and C. E. Price, Nucl. Phys. **A505**, 123 (1989).

[33] A. R. Bodmer, Nucl. Phys. **A526**, 703 (1991).

[34] C. J. Horowitz and B. D. Serot, Nucl. Phys. **A399**, 529 (1983).

[35] R. J. Furnstahl, B. D. Serot, and H.-B. Tang, Nucl. Phys. **A598**, 539 (1996).

[36] R. J. Furnstahl and B. D. Serot, e-print nucl-th/9911019.

[37] J. M. Lattimer and D. G. Ravenhall, Ap. J. **223**, 314 (1978).

[38] L. E. Reichl, *A Modern Course in Statistical Physics* (University of Texas Press, Austin, 1980).

[39] K. Huang, *Statistical Mechanics* (Wiley, New York, 1963).

[40] J. A. Lopez and P. J. Siemens, Nucl. Phys. **A431**, 728 (1984).

[41] R. J. Furnstahl and B. D. Serot, Phys. Rev. C **41**, 262 (1990).

[42] G. Fai and J. Randrup, Nucl. Phys. **A404**, 551 (1983).

[43] W. Bauer, D. R. Dean, U. Mosel, and U. Post, Phys. Lett. **150B**, 53 (1985).

[44] W. Bauer, U. Post, D. R. Dean, and U. Mosel, Nucl. Phys. **A452**, 699 (1986).

Isotopic Spin Dependent Lattice-Gas Model for Nuclear Fragmentation

S. Das Gupta* and S. K. Samaddar*†

Abstract

We use an isotopic spin dependent lattice gas model which is useful for gaining an understanding of many gross properties of nuclear fragmentations as seen in intermediate energy heavy ion reactions. In heavy systems, in addition to using the isotopic spin dependent nearest neighbour interaction, one needs to include the Coulomb interaction between protons. The major effect of the Coulomb force can be included with little extra effort. Several applications are studied.

1 Introduction

The basic characteristics of intermediate energy heavy ion collisions is the break-up of the colliding system into a number of fragments having different sizes. The role played by fluctuations in this disassembling process is undebatable. The general picture of the fragmentation process is as follows. Depending on the initial conditions like bombarding energy and impact parameter the system may compress at the early stage of collision. Then it expands and during this expansion phase density fluctuations give birth to various fragments. The fragmentation pattern stabilizes at certain stage called the 'freeze-out' stage and the volume at this stage is referred to as the 'freeze-out volume'. This volume is a few times the normal nuclear volume. Among the various statistical models employed to study the fragmentation phenomena, the lattice gas model (LGM) is the only model where the fragments are formed directly out of the density fluctuations mentioned above.

With the realisation that in an easily implemented extension of the standard lattice gas model [1, 2] it is possible to calculate cluster distributions [3, 4] the LGM has become a favourite theoretical tool to study fragmentation in intermediate energy heavy ion collisions [5-15]. The attractive part of the model, besides its simplicity, is that the model gives an equation of state as well as readily measurable cluster distributions. Influence of the equation of state on different observables can be studied.

Many properties can be studied in a simplified model of only one kind of particles. In fact this is the model that has been mostly studied. There are however several features of heavy ion collisions which require taking into account explicitly the fact that there are two kinds of particles and that nuclear interaction between like particles is different from the interaction between unlike particles. For large systems the Coulomb force between protons has significant effects on observables. This article focusses on the isotopic spin dependence aspect. We talk of general theory also but only for completeness, i.e., we will

*Physics Department, McGill University, 3600 University St., Montréal, Québec Canada H3A 2T8
†Saha Institute of Nuclear Physics, 1/AF Bidhannagar, Calcutta 700064,India

not fully explore all the aspects. Isotopic spin dependence and/or effects of coulomb on LGM predictions were dealt with in [5,11,12,15].

As far as we know the first qualtitative application of the LGM for nuclear fragmentation was made in 1986 [16]. The equation of state was not discussed and contiguous sites which were occupied by nucleons were considered to form a cluster. These clusters have been labelled Ising clusters in [13]. We now think this prescription overestimates sizes of clusters. From a dynamical point of view it is natural to enunciate a different rule for clusterisation. Since this formulation [3, 4] there has been a large amount of activity in the model.

The contents of the article are as follows. Section II describes the basic features of the LGM. The derivation of the clusterisation rule is given in section III. In section IV the LGM is brought in contact with the percolation model. The mean-field approximation of the LGM is made in section V with two types of bonds. The generation of an event is described in some details in section VI. Our treatment of the Coulomb force is included in section VII. In section VIII a fit to the nuclear masses calculated in the LGM with two types of bonds and Coulomb is done with a simple liquid-drop-model formula. The comparison of these masses with the experimental ones is also made here. The role of the isotopic spin dependence and the Coulomb force on several physical quantities is considered in section IX. In sections X and XI we present the results of our calculations for several observables and some of them are compared with the experimental data. Finally the summary and discussion are given in section XII.

2 The Model

We consider the participant zone in collisions of heavy ions and conjecture that because of nucleon-nucleon collisions the system reaches thermal equilibrium. We assume that the system is adequately described by classical statistical mechanics (more about this later). In that case the canonical partition function of an A particle system can be written in a separable form: $Z_{can} \propto Z_p(can)Z_r(can)$ where

$$Z_p(can) = \int \exp[-\beta \sum_{1}^{A} p_i^2/2m] d^3p_1....d^3p_A, \tag{1}$$

with β as the inverse temperature. The other part of the partition function is

$$Z_r(can) \propto \int \exp[-\beta \sum_{i<j} v(r_{ij})] d^3r_1.....d^3r_A. \tag{2}$$

Here $v(r_{ij})$ is the interaction energy between particles i and j. We approximate the configuration part of the partition function by the partition function of the LGM. In this model each lattice site may be occupied by at most one nucleon. At the moment we consider one kind of particles only. The number of lattice sites, N_s, gives a measure of the volume of the system and is usually larger than A. When the nucleus is in its ground state we have $A = N_s$. Thus the model is limited to normal nuclear volume or larger. Because cluster formation presumably takes place in a volume significantly larger than the normal volume, this restriction is not debilitating. If two nearest neighbouring sites are filled by nucleons, they will interact. The energy of interaction is denoted by ϵ where ϵ is negative. The exclusion of the possibility of two nucleons occupying the same site mimics a short range repulsion. It may also mimic some features of the Pauli principle. The attractive nearest neighbour interaction simulates the attractive nucleon-nucleon interaction which is also short range (but range longer than the short range repulsive interaction).

The part $Z_p(can)$ of the canonical partition function (eq.2.1) is simple and analytic. We need to calculate $Z_r(can)$. Let N_{nn} be the number of nn bonds in a specific lattice configuration. The energy carried by these bonds is then ϵN_{nn}. Thus $Z_r(can)$ is given by

$$Z_r(can) = \sum_{N_{nn}} g(N_s, A, N_{nn}) e^{-\beta \epsilon N_{nn}}. \qquad (3)$$

Here $g(N_s, A, N_{nn})$ is the number of configurations which have N_{nn} nearest neighbour bonds and which can be formed from A nucleons in N_s lattice sites. This is not analytically solvable and will require Monte-Carlo simulations. The grand canonical ensemble of the LGM can be mapped on to the Ising model [1],[2]. The latter has been well studied in the Monte-Carlo method and many of the known results can be transferred to the LGM. However one still needs to do fresh calculations to compare with data. The Ising model results are known in the $N_s \to \infty$ limit and in the nuclear case N_s is never very large. Also typically we will want to know the answer for a fixed A rather than over an average A. Moreover, some of the questions we will want to answer are irrelevant for Ising model and to our knowledge, were not studied.

The LGM can be solved in the mean field approximation. We will deal with this approximation but after we have introduced isotopic spin dependence in the model. For the moment, let us turn our attention to how clusters can be calculated in the model.

3 The Clusterisation Formula

An event is a lattice configuration of A nucleons in N_s lattice sites with $N_s \geq A$. The A nucleons also have momenta. After the nucleons have been placed in lattice configuration in a Monte-Carlo procedure the momenta of the A nucleons are generated by Monte-Carlo sampling of a Maxwell-Boltzmann distribution at temperature T. Because the partition function separates into two different parts (eqs 2.1 and 2.2) the two Monte-Carlo samplings (one for configuration and one for the momenta) are independent.

In a particular configuration there may be some isolated nucleons. They do not have any interacting partners and are therefore singles. The next case is when there is cluster of two nucleons which are nearest neighbours of each other. Clearly this will form a bound cluster if the kinetic energy of relative motion is insufficient to overcome the attraction between the nucleons, i.e., $p_r^2(1,2)/2\mu + \epsilon < 0$. Here $\vec{p}_r(1,2) = \frac{1}{2}(\vec{p}_1 - \vec{p}_2)$ and $\mu = m/2$ (m=mass of one nucleon).

It turns out this prescription which is rigorously correct for a cluster of two is also expected to work *statistically* for larger clusters. That is we can formulate the rule that independent of other neighbours two nearest neighbours form part of the same cluster if the relative kinetic energy of the two is insufficient to overcome their attraction. It is obvious that this reduces the many body problem of recognising a cluster of many nucleons into a sum of independent 2-body problems. For brevity we call this the PD recipe. To see why this works *statistically* even though not individually let us specifically consider a three-body cluster. The generalisation to higher clusters can be done.

For three particle clusters nearest neighbours are either linear or L-shaped. In either case there is only one particle which has two bonds (label this particle number 2) and two others (label them 1 and 3) which have one bond each. According to our "simple" prescription this will form a 3-body cluster if $p_r^2(1,2)/2\mu + \epsilon < 0$ and $p_r^2(2,3)/2\mu + \epsilon < 0$. To check if the particle 3 is part of a three-body cluster (similar arguments will be needed for particles 1 and 2) we should instead verify if $p_r^2(12,3)/2\tilde{\mu} + \epsilon < 0$. Here $\vec{p}_r(12,3)$ is the relative momentum between the centre of mass of (1+2) and 3; $\tilde{\mu} = (2/3)m$ is the reduced mass for this relative motion. Thus there may be cases where with our simple presription we get a three-body cluster though in reality the third one will separate; but also there will be cases where with our prescription we will deem that the third one will separate whereas in reality it stays attached. Statistically overestimation will cancel out underestimation because for a Maxwell-Boltzmann distribution all relative motions are also Maxwellian. That is, in Monte-Carlo simulations, $p_r^2(12,3)/2\tilde{\mu}$ will be as many times below the value of $-\epsilon$ as $p_r^2(1,2)/2\mu$ will be.

The same argument applies to particle 1. For particle 2, It can be verified that if $p_r^2(1,2)/2\mu + \epsilon < 0$ and $p_r^2(2,3)/2\mu + \epsilon < 0$ then $p_r^2(13,2)/2\tilde{\mu} + 2\epsilon < 0$ is always satisfied.

The generalisation to more complicated cases is straightforward. It only hinges on the fact that with a Maxwell-Boltzmann distribution relative motions are also Maxwellian.

This shows that in a statistical generation of different composites with this prescription we do not need to worry about subsequent evaporation as one needs to do in some other models [17]. Evaporation was already taken into account when we applied the formula. One does not take the size of the cluster to be given by just the number of nucleons which are connected to each other through nearest neighbour interaction. Some of these will fly away. The rest that remain and are counted, are particle stable.

A different derivation of the PD recipe is given in [9].

4 Percolation Characteristics and the Coniglio-Klein Formula

In the past, the bond percolation model was often used as an aid to understanding cluster distributions in heavy ion collisions [18], [19]. In the bond percolation model the number of nucleons is equal to the number of lattice sites: $N_s = A$. The fraction that the bond between two nearest neighbours is active is given by a probability p the value of which is between 0 and 1. If p=1, then we have one nucleus which stays together (low excitation). If p=0 (high excitation) then we have only monomers(nucleons). When the value of p is taken to be in between these extremes one obtains a distribution of clusters. There is one interesting value of $p = p_c$ at which, in an infinite system, a percolating cluster is obtained. A percolating cluster is one which extends from boundary to boundary. We call p_c the percolation point; p_c is the critical point in the model. Below p_c there is no percolating cluster. Above p_c there is always one such cluster. We refer to a monograph [20] for a review of the percolation model, however, some details necessary for this article, are provided in this section.

At p_c, the critical point (this is continuous phase transition as opposed to a first order phase transition), other interesting things happen. Let us denote by $Y(a)$ the number of clusters which has a nucleons. At p_c, the yield $Y(a)$ is given by a power law:

$$Y(a) \propto a^{-\tau}. \tag{4}$$

At the critical point fluctuation is the maximum. This means that if in the neighbourhood of p_c the yield $Y(a)$ is fitted by a power law, the exponent τ will be a minimum at p_c. The second moment is defined as $S_2 = \sum a^2 Y(a)/A$ where the sum excludes the largest cluster (defined henceforth as a_{max}). In the thermodynamic limit S_2 diverges at p_c and obeys a power law:

$$S_2 \propto |p - p_c|^{-\gamma}. \tag{5}$$

For finite systems S_2 will go through a maximum but not an infinity. The order parameter given by $\lim_{A\to\infty} a_{max}/A$ is zero for $p < p_c$ and increases with p for $p > p_c$ according to

$$a_{max}/A \propto |p - p_c|^{\beta}, \tag{6}$$

in the vicinity of p_c. The values of τ, β and γ are 2.2,0.41 and 1.8 for $N_s \to \infty$ and are obtained numerically from Monte-Carlo simulations [20].

The LGM also has phase transition and this phase transition is quite distinct from that of the percolation model. In fact the phase transition in the percolation model is special and the one in the lattice gas model is the more traditional one. The properties of this phase transition are known from studies of the three dimensional Ising model, the model often used for liquid-gas phase transition. For example it is well-known that the critical temperature is given by $T_c = 1.1275\epsilon$ and the critical density is given by $\rho_c = 0.5\rho_0$, that is, $A/N_s = 0.5$. For thermodynamic phase transition one does not need to invoke a clusterisation formula. However as soon as we invoke a clusterisation formula, such as the PD recipe, percolation model type predictions can be made. Indeed with a clusterisation formula, the bond percolation model is just a special case of the lattice gas model, namely the case $A = N_s$ [8].

In order that the thermodynamic critical point also becomes a point at which percolation just sets in, Coniglio and Klein used the following temperature dependent formula: the probability that two nearest neighbours have an active bond between them is given by [21]

$$p = 1 - \exp(-\beta|\epsilon|/2). \tag{7}$$

With this definition, these authors, using renormalisation group, proved that at (ρ_c, T_c) percolation just sets in. However, percolation sets in not just at the thermodynamic critical point; rather along a continuous line in the ρ, T plane (the dotted line in Fig.1). This was studied in [22] and the line is called the Kertész line. Hence the exponents τ, β and γ are meaningful not only at the critical point but along the entire line.

It turns out that the PD recipe derived in section III is very close to Coniglio-Klein formulation. The PD formula for p (using the fact that in a Maxwell-Boltzmann distribution the relative motion is also Maxwell-Boltzmann) is

$$p = 1 - \frac{4\pi}{(2\pi\mu T)^{3/2}} \int_{\sqrt{2\mu|\epsilon|}}^{\infty} \exp(-p_r^2/2\mu T) p_r^2 dp_r = 1 - \frac{2}{\sqrt{\pi}} \int_{|\epsilon|/T}^{\infty} e^{-q} q^{1/2} dq.$$

A comparison of p calculated according to the above formula to the Coniglio-Klein p (eq.4.4), taken from [4], is shown in Fig.2. They are very close. In this work we use the PD recipe.

5 Simple Mean-field Theory with Two Kinds of Bonds

The argument above for clusterisation immediately shows the need for distinguishing between interactions between unlike particles and interactions between like particles. The bond between two neutrons or between two protons can not be attractive because then simulations would give bound dineutrons and diprotons. We therefore need $\epsilon_{np} < 0$ and $\epsilon_{nn} = \epsilon_{pp} \geq 0$. The grand canonical partition function of this generalised LGM can be mapped onto a spin 1 Ising model in the presence of a dipolar and quadrupolar interactions [23]. For a general choice of $\epsilon_{np}, \epsilon_{pp}$ and ϵ_{nn} a rich assortment of phenomena can be predicted but with the restriction $\epsilon_{nn} = \epsilon_{pp} \geq 0$ this richness disappears. We will now obtain the partition function $Z_r(can)$ in the simplest mean field approximation, the Bragg-Williams approximation, to study phase-transition properties [11]. From now on, ϵ_{nn} can mean either ϵ_{pp} or ϵ_{nn}.

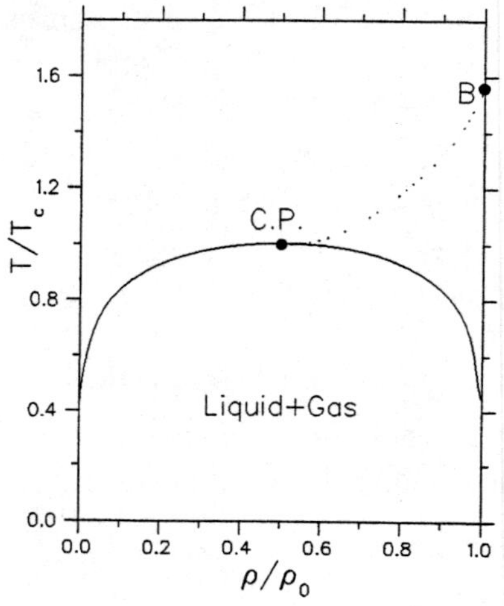

FIG.1. Phase diagram of three-dimensional lattice gas model. The full line is the coexistence curve. The dotted line is the percolation line.

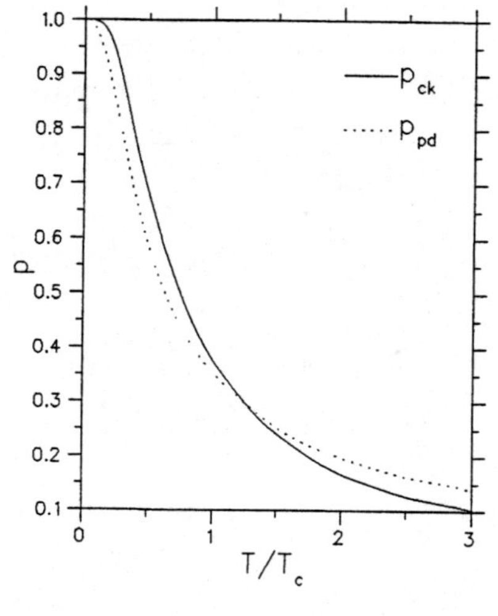

FIG.2. The bond probability p is plotted as a function of temperature. The solid line (p_{ck}) corresponds to Coniglio-Klein formula(eq.4.4) and the dotted line(p_{pd}) is the PD-recipe.

We denote by γ the number of nearest neighbours a particle can have. In 3 dimensions, the only dimension we use, γ is 6. There are N_s boxes and Z protons and N neutrons. Let one of the boxes be occupied by a proton. Then, in the Bragg-Williams approximation, among its nearest neighbours, on the average, $\gamma Z/N_s$ will be occupied by protons and $\gamma N/N_s$ by neutrons. The number of $n-p$ bonds will be $\gamma ZN/N_s$ and the number of $p-p$ bonds will be $(1/2)\gamma ZZ/N_s$ where the factor 1/2 remedies the double counting of the proton-proton bonds. Similarly, starting with a box occupied by a neutron we come up with the same number of neutron-proton bonds and the number of neutron-neutron bonds is $(1/2)\gamma NN/N_s$. Thus the interaction energy when there are Z protons, N neutrons placed in N_s boxes is $E=\gamma[\epsilon_{np}ZN+\epsilon_{nn}(N^2+Z^2)/2]/N_s$ and the partition function (eq.2.3) is

$$Z_r(N_s,Z,N)=\frac{N_s!}{(N_s-Z-N)!Z!N!}\exp(-\beta E). \quad (8)$$

We now find pressure P from the equation $P=T[\partial lnZ/\partial V]_T=T[\partial lnZ_r/\partial V]_T$. We use $V=a^3N$ where $a^3=1/\rho_0$ is the volume of one box; ρ_0 is the normal nuclear density= 0.16 fm^{-3}. Using Stirling's formula one arrives at

$$P=\rho_0 Tln\frac{N_s}{N_s-Z-N}+\rho_0\gamma\epsilon_{np}(Z/N_s)(N/N_s)+\rho_0\gamma\epsilon_{nn}[(N/N_s)^2+(Z/N_s)^2]/2. \quad (9)$$

We introduce an asymmetry parameter $\eta=(N-Z)/(N+Z)$ which takes the value 1 for neutron matter, 0 for nuclear matter and -1 for proton matter. We can then write

$$P=\rho_0 Tln\frac{V}{V-V_0}+\frac{1}{2}\rho_0\gamma\frac{V_0^2}{V^2}[\frac{\epsilon_{np}+\epsilon_{nn}}{2}+\frac{1}{2}\eta^2(\epsilon_{nn}-\epsilon_{np})]. \quad (10)$$

We determine the critical point from $\partial P/\partial\rho=\partial^2P/\partial\rho^2=0$. This gives the critical density $\rho_c=0.5\rho_0$ and the critical temperature $-(\gamma/4)[(\epsilon_{np}+\epsilon_{pp})/2+(1/2)\eta^2(\epsilon_{nn}-\epsilon_{np})]$. We know ϵ_{np} has to be attractive and ϵ_{nn} either zero or repulsive. If we take ϵ_{nn} to be zero (this will be justified in a subsequent section), then the critical temperature for nuclear matter (which has η=0) would be highest at $-(\gamma/4)(\epsilon_{np}/2)$ and would fall off quadratically with η to 0 at neutron or proton matter.

The Bragg-Williams approximation is the simplest one and is useful for qualtative estimates. Two of its shortcomings can be mentioned. In one dimension it predicts a phase transition whereas exact calculations which are analytic in one dimension shows that this is not correct. For nuclear matter at normal density ($Z=N=N_s/2$) it predicts a huge difference in binding energy for the choice $\epsilon_{np}=\epsilon_{nn}=\epsilon_{pp}$ =-5.33 MeV and the case $\epsilon_{pp}=\epsilon_{nn}=0;\epsilon_{np}=-5.33$ MeV. In the first case it predicts a binding energy of 16 MeV (which is the correct answer) but in the second case it predicts 8 MeV which is a gross underestimation. The correct answer in the second case also is 16 MeV. It is just that sites will be alternately populated by neutrons and protons so that all nearest neighbour bonds are of neutron-proton type. In a similar fashion the Bragg-Williams method predicts analytically that the critical temperature in the second case is half the value of the first case but essentially exact though numerical results show that the difference is much less, only about ten percent.

An improved mean-field calculation is the Bethe-Peierls approximation [2]. For an isospin dependent lattice gas model this has been worked out in [11]. Before we leave this section we mention mean field calculation of a different version of the isospin-dependent lattice gas model [24]. These authors, in addition to the nearest neighbour interaction, include a term in the Hamiltonian, modelled from a Fermi gas theory. For A fermions at zero temperature the total contribution to kinetic energy is $C_1 A\rho^{2/3} = C_1 A\rho_0^{2/3}(\rho/\rho_0)^{2/3} = C_2 N_s(A/N_s)^{5/3}$. This is added to the Hamiltonian of nearest neighbour interaction. There are difficulties of getting observables from this model. First of all, conceptually, the Fermi gas model arises from filling up $\exp(i\vec{k}.\vec{r})$ orbitals which by their construct, occupy the whole available configuration space. However the lattice gas model assumes that each nucleon occupies a lattice site, hence, is highly localised. To get some quantum mechanics into the model one could try using localised wave-packets, but this leads to difficult complications which are very hard to handle. Secondly one might question the justification of using a kinetic energy term whose value depends only upon the density but not on the temperature. All theories, including the Fermi gas theory, say that the value of kinetic energy also depends upon temperature. The most difficult part however is that there is no obvious way to go beyond mean field theory. Since practically no observable is calculable from a mean field theory, this model has not been used for any practical predictions.

6 Generating an Event

We now state in detail how events are generated. The principle is that of Metropolis algorithm [25]. Beginning with all lattice sites empty, an initial configuration is built according to the following procedure. By a random choice, a neutron or a proton is placed at the centre of the N_s lattice sites. The remaining $N_s - 1$ sites each has a probability of occupation by a proton or a neutron. If the first occupant was a proton, the six nearest neighbour lattice sites will have probability proportional to $\exp(-\beta(\epsilon_{pp} + u_c))$ for occupation by a proton and $\exp(-\beta\epsilon_{np})$ for occupation by a neutron. Other sites have probability proportional to $\exp(-\beta u_c)$ for occupation by a proton and probability proportional to 1 for occupation by a neutron. Here u_c is the appropriate coulomb energy between protons. If instead the first occupant was a neutron, then the six nearest neighbours will have a probability proportional to $\exp(-\beta\epsilon_{np})$ for occupation by a proton and probability proportional to $\exp(-\beta\epsilon_{nn})$ for occupation by a neutron. All other sites will have a probability of occupation proportional to 1 for both protons and neutrons. The second site therefore can be filled by Monte-Carlo sampling and it is clear how one can proceed to fill up N sites with N neutrons and Z sites with Z protons. This then forms the initial configuration from which we start Metropolis sampling. Switch is attempted between (1)occupied neutron and proton sites and (2) occupied and unoccupied sites. The switch is accepted if the switch causes the energy to go down or stay constant. If the energy goes up by an amount ΔE then the switch is accepted with probability $\exp(-\beta\Delta E)$. Coulomb energy difference is included in computing ΔE. After 10,000 attempts we accept a configuration as an event. A Monte-Carlo sampling is now made for momenta appropriate for a Maxwell-Boltzmann distribution at the appropriate temperature. Clusters from this event can now be calculated. The number of events has varied between 5,000 and 50,000, the criterion being that the statistical fluctuation should be acceptable.

The Monte-Carlo procedure can be used to calculate average energy, average yield of a particular composite etc. It can also be used to calculate the partition function. A very neat algorithm was used in [13].

7 Treating the Coulomb Force

We add this small section to emphasize that we have found that the time consuming approach of previous work where coulomb force was included [5],[11] can be given up for a more standard approach with negligible difference. The coulomb force is important when the disassembling system is large (in later sections we will further quantify this statement). In previous work this was demonstrated by (1) computing observables with the LGM without any coulomb and (2) generating the initial lattice gas configuration including the coulomb force (as described in VI) and then propagating the particles with molecular dynamics with coulomb and short range nuclear forces (such as would justify the nearest neighbour interactions). We find that it is important to include the coulomb force in the initialisation, but once this is done we can simply use the PD rule to obtain clusters. There is no need to follow this up with a molecular dynamics calculation since for coulomb fields met in heavy ion collisions the difference between application of PD rules and the full blown molecular dynamics calculation after initialisation is found to be small. This is a great saving in computer time. Although including the coulomb in the Metropolis algorithm does increase computing time, this is negligible compared with the molecular dynamics calculation that followed since molecular dynamics calculations had to be continued to large times to unambiguously identify clusters.

8 Empirical Mass Formula and the Isospin Dependent Lattice Gas Model

The value of the parameter ϵ_{np} is fixed at -5.33 MeV by the requirement that nuclear matter binding energy be 16 MeV per particle. We also have the condiition that $\epsilon_{nn} = \epsilon_{pp} \geq 0$. We may try to fix the value of this parameter by a "best" fit to the symmetry energy for medium to heavy mass nuclei. Of course nuclear binding energies also have surface tension term but once we fix ϵ_{nn} from symmetry energy considerations, we have no free parameter left.

We calculate binding energies for sevaral nuclei in the mass range $16 \leq A \leq 252$ taking $\epsilon_{nn} = 0$. The previous section described how to generate an event at a finite temperature. The method is readily adaptable to the case $T \to 0$ which then produces the ground state. The ground state energies for a number of nuclei so obtained are fitted to a simple liquid-drop model mass formula [26] given by

$$E/A = -a_v(1-\kappa I^2) + a_s(1-\kappa I^2)A^{-1/3} + a_c\frac{Z^2}{A^{4/3}} \tag{11}$$

Here $I = (N-Z)/A$, is the neutron-proton asymmetry of the nucleus. This mass formula contains four parameters, namely, a_v, a_s, a_c and κ. The volume energy coefficient a_v has already been fixed at 16 MeV through the choice of ϵ_{np}. The remaining three parameters are obtained from a least-squares fit to the calculated ground state energies. The per particle binding energies obtained in the LGM, their fitted values obtained from the mass formula and the experimental binding energies for a number of nuclei are shown in Table 1. It is seen that the fitted values are within 1% of their actual calculated values. This indicates that the LGM binding energies closely follow a liquid-drop mass formula. The comparison of the calculated binding energies with the experimental ones shows that for heavy nuclei they

Table 1: Calculated,fitted and experimental B.E./A

System	Calculated B/A	Fitted B/A	Expt. B/A
^{16}O	8.505	8.452	7.976
^{27}Al	8.899	9.064	8.331
^{40}Ca	9.169	9.137	8.562
^{56}Fe	9.455	9.321	8.790
^{64}Zn	9.304	9.269	8.736
^{90}Zr	9.067	9.124	8.710
^{114}Sn	8.839	8.903	8.523
^{150}Sm	8.589	8.496	8.244
^{197}Au	8.039	8.063	7.915
^{208}Pb	7.826	7.940	7.867
^{230}Th	7.803	7.737	7.631
^{238}U	7.702	7.649	7.635
^{252}Cf	7.501	7.539	7.465

Table 2: Lattice-gas and phenomenological liquid drop model parameters

model	a_v	a_s	κ	a_c
Lattice-gas	15.99	16.03	2.14	0.746
Phenomenological	15.677	18.56	1.79	0.717

agree within 1% which, we think, is an excellent agreement keeping in mind the simplicity of the LGM. However, for the relatively lighter nuclei the deviations in some cases are as high as 10%. The liquid-drop model parameters obtained from the least-squares fit to the calculated LGM energies and the corresponding phenomenological values [26] are listed in Table 2. We note that the asymmetry energy coefficient κ in the LGM is somewhat larger compared to the phenomenological one. A lower value for κ can be obtained only by making ϵ_{nn} negative. This is not permitted as then unphysical clusters like dineutron or diproton would be produced as mentioned earlier. So we fix the value of ϵ_{nn} at zero. Before leaving this section we would like to make a comment on the surface energy coefficient a_s. It can be shown that for number of nucleons equal to number of lattice points ($A = N_s$) the number of missing bonds at the surface is $3A^{2/3}$. For a symmetric nucleus ($N = Z$) the surface energy then becomes $3\epsilon_{np}A^{2/3}$ giving $a_s = 16.0$ MeV. It is seen from Table 2 that the value of a_s obtained in the LGM is very close to this value. Lastly we find the agreement between the phenomenological liquid-drop parameters and the LGM parameters very satisfying.

9 A Measure of the Influence of the Isotopic Spin Dependence and the coulomb force

In the framework of the LGM for finite nuclei at finite temperature how important is the feature that ϵ_{np} is different from ϵ_{nn} and that there is coulomb interaction? One measure is how much the one-body densities ρ_n and ρ_p change because one has, instead of one kind of bond ($\epsilon_{np} = \epsilon_{pp}$ =-5.33 MeV and no coulomb), two kinds of bonds (ϵ_{np} =-5.33 MeV,ϵ_{pp}=0), and two kinds of bonds plus coulomb interaction between protons. Here ρ_n is the one-body neutron density and ρ_p is the one-body proton density. One might attempt to get this

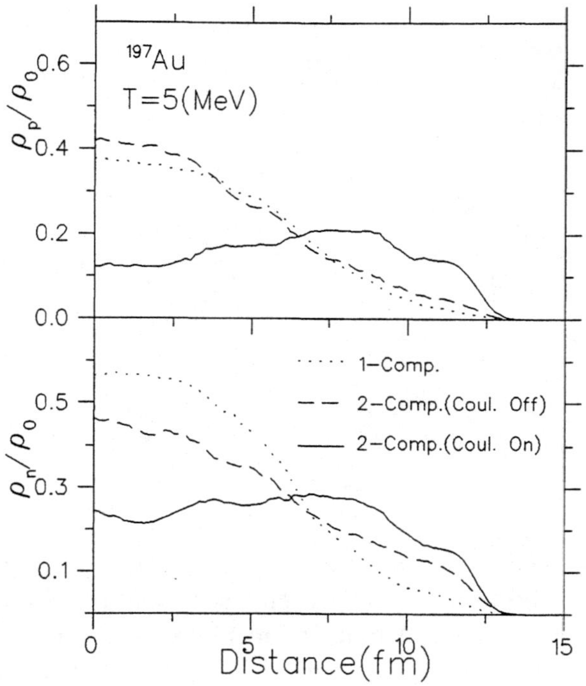

FIG.3. The proton(upper panel) and neutron(lower panel) one-body densities for ^{197}Au at a temperature of 5 MeV. The dotted lines correspond to calculations with one-type of bond($\epsilon_{nn} = \epsilon_{pp} = \epsilon_{np} = -5.33MeV$). The dashed lines represent calculations with two types of bond($\epsilon_{nn} = \epsilon_{pp} = 0, \epsilon_{np} = -5.33MeV$) without Coulomb interaction and the full lines are the same with the inclusion of the Coulomb interaction among the protons.

from mean-field theories (section V) but these give grossly wrong answers [27] and we compute these averaging over events generated by Monte-Carlo. In each event the centre of mass of the fragmenting system is computed and density is calculated as a function of distance from it. An unoccupied box has zero density throughout the box and an occupied box has density ρ_0 throughout the box. The density is usually very rugged in an event. A smoother density results after the canonical ensemble averaging. This is the "exact" one body density we are seeking.

In Fig. 3 we show the one body density for ^{197}Au at a temperature of 5 MeV. It is clear that for a nucleus of this size the isotopic spin dependence has considerable effect on the one-body density. What is also clear is that the coulomb interaction is very important. Unfortunately how this difference will affect the production of clusters, the real experimental observable, is hard to guess and one has to plot the data to see the difference it generates.

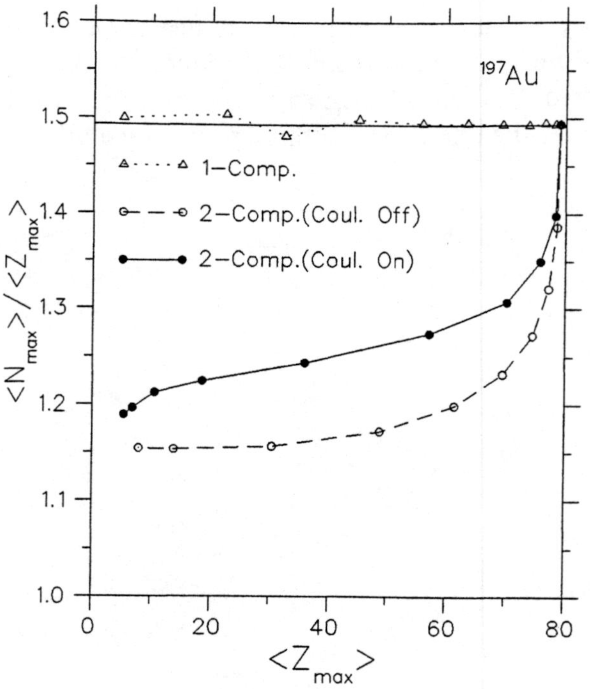

FIG.4. The evolution of the N/Z ratio of the largest fragment as a function of the atomic number of the fragment produced from the fragmentation of the ^{197}Au nucleus in the LGM. The different lines have the same meanings as in Fig.3. The temperature increases by 0.5 MeV between two successive symbols as we move from right to left,the first ones correspond to a temperature 2.5 MeV. The symbol $' <>'$ stands for the ensemble averaging.

However one can readily understand how the N/Z ratio of the largest cluster (we called this a_{max} in section IV) gets affected. This is shown in Fig. 4 where, as in Fig. 3, the disassmbling system has $Z = 79$ and $N = 118$. The average value of $< Z_{max} >$ (the average value of Z of the largest cluster) of course depends on the temperature used in the simulation. At low temperature $< Z_{max} >$ is close to 79 but will drop in value as the temperature gets raised. It is easy to understand that if only one kind of bond is used the $< N_{max} > / < Z_{max} >$ ratio stays close to that of the disassembling system. With two kinds of bonds but no coulomb the ratio of $< N_{max} > / < Z_{max} >$ is driven towards 1 because this is what the symmetry energy term prefers. However, the coulomb term will drive the largest cluster towards a higher $< N_{max} > / < Z_{Max} >$ ratio.

In the next section we turn to some typical predictions of the lattice gas model.

10 Some Salient Features of Intermediate Energy Heavy Ion Reactions: Experiment and Theory

We describe some of the broad features seen in heavy ion reactions. These features are all reproduced by the LGM. We have shown some representative calculations (Fig. 5) when the disintegrating system is ^{197}Au. The lattice size is 8^3 so the freeze-out density is $0.38\rho_0$. The following features noted in Fig.5 are also seen in experiments in general. At low bombarding energies (corresponding to low T), the yield as a function of A or Z is U shaped. In a given event typically there will be one large cluster and a bunch of lighter nuclei. As the temperature rises the height of the maximum of the avaraged yield at high Z goes down till it entirely disappears. At higher temperature the yield $Y(Z)$ against Z falls more steeply. The case when the maximum in the yield $Y(Z)$ just disappears (case T_B in Fig. 5) is particularly interesting. At this time $Y(Z)$ against Z is very nearly a power law ($Y(Z) \propto Z^{-\tau}$) if, in the nuclear case, one restricts oneself from $Z = 2$ to $Z \approx 20$.

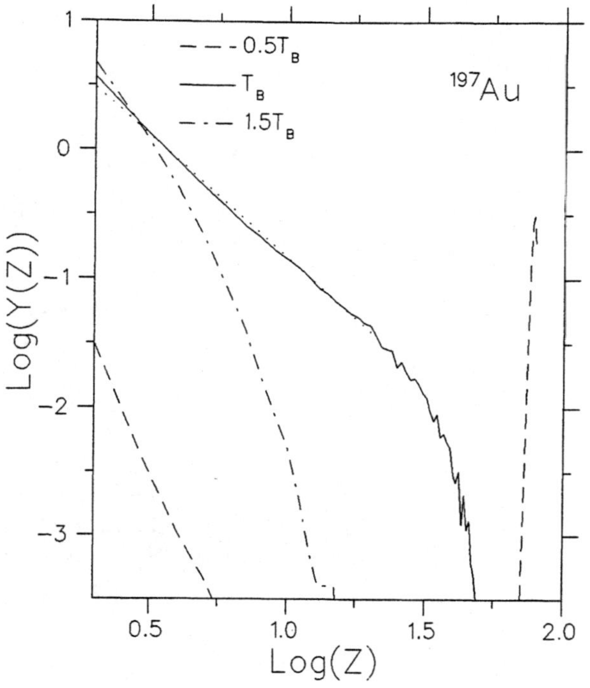

FIG.5. The log-log plot of the charge distributions obtained in the LGM from fragmentation of ^{197}Au at different temperatures as indicated in the figure. The calculations are with two types of bonds and Coulomb included. The dotted line corresponds to a power law fit to the calculated charge distribution at the temperature T_B(=4.5 MeV).

In the early eighties, in a pioneering experiment, the Purdue group saw a power law in the fragment distribution resulting from high energy proton-nuccleus collisions [28]. Since then the power law has been verified in many heavy ion collision experiments [29]. In experiments of gold on C, Al and Cu, the extracted value of τ was plotted against energy deposited per nucleon. The minimum was about 2. In calculations involving finite number of particles in a finite lattice size an approximate power law is obtained over a range of temperatures. In such cases one can derive a value of τ from

$$\frac{\sum_{z1}^{z2} ZY(Z)}{\sum_{z1}^{z2} Y(Z)} = \frac{\sum_{z1}^{z2} ZZ^{-\tau}}{\sum_{z1}^{z2} Z^{-\tau}}. \tag{12}$$

It is not unexpected that when this procedure is followed the minimum value of τ coincides rather remarkably well with the disappearence of the maximum in the high Z side. This τ_{min} in Fig. 5 is ≈ 2.

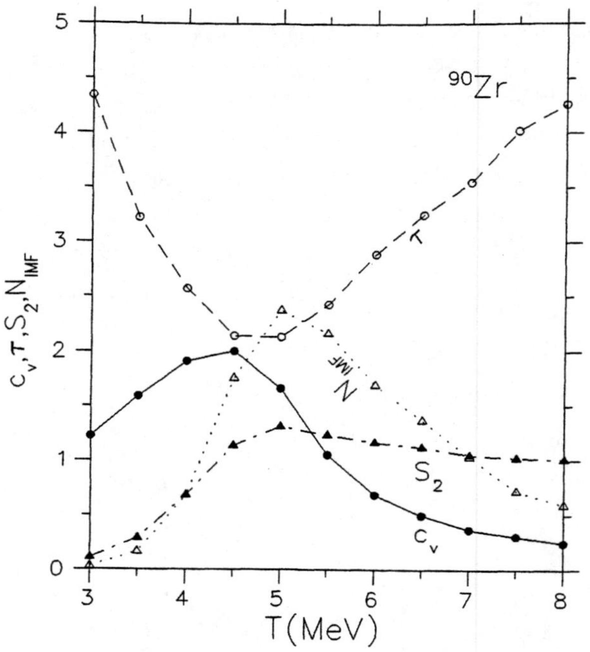

FIG.6. The specific heat C_v, the exponent τ for the power-law fit to the charge distriburion, the second moment S_2 for the charge distribution and the IMF yield are shown as a function of temperature for the fragmentation of ^{90}Zr in the LGM. The calculations are done in a 6^3-lattice with two types of bonds and Coulomb included.

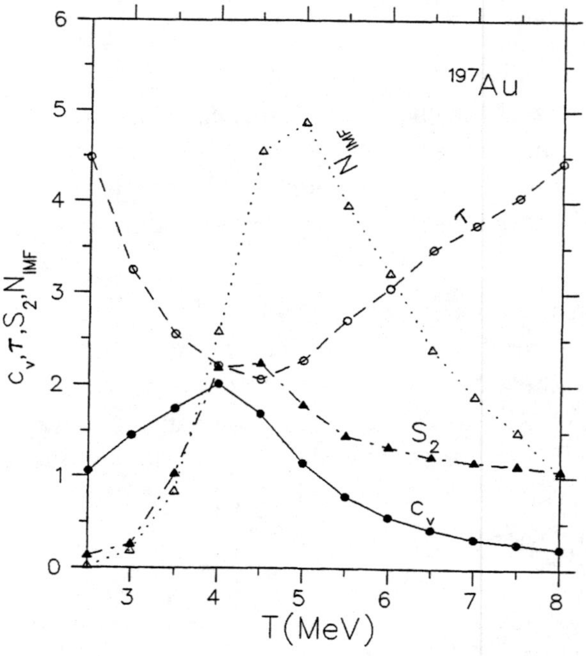

FIG.7. The same as in Fig.6 for ^{197}Au with 8^3-lattice.

In Figs. 6 and 7 we plot several quantities computed in the LGM when the disintegrating systems are ^{90}Zr and ^{197}Au respectively. In heavy ion physics one commonly defines IMFs (intermediate mass fragments) as those which have Z between and including 3 and 20. Experimentally, there is clear evidence that starting from low excitation energy in the disintegrating system N_{IMF} first rises, reaches a maximum and then falls again. The same feature is found in the lattice gas model. For ^{197}Au we find that our highest value for N_{IMF} is nearly 5, but experiments of Au on C, Al and Cu at GSI found a lower value, about 4; EOS experiment of Au on C found the maximum at about 3.5 [30]. In our calculation we used $Z = 79$ and $A = 197$; the disintegrating system in experiment (pre-equilibrium emission will reduce the number of nucleons which reach equilibrium) may be smaller which will explain the difference. In Figs. 6 and 7 we also have shown the value of S_2, the second moment for charge distriubutions mentioned in section IV. As a function of T a maximum is clearly discernible. We also plot C_v as a function of temperature which shows a maximum in the temperature range of the figures. In experimental data there is evidence of a maximum in specific heat [31]. This appears to occur around 5 MeV temperature.

It is clear from Figs. 6 and 7 that the maxima of C_v, N_{IMF}, S_2 and the minimum of τ all occur at nearly the same temperature. Similar results were also obtained by Ma et al. [15]. One can regard this as a signature of what would have been a phase transition in an infinite system with coulomb switched off. From the point of view of the LGM this would be a signature of first order phase transition [10]. The arguments go like this. The freeze-out density is very likely less than $0.5\rho_0$. A good fit of data in the model requires this [3]. All other reasonable models of fragmentation also demand this. Therefore we are on the low density side of the critical point. Now the percolation line of Fig. 1 extends not only from B to C.P. but extends on the low density side also. On the low density along the coexistence line there will be an extremum in the value of C_v as one crosses the coexistence line. We may not be able to measure this extremum value in the specific heat but in the lattice gas model other easier measurements such as a minimum in τ or a maximum in N_{IMF} will signify that we have crossed the coexistence curve. Incidentally according to this model it will be very difficult to ever reach critical point. To reach the critical point one needs the correct values for both the temperature and the freeze-out density. By varying the excitation dumped in the system (either by suitable selection in impact parameter or selection of the beam energy) one may be able to control the temperature reached in the system. But it is difficult to see how one can tune the freeze-out density. One however has second order type signatures even in the case of a first order transition here. The exponents τ, β and γ exist not only at the critical point but along an entire contour. This was pointed out in [13].

Finally we will mention the roles that the isotopic spin dependence of the nearest neighbour interaction and the coulomb interaction play in these computations. In Figs. 8 and 9 we plot for ^{197}Au, τ and C_v with temperature for (a) no isospin dependence (b) isospin dependent nearest neighbour interaction and (c) the case (b) plus coulomb. In shifts of the minimum of τ and the maximum of C_v the coulomb plays a significant role. We verified that

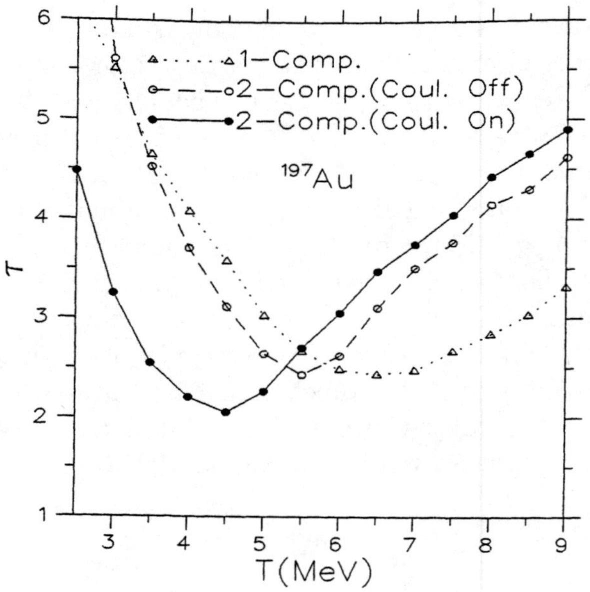

FIG.8. The exponent τ for the power-law fit to the charge distribution as a function of temperature for fragmentation of ^{197}Au for the cases: (a) no isospin dependence in the nearest neighbour interaction (dotted line with triangles), (b) with isospin dependent nearest neighbour interaction (dashed line with open circles) and (c) the same as in the case (b) with the Coulomb interaction (full line with filled circles).

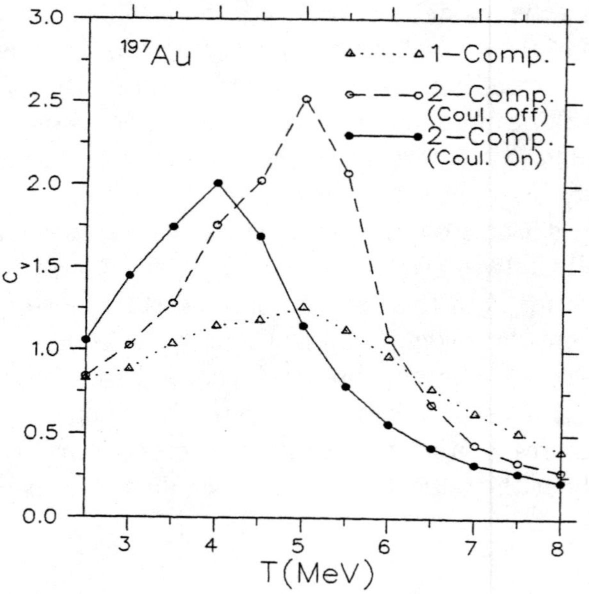

FIG.9. Same as in Fig.8 for the specific heat C_v.

as the disintegrating system gets larger with a corresponding growth in the value of Z, the proton number, the minimum of τ falls to lower and lower temperature finally disappearing altogether around Z=120. It is well-known that for central Au on Au collisions there is no minimum in τ [32]. The explanation of this was an early triumph of the LGM with coulomb interaction. As mentioned in section VII, we can do this without having to supplement the lattice gas part with molecular dynamics calculation.

11 Further Tests of the Model

We have seen above that the isospin dependent LGM reproduces the gross features of intermediate energy heavy ion data. In this section we examine some data which are more exclusive. In an experiment which highlighted the role of the isospin, average neutron multiplicity was measured as a function of charged-particle multiplicity in $^{124}Sn+^{124}Sn$ and $^{112}Sn + ^{112}Sn$ collisions [33]. For example, in the experiment, on the average, about 33 neutrons are emitted in $^{124}Sn+^{124}Sn$ collisions but only about 21 in $^{112}Sn+^{112}Sn$ collisions when the charged-particle multiplicity is 25. In Fig.10 our calculations for $< N_n >$ against $< N_c >$ for the two collisions once without any isotopic spin dependence and once including isotopic spin are shown. We input different tempeartures to change the average charged-particle multiplicities $< N_c >$ and at each temperature we calculate the average neutron multiplicity $< N_n >$. This of course does not correspond to the experimental situation identically. To correspond to the experimental situation additional assumptions need to be invoked, like variation of temperature with impact parameter etc. However, it is clear that an isotopic spin dependent LGM is needed to explain the data. In Fig 11 we fix temperature at 3 MeV and assume all N_c's between 10 and 25 arise from this temparature. We then compare the calculations with the data. Here also, in actuality, other temperatures arising from different impact parameters will also contribute to these N_c's but at least, there are no obvious large contradictions between the data and the isotopic spin dependent LGM.

This calculation also highlights one of the advantages of the LGM. Since the PD recipe

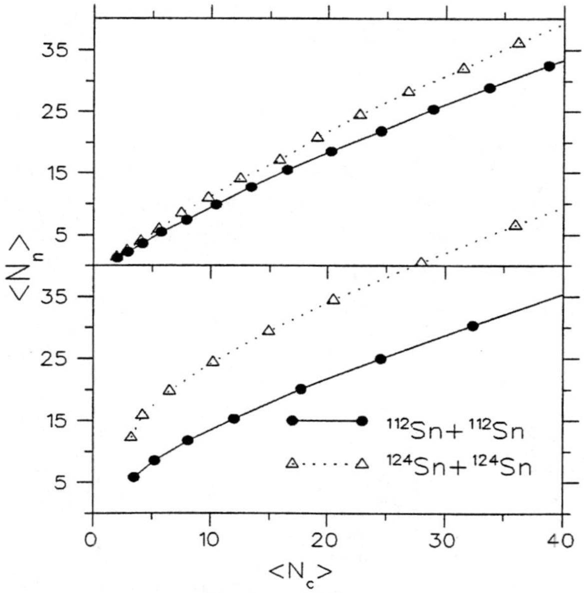

FIG.10. Correlation between average number of neutrons and charged particles produced at different temperatures for the reactions indicated. The lowest temperature is 2.5 MeV and T increases by 0.25 MeV between two successive symbols. The upper panel corresponds to calculations with one type of bond and the lower panel represents that with two types of bond including the Coulomb interaction. The calculations are done in a 9^3 lattice.

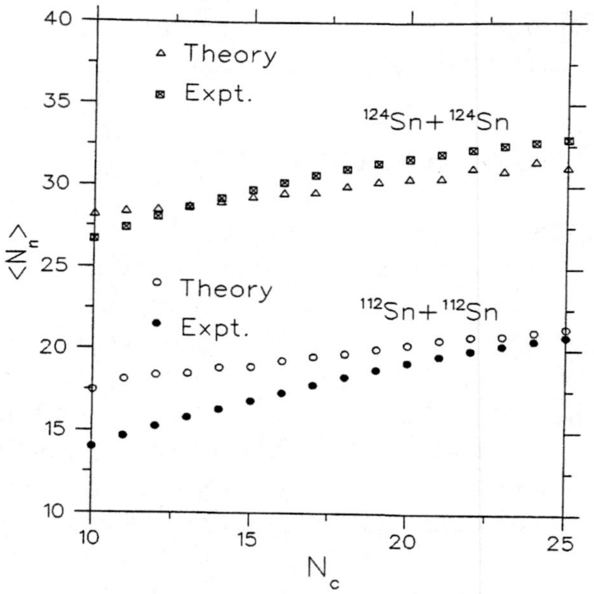

FIG.11. The same as in Fig.10 at a fixed temperature T=3 MeV with two types of bond (N_c not averaged). The experimental data are also included.

gives only particle stable nuclei, calculation of emitted neutrons is straightforward. This is, for example, not true in SMM [17] or other types of thermodynamic models [34]. Indeed in these models it is very difficult to compute the number of neutrons since they will largely arise from decay of particle unstable nuclei.

For future use, we mention other possible measurements which may explicitly test the isotopic spin dependent part of the LGM. We refer back to fig.4 where for ^{197}Au we have shown the N/Z of the largest cluster. The results are significantly different whether we use one kind of bond (no isotopic spin dependence) or two kinds of bonds. This is easily understandable as with proper isotopic spin dependence the largest cluster is driven towards N/Z=1 (coulomb will slightly compensate this as coulomb drives the ratio towards higher than 1). A corollary follows. For the rest of the system i.e., ^{197}Au$-$largest cluster, the N/Z ratio is much larger than that of ^{197}Au. Measured $Y(\text{n})/Y(\text{p})$ or $Y(\text{t})/Y(^3\text{He})$ will reflect the high N/Z of this remaining system. Therefore the ratio of free nuetrons to free protons (monomers) or tritons to ^{3}He is expected to be higher than that of the dissociating system.

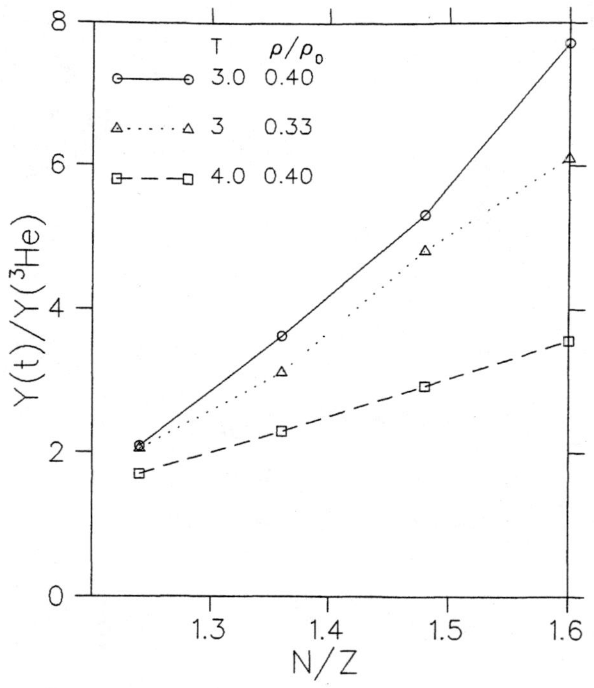

FIG.12. The ratio of yields of triton to ^{3}He as a function of the neutron(N) to proton(Z) ratio of the fragmenting system. The different curves correspond to different temperatures and freeze-out densities as indicated in the figure.

We test this idea by doing a calculation in an 8^3 lattice. We expect the ratio $Y(\mathrm{t})/Y(^3\mathrm{He})$ to be dependent upon the input N/Z, the temperature and possibly the freeze-out density. We expect (also verified numerically) the ratio to be much less dependent on the actual number $N + Z$. We therefore fix N and Z by imposing an N/Z value and $\rho/\rho_0 = (N + Z)/N_s$=0.4 and 0.33. Without any isotopic spin dependence the ratios are indistinguishably same as the input N/Z but the ratios are much higher with an isotopic spin dependent LGM. This is shown in Fig. 12. We again want to emphasize that such an observable is very difficult to calculate in a standard thermodynamic model because many tritons and ^{3}He will arise from decay of hot nuclei.

While the calculated $Y(\mathrm{t})/Y(^3\mathrm{He})$ might be dependable in spite of the lack of shell effects in the model (they may largely cancel out when ratios are taken) we do not expect isotope productions to be given accurately in the model because there shell efects etc are expected to be large. However, at least, the liquid drop part of the binding energy is reproduced by the model hence we calculated, more as a benchmark, isotope productions in the isotopic spin dependent LGM for ^{32}S+^{107}Ag collisions. The experimental data are taken from [35]. In Fig. 13 we also show results from a thermodynamic model calculation [34] where subsequent evaporation from hot nuclei are included. Not unexpectedly for these data the thermodynamic model does much better. Thermodynamic models have their own failures however. A power law does not emerge; if a value of τ is nonetheless extracted from a least square fit it can fall below the measured value 2. Also a minimum in τ continues to be obtained even for large coulomb field, in contrast with experimental finding [36].

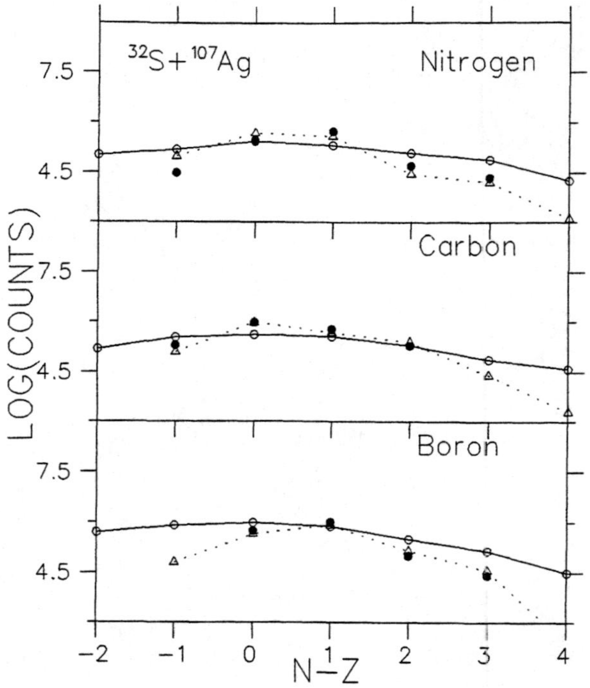

FIG.13. Isotopic yields for the elements B, C and N from the reaction $^{32}S+^{107}Ag$ at 22.3A MeV, the abscissa being the difference between the neutron and proton numbers of the isotopes. The experimental data(filled circles) are from [35]. The full lines with empty circles are LGM calculations while the dotted lines with triangles refer to calculations in the thermodynamic model.

12 Summary and Discussion

An isotopic spin dependent lattice gas model is a very useful tool for understanding gross features of nuclear fragmentation. It is easy to implement and reproduces important features of power law, peaks in specific heat, rise and fall of the average number of intermediate mass fragments etc. It is the only workable model where interactions between different fragments are explicitly in and accounted for.

Phase transition can be studied in this model in an exact (though numerical) fashion. We did not pursue the phase transition aspect in any detail in this article as there is another article on this aspect in this volume. We looked at phase transition only in the mean field approximation.

This article has emphasized the inclusion of the coulomb field for large fragmenting systems. If fitting data is the main goal, this must be included. This is not hard since most of the effects are included by incorporating the coulomb in the Monte-Carlo simulation of events. This probably takes care of the coulomb force in a more rigorous fashion than the Wigner-Seitz approximation [17] which ignores all fluctuations in the coulomb energy.

The main deficiency of the model is in the lack of shell effects of the composites and loss of spherical symmetry. We know of no easy ways to rectify these.

13 Acknowledgment

This work is supported in part by the Natural Sciences and Engineering Research Council of Canada and by *le Fonds pour la Formation de Chercheurs et l'Aide à la Recherche du Québec*. SKS thanks the physics department, McGill University for hospitality during the course of this work. SDG thanks Betty Tsang for generous help for clarification of experimental data. The authors thank Abhijit Majumder for many helpful discussions.

References

[1] T. D. Lee and C. N. Yang, Phys. Rev **87**, 404 (1952).

[2] K. Huang, *Statistical Mechanics*(John Wiley, New York,1987), chapter 14.

[3] J. Pan and S. Das Gupta, Phys. Lett **B344**, 29 (1995).

[4] J. Pan and S. Das Gupta, Phys. Rev **C51**, 1384 (1995).

[5] S. Das Gupta and J. Pan, Phys. Rev **C53** , 1319 (1996).

[6] L. Beaulieu et. al. Phys. Rev **C54**, R973 (1996).

[7] S. Das Gupta, J. Pan and M. B. Tsang, Phys. Rev **C54**, R2820 (1996).

[8] S. Das Gupta, J. Pan, I. Kvasnikova and C. Gale, Nucl. Phys. **A621**, 897 (1997).

[9] X. Campi and H. Krivine, Nucl. Phys **A620**, 46 (1997).

[10] J. Pan, S. Das Gupta and M. Grant, Phys. Rev. Lett. **80**, 1182 (1998).

[11] J. Pan and S. Das Gupta, Phys. Rev **C57**, 1839 (1998).

[12] Ph. Chomaz and F. Gulminelli, Phys. Lett **B447**, 221 (1999).

[13] F. Gulminelli and Ph. Chomaz, Phys. Rev. Lett., **82**, 1402 (1999).

[14] Y. G. Ma et. al., Euro. Phys. J. **A4** , 217 (1999).

[15] Y. G. Ma et al., Phys. Rev **C60**, 024607, (1999).

[16] T. S. Biro, J. Knoll and J. Richert, Nucl. Phys. **A459** , 692 (1986).

[17] J. P. Bondorf, A. S. Botvina,, A. S. Ijilinov, I. N. Mishustin, and K. Sneppen, Phys. Rep. **257**, 133 (1995).

[18] W. Bauer, Phys. Rev. **C38**, 1297 (1988).

[19] X. Campi, Phys. Lett B208, 351 (1988).

[20] D. Stauffer and A. Aharony, *Introduction to Percolation Theory* (Taylor and Francis, London, 1992).

[21] A. Coniglio and W. Klein, J. of Phys. **A13**, 2775 (1980).

[22] J. Kertész, Physica, **A161**, 58 (1989).

[23] J. Sivardiere and J. Lajzerowicz, Phys. Rev **A11**, 2090 (1975).

[24] S. Ray, J. Shamanna and T. T. S. Kuo, Phys. Lett. **B392**, 7 (1997).

[25] N. Metropolis, A. W. Rosenbluth, M. N. Rosenbluth, A. H. Teller, and E. Teller, J. Chem. Phys. **21**, 1087 (1953).

[26] W. D. Myers and W. J. Swiatecki, Nucl. Phys. **81**, 1 (1966).

[27] S. K. Samaddar, S. Das Gupta, J. N. De, B. K. Agrawal, and T. Sil, Phys. Lett **B459**, 8 (1999).

[28] J. E. Finn et. al. Phys. Rev. Lett. **49**, 1321 (1982).

[29] C. A. Ogilvie et. al. Phys. Rev. Lett. **67**, 1214 (1991).

[30] J. B. Elliott, Ph. D. Thesis, Purdue University, 1995, unpublished.

[31] J. Pochodzalla et. al. Phys. Rev. Lett. **75**, 1040 (1995).

[32] M. D'Agostino et al., Phys. Rev. Lett **75**, 4373 (1995).

[33] G. J. Kunde et al., Phys. Rev Lett **77**,2897 (1996).

[34] A. Majumder and S. Das Gupta, to be published in Phys. Rev C.

[35] T. K. Nayak et al., Phys. Rev **C45**, 132 (1992).

[36] P. Bhattacharyya, S. Das Gupta and A. Z. Mekjian, to appear in Phys. Rev C.

Phase Transition with an Isospin Dependent Lattice Gas Model

Philippe Chomaz*, Francesca Gulminelli+ and Véronique Duflot *†

1 Introduction

One of the most important challenges of heavy ion physics at intermediate bombarding energies is the identification and characterization of the nuclear liquid-gas phase transition [40]. With the increasing availability of quasi complete and exclusive data on the deexcitation of well identified single nuclear sources in heavy ion reactions [26], different indirect evidences of such a transition are rapidly accumulating. On one side, the sudden opening of the multifragmentation [2, 38] and vaporization [37] channels can be interpreted as the signature of the boundaries of phase mixture [16, 17]. This is reinforced by the fact that at the corresponding excitation energy values the caloric curve [35] shows a structure similar to a first order phase transition[3, 17]. On the other side, the observation of critical exponents like power laws in the charge distribution of the multifragmenting system [22] can be interpreted as an evidence of the transition.

An important property of nuclear matter is the fact that it is built out of two different fluids: protons and neutrons. This new degree of freedom is important to understand the properties of the equation of state. The first theoretical calculations of the phase transition within finite systems were performed [6, 32, 7] in the isospin symmetric case. It is well known that in a system with two conserved charges (baryon number and isospin, in the nuclear case) the nature itself of the liquid-gas phase transition is modified: the transition is continuous, phase separation is driven by isovector as well as isoscalar instabilities, the value of critical parameters is modified [28, 36]. Since most of the multifragmenting systems produced by heavy ion collisions are far from being isospin symmetric, the comparison of experimental charge distributions with theoretical size distributions can be misleading. Using an exactly solvable model of the liquid-gas phase transition, in this chapter we shall show that if one drops out the mean field approximation, the critical parameters of the phase transition depend only slightly on the asymmetry and this dependence is probably hardly detectable for experimentally available N/Z ratios. However a phenomenon analogous to distillation occurs in the fragmentation of a finite piece of matter containing two fluids. This distillation of neutron matter characterizes the transition. This means that the isospin degree of freedom can be used as a new observable to study the phase transition.

The plan of the chapter is as follows: in the next section the isospin dependent Lattice Gas model will be introduced and its thermodynamics calculated from an exact evaluation of the various entropies and partition sums associated with the different ensembles; section 3 will be devoted to a discussion of the different ways to extract thermodynamical signals,

*GANIL (DSM-CEA/IN2P3-CNRS), B.P. 5027, F-14076 Caen cédex, France
†LPC Caen, IN2P3-CNRS, ISMRA et Université , F-14050 Caen cédex, France

in particular using fluctuations. Our main results on the thermodynamics of the lattice gas model are summarized in section 4. The cluster observables at subcritical and supercritical densities with a special emphasis on finite size distortions of the critical parameters and on the effect of the isospin degree of freedom will be presented in section 5; The distillation of isospin asymmetric systems will be discussed in section 6 and finally section 7 will be devoted to conclusions and outlooks.

Further results obtained within this same model can be found in the contribution of S.Das Gupta to this book.

2 Thermodynamics of the lattice gas model

The Lattice Gas Model of Lee and Yang [42] has successfully described the liquid-gas phase transition for atomic systems. This same model has already been applied to nuclear physics for isospin symmetric systems in the gran-canonical ensemble [7, 6], with an approximate sampling [29] of the canonical ensemble [32], and for isospin asymmetric systems in the mean field approximation [36, 33]. The isospin dependent Lattice Gas Model is isomorphous to a spin 1 Ising model with dipolar and quadrupolar interactions [39]. The N sites of a lattice are characterized by an occupation number τ which is defined as $\tau = 0$ for a vacancy, and $\tau = 1(-1)$ for a proton(neutron). Particles occupying nearest neighboring sites interact with an energy ϵ_{τ_i,τ_j}. The Hamiltonian is given by

$$\hat{\mathbf{E}} = \sum_{i=1}^{N} \frac{p_i^2}{2m}|\tau_i| + \frac{1}{2} \sum_{i,j \in \mathcal{N}_i} \epsilon_{\tau_i,\tau_j} \tau_i \tau_j \tag{1}$$

where the second sum runs only on the neighboring sites $\mathcal{N}_i$ of the site i. The coupling constants ϵ_{τ_i,τ_j} are fixed such as to reproduce the volume and symmetry part of the liquid drop model parametrization, $\epsilon_{1,1} = \epsilon_{-1,-1} \equiv \epsilon_d = 0.$ MeV, $\epsilon_{1,-1} \equiv \epsilon_s = -5.5$ MeV. When no isospin dependence is introduced we use $\epsilon_d = \epsilon_s = -5.5$ MeV. In all the calculations shown below the numerical realization of the model is a three-dimensional cubic lattice of size $S = L^3$.

With the finite range interaction defined above the thermodynamical limit does exist for the infinite system. In this limit all the different statistical ensembles converge to a unique thermodynamics. However, for finite systems the various statistical ensembles are not equivalent because fluctuations cannot be neglected. Therefore, the equations of state will be different from one ensemble to the other. Then we are forced to carefully discuss the various statistical ensembles and their adequacy to the nuclear physics experimental situation.

2.1 Canonical isochore ensemble

In the liquid-gas phase transition, since the order parameter is the density difference between the two phases, the volume is essential in determining thermodynamical properties. A constant volume ensemble can be trivially generated by considering a fixed constraining cubic box. Many studies have been performed considering periodic boundary conditions in order to avoid the effects of the surface [9]. Systems in a fixed cubic volume have also been investigated [5]. In this section periodic boundary conditions will be systematically applied, while the discussion of surface effects will be postponed to section 2.5 where more realistic spherical systems will be considered.

In the canonical isochore ensemble the system is characterized by a linear size L, a fixed number of particles $A = N + Z$ (or equivalently a density $\rho/\rho_0 = A/L^3$), and a temperature β^{-1}. Statistical averages are taken over events obtained with a standard Metropolis sampling [27] of the lattice in the canonical ensemble. In this section, we shall limit our analysis to isospin symmetric systems $N = Z = A/2$.

The canonical thermodynamics of the model is calculated from a direct evaluation of the partition sum Z via an iterative procedure. At a temperature β^{-1}, the occupation probability of an energy state E is

$$P_\beta(E) = \frac{1}{Z_\beta} W(E) e^{-\beta E} \tag{2}$$

where $W(E)$ is the degeneracy of the state. In order to have an exact calculation of the equations of state and of the partition properties of the system we use a standard mass conserving Metropolis method to sample events according to the canonical probability eq.(2). Starting from a lattice configuration characterized by an energy E, the occupations of two randomly selected sites are exchanged leading to a new configuration of energy E' with a probability $exp(-\beta \Delta E)$ with $\Delta E = E' - E$. A large ensemble of preliminary events (typically $2 \cdot 10^7$) is first generated in order to reach statistical equilibrium. Then many configurations (typically $2 \cdot 10^4$) are produced using the same Metropolis algorithm before a single one is stocked to avoid event-to-event correlations and the occupation probability distribution is constructed via

$$P_\beta(E) = \frac{N_\beta(E)}{N_0} \tag{3}$$

Here, N_β is the number of events falling in the energy bin of size ΔE around E, and N_0 is the total number of sampled events. The bin size ΔE is chosen small enough that the results are independent of ΔE.

We have checked that an inversion of eq.(2) leads to an a posteriori estimation of the temperature strictly equal to the input one[20]. From the comparison of the occupation probabilities at two different temperatures β_1, β_2 one then has

$$Z_1 = Z_2 \frac{N_{\beta_1}(E)}{N_{\beta_2}(E)} e^{-E(\beta_1 - \beta_2)} \tag{4}$$

Since this equation is valid for every energy E it can be used to control the algorithm. Then a best fit of Z_1 can be derived. This can be done iteratively with an initial normalization to the infinite temperature limit where the partition sum is analytical. Once the partition sum is known, all thermodynamical quantities can be computed from the standard statistical definitions in the canonical ensemble.

As an example, in figure 1 a few isotherms are shown in the chemical potential versus pressure plane for a lattice of size $L = 6$ with a number of particles varying from $A = 10$ to $A = 210$. Even for such small systems, a very clean liquid and gas branch is seen for temperatures up to about 6.5 MeV. Since the crossing point of the two branches represents the coexistence boundary, the coexistence line can be evaluated very precisely from figure 1 leading to a critical temperature $T_c = 6.7$ MeV and a critical exponent $\beta = 0.31$ for the temperature dependence of the order parameter. This value has to be compared with the mean field approximation [32] (shown as dotted line in figure 1) $\beta = 0.5, T_c = 8.1$ and with the expected value in the thermodynamical limit $\beta^\infty = 0.33, T_c^\infty = 6.16$. It is clear that the finite size of accessible nuclear systems does not imply a drastic deformation

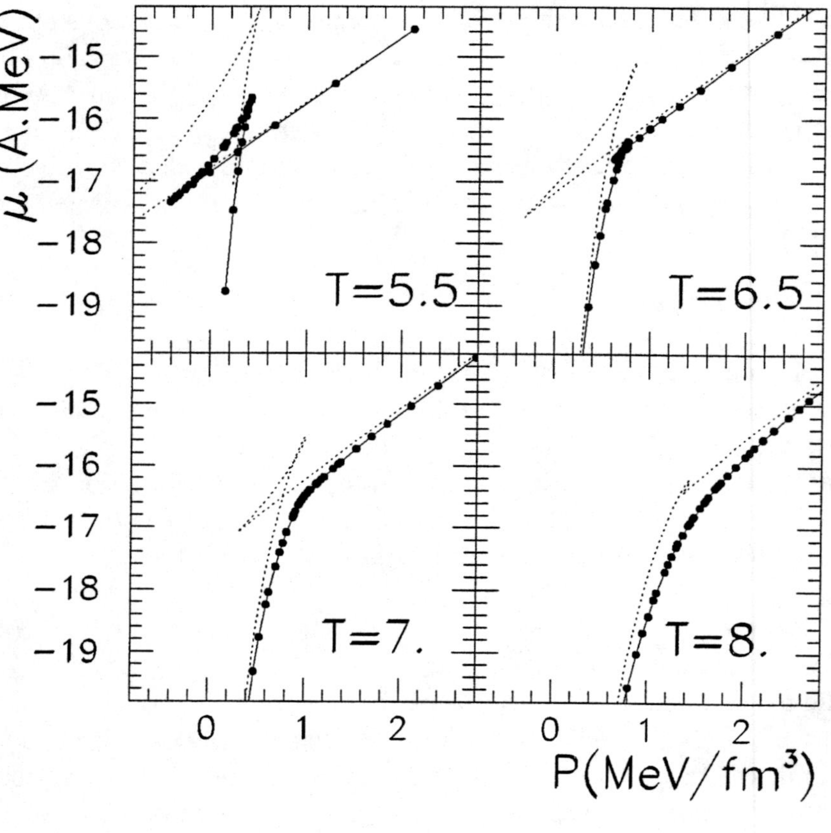

Figure 1: Chemical potential as a function of pressure at different temperatures for a cubic lattice of size $L = 6$ calculated from the exact partition sum (circles) and with the mean field approximation (dotted line).

of thermodynamical parameters (the corresponding values for $L = 8$ are $T_c = 6.6$ and $\beta = 0.31$). However the speed of convergence towards the thermodynamical limit strongly depends on the observable studied [6].

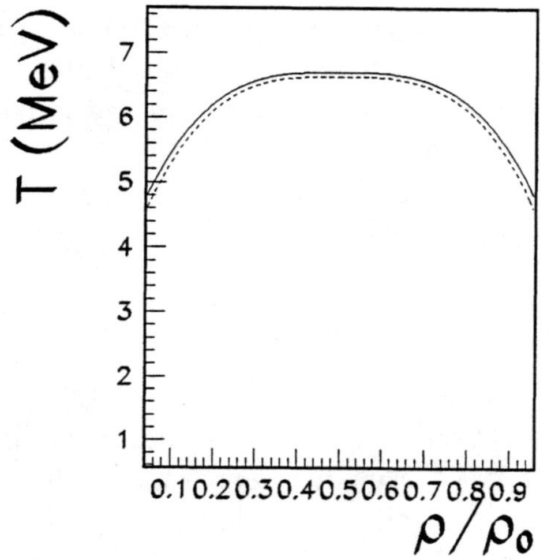

Figure 2: Coexistence zone computed from the anomaly in the pressure versus chemical potential diagram for a linear size of the lattice L=6 (solid line) and L=8 (dashed line).

2.2 Microcanonical isochore ensemble

A more fundamental statistical framework for actual applications is the microcanonical ensemble. This ensemble is directly related via its entropy to the elementary density of states. Moreover the microcanonical ensemble is relevant for the analysis of experimental data because of the absence of a heat bath and since using calorimetry techniques the excitation energy can be measured on an event-by-event basis; the events can thus be sorted in constant energy bins, i.e. in microcanonical ensembles. An easy and exact way to access microcanonical quantities is to sort the canonical events according to their total energy. It is important to notice that this procedure may be in some cases numerically time consuming but it is always an exact method to generate constant energy events with the correct microcanonical weight (see eq.(5) below). At a given canonical temperature β^{-1}, the number N_β of events falling in the energy bin E is proportional to the canonical distribution (2), $P_\beta(E)$, and to the total number of events N_0 :

$$\frac{N_\beta(E)}{N_0} = \frac{W(E)}{Z_\beta} \exp(-\beta E) \tag{5}$$

Equation (5) can be inverted in order to extract the density of states $W(E)$ leading to the entropy

$$S(E) \equiv \log(W(E)) \tag{6}$$

This allows a direct estimation of the microcanonical caloric curve

$$T^{-1}(E) \equiv \frac{\partial S(E)}{\partial E} = \beta + \frac{\partial \log N_\beta(E)}{\partial E} \tag{7}$$

which is valid for every β. With a single canonical sampling at an arbitrary β it is in principle possible to directly compute the whole microcanonical caloric curve (7). However for a given β the energies far away from the canonical average $\bar{E}(\beta)$ are hardly sampled in the canonical ensemble. To minimize numerical efforts and statistical errors it is therefore more convenient to perform many canonical samplings at different β associated with average energies scanning the energy region under study.

2.3 Isobar ensemble: enthalpy properties

In actual nuclear physics experiments, even if different sources with different excitation energies can be prepared, the other thermodynamical parameters can not be independently controlled on an event by event basis even if in some cases they can be measured. In particular the volume is not defined through boundary conditions but is an experimental observable known at best in average. For example the average radius of a selected ensemble of events can be estimated through interferometry or through comparisons with statistical models. From a theoretical point of view one is therefore forced to consider a statistical ensemble for which the volume can fluctuate from event to event. In terms of information theory this implies that at equilibrium the entropy of the system should be maximized under the constraint of a specific value for the average volume. In the absence of a prefered direction, an average volume can be defined through the one-body observable

$$\hat{\mathbf{V}} = \frac{4\pi}{3} \sum_{i=1}^{N} r_i^3 \tau_i \tag{8}$$

where r_i is the distance to the center of the lattice.

Introducing first a canonical description in which the energy observable $\hat{\mathbf{E}}$ as well as the volume $\hat{\mathbf{V}}$ are known in average we have to introduce the associated Lagrange multipliers β, λ so that the partition function reads

$$Z_{\beta,\lambda} = \sum_{(n)} \exp\left(-\beta E^{(n)} - \lambda V^{(n)}\right) \tag{9}$$

where the sum runs over all possible partitions. Here, $E^{(n)}$ and $V^{(n)}$ are the expectation values of the operators $\hat{\mathbf{E}}$ and $\hat{\mathbf{V}}$ in the nth event, β is the inverse of the canonical temperature, and the quantity $P = -\lambda/\beta$ has the dimension of a pressure. In other words, the experimental fact that the volume is known only in average means that the pressure, interpreted as the Lagrange multiplier associated with the volume observable, can be considered as the relevant state variable. The associated statistical ensemble of events can easily be generated through a canonical sampling using a constrained Hamiltonian $\hat{\mathbf{H}} = \hat{\mathbf{E}} - P\,\hat{\mathbf{V}}$ where $P\,\hat{\mathbf{V}}$ can be considered as a constraining one-body external field. The constrained energy $H = E - PV$ can be interpreted as an enthalpy. The actual value of the pressure parameter must be defined to get the desired average volume.

When the pressure is fixed we can sort the events as a function of the constrained enthalpy H. Then at a given canonical temperature β, the canonical distribution of enthalpies reads

$$P_\beta (H) = \frac{N_\beta (H)}{N_0} = \frac{W (H)}{Z_\beta} \exp(-\beta H) \tag{10}$$

where $N_\beta (H)$ is the number of events falling in the enthalpy bin H out of the total number N_0 of sampled events. Using the general relation $S \equiv \log W$ we can estimate the microcanonical caloric curve via $T^{-1} (H) = \partial_H S (H)$ as in section 2.2 above.

In the calculations shown in this section and in section 2.4 below an isospin symmetric system is considered with a fixed $A = 216$ number of particles. The qualitative results of our analysis are not modified by varying the size of the system; however the actual values of the temperature and heat capacity do depend on A due to finite size effects[19]. For these constrained calculations the size of the lattice is chosen large enough so that the boundary conditions do not affect the numerical sampling. For the results displayed in this section $S = 8000$ is used while in section 2.4 where weaker pressures are considered a lattice size as large as $S = 125000$ is needed. The numerical technique is the same as described in section 2.1 but the lattice occupations are sampled according to the partition function (9).

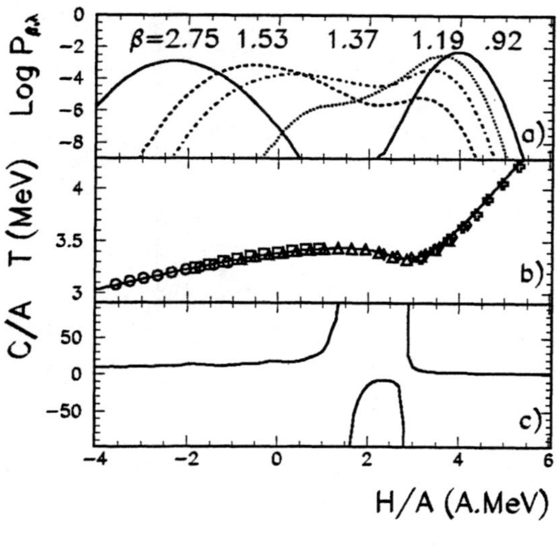

Figure 3: a) Total constrained enthalpy distributions associated with various canonical β . b) Microcanonical temperature as a function of the enthalpy per particle from various canonical samplings at the different β shown above (symbols) and the corresponding average (line). c) Microcanonical specific heat.

To illustrate a first order phase transition the pressure P is first chosen in such a way that the isobar crosses the canonical coexistence line at about the half of the critical temperature. Figure 3a presents the energy distributions at different β. According to equation (10) the logarithm of these probabilities directly gives the entropy apart from a linear correction. Far from the coexistence region the distribution presents the expected Gaussian behavior. When we get close to the coexistence region the observed double bump is a direct evidence of the convex intruder expected in the entropy in the case of a finite microcanonical system undergoing a first order phase transition. This can be better seen on the associated microcanonical caloric curve presented in Figure 3b. For each canonical ensemble the total energy exhibits large fluctuations allowing the calculation of the microcanonical temperature over a wide range of energies using equation (7). The fact that the microcanonical temperatures defined from various canonical ensembles at different β all agree within the statistical error bars is a demonstration of the accuracy of the Metropolis sampling. The microcanonical caloric curve presents the expected back-bending in the coexistence region due to the anomalous convexity of the entropy. This induces a negative branch in the microcanonical heat capacity[24] $C = (\partial_E T)^{-1}$ as shown in Figure 3c. The anomalous curvature of the thermodynamical potential is a clear signature of a first order phase transition, the distance between the two poles being a direct measure of the latent heat. Increasing the pressure, the two poles get closer up to the critical point where they merge before they disappear.

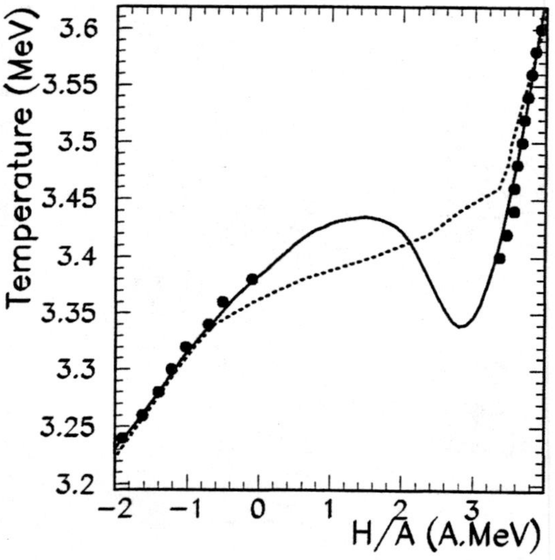

Figure 4: Microcanonical caloric curve (full line) compared with the most probable (dots) and average (line) canonical enthalpy for each β.

It is interesting to compare the microcanonical and canonical caloric curves as shown in Figure 4. In finite systems two canonical caloric curves can be defined, corresponding to the average and most probable energy associated to a given β. In infinite systems these two energies are equal because fluctuations can be neglected. Far from the phase transition the canonical and microcanonical curves agree. Indeed as seen in Fig. 3*a* the canonical energy distribution is Gaussian showing that the two ensembles only differ through minor fluctuation corrections. Moreover, from equation (7) we can see that the most probable canonical energy is characterized by the equality of the microcanonical and canonical temperatures. In the coexistence region however the predictions of the two ensembles differ in an noticeable manner. The canonical caloric curves are by definition monovaluated while this restriction does not apply to the microcanonical case: in the back-bending region the canonical caloric curve associated with the most probable energy presents a discontinuity equivalent to the Maxwell construction. The observed energy jump is directly related to the latent heat of the first order phase transition. Because of fluctuations the average energy presents a smoother behavior with however a clear slope change in the transition region. It should be noticed that the partitions which fall in the energy region corresponding to the canonical temperature jump are hardly sampled by the canonical ensemble, but are accessible in the microcanonical ensemble. Therefore, the microcanonical sorting of events allows to study regions of the phase diagram which are forbidden in the canonical formalism. These regions are characterized by specific properties such as negative heat capacities which will be studied in more detail in the next section. In particular it is important to identify experimental observables which can directly inform us about these peculiar properties.

2.4 Isobar ensemble: volume and energy properties

In the previous section we have concentrated the discussion on the enthalpy properties. However, if the pressure is not fixed, as it is most probably the case in the experimental situation, one should discuss energy and volume independently. The corresponding microcanonical statistical ensemble is defined through the partition function

$$Z_\lambda(E) = \sum_V W_V(E) \exp(-\lambda V) \tag{11}$$

with $W_V(E) = \exp(S(E,V))$ the density of states with an energy E and a volume V . This partition function of the isobar microcanonical ensemble can be computed in an exact way from a standard Metropolis sampling of the corresponding canonical ensemble as in the previous sections.

In the statistical ensemble with fluctuating volume introduced above the energy E and the Lagrange conjugate of the volume λ represent the two state variables of the system. It should be noticed that λ can still be formally related to a pressure via $P = T_\lambda \lambda$. They are associated to two equations of states, giving the microcanonical temperature $T_\lambda^{-1} \equiv \partial_E \log Z_\lambda$ and the average volume $< V >= -\partial_\lambda log Z_\lambda$ as a function of E and λ. The temperature can be evaluated from the energy derivative of the logarithm of the canonical energy distribution $N_{\beta,\lambda}(E)$

$$N_{\beta,\lambda}(E) = \frac{N_0}{Z_\lambda(\beta)} Z_\lambda(E) \exp(-\beta E) \tag{12}$$

where N_0 is the number of states canonically sampled with a given constraining λ, for any thermal Lagrange multiplier β, and the temperature is given by

$$T_\lambda^{-1}(E) \equiv \frac{\partial S_\lambda(E)}{\partial E} = \beta + \frac{\partial \log N_{\beta,\lambda}(E)}{\partial E} \tag{13}$$

Again since this equation is valid for every β, we can use many canonical samplings at different β to derive the same microcanonical caloric curve. The agreement between the different curves provides a strong test of the numerical sampling[21]. $< V >$ is calculated by averaging at a given λ over microcanonical events, which are obtained through an energy sorting of canonical partitions.

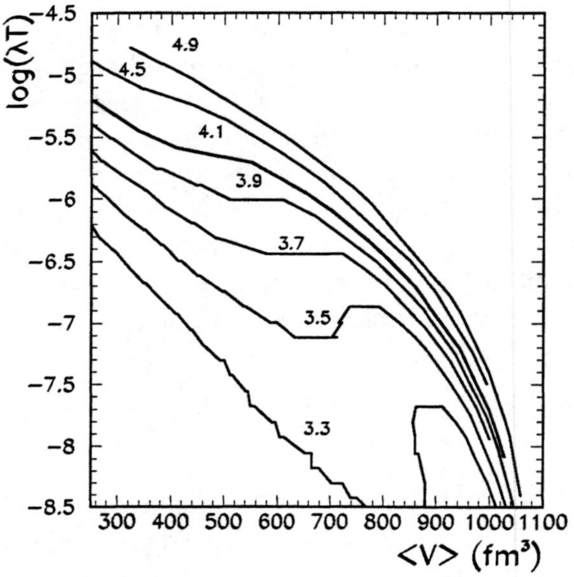

Figure 5: Correlation between pressure and volume for a system of 216 particles in the microcanonical ensamble with fluctuating volume. The microcanonical temperature is indicated on each isotherm. The thick line gives the critical isotherm.

The $(\lambda T, < V >)$ diagram of the system displayed in Figure 5 shows a striking similarity with the classical (P-V) equation of state of a Van der Waals gas. Far from the coexistence region the curves $P(< V >)$ at constant temperature are monotonous. However, when we get close to the coexistence region we observe an anomalous backbending in the EOS. This induces a negative branch in the compressibility which has already been observed in canonical lattice gas calculations at constant volume [19, 20]. The physical origin of this anomaly is completely different from the backbendings systematically found in the mean field approximation which simply reflect the instability of the homogenous system respect to phase separation. In our exact calculation which naturally includes inhomogeneous partitions conversely this feature corresponds to equilibrium under specific conservation laws (here mass number and energy) [19, 20]. In the microcanonical case the same anomaly is apparent in the equation of state providing the temperature T as a function of the state variables (λ, E) as shown in the two-dimensional caloric curves displayed in figure 6. The backbending of the temperature surface induces a negative branch in the microcanonical heat capacity at constant $\lambda : C_\lambda^{-1} = \partial T_\lambda (E) / \partial E$.

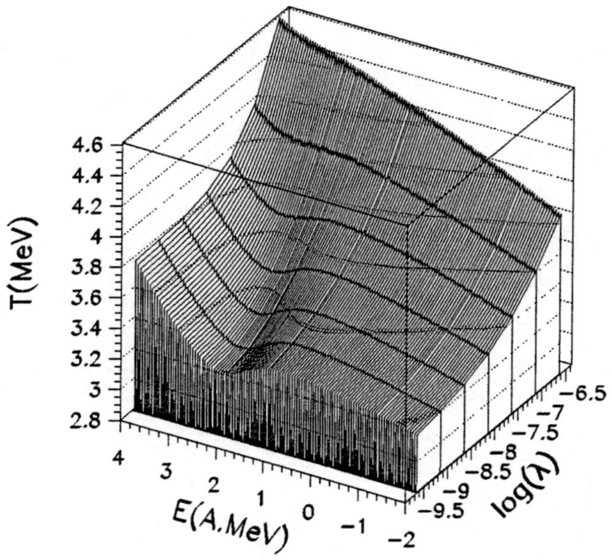

Figure 6: Temperature as a function of the energy per particle and the Lagrange parameter for a system of 216 particles in the microcanonical ensamble with fluctuating volume.

2.5 Discussion of the different shapes

The use of the Lagrange multipliers λ in the constrained calculation introduced above leads to a diffuse spherical density distribution shown in figure 7 for two different temperatures and average volumes[13]. In order to study the role of the surface conditions, we can compare the constrained calculation with a sharp boundary condition imposed on a spherical surface. The lines in figure 7 represent the respective density profiles for a boundary condition adjusted in such a way that the two calculations lead to identical mean volumes. In the upper part of figure 7 the volume of the sharp sphere is defined to be twice the mass considered so that at high temperature the value of the density corresponds to the critical density $\rho = \rho_c = \rho_0/2$. In the lower part of the figure the volume is doubled to consider a more realistic freeze out condition for nuclear fragmentation. In all figures r_0 represents the elementary mesh size of the lattice. At $T = 3$ MeV the densities are surprisingly close. This demonstrates the preeminence of the interaction on the boundary conditions. Indeed, the fluid forms a drop of liquid which develops a self consistent density profile independently of the surface conditions. At higher temperature the boundary conditions start to be important. At $T = 15$ MeV the system with an imposed sharp boundary finally becomes uniform.

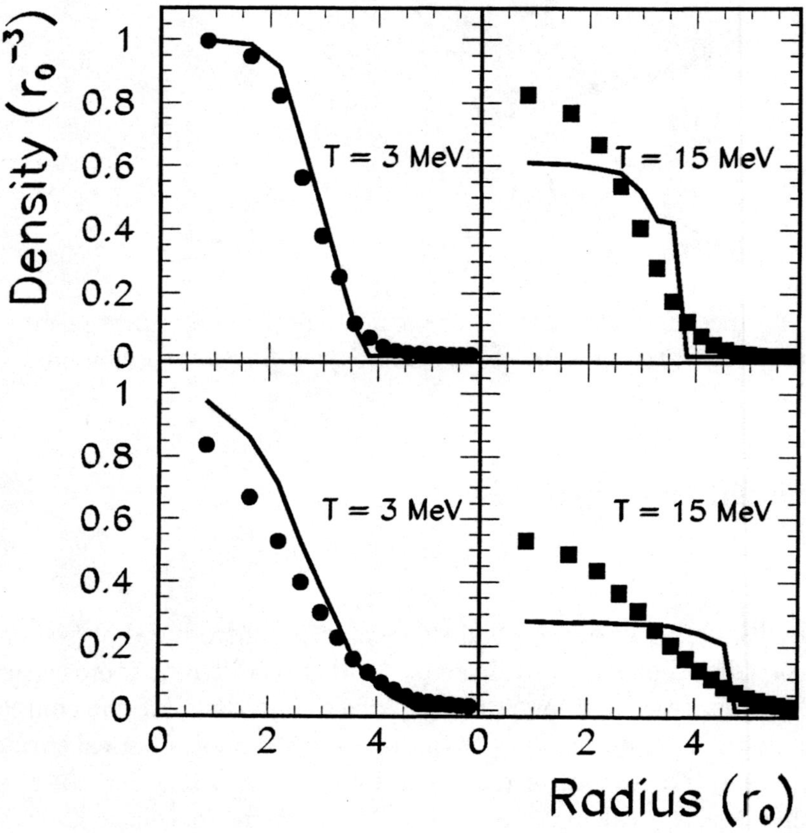

Figure 7: Density profiles for a system of 108 particles at two temperatures $T = 3$ MeV (left) and $T = 15$ MeV (right). Symbols: Metropolis sampling with a Lagrange pressure such that $< V >= 108$ (upper part) and $< V >= 216$ (lower part). Full lines: sharp boundary conditions in spherical symmetry for the same values of $T, < V >$ as the symbols.

3 Phase transition signals in the isobar microcanonical ensemble

It has been recently proposed that the microcanonical latent heat can be measured using partial energy fluctuations[10]. The total energy E of the considered system can be decomposed into two independent components, its kinetic and potential energy: $E = E_k + E_p$. In a microcanonical ensemble with a total energy E the total degeneracy factor $W(E) = \exp(S(E))$ is thus simply given by the folding product of the individual degeneracy factors $W_i(E_i) = \exp(S_i(E_i))$ of the two subsystems $i = k, p$. One can then define for the total system as well as for the two subsystems the microcanonical temperatures T_i and the associated heat capacities C_i. Since we are dealing with a classical gas the kinetic equation of state is simply given by $E_k = 3AT_k/2$ and the kinetic heat capacity is a constant, $C_k = 3A/2$. If we now look at the kinetic energy distribution when the total energy is E we get

$$P_k^E(E_k) = \exp(S_k(E_k) + S_p(E - E_k) - S(E)) \tag{14}$$

Using Eq.(14) we directly get that the most probable partitioning of the total energy E between the potential and kinetic components is characterized by a unique microcanonical temperature $\bar{T} \equiv T_k\left(\bar{E}_k^E\right) = T_p\left(E - \bar{E}_k^E\right)$. Therefore the most probable kinetic energy $\bar{E}_k^E$ can be used as a microcanonical thermometer. Using a Gaussian approximation for $P_k^E(E_k)$ the kinetic energy variance can be calculated as [31]

$$\sigma_k^2 = \bar{T}^2 \frac{C_k C_p}{C_k + C_p} \tag{15}$$

where C_k and C_p are the microcanonical heat capacities calculated for the most probable energy partition. Equation (15) can be inverted to extract from the observed fluctuations the heat capacity

$$C \simeq C_k + C_p = \frac{\bar{T}^2 C_k^2}{\bar{T}^2 C_k - \sigma_k^2} \tag{16}$$

When C_p diverges and then becomes negative, σ_k^2 remains positive but overcomes the canonical expectation $\sigma_k^2 = \bar{T}^2 C_k$. This anomalously large kinetic energy fluctuation is a signature of the first order phase transition. This means that kinetic energy fluctuations are an experimentally accessible measure of the heat capacity which indicates divergences and negative branches characteristic of the phase transition.

If we now turn back to the fact that the nuclear EOS is two dimensional, two conjugated pairs of thermodynamical variables are needed, one describing the energy part and the other one the volume. The normalized fluctuations σ_K^2/T^2 obtained in the microcanonical ensemble with a constrained volume are shown in the energy-λ plane in figure 8. From figures 6 and 8 it is apparent that the phase transition signal is visible in the temperature as well as in the fluctuation observable.

However the experimentally measured caloric curves are not bidimensional. Indeed, since experimentally the freeze out volume cannot be measured on an event by event basis, fragmentation data can be sorted in excitation energy bins but not in volume bins. This means that experiments are sampling a monodimensional curve on the equation of state

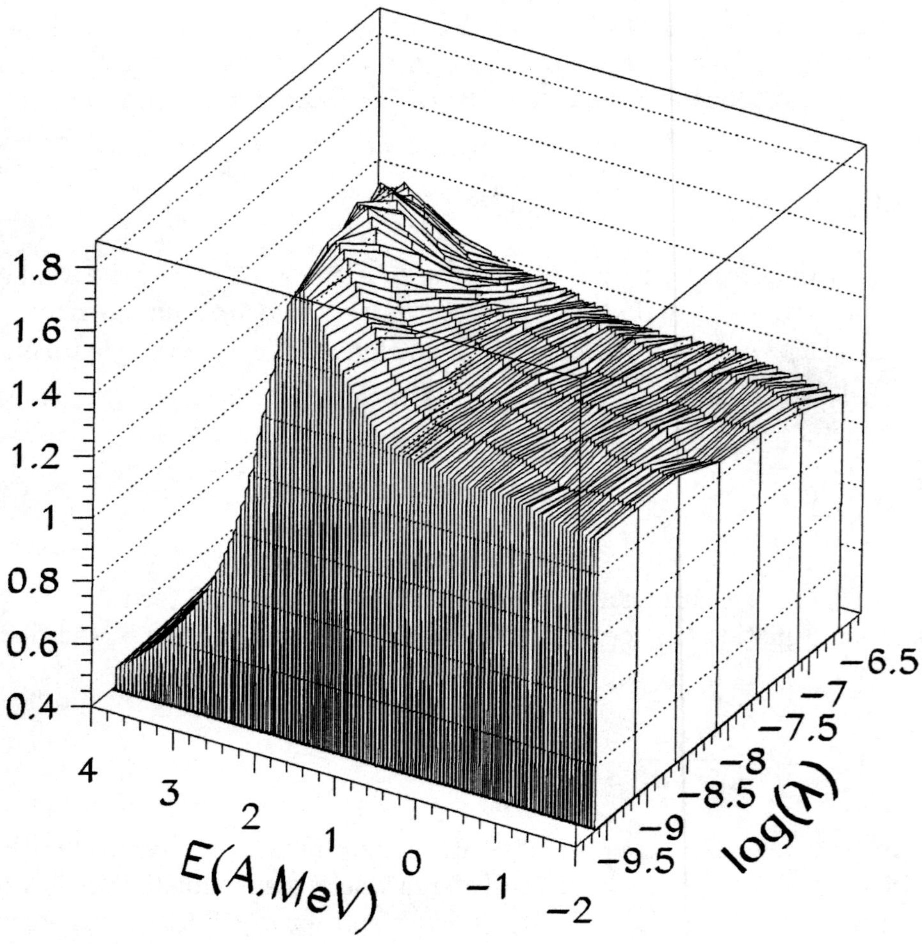

Figure 8: Normalized kinetic energy fluctuations as a function of the energy per particle and the Lagrange parameter for a system of 216 particles in the microcanonical ensamble with fluctuating volume.

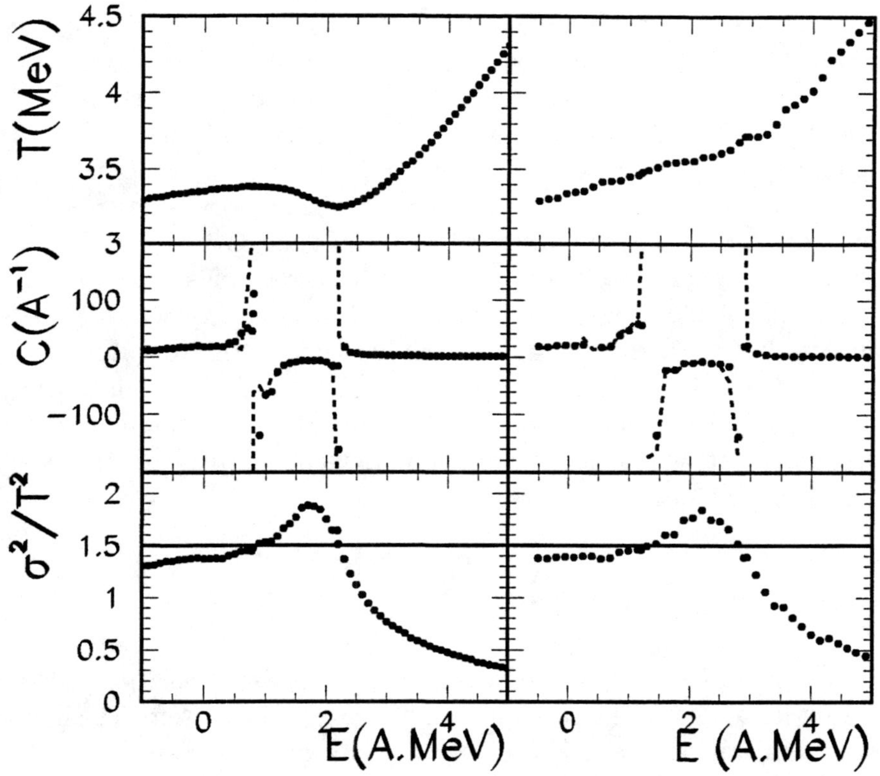

Figure 9: Thermodynamical quantities in the microcanonical ensemble for a transformation at constant pressure (left part) and at constant volume (right part). Upper panels: caloric curve. Lower panels: normalized kinetic energy fluctuations compared to the canonical expectation (lines). Medium panels: specific heat (symbols) compared to the estimation through eq.(16) (lines).

surface. The resulting caloric curve therefore depends on the actual transformation i.e. on the actual path followed in the thermodynamical parameters plane. As an example the behavior of the temperature as a function of energy at a constant pressure or a constant average volume in the subcritical region are displayed in the upper part of figure 9. At constant pressure the caloric curves are steeper than the ones at constant λ presented in figure 6 when the system is in the liquid or in the vapor phase; in the coexistence region the isobars are almost identical to the iso-λ's since P and λ differ only by the temperature which is almost constant in the phase transition region, and a backbending is clearly seen. On the other hand at constant volume a smooth behavior is observed with a slope change entering the gas phase, as expected from general thermodynamics (see also [6, 8]) . This is due to the fact that the λ parameter varies rapidly in the coexistence region (see fig.5). From these examples one clearly sees that the various transformations leads to very different caloric curves. More generally, it is clear that the backbending of the temperature surface can be avoided depending on the path of the considered transformation and the phase transition signal can be hidden in the observation of the caloric curve.

On the other side partial energy fluctuations are a state variable which does not depend on the transformation from one state to another and can directly give access to the equation of state. From figure 8 we can see that in the whole phase transition region the microcanonical fluctuations present a strong maximum which exceeds the canonical value: an anomalously large fluctuation signal will be always seen if the system undergoes a first order phase transition, independent on the path. As an example the lower part of figure 9 shows a constant p or $< V >$ cut of the bidimensional fluctuation surface shown in figure 8. The quantitative behavior of the specific heat as a function of energy will depend on the specific transformation, but at each point the specific heat extracted from fluctuations will be a direct measure of the underlying equation of state. This is clearly demonstrated in the medium part of figure 8 where the symbols which represent the specific heat C_λ extracted from the equation of state eq.(13) are in very good agreement with the lines which correspond to the fluctuation estimation eq.(15).

4 Conclusions about thermodynamical signals

To summarize the two preceding sections, when finite systems are discussed the statistical ensemble of events is essential to determine the equation of state of the system. In all cases a first order phase transition can be unambigously signed by a convexity anomaly of the relevant thermostatistical potential [15]. This anomaly can be accessed from the study of specific observables depending on the sorting variable which defines the statistical ensemble. In particular, a negative heat capacity is a well defined signal of a first order phase transition only if events are sorted in constant excitation energy bins. In the case of the liquid gas phase transition one is forced to introduce a second thermodynamical variable in order to specify the volume of the system. Then a monodimensional curve such as the measured caloric curves provides insufficient informations to measure the equation of state and can be misleading. On the other hand, kinetic energy fluctuations are independent of the actual trajectory followed in the parameters space and represent a direct measure of the EOS. Considering a statistical ensemble of states characterized by total energy and an average volume we have shown that abnormal fluctuations are a robust signal of a first order phase transition. In this case kinetic energy fluctuations in the phase transition region are indeed related to the isobar heat capacity C_P which is known to diverge[30].

The connection to experimental fragmentation data is straightforward. The microcanonical ensemble is relevant for the analysis of experimental data because of the absence of a heat bath and since using calorimetry techniques the excitation energy can be measured on an event-by-event basis. As far as the freeze-out volume is concerned, the absence of a constraining box implies that this variable can be known at best in average. Therefore, we expect that the partial energy fluctuations will present a strong anomaly if the multifragmenting nuclear system is undergoing a liquid gas phase transition[1].

5 Finite size scaling

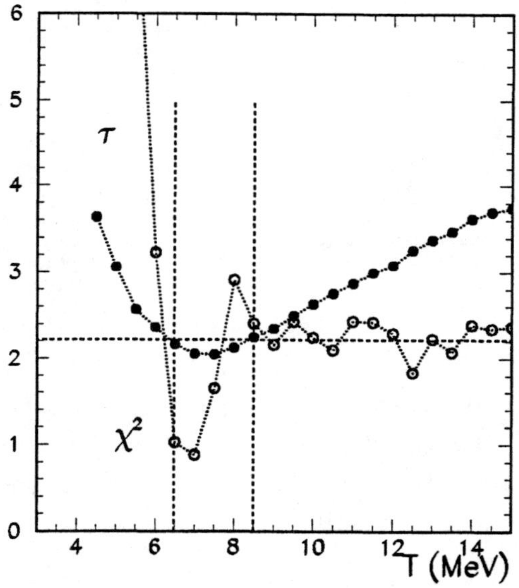

Figure 10: Slope (filled circles) and corresponding χ^2 (open circles) of the size distribution as a function of temperature for two different system sizes at the critical density. Dashed lines: τ_{max} extracted from the maximum production yields (see text) and corresponding critical temperature interval.

Following the well known example of critical opalescence for macroscopic fluids, we can expect that also in the nuclear case important density fluctuations will characterize the proximity of the critical point leading to a large variety of multifragment configuration patterns. Then the properties of fragment partitions should carry important informations on the presence of a phase transition.

The definition of clusters in the lattice gas combines a site and bond percolation algorithm: two particles are bonded if they

- a) occupy neighbor sites

- b) the kinetic energy of their relative motion $\vec{p}_{\tau}^{2}/2\mu$ does not exceed the binding energy ϵ_{τ_i,τ_j}.

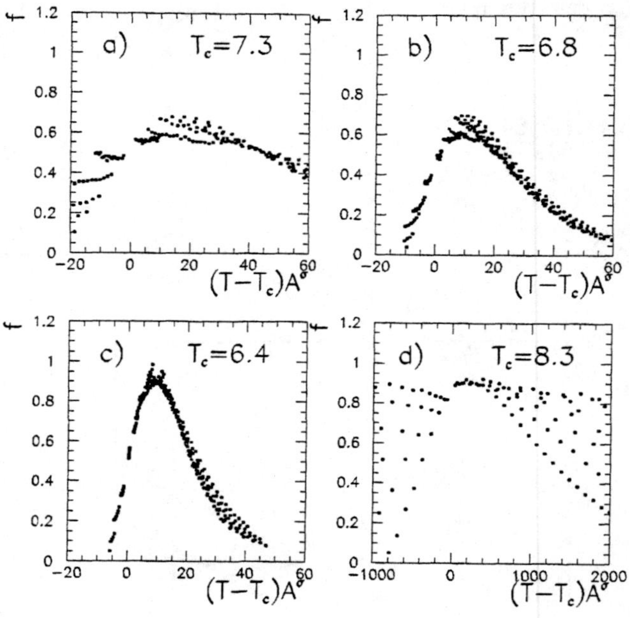

Figure 11: Scaling function for clusters of size ranging from 4<A<30 and temperature 2<T<20 obtained from eq.(17) with critical exponents extracted with methods 1 (a), 2 (b), and 3 (c,d) described in the text. In figure 3c (d) T_c is given by the lower (upper) bound of the critical temperature region of figure 10.

For details about the definition of clusters we report the reader to refs.[6, 32]. Remark that condition *b)* implies that the fragments produced are relatively cold, *i.e.* their thermal excitation energy is lower than the threshold of particle emission; their finite excitation energy is given by their shape rather than by internal degrees of freedom. This definition of clusters guarantees that the thermodynamical critical point exhibits critical fragment size distributions[11] and that in the proximity of the critical point the size distribution scales as [41, 14, 4]

$$\frac{dN}{dA}(A,T) = A^{-\tau} f(A^{\sigma}(T-T_c)) \tag{17}$$

where f is a universal scaling function and τ, σ are critical exponents.

It has already been observed that power laws in the size distribution are not characteristic of the critical point solely, but occur also at supercritical densities along the Kertész line [25, 6] and at some subcritical densities at lower temperatures [34]. In order to better quantify this statement and study the distortions due to finite size effects, we have checked the validity of eq.(17) for different densities and lattice sizes. Different techniques can be proposed to extract the value of the τ exponent. A first possibility[22] (method 1) is to plot the slope of $ln(dN)$ versus $ln(A)$ as a function of temperature; the minimum of this curve (τ_{min}) gives the exponent τ and the corresponding temperature is identified with T_c. Alternatively[32, 34] (method 2) one can use the fact that at $T = T_c$ the χ^2 of the power law fit should be minimum, and the corresponding slope τ_χ of the fit gives the value of τ. Another method (method 3) consists in plotting the maximum production yield $dN_{max}(A)$ of a species of size A as a function of A; one can see from eq.(17) that this quantity should behave as a power law of exponent τ_{max}. Then T_c can be obtained as the temperature at which the power law fit to the size distribution $dN/dA(A)$ gives an exponent $\tau = \tau_{max}$. Note that two solutions may fulfil this requirement.

5.1 Scaling in the isochore ensemble

These three methods lead to approximately the same value of τ, as it is shown in figure 10; however the corresponding T_c (and consequently the value of σ) are quite different. The values of τ and T_c found systematically lie inside the "critical region" (bounded by the dashed vertical lines of figure 10) obtained with the third method introduced above. This is not a smoothing effect due to the finite size of the system (a well defined critical point of an infinite system being replaced by a wider critical region in the finite one) but it is rather due to the intrinsic imprecision of the methods. As a matter of fact finite size scaling is clearly violated by the use of method 1 (figure 11) implying that this method cannot be used to define the critical point. By requiring that the scaling is verified, the critical temperature can be determined within 0.5 MeV even for a system of a typical nuclear size.

Finite size scaling holds with the same quality at different freeze out densities and lattice sizes. A critical behavior is clearly seen at supercritical densities as well as at subcritical densities. Figure 12 presents the evolution of the critical temperature as a function of the density for a lattice size $L = 8$. This line passes close to the thermodynamical critical point. At high density $\rho \approx 1$ this line reaches the bond percolation critical point. Below the critical point a critical behavior is also seen for a first order phase transition. In this sense there is no contradiction between the scenario of fragmentation at low density inside the coexistence or the spinodal region (first order phase transition) and the observation of critical signals characteristic of a second order phase transition. This property is specific of small systems because it disappears when the volume goes to infinity. Indeed, when the system is very large an infinite fragment is always present in the system for all temperatures up to the coexistence line.

If the critical temperature is extracted from data by verifying that finite size scaling is respected, critical parameters depend only very slightly on the mass for typical nuclear sources ranging from $A \approx 50$ to $A \approx 300$ and relevant thermodynamical informations can be extracted in spite of finite size effects (figure 13). It should be noticed that at low density in the (experimentally inaccessible) limit of very large systems, scaling becomes very poor. The extracted critical temperature gets close to the coexistence line. This means that in the coexistence region the apparent power laws are not connected to the existence of critical fluctuations of all sizes but rather to the disappearance of the infinite cluster on the coexistence line. Finite size scaling at low densities is therefore a specific property of small systems perfectly compatible with the presence of a first order phase transition. In the thermodynamical limit this scaling is violated as it can be seen by directly looking at the fragment size distribution and as it is shown by the fact that the critical exponents deviate from the expected value of their universality class[12] (figure 13a, 13b).

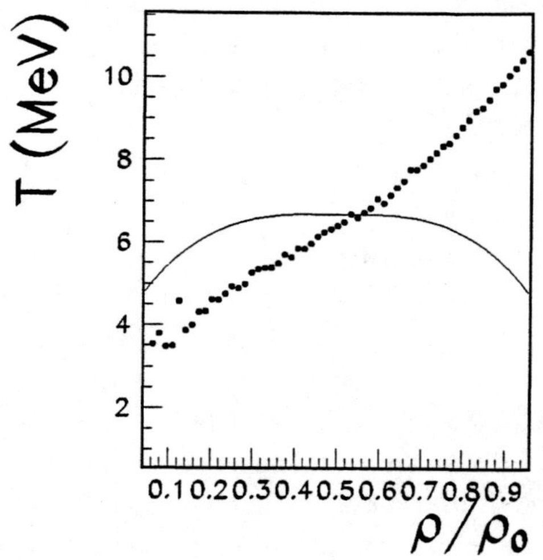

Figure 12: Full line: coexistence line from thermodynamics (see section 1); symbols: critical curve from fragment size distributions for a cubic lattice of linear dimension $L = 8$.

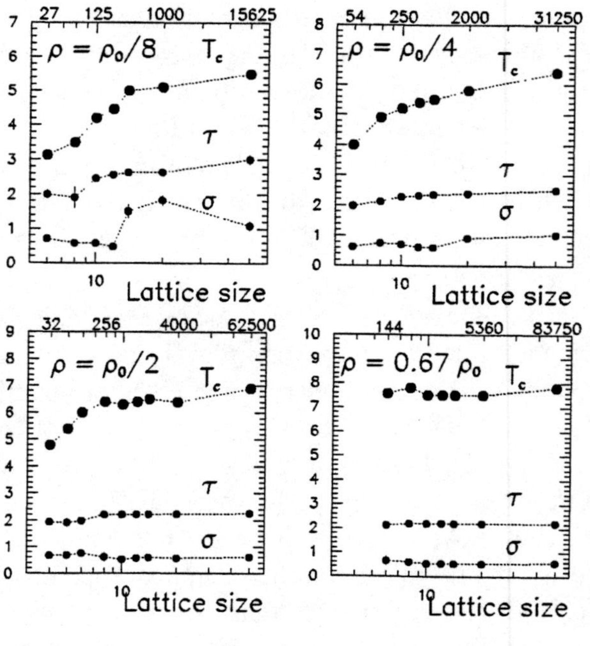

Figure 13: Critical parameters as a function of the linear size of the lattice (lower scale) or of the mass of the system (upper scale) at different freeze out densities.

5.2 Role of the volume definition

Besides the role of the finite size, one may wonder about the effects of the surface on the cluster partitions. In order to study the influence of the surface conditions on critical behaviors we have compared the following different systems with the same number of particles :

- a finite cubic system with periodic boundary conditions at half normal density corresponding to the critical density ;
- a finite cubic system without periodic boundary conditions with the same volume as in (1) ;
- a finite sphere with the same volume as in (1) and sharp edges ;
- a finite sphere with a diffuse surface obtained using a constraint on the average volume, the volume being fixed to be the same as the average one of case (3).

The universality of the critical behavior is tested for the diffuse sphere calculation in figure 14. One can see that all the different data points collapse on a single curve demonstrating the validity of the universal scaling. As already observed above, using the requirement that around the critical point fragment production should exhibit a finite size scaling allows to define unambiguously the critical temperature within less than 0.5 MeV even for a finite system of a typical nuclear size.

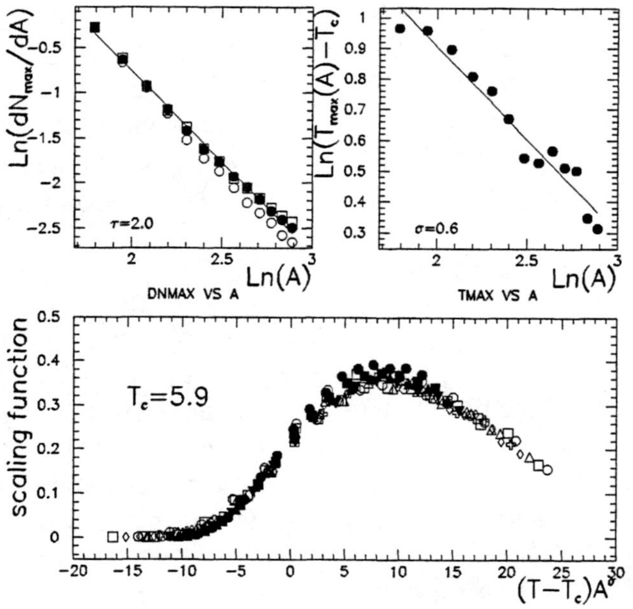

Figure 14: Scaling function f (see text) extracted from the finite size scaling analysis for a spherical system with $A_0 = 108$ with a diffuse surface for temperatures $3 \leq T \leq 10$ MeV and fragment masses $4 \leq A \leq 20$.

The quality of the critical exponents fit is comparable for all the surface conditions examined and the resulting critical parameters are presented in table 1. The slight variation of the critical temperature can be related to the variations of the density profiles inside the system. If the density of the system is varied between $\rho_0/10$ and ρ_0, the critical exponents remain almost unchanged as well as the agreement between the different surface conditions; only the critical temperature extracted from scaling analysis is seen to decrease monotonically with increasing volume as in the case of periodic boundary conditions (see figure 12).

System	τ	T_c(MeV)	σ
Cubic periodic	2.0 ± 0.1	6.5 ± 0.4	0.8 ± 0.2
Cubic non periodic	2.0 ± 0.1	6.0 ± 0.4	0.7 ± 0.2
Sharp sphere	1.9 ± 0.1	6.2 ± 0.4	0.7 ± 0.2
Diffuse sphere	2.1 ± 0.1	6.1 ± 0.4	0.6 ± 0.2

Table 1: Critical parameters for the different surface conditions of a system of 108 particles at the critical density

In other words, it is apparent that the change of the surface shape only slightly affects the detailed structure of the scaling function and does not change qualitatively the critical behavior. Power laws in the fragment size distributions are also found in the microcanonical ensemble, at the energy corresponding to the peak of the kinetic energy fluctuations (i.e. at approximately the middle of the negative branch of the heat capacity). Due to the different sorting of events in the microcanonical ensemble however the value of τ is slightly deformed respect to the expected critical exponent of the liquid-gas universality class [21].

For a realistic application to fragmentation data however other physical ingredients have to be added to the model. In particular the persistence of signals must be verified respect to the long range Coulomb interaction. The fact that minor violations to the scaling in the size of the largest cluster are observed in the macroscopic microcanonical statistical model MMMC[23] which includes the Coulomb field is in this sense very encouraging.

5.3 Protons and neutrons partitions

The results presented in the preceding section have been obtained by assuming an isospin symmetric system, $y = (N - Z)/A = 0$ and considering size distributions. Of course the measurable quantity is the charge rather than the mass distribution. Figure 15 shows that mass, neutron number and proton number distributions exhibit a power law behavior at the same critical temperature. We can also study the proton or neutron distribution for isospin symmetric and asymmetric systems. For an $y \neq 0$ system in the presence of a symmetry energy, *i.e.* a different coupling for isoscalar and isovector pairs, the nature itself of the phase transition is modified [28]. Just to mention a few well known features from macroscopic thermodynamics with two conserved charges, the coexistence curve is transformed into a surface which implies that no discontinuity is observed passing through the phase transition, i.e. as far as transformations are concerned the transition is second order instead of first order; phase separation is triggered by isovector as well as isoscalar instabilities; the isotopic composition of the liquid and gas phase differ. Figure 15 shows that critical behaviors are observed for all the different isotopic composition of the source.

The asymmetry dependence of the critical parameters at the critical density is presented in figure 16 for a system of mass $A = 256$. The critical temperature has been obtained with

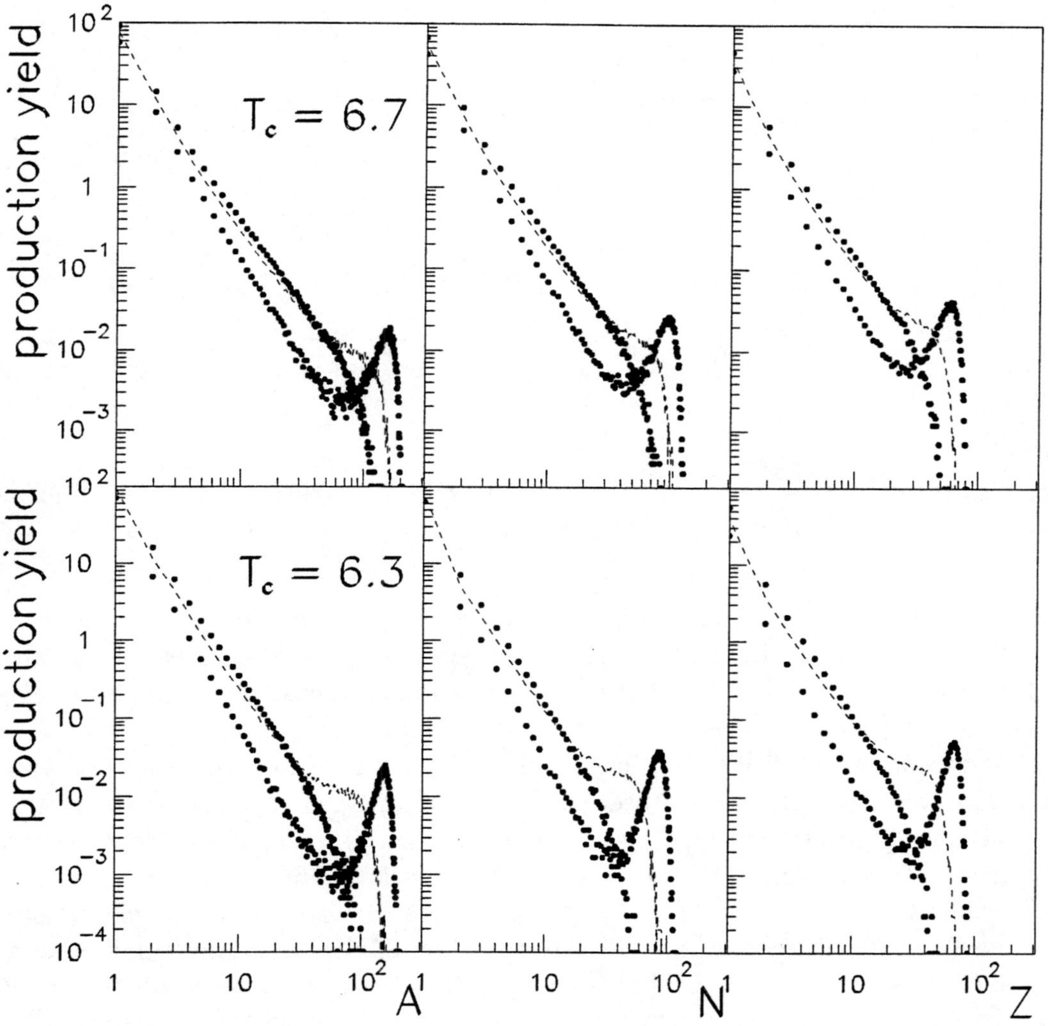

Figure 15: Fragment mass, neutron number and charge distributions around the critical temperature at half the saturation density for a 8x8x8 cubic lattice with an isospin asymmetry $y = 0$ (upper part) and $y = 0.25$ (lower part).

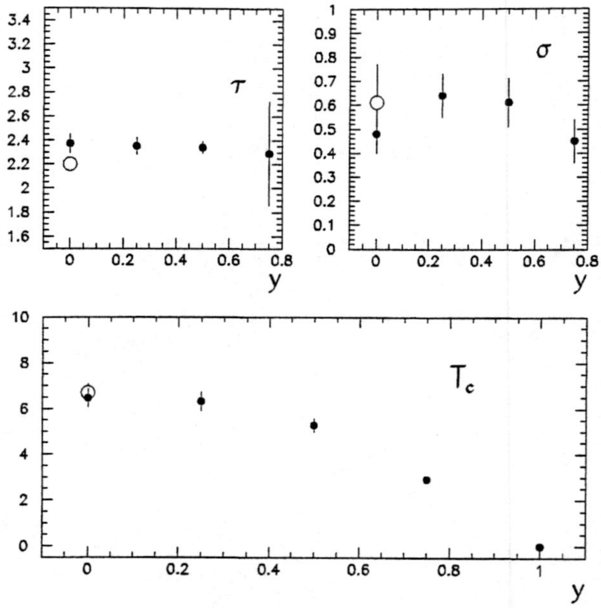

Figure 16: Critical parameters as a function of the isospin asymmetry of the system for a 8x8x8 cubic lattice at critical density. Open symbols: isospin independent interaction.

the method 3 of section 4. For comparison, the values obtained in a system of the same size with isospin independent interactions $\epsilon_s = \epsilon_d = -5.5 MeV$ are also given. The same results within error bars are obtained if the parameters are extracted using instead of the total mass of the fragment the total charge or neutron number.

As expected from universality the critical exponents do not depend on the isospin asymmetry. The observed deviation of τ respect to the isospin independent case is due to the different influence of finite size effects. This can be understood since in an isospin dependent case the fragments lie close to the bottom of the stability valley $N \approx Z \approx A/2$ while in the isospin independent case no restrictions are expected on the partitioning of A nucleons into Z protons and N neutrons. For $A/A_0 \ll 1$ this reduction of the partitions number is an increasing function of A which means that from a combinatorial point of view it is harder to make a big cluster in the isospin dependent system even if the system is isospin symmetric. The difference vanishes as $A \to \infty$ but for the finite system IMF size distributions are naturally steeper, *i.e.* τ is larger than the expected value of the Ising universality class. This result is qualitatively in agreement with the findings of ref.[33].

The critical temperature shows an approximate quadratic dependence on asymmetry (lower part of figure 16) similar to the mean field calculations of refs. [28, 36] $T_c(y) - T_c(0) \propto y^{\xi}$ with $\xi \approx 2$. It should be noticed that since y does not correspond to an order parameter in the considered case the coefficient ξ should not be considered as a critical exponent. Considering the limited variation of y in experiments, even for very exotic nuclei and extremely precise measurements the asymmetry dependence of the critical temperature would be barely visible. On the other hand, the results shown in figure 16 imply that fragmentation data obtained from systems with different asymmetries can be analysed together for a better determination of critical parameters, isospin observables being used to pin down the phase transition and measure thermodynamical quantities.

6 Distillation of isospin asymmetric systems

Isospin can be not only viewed as a degree of freedom but also as an observable: we can hope that next generation detectors for multifragmentation, measuring masses as well as charges of the reaction products, will supply us with a richer two dimensional information reducing the ambiguities of the existing signals for the phase transition. One intriguing possibility is the expectation that for an $y \neq 0$ system inside the coexistence region the vapor phase should be more asymmetric than the liquid phase [28].

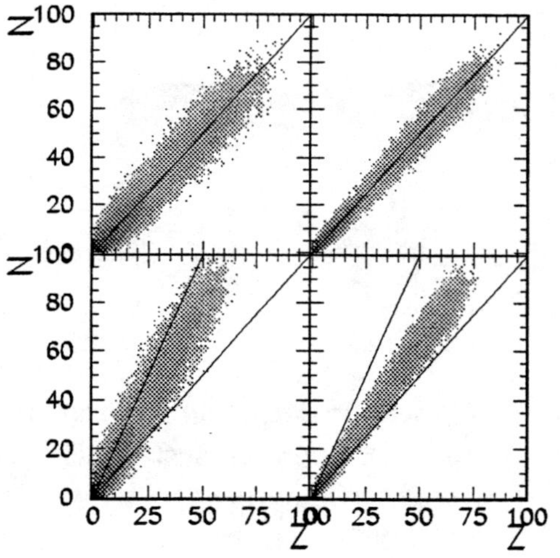

Figure 17: Distribution of fragments around the critical temperature as a function of their neutron and proton number. Top (bottom) part: $y = 0$ ($y = 0.25$). Left (right) part: $\epsilon_d = \epsilon_s = -5.5$ MeV ($\epsilon_d = 5.5$ MeV, $\epsilon_s = 0$ MeV). Lines: stability $N = Z$ line and isotopic ratio of the initial source.

Figure 17 presents the isobaric distribution of fragments in the proximity of the critical temperature as a function of their neutron and proton number. For an isospin symmetric source the distribution is centered around the $N = Z$ line. When the interaction does not depend on the isospin the observed isobaric distribution is only due to a combinatorial effect. For an isospin dependent interaction the distribution is narrower because the different isobars do not have the same binding energy. In fact the bottom of the stability valley in our simple model which neglects Coulomb effects lies on the $N = Z$ line. Therefore, in addition to the simple combinatorial coefficient the statistical weight due to the variation of the binding energy along an isobaric line leads to a narrowing of the distribution. In the case of an asymmetric source the fragments appear to have the same N/Z ratio as the initial source when the interaction does not distinguish between protons and neutrons. However, when the isospin dependent interaction is switched on we observe that the heavy fragments get closer to the bottom of the stability valley while the light ones present a stronger asymmetry.

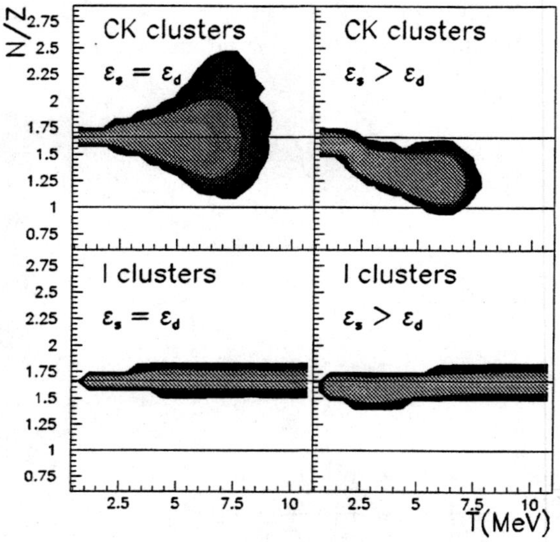

Figure 18: Ratio between neutron N and proton Z number of produced clusters with $N > 40$ as a function of temperature for a fragmenting system with $N_0 = 160$ and $Z_0 = 96$ at $\rho = \rho_0/2$. Upper (lower) part: Coniglio-Klein (Ising) clusters. Left part: interaction energy $\epsilon_s = \epsilon_d = -5.5$ MeV. Right part: interaction energy $\epsilon_s = -5.5$, $\epsilon_d = 0.$ MeV.

This feature is better visible in figure 18 which displays the isotopic composition of heavy fragments (*i.e.* of the liquid fraction) as a function of temperature for a system with $y = 0.25$. If the interaction energy is assumed independent of isospin, at all temperatures the distribution is centered on the N/Z of the source. This effect is again purely combinatorial and the broadening of the distribution approaching the critical temperature reflects the maximum of fluctuations at the critical point. If symmetry energy is taken into account, the distribution is centered on a value the closer to stability ($N/Z = 1$ in the model, independent on the mass) the higher is the temperature.

The algorithm for cluster definition in the lattice gas model explained in section 5 guarantees that the excitation energy per nucleon is lower than the threshold of particle emission[6, 32], *i.e.* the fragments produced are relatively cold and secondary evaporation is (in average) taken into account. One can then wonder if the observed proton richness (relative to the isotopic composition of the source) of the fragments does not trivially reflect secondary decay. If we construct clusters simply by connecting neighbour occupied sites (Ising clusters), then the average isotopic distribution is given by the lower part of figure 18, which is centered on the combinatorial expectation except at very low temperature.

Therefore, even if the cluster proton richness in the model is a genuine signal of phase coexistence one can argue that any model including secondary decay should exhibit the same feature and it may be difficult to disentangle the two effects.

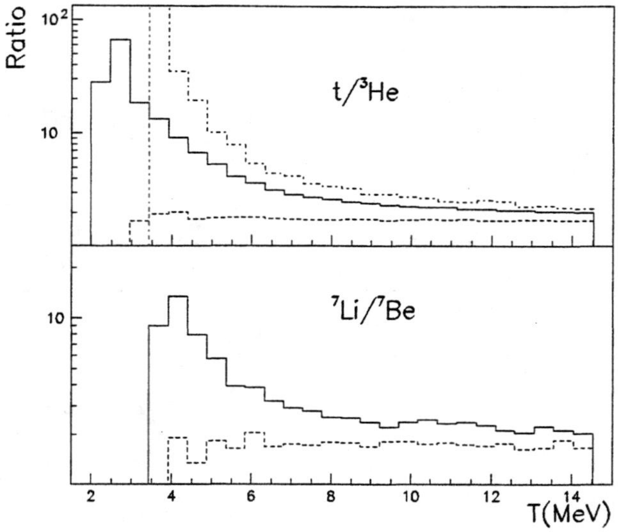

Figure 19: Relative production yield of light isobars as a function of temperature for a fragmenting system with $N_0 = 160$ and $Z_0 = 96$ at $\rho = \rho_0/2$. Full (dashed-dotted) lines: interaction energy $\epsilon_s = -5.5$, $\epsilon_d = 0.$ MeV; Coniglio-Klein (Ising) clusters. Dashed lines: interaction energy $\epsilon_s = \epsilon_d = -5.5$ MeV; Coniglio-Klein clusters.

A clearer signal of phase coexistence with two conserved charges can be obtained from the analysis of the vapor isotopic composition. Figure 19 shows the ratio between the production yields of different light isobars as a function of temperature with and without isospin dependent interactions. Only at supercritical temperatures the combinatorial value $R = N/Z = 2.1$ is achieved, and the trend is opposite respect to the expected effect of secondary decay. Figure 20 shows the isobaric ratio for different light masses as a function of the asymmetry of the fragmenting system. Above the critical point one can see that this ratio tends toward the isotopic ratio of the source while below the critical point the initial isotopic ratio is strongly enhanced. This very strong enrichment can be related to the existence of two conservation laws, one acting on the proton and the other one on the neutron number. In the presence of a large fragment, i.e. below the critical point, very few protons are free to be captured by the light particles. Indeed, a heavy fragment as close as possible to the bottom of the stability valley is strongly favored from the energetic point of view. Then, as shown by figure 18 above starting with a neutron rich source we get an almost equilibrated large cluster. Only a few protons are left to be shared between the light fragments leading to the observed strong enhancement of neutron rich "gas"-particles. This is the distillation of neutron matter. It should be noticed that this is a clear signature of the thermodynamical nature of the phase transition.

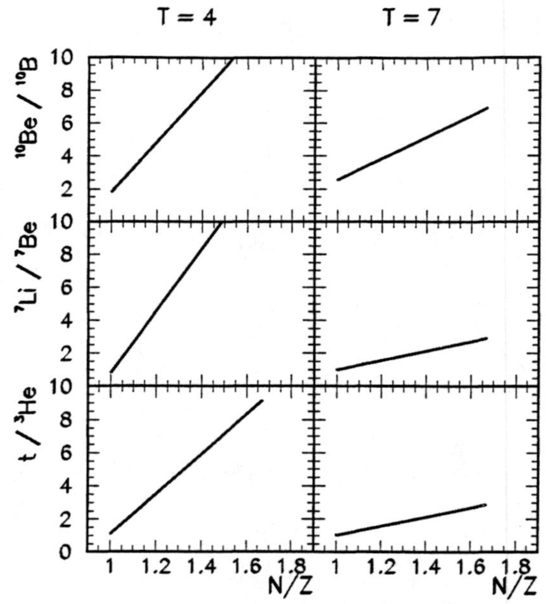

Figure 20: Relative production yield of light isobars as a function of the N/Z of the fragmenting source and for two different temperatures.

7 Conclusions

In this chapter we have looked for relevant signals of the nuclear liquid-gas phase transition within an isospin dependent Lattice Gas model. An exact calculation of the microcanonical and canonical partition function has allowed us to compute the critical point and coexistence line exactly in finite systems. We have also shown that the study of energy partition fluctuations can be a source of information and can provide a measure of the heat capacity. In particular abnormal fluctuation are related to negative heat capacities which are a clear signature of first order phase transition. The critical temperature is also accessible from the fragment production yield, as well as the universal critical exponents of the transition. A full agreement between thermodynamical and fragment variables is obtained if the parameters are extracted from a systematic analysis of size distributions at different temperatures. In this case finite size scaling is remarkably verified over a wide range of sizes and temperatures. The presence at the same time of first and second order transition signals can be understood within a detailed analysis of phase transitions in finite systems. In particular we have demonstrated that a critical behavior in fragment observables can be consistent with the thermodynamics of phase coexistence and the occurrence of a low freeze out density due to finite size effects. Indeed, at variance with the supercritical regime, critical behavior at subcritical densities is a side effect of the finite size of the system which disappears in the infinite limit. However, the dependence with mass being smooth, the observation of critical signals from multifragmenting systems at low densities can provide useful informations on the characteristics and the parameters of the transition. The fact that the critical behavior is not confined to a single thermodynamical point, but can be seen along a whole critical line implies that the temperature and density of the multifragmenting source have to be inferred at the same time from experimental data. Then an interesting observable to better localize the system in the $T - \rho$ plane is given by the isotopic composition of the fragments. We have shown that in the coexistence region the vapor fraction is more asymmetric than the liquid fraction and the signal is not washed out by secondary decay. We have also shown that critical parameters are presenting only a weak dependence on nuclear asymmetry demonstrating that meaningful informations can be extracted from the multifragmentation study of different N/Z sources. Moreover both charge and mass distributions can be used to extract critical behaviors and to sign the distillation phenomenon of asymmetric nuclear matter.

Even if qualitatively the neutron distillation signal has already been experimentally observed [43], for a realistic application to fragmentation data the persistence of signals must be verified respect to the Coulomb interaction.

References

[1] M. D'Agostino et al., nucl-ex/9906004 and to be published in Phys.Lett.B.

[2] G. Bizard et al., Phys. Lett. B 302 (1993) 162.

[3] J.P.Bondorf, A.S.Botvina, I.N.Mishustin, Phys. Rev. C, submitted.

[4] L. Cardy, Current physics - Sources and Comments : vol. 2 : Finite size scaling (North-Holland, Elsevier Science Pub. 1988).

[5] X. Campi and H. Krivine, Nucl. Phys. A 620 (1997) 46; J. Pan and S. Das Gupta, Phys.Rev.Lett. 80 (1998)1182.

[6] X.Campi and H.Krivine, Nucl. Phys. A 620 (1997) 46 and Physica A, in press.

[7] J.M.Carmona, J.Richert, A.Tarancon, Nucl. Phys. A 643 (1998) 115.

[8] J.M.Carmona, N.Michel, J.Richert, P.Wagner, nucl-th/9910061.

[9] J.M.Carmona, J.Richert, A.Tarancon, Nucl. Phys. A 643 (1998) 115; F.Gulminelli and P.Chomaz, Phys. Rev. Lett. 82 (1999)1402; Phys. Lett. B 447 (1999) 221.

[10] P.Chomaz and F.Gulminelli, Nucl. Phys. A647, 153 (1999).

[11] A. Coniglio and W. Klein, J. of Phys. A13 (1980) 2775.

[12] B. Diu et al., Physique Statistique, Hermann (1989) Paris.

[13] V.Duflot, Ph.Chomaz and F.Gulminelli, preprint GANIL P 99-29 and Phys.Lett.B, in press.

[14] M.E. Fisher, Physics vol. 3-5 (1967) 255; M.E. Fisher and M.N. Barber, Phys. Rev. Lett. 28 (1972) 1516.

[15] D. H. E. Gross and P.A.Hervieux, Z.Phys. D35 (1995) 27; D.H.E. Gross and E.Votyakov, cond-mat 9911257 .

[16] D.H.E. Gross, Phys. Rep. 279(1997) 119.

[17] F. Gulminelli and D. Durand, Nucl.Phys. A 615 (1997) 117.

[18] F.Gulminelli and Ph.Chomaz, invited paper to the XXXVI Int.Winter Meeting on Nucl.Phys., Bormio, I.Iori ed., (1998) and invited paper to the XIV Winter Wokshop on Nucl.Dynamics, Snowbird, W.Bauer ed., (1998).

[19] F.Gulminelli and Ph.Chomaz, Phys. Rev. Lett. 82 (1999) 1402.

[20] F.Gulminelli and Ph.Chomaz, Int. Journ. Mod. Phys. E, vol.8,n.6 (1999)1.

[21] F.Gulminelli, P.Chomaz and V.Duflot, preprint LPCC 99-17 and to be published in Europhys.Lett.

[22] J.A. Hauger et al., Phys.Rev.Lett. 77 (1996) 235; J.B. Elliott et al., Phys. Lett. B 381 (1996) 35; Phys.Lett. B 418 (1998) 34.

[23] H.R.Jaqaman and D. H. E. Gross, Nucl.Phys A 524 (1991) 321.

[24] W. Thirring, Zeit. Phys. 235 (1970) 339.

[25] J.Kertész, Physica A 161 (1989) 58.

[26] N. Marie et al., Phys.Lett. B 391 (1997) 15, and references therein.

[27] N. Metropolis et al., J. Chem. Phys. 21 (1953) 1087.

[28] H.Mueller and B.D. Serot, Phys.Rev. C 52 (1995) 2072.

[29] W.F. Mueller, Phys.Rev. C 56 (1997) 2873.

[30] D.Lynden-Bell, Proc. of the XXth IUPAP Int.Conf on Stat. Phys., Paris, July 20-24 (1998) and cond-mat/9812172.

[31] J.L. Lebowitz et al., Phys. Rev. 153 (1967) 250.

[32] J.Pan and S. Das Gupta, Phys.Lett. B 344 (1995) 29; Phys.Rev. C 51 (1995) 1384; Phys.Rev. C 53 (1996) 1319; Nucl.Phys. A 621 (1997) 897.

[33] J.Pan and S. Das Gupta, Phys.Rev. C 57 (1998) 1839.

[34] J.Pan and S. Das Gupta, Phys.Rev.Lett. 80 (1998)1182.

[35] J. Pochodzalla et al., Phys.Rev.Lett. 75 (1995) 1040; Y. G. Ma et al, Phys.Lett. B 390 (1997) 41; M.D'Agostino et al., Nucl. Phys. A650 (1999) 329.

[36] S.Ray, J.Shamanna, T.T.S. Kuo, Phys.Lett. B 392 (1997) 7.

[37] M. F. Rivet et al., Phys. Lett. B 388 (1996) 219; B. Borderie et al.,Phys. Lett. B 388 (1996) 224; Eur. Phys. Jour. A6 (1999) 197.

[38] A. Schuttauf et al., Nucl.Phys. A 607 (1996) 457.

[39] J.Sivardière and J.Lajzerowicz, Phys.Rev. A 11 (1975) 2090.

[40] P.J. Siemens, Nature 305 (1983) 410; M.W. Curtin et al., Phys. Lett. B 123 (1983) 289.

[41] D.Stauffer and A.Aharony, Introduction to percolation theory (Taylor and Francis, London 1992).

[42] C.N. Yang and T.D. Lee, Phys.Rev. 87 (1952) 410.

[43] S.J.Yennello et al., Phys. Lett. B 321 (1994) 15; Phys. Lett. B 371 (1996) 186.

Chemical Versus Mechanical Instability and Isospin Fractionation in Asymmetric Nuclear Matter

Bao-An Li * C. M. Ko †

Abstract

Based on phenomenological nuclear equations of state, we discuss the chemical and mechanical instabilities in a hot isospin-asymmetric nuclear matter. Although both instabilities depend strongly on the isospin asymmetry of the nuclear matter, the boundary of chemical instability in the pressure-density space is more extended than that of mechanical instability. Locations of these boundaries further depend sensitively on both the bulk compressibility of nuclear matter and the density dependence of the symmetry energy. For an asymmetric nuclear matter that is chemically and mechanically stable, a lower neutron excess exists in the liquid phase than in the gas phase. The relation of such isospin fractionation in the nuclear liquid-gas phase transition and the enhanced production of neutron-rich light particles observed in heavy ion collisions is also discussed.

1 Introduction

During the last decade, there have been extensive studies on multifragmentation in nuclear reactions, which is characterized by the emission of intermediate-mass fragments that follow a power law distribution; see, e.g., the contributions by Pochodzalla, Sobotka, Schröder, and Tsang *et al.* in this book. It has been suggested that this phenomenon is due to various mechanical instabilities of the nuclear matter, such as the volume instability, the surface instability of the Rayleigh kind, and the Coulomb instability [1]. At currently available radioactive ion beam facilities and the future Rare Isotope Accelerator (RIA), nuclear matter with a large isospin asymmetry can be created transiently during reactions induced by neutron-rich nuclei. This offers the possibility to study also the chemical instability associated with the isospin asymmetry of nuclear matter. Indeed, recent experiments using projectile-target combinations of different neutron to proton ratios have shown that products from nuclear multifragmentation depends strongly on the isospin asymmetry of the colliding nuclei [2, 3, 4, 5].

Nuclear matter is not thermodynamically stable at all densities, ρ, temperatures, T, and neutron excesses,

$$\delta \equiv (\rho_n - \rho_p)/(\rho_n + \rho_p), \tag{1}$$

*Department of Chemistry and Physics, Arkansas State University P.O. Box 419 State University, AR 72467-0419, USA

†Cyclotron Institute and Department of Physics, Texas A&M University College Station, TX 77843-3366, USA

where ρ_n and ρ_p are the neutron and proton densities, respectively. The necessary and sufficient conditions for the stability of an isospin-asymmetric nuclear matter are given by the following inequalities [6, 7, 8]:

$$\left(\frac{\partial E}{\partial T}\right)_{\rho,\delta} > 0, \tag{2}$$

$$\left(\frac{\partial P}{\partial \rho}\right)_{T,\delta} \geq 0, \tag{3}$$

$$\left(\frac{\partial \mu_n}{\partial \delta}\right)_{P,T} \geq 0, \tag{4}$$

where E, P and μ_n are the energy per nucleon, pressure, and neutron chemical potential, respectively. The first condition ensures the thermodynamic stability and is satisfied by any reasonable equation of state; the second condition prevents the mechanical instability against the growth of density fluctuations; and the last one protects the diffusive stability against the development of neutronization. The last two conditions might be violated in certain regions of the (ρ, T, δ) configuration space.

To understand the respective roles of chemical and mechanical instabilities in the dynamic properties of nuclear matter, it is important to study their relative locations in the (ρ, T, δ) configuration space. Since the adiabatic spinodal is always enclosed by the isothermal spinodal (ITS), it is sufficient to compare the diffusive spinodal (DS) and ITS. Obviously, boundaries of these instabilities depend on the equation of state. Using the Relativistic Mean Field (RMF) theory, Müller and Serot (see their contribution in this volume) have found that the diffusive spinodal encloses more of the configuration space than the isothermal one. For currently reachable regions of isospin asymmetry in nuclear reactions induced by neutron-rich nuclei, we expect that an appreciable separation between the isothermal and diffusive spinodals in the pressure-density configuration space.

As discussed in detail in previous chapters (see, e.g., the contribution by Lombardo and Zuo in this volume), the equation of state of asymmetric nuclear matter varies appreciably among theoretical models. In particular, the RMF theory predicts a characteristic linear density dependence of the symmetry energy. It is therefore of interest to examine how the conclusions based on the RMF theory might change when other equations of state for asymmetric nuclear matter are considered [9, 10, 11]. In this chapter, we first present a phenomenological model for isospin asymmetric nuclear matter and then investigate the dependence of the chemical and mechanical instabilities on the isospin-dependent nuclear equation of state.

2 A phenomenological model of hot asymmetric nuclear matter

For a gas of neutrons and protons at temperature T, the density ρ_q of neutrons $(q = n)$ or protons $(q = p)$ is

$$\rho_q = \frac{1}{\pi^2}\int_0^\infty k^2 f_q(k)dk, \tag{5}$$

where

$$f_q(k) = [exp(e_q - \mu_q)/T + 1]^{-1} \tag{6}$$

is the nucleon Fermi distribution function, and e_q is the single particle kinetic energy. At modestly high temperatures ($T \geq 4$ MeV), this equation can be inverted to give the chemical potential [12, 13, 14]

$$\mu_q = V_q + T\left[\ln(\frac{\lambda_T^3\rho_q}{2}) + \sum_{n=1}^{\infty}\frac{n+1}{n}b_n(\frac{\lambda_T^3\rho_q}{2})^n\right], \tag{7}$$

where

$$\lambda_T = \left(\frac{2\pi\hbar^2}{mT}\right)^{1/2} \tag{8}$$

is the thermal wavelength of nucleon, and b_n's are the virial coefficients for an ideal Fermi gas. The first five coefficients are [13]

$$\begin{aligned} b_1 &= \frac{\sqrt{2}}{8} \approx 0.177, \\ b_2 &= \frac{1}{8} - \frac{2\sqrt{3}}{27} \approx -3.30 \times 10^{-3}, \\ b_3 &= \frac{3}{32} + \frac{5\sqrt{2}}{64} - \frac{\sqrt{6}}{12} \approx 1.11 \times 10^{-4}, \\ b_4 &= \frac{317}{1728} + \frac{\sqrt{2}}{8} - \frac{\sqrt{3}}{6} - \frac{4\sqrt{5}}{125} \approx -3.54 \times 10^{-6}, \\ b_5 &= \frac{35}{128} + \frac{1687\sqrt{2}}{6912} - \frac{5\sqrt{3}}{72} - \frac{5\sqrt{6}}{36} - \frac{\sqrt{10}}{20} \approx 8.30 \times 10^{-8}. \end{aligned} \tag{9}$$

In Eq. (7), V_q is the single particle potential energy

$$V_q(\rho, \delta) = V_0(\rho) + V_{\mathbf{asy}}^q(\rho, \delta), \tag{10}$$

where V_0 and $V_{\mathbf{asy}}$ are the isospin-independent and dependent parts, respectively. With a Skyrme-type effective interaction, the first term can be parameterized as

$$V_0 = a(\rho/\rho_0) + b(\rho/\rho_0)^\sigma. \tag{11}$$

The parameters a, b, and σ are determined by the saturation properties and compressibility K of symmetric nuclear matter, i.e.,

$$a = -29.81 - 46.90\frac{K + 44.73}{K - 166.32} \text{ (MeV)}, \tag{12}$$

$$b = 23.45\frac{K + 255.78}{K - 166.32} \text{ (MeV)}, \tag{13}$$

$$\sigma = \frac{K + 44.73}{211.05}. \tag{14}$$

Varying the value of K allows one to study its effects on the chemcial and mechanical instabilities of nuclear matter. The isospin-independent part of V_q should also contain a momentum-dependent part. Although the latter is important for some dynamical observables in heavy ion collisions such as the collective flow, it is not essential for observables such as the neutron to proton ratio of preequilibrium nucleons [11]. A recent study has also shown that the momentum-dependent interaction has rather small effects on the global properties of nuclei [15, 16]. For the isospin-asymmetric potential $V_{\mathbf{asy}}^q$, we postpone the discussion to Section 3.

Using the Gibbs-Duhem relation,

$$\left(\frac{\partial P}{\partial \rho}\right)_{T,\delta} = \frac{\rho}{2}\left[(1+\delta)\left(\frac{\partial \mu_n}{\partial \rho}\right)_{T,\delta} + (1-\delta)\left(\frac{\partial \mu_p}{\partial \rho}\right)_{T,\delta}\right], \tag{15}$$

the mechanical stability condition can be obtained from the neutron and proton chemical potential, i.e.,

$$\begin{aligned}\left(\frac{\partial P}{\partial \rho}\right)_{T,\delta} &= T + a\frac{\rho}{\rho_0} + b\sigma\left(\frac{\rho}{\rho_0}\right)^{\sigma} \\ &+ \frac{1+\delta}{2}\left(\rho\frac{\partial V^n_{\rm asy}}{\partial \rho} + T\sum_{n=1}^{\infty}(n+1)b_n\left(\frac{\lambda_T^3(1+\delta)\rho}{4}\right)^n\right) \\ &+ \frac{1-\delta}{2}\left(\rho\frac{\partial V^p_{\rm asy}}{\partial \rho} + T\sum_{n=1}^{\infty}(n+1)b_n\left(\frac{\lambda_T^3(1-\delta)\rho}{4}\right)^n\right).\end{aligned} \tag{16}$$

The first two terms are contributions from the kinetic energy and isospin-independent potential while the last two terms are those from the isospin-dependent kinetic energy and potential.

The condition for the chemical stability (i.e., Eq.(4)) can be obtained by using the following transformation:

$$\begin{aligned}\left(\frac{\partial \mu_n}{\partial \delta}\right)_{T,P} &= \left(\frac{\partial \mu_n}{\partial \delta}\right)_{T,\rho} - \left(\frac{\partial \mu_n}{\partial P}\right)_{T,\delta}\cdot\left(\frac{\partial P}{\partial \delta}\right)_{T,\rho} \\ &= \left(\frac{\partial \mu_n}{\partial \delta}\right)_{T,\rho} - \left(\frac{\partial \mu_n}{\partial \rho}\right)_{T,\delta}\cdot\left(\frac{\partial P}{\partial \rho}\right)^{-1}_{T,\delta}\cdot\left(\frac{\partial P}{\partial \delta}\right)_{T,\rho}.\end{aligned} \tag{17}$$

To evaluate the above equation, we use Eq.(16) for $\left(\frac{\partial P}{\partial \rho}\right)_{T,\delta}$ and the following expressions for the other three partial derivatives:

$$\begin{aligned}\left(\frac{\partial \mu_n}{\partial \delta}\right)_{T,\rho} &= \left(\frac{\partial V^n_{\rm asy}}{\partial \delta}\right)_{T,\rho} \\ &+ T\left[\frac{1}{1+\delta} + \sum_{n=1}^{\infty}(n+1)b_n\left(\frac{\lambda_T^3\rho}{4}\right)^n(1+\delta)^{n-1}\right],\end{aligned} \tag{18}$$

$$\begin{aligned}\left(\frac{\partial \mu_n}{\partial \rho}\right)_{T,\delta} &= \frac{a}{\rho_0} + b\sigma\frac{\rho^{\sigma-1}}{\rho_0^\sigma} + \left(\frac{\partial V^n_{\rm asy}}{\partial \rho}\right)_{T,\delta} \\ &+ T\left[\frac{1}{\rho} + \sum_{n=1}^{\infty}(n+1)b_n\left(\frac{\lambda_T^3(1+\delta)}{4}\right)^n\cdot\rho^{n-1}\right],\end{aligned} \tag{19}$$

$$\left(\frac{\partial P}{\partial \delta}\right)_{T,\rho} = \left(\frac{\partial P_{\rm asy}}{\partial \delta}\right)_{T,\rho} + \frac{T\rho}{2}\sum_{n=1}^{\infty}(n+1)b_n(\frac{\lambda_T^3\rho}{4})^n\left[(1+\delta)^n - (1-\delta)^n\right]. \tag{20}$$

We note that the total pressure also be obtained from Eq.(15), and it has contributions from the kinetic motion, the isospin-independent potential, and the isospin-dependent potential, i.e.,

$$P = P_{\rm kin} + P_0 + P_{\rm asy}. \tag{21}$$

The kinetic contribution $P_{\rm kin}$ is given by

$$P_{\rm kin} = T\rho\left[1 + \frac{1}{2}\sum_{n=1}^{\infty}b_n(\frac{\lambda_T^3\rho}{4})^n\left((1+\delta)^{n+1} + (1-\delta)^{1+n}\right)\right]. \tag{22}$$

The contribution from the isospin-independent nuclear interaction is

$$P_0 = \frac{1}{2}a\rho_0(\frac{\rho}{\rho_0})^2 + \frac{b\sigma}{1+\sigma}\rho_0(\frac{\rho}{\rho_0})^{\sigma+1}. \tag{23}$$

$P_{\rm asy}$ is the contribution from the isospin-dependent part of the nuclear interaction $V_{\rm asy}^{n/p}$, i.e.,

$$P_{\rm asy} = \frac{1+\delta}{2}\int \rho\left(\frac{\partial V_{\rm asy}^n}{\partial\rho}\right)_\delta d\rho + \frac{1-\delta}{2}\int \rho\left(\frac{\partial V_{\rm asy}^p}{\partial\rho}\right)_\delta d\rho. \tag{24}$$

3 Symmetry energy and potential

To proceed with the explicit determination of stability conditions, we need to the neutron and proton symmetry potentials $V_{\rm asy}^{n/p}$ and their dependence on ρ and δ. Studies based on microscopic many-body theories have shown that the energy per nucleon in asymmetric matter $e(\rho,\delta)$ can be approximated by a parabolic function [17, 18], i.e.,

$$e(\rho,\delta) = e(\rho,0) + S(\rho)\delta^2, \tag{25}$$

where $e(\rho,0)$ is the EOS of symmetric nuclear matter and $S(\rho)$ is the symmetry energy. The latter is normally extracted (see the contribution by Lombardo *et al.* in this book) from the energy difference between pure neutron matter and symmetric matter, i.e.,

$$S(\rho) = e(\rho,1) - e(\rho,0). \tag{26}$$

The symmetry energy can be separated into a kinetic and a potential contribution according to

$$S(\rho) = T_F(\rho,1) - T_F(\rho,0) + S_{\rm pot}(\rho). \tag{27}$$

In the above,

$$T_F(\rho,\delta) = \frac{3\hbar^2}{20m}\left(\frac{3\pi^2\rho}{2}\right)^{2/3}\left[(1+\delta)^{5/3} + (1-\delta)^{5/3}\right] \tag{28}$$

is the Fermi-gas kinetic energy per nucleon and $S_{\rm pot}(\rho)$ is the potential energy per nucleon in asymmetric nuclear matter due to the isospin-dependent interaction. Inserting Eq.(28) into Eq.(27) gives

$$S(\rho) = (2^{2/3}-1)\frac{3}{5}E_F^0u^{2/3} + S_{\rm pot}(\rho), \tag{29}$$

where E_F^0 is the Fermi energy at normal nuclear matter density and $u = \rho/\rho_0$. The above expression allows us to discuss the differences among the theoretically calculated density dependence of the symmetry energy through parameterizing the potential term $S_{\rm pot}(\rho)$ as

$$S_{\rm pot}(\rho) = (S(\rho_0) - (2^{2/3}-1)\frac{3}{5}E_F^0)F(u). \tag{30}$$

In the above, $S(\rho_0)$ is the symmetry energy at normal nuclear matter density ρ_0 and is known to have values in the range of 27-36 MeV [19]. We consider here three typical forms of $F(u)$, i.e., $F(u) = u^2$, u and $u^{1/2}$, which approximately describe the large range of predicted density dependence of the symmetry energy by various many-body approaches

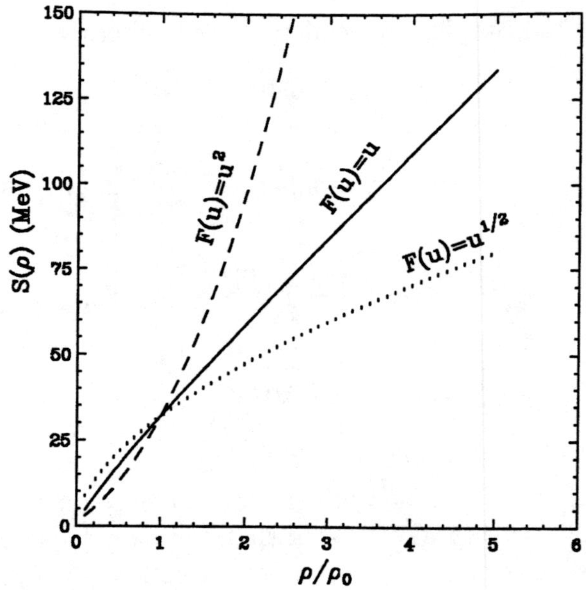

Figure 1: Symmetry energy as a function of density for different $F(u)$.

[9, 10]. Shown in Fig. 1 is the symmetry energy as a function of density obtained from Eq.(29) using the three parameterizations for $F(u)$ and $S(\rho_0) = 32$ MeV.

Although all parameterizations give the same symmetry energy at normal nuclear matter density, they behave differently at both lower and higher densities. One way of characterizing their differences is through their slope L and curvature $K_{\rm sym}$ at normal density, i.e.,

$$L \equiv 3\rho_0 \frac{\partial S}{\partial \rho}|_{\rho=\rho_0}, \quad (31)$$

$$K_{\rm sym} \equiv 9\rho_0^2 \frac{\partial^2 S}{\partial^2 \rho}|_{\rho=\rho_0}. \quad (32)$$

From Eqs.(29) and (30), one finds $L = 141.2$, 83.3, and 54.3 MeV and $K_{\rm sym} = 321.84, -25.6$, and -69.0 MeV for $F(u) = u^2$, u and $u^{1/2}$, respectively. These values are within the range of predictions based on many-body theories. As shown in Table 1, $K_{\rm sym}$ ranges from positive 466 MeV to negative 400 MeV as predicted by both non-relativistic and relativistic many-body approaches using different interactions [21]. From the most recent data on giant monopole resonances, the value of $K_{\rm sym}$ is found between -566 ± 1350 MeV and 34 ± 159 MeV [20], depending on the mass region of nuclei and the number of parameters used in parameterizing the compressibility of finite nuclei. The $K_{\rm sym}$ data thus still cannot distinguish these model predictions.

In terms of the parameterized symmetry energy, the symmetry potential $V^q_{\rm asy}$ is obtained through

$$V^q_{\rm asy}(\rho, \delta) = \partial w_{\rm asy}(\rho, \delta)/\partial \rho_q, \quad (33)$$

where $w_{\rm asy}(\rho, \delta)$ is the contribution of nuclear interactions to the symmetry energy density

$$w_{\rm asy}(\rho, \delta) = e_a \cdot \rho F(u)\delta^2, \quad (34)$$

and

$$e_a \equiv S(\rho_0) - (2^{2/3} - 1)\,\tfrac{3}{5}E_F^0 \approx 19.3\ MeV. \quad (35)$$

Table 1: The slope and curvature of symmetry energy in units of MeV at normal nuclear density.

Force	Paris	SKM*	SI′	SIII	DHF (b)	DHF (e)
L	68.8	45.78	35.34	9.91	132	138
K_{sym}	37.56	-155.9	-259.1	-393.7	466	276

For $F(u) = u^2$, u and $u^{1/2}$, the symmetry potentials are thus

$$V_{\rm asy}^{n(p)} = \pm 2e_a u^2\delta + e_a u^2\delta^2, \tag{36}$$

$$V_{\rm asy}^{n(p)} = \pm 2e_a u\delta, \tag{37}$$

and

$$V_{\rm asy}^{n(p)} = \pm 2e_a u^{1/2}\delta - \frac{1}{2}e_a u^{1/2}\delta^2, \tag{38}$$

respectively. In the above, the $\pm$ sign is for the neutron and proton, respectively. Since the isospin asymmetry involved in heavy-ion collisions is not very large, the δ^2 term can be neglected. In this case, the symmetry potential increases linearly with the neutron excess δ. Among the three expressions given above for $V_{\rm asy}^{n/p}$,the second one has been widely used in dynamical models for nuclear reactions [21].

The isospin-dependent part of the pressure $P_{\rm asy}$ are then given by

$$P_{\rm asy} = \frac{4}{3}e_a\rho_0(\frac{\rho}{\rho_0})^3\delta^2, \tag{39}$$

$$P_{\rm asy} = e_a\rho_0(\frac{\rho}{\rho_0})^2\delta^2, \tag{40}$$

and

$$P_{\rm asy} = \frac{2}{3}e_a\rho_0(\frac{\rho}{\rho_0})^{3/2}\delta^2 \tag{41}$$

for $F(u) = u^2$, u and $u^{1/2}$, respectively.

The above expressions for $V_{\rm asy}^{n(p)}$ and $P_{\rm asy}$ allow us to express analytically both the mechanical and chemical stability conditions in the ρ, T, and δ configuration space and to study the region of chemical and/or mechanical instability as well as their dependence on the nuclear equation of state.

4 Chemical and mechanical instabilities in asymmetric nuclear matter

First, we show in Figs. 2 and 3 $\left(\frac{\partial P}{\partial \rho}\right)_{T,\delta}$ and $\left(\frac{\partial \mu_n}{\partial \delta}\right)_{P,T}$ as functions of ρ for various T and δ using $F(u) = u$ and a nuclear compressibility of $K = 210$ MeV. For comparisons, results for symmetric nuclear matter are also shown in the left window of Fig. 2. In the latter case there is no chemical instability as the quantity $(\partial\mu_n/\partial\delta)_{T,P}$ increases with both density and temperature. To see this, we note that

$$\lim_{\delta\to 0}\left(\frac{\partial P}{\partial\delta}\right)_{T,\rho}=0. \tag{42}$$

Then, it is easy to show that

$$\begin{aligned}\lim_{\delta\to 0}\left(\frac{\partial\mu_n}{\partial\delta}\right)_{T,P} &= \lim_{\delta\to 0}\left(\frac{\partial\mu_n}{\partial\delta}\right)_{T,\rho}\\ &= 2e_aF(u)+T\left[1+\sum_{n=1}^{\infty}(n+1)b_n\left(\frac{\lambda_T^3\rho}{4}\right)^n\right],\end{aligned} \tag{43}$$

which is positive for symmetric nuclear matter and increases with both density and temperature.

The mechanical instability is known to occur at intermediate densities, i.e., in the mixed phase, between the gas and liquid phases at subcritical temperatures. This is clearly demonstrated by the ρ dependence of $(\partial P/\partial\rho)_{T,\delta}$. It is interesting to examine this quantity in the limit of low densities. From the Gibbs-Duhem relation of Eq. (15) and the expression for μ_q of Eq. (7), one has

$$\lim_{\rho\to 0}\left(\frac{\partial P}{\partial\rho}\right)_{T,\delta}=\lim_{\rho\to 0}(T+a\frac{\rho}{\rho_0})=T, \tag{44}$$

as expected for an ideal gas.

We note from Eq. (17) that the quantity $(\partial\mu_n/\partial\delta)_{T,P}$ is singular along the boundary for mechanical instability, where $(\partial P/\partial\rho)_{T,\delta}=0$. This singularity is indicated by the spikes in Figs. 2 and 3. However, their heights at the boundary of the gas and mixed phases is very small because the last derivative in Eq. (17) is close to zero, i.e.,

$$\lim_{\rho\to 0}\left(\frac{\partial P}{\partial\delta}\right)_{T,\rho}=\left(\frac{\partial P_{\text{asy}}}{\partial\delta}\right)_{T,\rho}=0. \tag{45}$$

Furthermore, in the ideal gas limit the nuclear matter is also chemically stable, i.e.,

$$\lim_{\rho\to 0}\left(\frac{\partial\mu_n}{\partial\delta}\right)_{P,T}=\frac{T}{1+\delta}>0. \tag{46}$$

As the system becomes neutron-rich, the chemical instability starts to develop, so the boundary for mechanical instability shrinks while that for chemical instability expands as δ increases. Also, in the mechanically unstable region the system is chemically stable, but at higher densities chemical instability can appear in the mechanically stable region. Therefore, in neutron-rich nuclear matter the chemical instability is a more relevant quantity in determining its properties. As shown in Fig. 4, where the pressure along the chemical (DS), and mechanical (ITS) spinodals are plotted as functions of density. One sees that both the isothermal pressure and the separation between the two spinodals increase with increasing δ, so not only the chemical but also the mechanical instability is strongly isospin-dependent.

The isospin-dependence of mechanical and chemical instabilities at a fixed temperature are shown in more detail in Figs. 5 and 6. It is seen that nuclear matter is both mechanically and chemically stable at low and high densities. At intermediate densities (e.g., $\rho/\rho_0=$

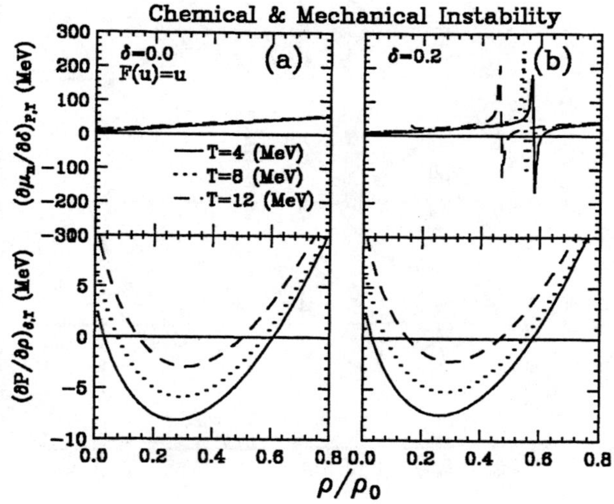

Figure 2: Chemical (upper window) and mechanical (lower window) stability conditions as functions of density at fixed temperatures T=4, 8 and 12 MeV for $\delta = 0.0$ (left panel)and $\delta = 0.2$ (right panel). Taken from Ref. [22].

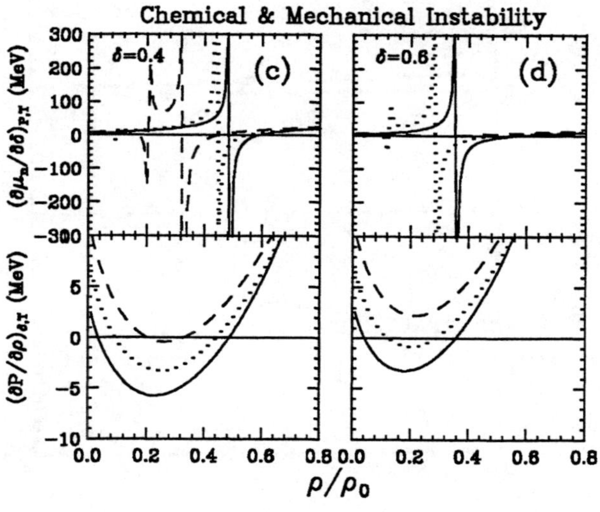

Figure 3: Same as Fig. 2 for δ =0.4 (left panel) and $\delta = 0.6$ (right panel). Taken from Ref. [22].

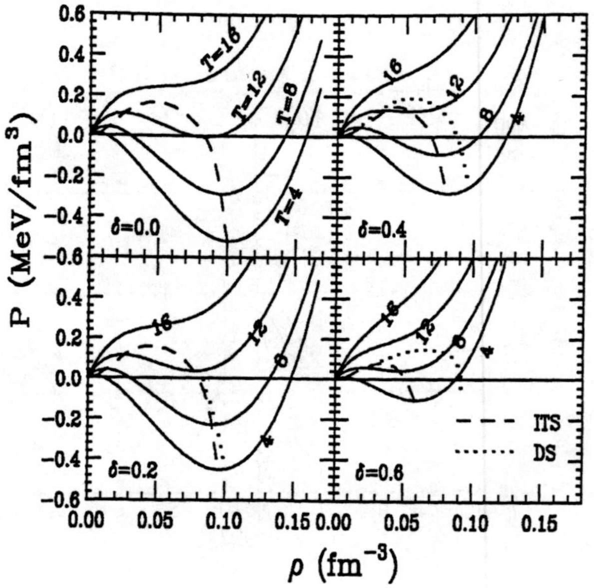

Figure 4: Pressure as a function of density at fixed temperatures T=4, 8, 12 and 16 MeV for δ =0.0, 0.2, 0.4 and 0.6. The isothermal and diffusive spinodals are shown by the dashed and dotted lines, respectively. Taken from Ref. [22].

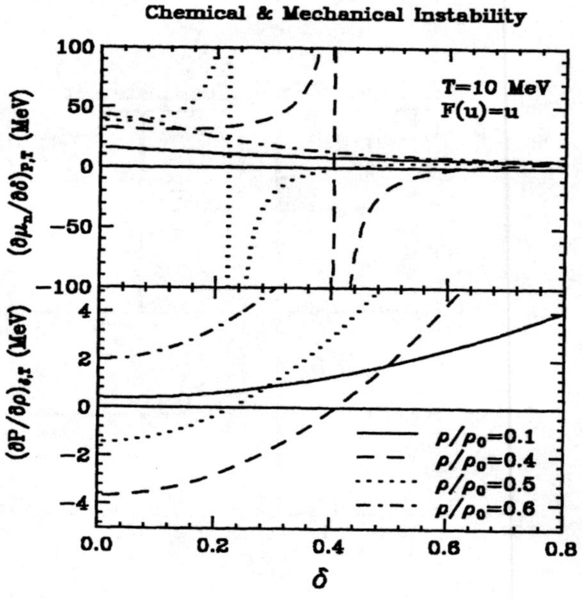

Figure 5: Chemical (upper window) and mechanical (lower window) instability conditions as functions of δ at a fixed temperature T = 10 MeV and various densities. Taken from Ref. [22].

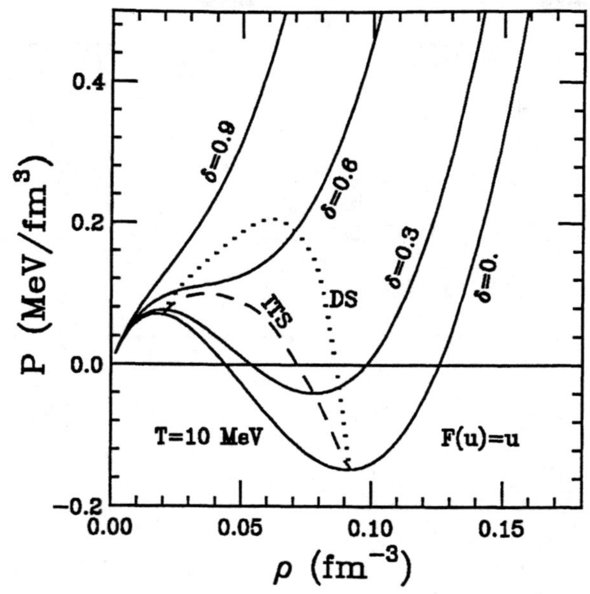

Figure 6: Pressure as a function of density at a fixed temperature $T = 10$ MeV at various δ. The isothermal and diffusive spinodals are shown by the dashed and dotted lines, respectively. Taken from Ref. [22].

0.4, 0.5) the system is mechanically unstable for small δ (e.g., $\delta \leq 0.2$ and $\delta \leq 0.4$ for $\rho/\rho_0 = 0.4$ and 0.5, respectively), and chemically unstable for intermediate δ (e.g. $0.2 \leq \delta \leq 0.4$ for $\rho/\rho_0 = 0.4$ and $0.4 \leq \delta \leq 0.6$ for $\rho/\rho_0 = 0.5$, respectively). The corresponding boundaries in the pressure-density space are shown in Fig. 6. Again, the diffusive spinodal is more extended than the isothermal spinodal. The dependence of pressure on isospin is also shown in this figure. Its strong dependence on δ is due to the significant contribution of the isospin-dependent potential to the total pressure.

5 Effects of nuclear EOS on chemical and mechanical instabilities and isospin fractionation

Since both chemical and mechanical instabilities depend strongly on the isospin asymmetry of the system, it is of interest to know how this dependence might be affected by the isospin-dependent and -independent parts of nuclear equation of state. This is illustrated in Figs. 7 and 8, where the pressure is shown as a function of density (left panels) and δ (right panels) along the diffusive spinodals (upper windows) and isothermal spinodals (lower windows) at a constant temperature of $T = 10$ MeV by using the three forms of $F(u)$ and a compressibility of 380 and 210 MeV, respectively. Several interesting observations can be made from these two figures. First, the diffusive spinodal is always more extended in the configuration space than the isothermal spinodal. However, their relative importance depends on their relative growth rates as already pointed out by Mueller and Serot in their chapter. To our best knowledge, there is no detailed study on the relative time scales for the chemical and mechanical stabilities. In the work by Baran *et al.* [23], the time scale of dynamical instability in asymmetric nuclear matter has been studied but no clear distinction between the growth rates of chemcial and mechanical instability has been made. Second, both DS and ITS depend on the density dependence of the symmetry energy; and this dependence is

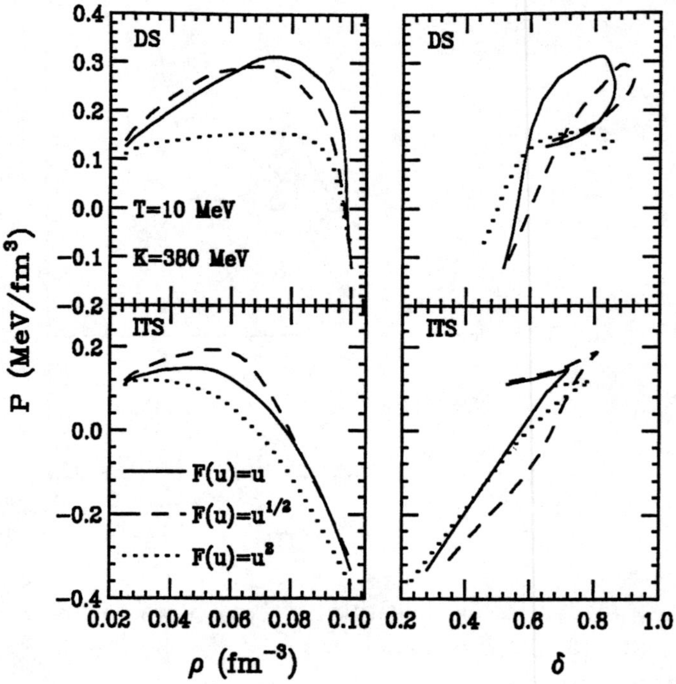

Figure 7: Pressure as a function of density (left panel) and isospin asymmetry (right panel) at a fixed temperature $T = 10$ MeV along the boundary of diffusive spinodals (upper windows) and isothermal spinodals (lower windows) by using the three forms of $F(u)$ and a compressibility of $K = 380$ MeV. Taken from Ref. [22].

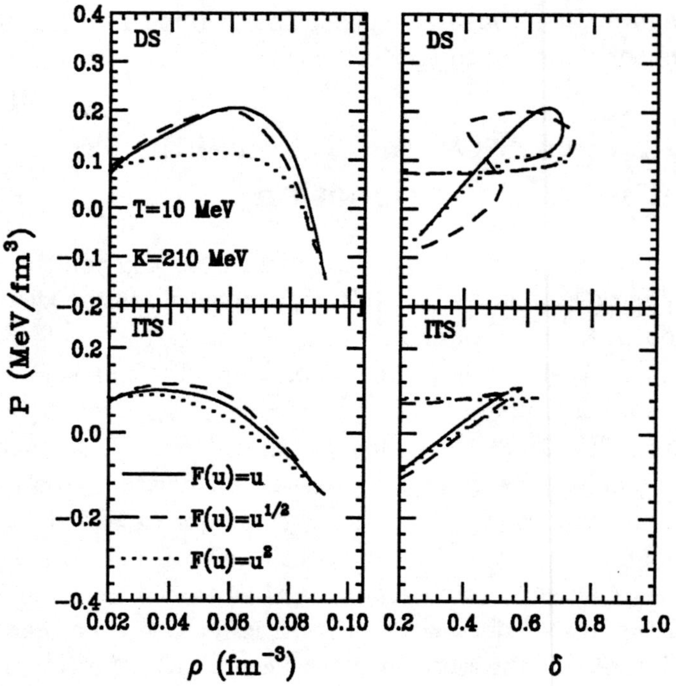

Figure 8: Same as Fig. 7 for a compressibility of $K = 210$ MeV. Taken from Ref. [22].

stronger for the stiff equation of state with $K = 380$ MeV. Moreover, both the DS and ITS boundaries are more affected by the compressibility in the $P - \delta$ than in the $P - \rho$ space. Third, from the right panels of the above two figures (especially the one with $K = 380$ MeV, one sees that it is both chemically and mechanically favorable for the nuclear matter to be less asymmetric (smaller δ) in the liquid phase (higher density) than in the gas phase (lower density). This observation is consistent with those from the Relativistic Mean Field model (see the chapter by Mueller and Serot) and the isospin-dependent lattice gas model (see the chapters by Das Gupta *et al.* and Chomaz *et al.* in this book). This common feature of isospin fractionation in the liquid and gas phases can be understood from the energy considerations. Since the equation of state for asymmetric nuclear matter contains a $S(\rho)\delta^2$ term, it is energetically favorable for asymmetric matter to separate into a liquid phase that is less asymmetric and a gas phase that is more asymmetric, rather than into two phases with equal isospin-asymmetry. In studies based on dynamical evolutions of particles which are initialized in the mechanical instability region in a box with periodic boundary conditions, a similar isospin fractionation has been found [23]. This phenomenon has been recently suggested as a possible signature of the liquid-gas phase transition in asymmetric nuclear matter (see, e.g., the contribution of Chomaz *et al.*). As we shall discuss in the next section, the isospin-dependent transport model for nuclear reactions also predicts a neutron-rich gas (single nucleons and light clusters) and more isospin-symmetric liquid (heavy residues and fragments). Therefore, the phenomenon of isopsin fractionation is not necessariely a unique property associated with the gas-liquid phase transition in nuclear matter.

6 Isospin fractionation from a dynamical point of view

In analyses of recent experimental data on particle yields from intermediate energy heavy ion collisions, it has been found that the gas phase is enriched with neutrons relative to the liquid phase that consists of bound nuclei [24]. This is consistent with the predicted isospin fractionation in liquid-gas phase transition using equilibrium models as discussed in the previous section. In this section, we discuss an alternative explanation of the observed isospin fractionation using a dynamical model for nuclear reactions. Instead assuming that an isospin-asymmetric system breaks up instantaneously into a neutron-rich gas and neutron-poor liquid phase, we show that a strong symmetry potential in nuclear reactions involving isospin asymmetric nuclei naturally leads to a more isospin symmetric heavy residue that is surrounded by a neutron-rich gas and subsequentlly decays by either particle emission or multifragmentation.

Denoting the neutron to proton ratio of the reaction system and the gas component by X and X_g, respectively, then the neutron to proton ratio of the liquid phase is given by

$$X_l = X + \frac{Z_g}{Z_l}(X - X_g), \tag{47}$$

where Z_g and Z_l are the proton number of the gas and liquid components, respectively. It has been shown in transport model calculations [11] and experiments [25, 26, 27, 28, 29] that X_g, particularly its preequilibrium component, is much larger than X, leading to a smaller X_l. As an example, we consider collisions of $^{112}\mathrm{Sn}+^{112}\mathrm{Sn}$, $^{124}\mathrm{Sn}+^{124}\mathrm{Sn}$, and $^{132}\mathrm{Sn}+^{132}\mathrm{Sn}$ at a beam energy of 40 MeV/nucleon using the isospin-dependent IBUU model [30, 31] with the neutron and proton distributions in the Sn isotopes calculated from the RMF theory. To identify the free nucleons, a phase-space coalescence model has been used at 200 fm/c after the initial contact of the two nuclei, when the quadrupole moment of the nucleon momentum distribution in the heavy residue almost vanishes, indicating the approach of thermal equilibrium. A nucleon is considered free if it is not correlated with other nucleons within a spatial separation of $\Delta r = 3$ fm and a momentum separation of $\Delta p = 300$ MeV/c. Otherwise, it is bounded in a cluster. These results are not sensitive to these parameters if they are varied by less than 30% around the above values.

The symmetry potential has the following effects on preequilibrium nucleons. First, it tends to make more neutrons than protons unbound. One thus expects that a stronger symmetry potential leads to a larger ratio of free neutrons to protons. Second, if both neutrons and protons are already free, the symmetry potential makes neutrons more energetic than protons. In Fig. 9, the ratio ($R_{n/p}(E_{\text{kin}}) \equiv dN_n/dN_p$) is shown as a function of kinetic en-

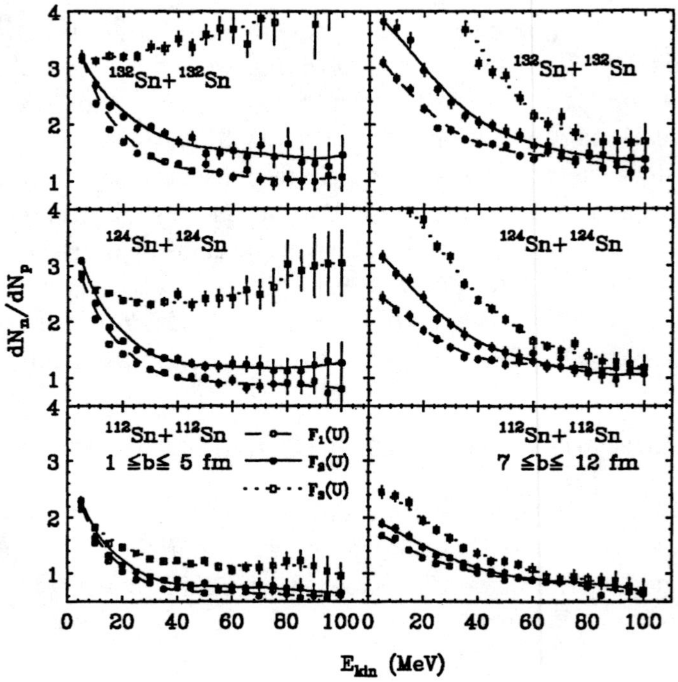

Figure 9: The ratio of preequilibrium neutrons to protons as a function of kinetic energy in central (left windows) and peripheral (right windows) reactions using the three forms of $F(u)$. Taken from [11].

ergy. These results are obtained by using the three forms of $F_1(u) \equiv 2u^2/(1+u)$, $F_2(u) \equiv u$ and $F_3(u) \equiv u^{1/2}$ for typical central (left windows) and peripheral (right windows) collisions of $^{112}\text{Sn} + ^{112}\text{Sn}$, $^{124}\text{Sn} + ^{124}\text{Sn}$ and $^{132}\text{Sn} + ^{132}\text{Sn}$, respectively. The increase of the ratio at lower kinetic energies in all cases is due to the Coulomb repulsion which shifts protons from lower to higher kinetic energies. On the other hand, the different ratios calculated using different $F(u)$'s reflect clearly the effect mentioned above, i.e., with a stronger symmetry potential the ratio of preequilibrium neutrons to protons becomes larger for more neutron rich systems. Between the central and peripheral collisions the ratios vary smoothly as a function of impact parameter. This is shown in Fig. 10 where the energy-dependence of the ratio $R_{n/p}(E_{\text{kin}})$ is shown for different impact parameters for the reaction of $^{124}\text{Sn} + ^{124}\text{Sn}$ using $F_3(u)$. The impact parameter dependence of the ratio of the total yield of neutrons to that of protons shown in Fig. 11 clearly demonstrates the effects of neutrons skins in more neutron-rich systems at larger impact parameters.

It is interesting to note that the effects due to different symmetry potentials are seen in

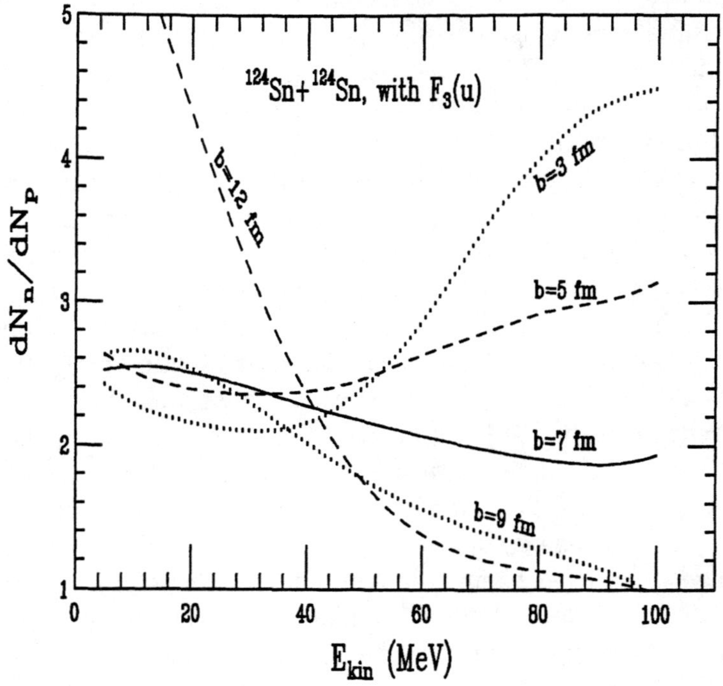

Figure 10: Impact parameter dependence of the ratio $R_{n/p}(E_{\text{kin}})$ as a function of kinetic energy for the reaction of $^{124}\text{Sn} +^{124}\text{Sn}$ using $F_3(u)$.

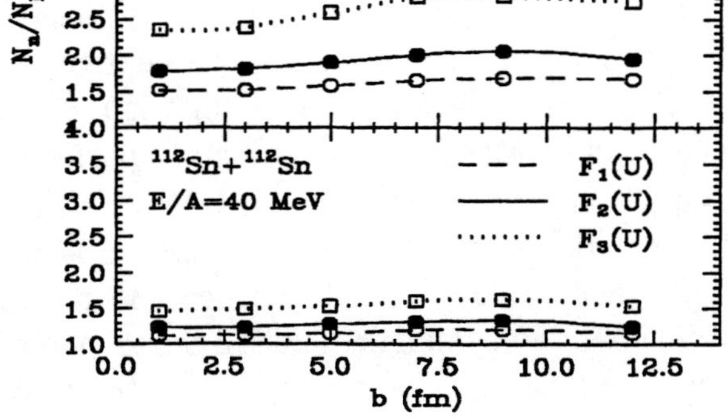

Figure 11: The ratio of preequilibrium neutrons to protons as a function of impact parameter using the three forms of $F(u)$.

different kinetic energy regions for central and peripheral collisions. This can be understood from the impact parameter dependence of $R_{n/p}(E_{\rm kin})$, shown in Fig. 10, and the reaction dynamics. In central collisions, effects of the symmetry potential are most prominent at higher kinetic energies. This is because most of finally observed free neutrons and protons are already unbound in the early stage of the reaction as a result of violent nucleon-nucleon collisions. The symmetry potential thus mainly affects the nucleon energy spectra by shifting more neutrons to higher kinetic energies with respect to protons. In peripheral collisions, however, there are fewer nucleon-nucleon collisions; whether a nucleon can become unbound depends strongly on its potential energy. With a stronger symmetry potential more neutrons (protons) become unbound (bound) as a result of a stronger symmetry potential, but they generally have smaller kinetic energies. As a result, in peripheral collisions effects of the symmetry potential show up chiefly at lower kinetic energies. For systems with more neutrons the effects of the symmetry potential are so strong that in central (peripheral) collisions different forms of $F(u)$ can be clearly distinguished from the ratio of preequilibrium neutrons to protons at higher (lower) kinetic energies. However, because of energy thresholds in detectors, it will be difficult to measure low energy nucleons, especially neutrons. Furthermore, the low energy spectrum also has appreciable contribution from emissions at the later stage when the system is in equilibrium. Therefore, the measurement of the ratio $R_{n/p}(E_{\rm kin})$ in central heavy-ion collisions for nucleons with energies higher than about 20 MeV is more suitable for extracting the EOS of asymmetric nuclear matter.

There are already experimental indications for the existence of a neutron-rich gas phase in heavy-ion collisions at intermediate energies. In several experiments around the Fermi energy [25, 26, 27, 28], Hilscher *et al.* have found that the neutron/proton ratio, $((N/Z)_{\rm free})$, of preequilibrium nucleons is consistently higher than that of the projectile-target system $(N_P + N_T)/(Z_P + Z_T)$ and cannot be explained by the Coulomb effect. For example, in the reaction of $^{12}{\rm C} +^{165}{\rm Ho}$ at a beam energy of 32 MeV/nucleon, they have measured the neutron and proton spectra at 14° and energies between 70 and 100 MeV, and found that the multiplicity of neutrons is larger than that of protons by a factor of $1.4 \pm 0.2, 1.7 \pm 0.3$ and 2.4 ± 0.3 for linear-momentum transfers of 52%, 73% and 93%, respectively. Therefore, the neutron to proton ratio is much higher than that of the reaction system $(N/Z)_{\rm cs}$=1.42 in central collisions corresponding to higher linear-momentum transfer. This result cannot be explained by the standard Fermi jet model of preequilibrium nucleon emission [25], but is in agreement with the BUU predictions discussed above. From Figs. 9 and 11, it is seen that with a large symmetry potential the n/p ratio for energetic nucleons can be much larger than that of the colliding system. In Fig. 12, the relative neutron excess of preequilibrium nucleons $\delta_{\rm pre-eq} = (N - Z)/(N + Z)$ is shown as a function of time for the reaction of $^{12}{\rm C} +^{165}{\rm Ho}$ at a beam energy of 35 MeV/nucleon and an impact parameter of 1.0 fm. The calculations are carried out with a symmetry potential $V_{\rm asy}^{n/p} = \pm c(\rho_n - \rho_p)/\rho_0$. In the early stage of the reaction, the neutron excess is seen to fluctuate due to violent collisions, but reaches a constant value after about 100 fm/c. The dashed line at 0.17 is the neutron excess of the reaction system. Without the symmetry and Coulomb potentials the final neutron excess of preequilibrium nucleons is about the same as that of the reaction system. Including the Coulomb but not the symmetry potential ($c = 0$) leads to the emission of more protons in the preequilibrium stage of the reaction due to the Coulomb repulsion. Furthermore, the ratio $(N/Z)_{\rm free}$ increases monotonically as the strength of symmetry energy increases. To have a ratio larger than the N/Z of the reaction system one needs to use a strength parameter $c \geq 20$ MeV that corresponds to $e_{\rm sym}(0) \geq 32$ MeV.

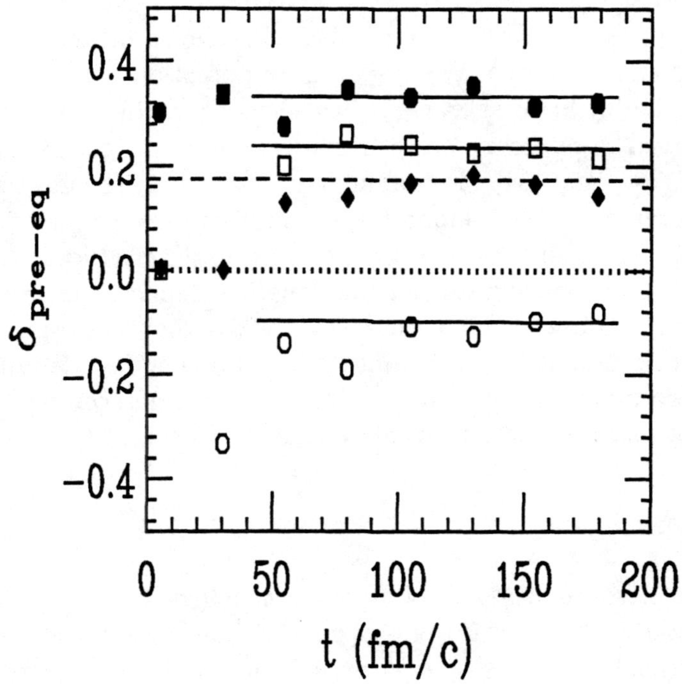

Figure 12: The ratio of preequilibrium neutrons to protons as a function of time for the reaction of $^{12}C + ^{165}Ho$ at a beam energy of 35 MeV/nucleon and an impact parameter of 1.0 fm. Solid circles: $c = 46$ MeV with Coulomb potential; open squares: $c = 32$ MeV with Coulomb potential; open circles: $c = 0$ MeV with Coulomb potential; solid diamonds: $c = 0$ MeV without Coulomb potential. Taken from [21].

More recently, Schröder *et al.* have studied systematically the spectra of preequilibrium neutrons and protons in collisions of both isospin symmetric and asymmetric nuclei [29]. It is found that the n/p ratio of preequilibrium nucleons does not scale with the N/Z ratio of the combined system. In particular, a preliminary analysis of the experimental data indicates that the ratio $(n/p)_{^{48}\mathrm{Ca}+^{112}\mathrm{Sn}}/(n/p)_{^{40}\mathrm{Ca}+^{112}\mathrm{Sn}}$ at a beam energy of 35 MeV/nucleon is about 4 to 12 depending on the impact parameter. This result is consistent with the existence of a strong symmetry potential as illustrated in the above studies based on the transport model.

Enhanced production of free neutrons than that predicted by the standard statistical model has also been reported in Refs. [4, 32]. In Ref. [33], it was pointed out that within the experimental and model uncertainties the production of neutron-deficient residues is much higher than that predicted by the statistical model. Because of charge conservation, this is equivalent to an enhanced neutron emission as shown in Refs. [3, 32]. Again, this is consistent with the existence of a strong symmetry potential.

In the contribution by Sobotka *et al.*, it has been shown that the abundance of neutron-rich isotopes observed in experiments is primarily due to the coalescence of clusters with neutrons from the gas phase [34]. Such abundant neutrons in the gas phase directly reflect its large isospin-asymmetry and support the picture that asymmetric nuclear reactions proceed by gradually increasing (decreasing) neutrons in the gas (liquid) component through particle emissions. The resulting isospin fractionation depends sensitively on the nuclear symmetry potential. To discern this picture from the one that an isospin asymmetric system instantaneously breaks statistically into a neutron-rich gas and neutron-poor liquid requires the use of correlation function techniques, as they allow the determination of the time scales for the emissions of neutron and intermediate mass fragments.

7 Summary

In this chapter, we have discussed using a phenomenological thermal model the chemical and mechanical instabilities in hot isospin-asymmetric nuclear matter. Both instabilities are found to depend strongly on the isospin asymmetry in the nuclear matter. For a chemically and mechanically stable asymmetric nuclear matter, a lower neutron excess is found in the liquid phase than in the gas phase. The boundary of chemical instability in the pressure-density space is found to be more extended than that of mechanical instability. However, locations of these boundaries depend sensitively on both the bulk compressibility of nuclear matter and the density dependence of the symmetry energy. The isospin fractionation found in this model is consistent with the predictions based on other more complicated equilibrium models. We have also pointed out that the experimentally observed neutron-enriched light particle production is not uniquely related to the isospin fractionation in a liquid-gas phase transition. Using a transport model, we have shown that there is a gradual dynamical enhancement of neutrons in the gas component. Furthermore, we have demonstrated that the isospin fractionation depends sensitively on the isospin-dependence of the nuclear equation of state.

The work of Bao-An Li is supported in part by the National Science Foundation under Grant No. PHY-0088934 and Arkansas Science and Technology Authority Grant No. 00-B-14. C.M. Ko is is supported in part by the National Science Foundation under the Grant PHY-9870038, the Welch Foundation under the Grant A-1358, and the Texas Advanced

Research Program under the Grants FY97-010366-0068 and FY99-010366-0081.

References

[1] L.G. Moretto and G.J. Wozniak, Ann. Rev. Nucl. Part. Sci. **43**, 123 (1993).

[2] S.J. Yennello, et al., Phys. Lett. **B321**, 15 (1994).

[3] G.J. Kunde *et al.*, Phys. Rev. Lett. **77**, 2897 (1996).

[4] J.F. Dempsey *et al.*, Phys. Rev. C**54**, 1710 (1996).

[5] M.L. Miller et al., Phys. Rev. Lett. **82**, 1399 (1999).

[6] J.M. Lattimer and D.G. Ravenhall, Astr. Jour., **223**, 314 (1978).

[7] M. Barranco and J. R. Buchler, Phys. Rev. C**22**, 1729 (1980).

[8] H. Müller and B.D. Serot, Phys. Rev. C **52**, 2072 (1995).

[9] M. Prakash, T.L. Ainsworth and J.M. Lattimer, Phys. Rev. Lett. **61**, 2518 (1988).

[10] J.M. Lattimer, C.J. Pethick, M. Prakash and P. Haensel, Phys. Rev. Lett. **66**, 2701 (1991).

[11] B.A. Li, C.M. Ko and Z.Z. Ren, Phys. Rev. Lett. **78** 1644 (1996).

[12] M. Brack, C. Guet and H.-B Hákansson, Phys. Rep., **123**, 275 (1985).

[13] H.R. Jaqman, A.Z. Mekjian and L. Zamick, Phys. Rev. C**27**, 2782 (1983); *ibid* C**29**, 2067 (1984).

[14] H.R. Jaqaman, Phys. Rev. C**39**, 169 (1988); *ibid* C**40**, 1677 (1989).

[15] L.P. Csernai, G. Fai, C. Gale and E. Osnes, Phys. Rev. C **46**, 736 (1992).

[16] V.K. Mishra, G. Fai, L.P. Csernai and E. Osnes, Phys. Rev. C**47**, 1519 (1993).

[17] I. Bombaci and U. Lombardo, Phys. Rev. C**44**, 1892 (1991).

[18] H. Huber, F. Weber and M.K. Weigel, Phys. Lett. **B317**, 485 (1993); Phys. Rev. C**50**, R1287 (1994).

[19] P.E. Haustein Atomic data and nuclear data tables, **39**, 185-395 (1988)

[20] S. Shlomo and D.H. Youngblood, Phys. Rev. C**47**, 529 (1993).

[21] B.A. Li, C.M. Ko and W. Bauer, Int. J. Mod. Phys. E**7**, 147 (1998).

[22] B.A Li and C. M. Ko, Nucl. Phys. **A618**, 498 (1997).

[23] V. Baran, M. Colonna, M. Di Toro and A.B. Larionov, Nucl. Phys. A**632**, 287 (1998).

[24] H.S. Xu et al., preprint MSUCL-1137, Oct. 1999.

[25] D. Hilscher *et al.*, Phys. Rev. C**36**, 208 (1987).

[26] D. Hilscher, in Proc. of Specialists's Meeting on Preequilibrium Nuclear Reactions, Semmering, Austria, 10-12th, Feb. 1988, Ed. B. Strohmaier (OECD, Paris, 1988), p. 245.

[27] D. Polster *et al.*, in Book of abstracts, International Nuclear Physics Conference, Wiesbaden, Germany, July 26-Aug. 1, 1992, p3.3.8.

[28] D. Polster *et al.*, Phys. Rev. C**51**, 1167 (1995).

[29] D.K. Agnihotri et al., in Proceedings of the 13th Winter Workshop of Nuclear Dynamics, Marathon, Florida, Feb. 1997, Eds. W. Bauer and A. Mignerey, Plenum Press, p67.

[30] B.A. Li and S.J. Yennello, Phys. Rev. C**52**, R1746 (1995).

[31] B.A. Li, Z.Z. Ren, C.M. Ko and S.J. Yennello, Phys. Rev. Lett. **76**, 4492 (1996).

[32] M. Gonin *et al.*, Phys. Rev. C**42**, 2125 (1990).

[33] K.A. Hanold *et al.*, Phys. Rev. C**52**, 1462 (1995).

[34] L.G. Sobotka, J.F. Dempsey, R.J. Charity, and P. Danielewicz, Phys. Rev. C **55**, 2109 (1997).

Collective Modes in Asymmetric Nuclei

Klaus Morawetz [*], Uwe Fuhrmann, and Rainer Walke [†‡]

Abstract

The collective motion of a finite nuclear system is investigated by numerical simulation and by linear response theory. Using a pseudo-particle simulation technique we analyze the giant resonances with a multipole decomposition scheme. We examine the energy and the damping of different giant collective modes and obtain the dependence of these quantities on the proton–neutron ratio. In simulations of finite nuclei including only mean fields the centroid energy of the resonance decreases with higher asymmetry due to the change of the compressibility and the damping increases. The collisional correlation in turn leads to a decreasing damping with increasing asymmetry.

Alternatively the giant collective modes in asymmetric nuclear matter are investigated within linear response theory including the collisional correlations via a dynamic relaxation time approximation. For a multicomponent system we derive a coupled dispersion relation and show that two sources of coupling appear: (i) a coupling of isoscalar and isovector modes due to the action of different mean-fields and (ii) an explicit new coupling in asymmetric matter due to collisional interaction. We show that the latter is responsible for a new mode arising besides isovector and isoscalar modes. The comparison with simulation results as well as experiments is performed. It is stressed that the surface effects and the collisional correlations are both essential in order to describe the correct damping behavior. A model is presented which allows to combine surface and collisional effects. Higher-order modes like the isoscalar dipole mode which where recently measured are discussed within this frame.

Collective motion beyond the linear regime is demonstrated for the example of large-amplitude isoscalar giant octupole excitations in finite nuclear systems. Depending on the initial conditions we observe either clear octupole modes or over-damped octupole modes which decay immediately into quadrupole ones. This dependence on initial correlations represents a behaviour beyond linear response.

1 Introduction

Giant resonances are currently enjoying a resurgence of attention as a tool for the investigation of many particle effects in finite quantum systems. The study of giant resonances offers the possibility to learn more about nuclear forces. Of special interest are modern experiments with exotic nuclei which broaden our knowledge in a new degree of freedom: the variability in the proton to neutron ratio. The treatment of collective modes in nuclear matter is documented in an enormous literature starting from their discovery [1] up to recent discussions [2] -[35] and citations therein.

We want to investigate the collective excitations in asymmetric nuclear matter [6, 7] by numerical simulation of the Vlasov equation and by comparison with the linear response theory. While the former leads to an insight into finite-size effects like surface, the latter

*LPC-ISMRA, Bld Marechal Juin, 14050 Caen and GANIL, Bld Becquerel, 14076 Caen, France
†Fachbereich Physik, Universität Rostock, 18051 Rostock, Germany
‡current address: Max-Planck-Institute for demographic research, Rostock, Germany

allows to consider collisional correlations in a control-able way. Both methods are contrasted and we demonstrate that the collisional correlations as well as the surface effects are important to describe the experimental damping of giant resonances.

The damping mechanisms of collective motions in excited nuclei are a topic of continuing debate. Mainly two lines of thought are pursued. In one line of thought it is assumed that collisions are the only physical reason for damping which is described via a Fermi liquid approach [6],[8]- [20, 36]. The other line of thought considers new features of the finite nucleus, such as surface oscillations and a level density with finite spacing. The investigations are performed without inertia [21]- [29, 37] or by including inertia [27],[30]-[35]; note that inertia is absent in infinite matter.

Both classes of models predict a comparable degree of damping necessary to reproduce the experimental data. Consequently, it is an open question which is the correct physical reason for damping. Of course, the correct description has to assume a finite nucleus consisting of nucleons which are bound via the mean field, through which the nucleons undergo mutual collisions and where the surface is formed by the particles themselves. These features are principally included in Boltzmann-Uehling-Uhlenbeck-(BUU) simulations [38, 39, 40] or in its nonlocal extensions [41, 42]. In full simulations, however, we will not gain a simple insight into the physical origins of the damping mechanism, in particular, how much is contributed by the surface and how much by collisions.

One aim of this article is therefore to compare both pictures in the frame of linear response theory. Within the collision-free Vlasov equation the linear response of finite systems is well known [43] and allows one to calculate the strength function of finite nuclei. The resulting damping, however, does not reproduce the experimental damping of giant resonances since collisions are absent. This motivates us to develop a linear response theory including collisions.

While most of the theoretical treatments of oscillations rely on the linear response method or RPA methods, large amplitude oscillations require methods beyond this level. In particular the question of the appearance of chaos has recently been investigated [44, 45, 46]. The hypothesis was established that the octupole mode is over-damped due to negative curved surface and consequent additional chaotic damping [47, 48, 49]. Here we want to discuss in which conditions one might observe octupole modes at least in Vlasov - simulations of giant resonances which will turn out to be dependent on initial conditions and are consequently an effect beyond linear response.

The outline is as follows: In the second chapter we will give the numerical results of pseudo-particle simulation of the Vlasov equation. This will provide us with some insight into the magnitude of finite size effects on the damping. In the third chapter we present the linear response result of the Vlasov equation including collisional correlations. We will consider a simplified picture of infinite matter response and will consider the Steinwedel Jensen picture. We discuss two possibilities to include surface contributions in the linear response formalism. This allows us to describe the experimental temperature dependence of the damping of giant dipole resonances as well as the structure function of the isovector dipole resonances. The surface consideration allows to describe the isoscalar dipole resonance as a higher order mode. The comparison between linear response results and simulations is performed and finally we discuss nonlinear effects beyond linear response for the example of giant octupole modes.

2 Pseudoparticle simulation

We will describe the giant resonance first by a kinetic equation using a pseudo-particle simulation [50]. The kinetic equation for the quasiclassical distribution function reads for neutrons (for protons analogously)

$$\dot{f}_n(\mathbf{p},\mathbf{r},t) + \frac{\mathbf{p}}{m}\partial_{\mathbf{r}} f_n(\mathbf{p},\mathbf{r},t) - \partial_{\mathbf{r}}(U_n + U_{\text{ext}})\partial_{\mathbf{p}} f_n(\mathbf{p},\mathbf{r},t) = I_{\text{corr}} \tag{1}$$

with the collisional integral I_{corr} discussed later and the self-consistent mean-field potential U given by a schematic Skyrme type [51]

$$U_{n/p}(\varrho, I) \;=\; a(\tfrac{\varrho}{\varrho_0}) + b(\tfrac{\varrho}{\varrho_0})^s \pm cI(\tfrac{\varrho}{\varrho_0}) \tag{2}$$

with $a = -356$MeV, $b = 303$MeV , $c = 54$MeV and $s = 7/6$. Here the neutron excess is $I = \frac{\varrho_n - \varrho_p}{\varrho_n + \varrho_p}$, the neutrons feel the +–sign potential, the protons the opposite one.

By multiplying the kinetic equation by $1, p$ or $E = \frac{p^2}{2m} + U$ respectively one obtains the balance for particle density ρ, momentum density u and energy density $\mathcal{E}$. Since the collision integrals vanish for density and momentum balance we get the usual balance equations

$$\begin{aligned} \frac{\partial \rho(\mathbf{r},t)}{\partial t} + \frac{\partial}{\partial \mathbf{r}} \int \frac{d\mathbf{p}}{(2\pi\hbar)^3} \frac{\partial E}{\partial \mathbf{p}} f(\mathbf{p},\mathbf{r},t) = 0 \\ \frac{\partial u_i(\mathbf{r},t)}{\partial t} + \frac{\partial}{\partial r_j} \int \frac{d\mathbf{p}}{(2\pi\hbar)^3} (p_i \frac{\partial E}{\partial p_j} f(\mathbf{p},\mathbf{r},t) + \mathcal{E}(\mathbf{r},t)\delta_{ij}) = 0 \end{aligned} \tag{3}$$

where the mean field energy of the system varies as

$$\begin{aligned} \delta\mathcal{E} &= \int \frac{d\mathbf{p}}{(2\pi\hbar)^3} \frac{\delta\mathcal{E}}{\delta f(\mathbf{p},\mathbf{r},t)} \delta f(\mathbf{p},\mathbf{r},t) \\ &= \int \frac{dp}{(2\pi\hbar)^3} (\frac{p^2}{2m} + U)\delta f(\mathbf{p},\mathbf{r},t) \end{aligned} \tag{4}$$

such that from (2) follows the total energy density as

$$\mathcal{E}(\mathbf{r},t) = \int \frac{d\mathbf{p}}{(2\pi\hbar)^3} \frac{p^2}{2m} f(\mathbf{p},\mathbf{r},t) + \frac{a}{2}(\tfrac{\varrho^2(\mathbf{r},t)}{\varrho_0}) + \frac{b\rho(\mathbf{r},t)}{s+1}(\tfrac{\varrho(\mathbf{r},t)}{\varrho_0})^{s+1} \pm \frac{c}{2} I(\tfrac{\varrho^2(\mathbf{r},t)}{\varrho_0}). \tag{5}$$

With the help of this quantity the balance of energy density reads from (1)

$$\frac{\partial \mathcal{E}(\mathbf{r},t)}{\partial t} + \frac{\partial}{\partial \mathbf{r}} \int \frac{d\mathbf{p}}{(2\pi\hbar)^3} E \frac{\partial E}{\partial \mathbf{p}} f(\mathbf{p},\mathbf{r},t) = -\frac{\partial}{\partial t} E_{\text{corr}}(\mathbf{r},t) \tag{6}$$

with the correlation energy [52, 53] arising from the collisional side. The collisional side we will consider later in linear response. Here for the simulation we will neglect the collisions and will restrict ourselves to the Vlasov kinetic equation. In this way we will learn what are the effects of finite size and what are the effects of collisions.

Let us now first look at the input for the simulation. The mass number dependence of binding energy $\mathcal{E}/A$ from (5) is shown in figure 1. One sees that with increasing asymmetry the binding energy becomes weaker and consequently the compressibility decreases. The

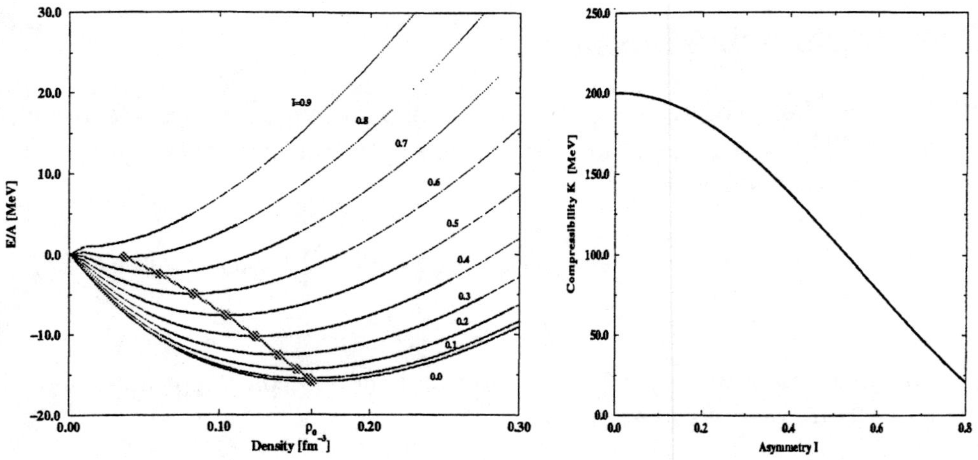

$$K = 9n^2 \frac{\partial(\mathcal{E}/A)}{\partial n^2}$$

Figure 1: Left: Mass number dependence of the energy for different neutron excess $I = \frac{\varrho_n - \varrho_p}{\varrho_n + \varrho_p}$. The dots denote the saturation densities. With increasing asymmetry the saturation density and the binding energy decrease. Right: Asymmetry dependence of compressibility modulus. With increasing asymmetry this modulus decreases.

compression modulus is defined as the derivative of the Energy per particle

$$K = 9n^2 \frac{\partial(\mathcal{E}/A)}{\partial n^2} \tag{7}$$

and took for $I = 0$ the value $K = 200$ MeV according to (5). Figure 1 shows the asymmetry dependence of K. One sees that for neutron matter no binding energy would occur and a very low compressibility.

We solve now the kinetic equation by representing the distribution function $f(\mathbf{p}, \mathbf{r}, t)$ by a sum of N pseudo-particle distributions

$$f(\mathbf{p}, \mathbf{r}, t) \approx f_0(\mathbf{p}, \mathbf{r}, t) \quad = \quad \sum_{i=1}^{AN} \frac{1}{N} f_S(\mathbf{p} - \mathbf{p_i}(t), \mathbf{r} - \mathbf{r_i}(t)) \tag{8}$$

and use Gaussian pseudo-particles

$$f_S(\mathbf{p} - \mathbf{p_1}, \mathbf{r} - \mathbf{r_1}) = c\ e^{-(\mathbf{p}-\mathbf{p_1})^2/2\sigma_p^2}\ e^{-(\mathbf{r}-\mathbf{r_1})^2/2\sigma_r^2} \tag{9}$$

at $\mathbf{r_1}$ with momentum $\mathbf{p_1}$ [54]. These pseudo-particles follow classical Hamilton equations

$$\dot{\mathbf{p}}_i \quad = \quad -\nabla U, \quad \dot{\mathbf{r}}_i = \frac{\mathbf{p}_i}{m}\ . \tag{10}$$

The only fluctuations introduced are due to some small unavoidable numerical noise[55]. We are using 300 pseudo-particles per nucleon and a pseudo-particle width of $\sigma_r = 0.466$ fm. They are adjusted both in such a way that the experimental energy of the ^{40}Ca giant monopole mode is reproduced. With these fixed two parameters the experimental behavior of centroid energy with mass number is than reproduced over the full range for giant monopole and giant dipole resonances. We have checked different numbers of test particles. The dependence of observables on the width is discussed in [56]. Numerically the ground states of nuclei are realized by Wood- Saxon shapes of density and Fermi spheres in momentum.

3 Multipole analysis

Provided we have now solved the kinetic equation we will have the distribution function represented by N pseudo-particles. Since we are interested in moments of the distribution function, the density, current and energy, we can expand these moments in the test particle representation as well. The momentum distribution of these moments can be obtained by spatial integration: $F_a(\mathbf{p}) = \int d\mathbf{r}\, a\, f(\mathbf{p}, \mathbf{r})$ with, for the mass distribution, $a = 1$, for isospin, $a = \tau$, for kinetic energy, $a = \frac{p^2}{2m}$, and for kinetic isospin energy, $a = \frac{\tau p^2}{2m}$ respectively. Decomposed into spherical coordinates they read

$$F_a(p, \vartheta, \varphi) = \sum_{i=1}^{AN} \frac{(2\pi)^3}{N} a_i \frac{\delta(p - p_i)}{p_i^2} \delta(\varphi - \varphi_i) \frac{\delta(\vartheta - \vartheta_i)}{\sin(\vartheta_i)} \,. \tag{11}$$

Radial integration determines now a spherical distribution

$$\bar{F}_a(\vartheta, \varphi) = \sum_{i=1}^{AN} \frac{1}{N} a_i \delta(\varphi - \varphi_i) \frac{\delta(\vartheta - \vartheta_i)}{\sin(\vartheta_i)} \tag{12}$$

which can be decomposed into spherical harmonics

$$\bar{F}_a(\vartheta, \varphi) = \sum_{l=0}^{\infty} \sum_{m=-l}^{l} a_{lm} Y_{lm}(\vartheta, \varphi) \,, \tag{13}$$

$$a_{lm} = \sum_{i=1}^{AN} \frac{1}{N} a_i Y_{lm}^{\star}(\vartheta_i, \varphi_i). \tag{14}$$

The observable distributions $\bar{F}_a(\vartheta, \varphi)$ are normalized to $\sqrt{4\pi}\ a_{00}$, i.e. to mass number A, total isospin T, kinetic energy E_{kin} and kinetic isospin energy E_{kinT}, respectively. The polar angular distribution of moments reads now from (12) and (13)

$$\hat{F}_a(\vartheta) = \sum_{i=1}^{AN} \frac{1}{N} a_i \frac{\delta(\vartheta - \vartheta_i)}{\sin(\vartheta_i)} \equiv \int_0^{2\pi} d\varphi\, \bar{F}_a(\vartheta, \varphi) = \sum_{l=0}^{\infty} a_{l0} \sqrt{\tfrac{2l+1}{4\pi}} P_l(\cos\vartheta) \tag{15}$$

with Legendre polynomials P_l. As a measure for the strength of the resonances we obtain now the coefficients a_{l0} of the corresponding moment from (15) as

$$a_{l0} \sqrt{\frac{2l+1}{4\pi}} = \frac{2l+1}{2} \sum_{i=1}^{AN} \frac{1}{N} a_i \frac{P_l(\cos\vartheta_i)}{\sin(\vartheta_i)}. \tag{16}$$

These amplitudes of multipole moments, $a_{l0}\sqrt{\frac{2l+1}{4\pi}}$, are displayed as a function of time in figures 2,5,8 and 9. The value a_{10} means the dipole moment, vanishing for isoscalar resonances, a_{20} is characterizing the quadrupole oscillations and a_{30} the octupole ones, etc.

4 Isoscalar monopoles

The first analyzed mode is the isoscalar giant monopole mode which plays an important role in the determination of nuclear compressibility. The connection of the compressibility and the energy of giant monopole resonances is discussed e.g. in [57].

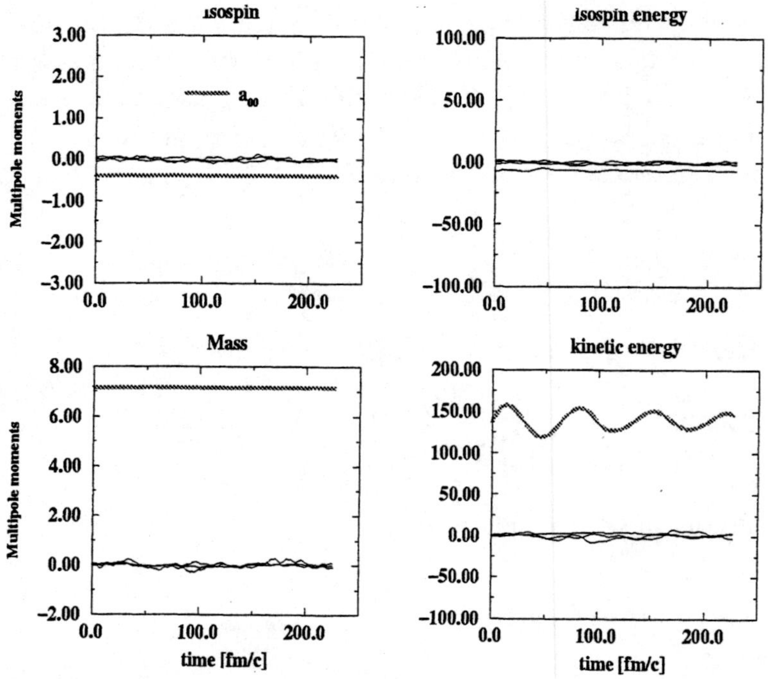

Figure 2: The picture shows for isoscalar monopole oscillations of ${}^{90}_{40}Zr$ the time evolution of multipole moments $\sqrt{\frac{2l+1}{4\pi}} a_{l0}(t), l = 0, 1, 2, 3$ corresponding to monopole, dipole, quadrupole and octupole oscillation for the distribution of isospin $a = \tau$, kinetic isospin energy $a = \frac{\tau p^2}{2m}$, mass $a = 1$ and kinetic energy $a = \frac{p^2}{2m}$.

energy E_{kin} and kinetic isospin energy E_{kinT}, respectively. The polar angular distribution of moments reads now f (2.12) and (2.13)

$$\hat{F}_a(\vartheta) = \sum_{i=1}^{AN} \frac{1}{N} a_i \frac{\delta(\vartheta - \vartheta_i)}{\sin(\vartheta_i)} \equiv \int_0^{2\pi} d\varphi \, \bar{F}_a(\vartheta, \varphi) \;=\; \sum_{l=0}^{\infty} a_{l0} \sqrt{\tfrac{2l+1}{4\pi}} P_l(\cos\vartheta) \tag{2.15}$$

The figure 2 shows simulation results for a monopole oscillation. Here the excitation has been performed by adding an extra momentum to the test-particles in the direction of the center of mass. We have excited in this way a clear monopole breathing mode which can be seen in the oscillation of the kinetic energy. The corresponding mean field energy performs the opposite oscillations that the total energy is constant. The fact that energy is oscillating between kinetic and correlational ones describes why we have here a compressional or breathing mode. All other modes remain unexcited. The finite value of the isospin and mass for $l = 0$ reflects the conservation of isospin and particle number.

With the previously chosen fixed width of test-particles we see from figure 3 that the experimental mass number dependence of monopole oscillation is well reproduced over the whole range of mass numbers. However, the experimental damping can be seen to be largely underestimated by the Vlasov simulation which yields about 2 MeV. This is a first indication that the mean field, even for a finite system, cannot account for the whole damping and dissipation. Instead we have to take into account collisional correlations which will be performed later.

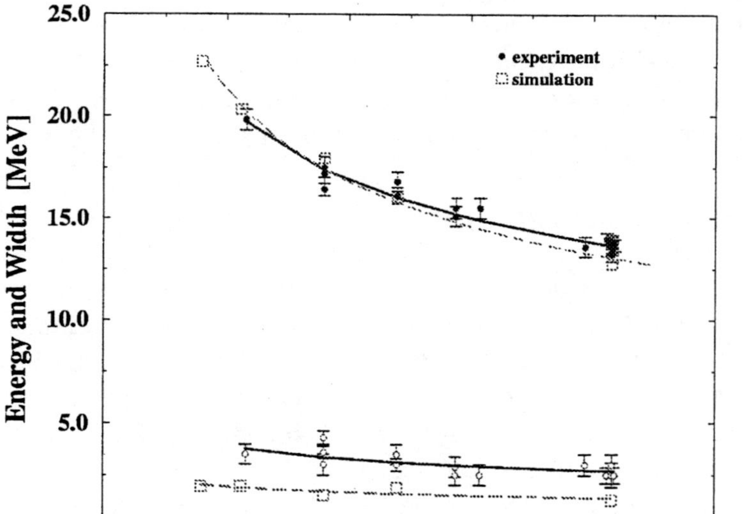

Figure 3: The mass number dependence of giant monopole oscillation energy and damping width. The solid line denotes the experimental values while the dashed line describes the Vlasov simulation results.

It is now instructive to examine the isospin dependence of the isoscalar monopole. Since this is the experimental value which determines the nuclear compressibility, the isospin dependence is of direct importance. In figure 4 the dependence on asymmetry of the isoscalar monopole energy is plotted for a hard as well as a stiff equation of state. Of course, the hard equation of state leads to a higher monopole energy corresponding to a higher compressibility. Analogously to figure 1 the monopole energy decreases with the asymmetry. The hard equation of state leads to a more linear decrease while the soft equation of state remains almost unchanged up to a certain asymmetry and then decreases faster.

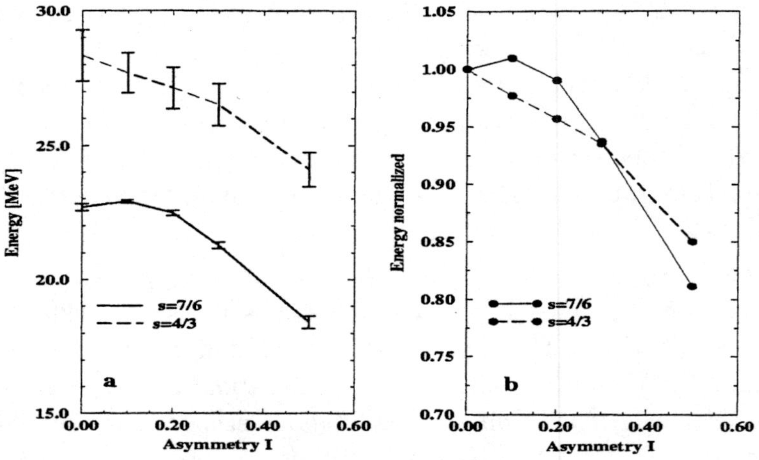

Figure 4: The giant monopole energy (a) and normalized to the symmetric case (b) versus asymmetry for two different parameterizations of the mean field according to (6). The stiff equation of state (black line) leads to higher monopole energy and shows a more linear decrease while the soft equation of state (grey line) shows a weak dependence for small asymmetry and decreases faster for higher asymmetries.

5 Isovector dipole resonances

On the next figure 5 the multipole analysis for $^{90}_{40}Zr$ oscillations is performed. The excitation is now chosen as an isovector dipole one created by a shift of proton against neutron spheres. One recognizes that a clear dipole a_{10} mode is excited which is seen in the oscillation of the isospin and isospin energy. The kinetic energy now remains constant in contrast to the isoscalar mode because we have no compression mode. Like the monopole case we now investigate the mass number dependence. With the same parameterization as for the monopoles, the experimental mass dependence of the energy is well reproduced, figure 6. The damping is again under-predicted. The Vlasov simulation leads to a nearly constant width of 2 MeV. This is again a hint that collisions cannot be neglected. The isospin dependency of the dipole mode is plotted in figure 7 and is characterized by an increasing width and decreasing centroid energy with increasing asymmetry.

6 Quadrupole resonances

Let us now investigate the quadrupole resonances. For these we distinguish isoscalar and isovector modes. The quadrupole resonances are excited by dividing the spatial distribution into two equal pieces which are accelerated in opposite directions. In the first example the neutrons and protons are in phase giving rise to an isoscalar mode.

We see from the kinetic energy in figure 8 that besides the clear isoscalar quadrupole mode there is also a weak isoscalar monopole mode excited. The isospin dependency of the isoscalar quadrupole mode is plotted in figure 7. As in the case of the isovector dipole mode one sees a slight decrease of the centroid energy and an increase of damping with increasing asymmetry.

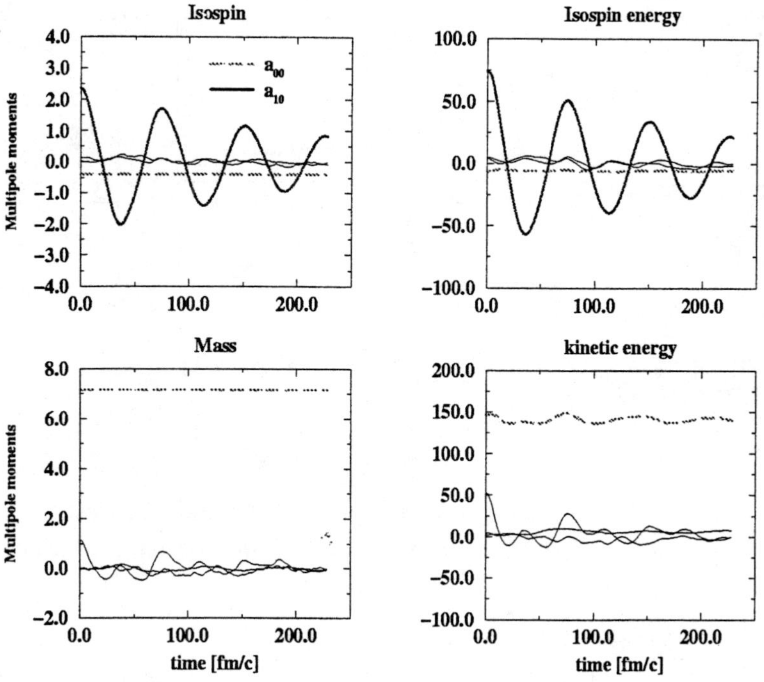

Figure 5: The time evolution of multipole moments for the isovector dipole excitation of ${}^{90}_{40}Zr$ analogously to figure 2.

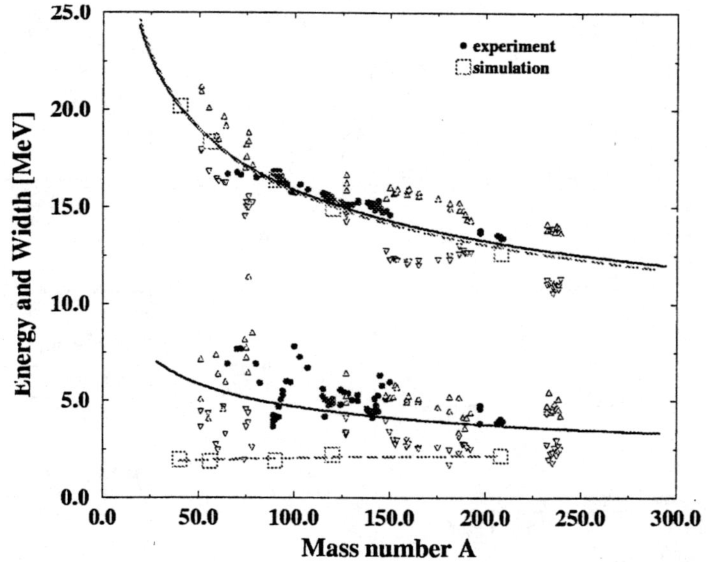

Figure 6: The experimental mass number dependence of isovector giant dipole oscillations (upper curve) is compared with the Vlasov simulation results. The oscillations in the experimental damping width are due to shell closures.

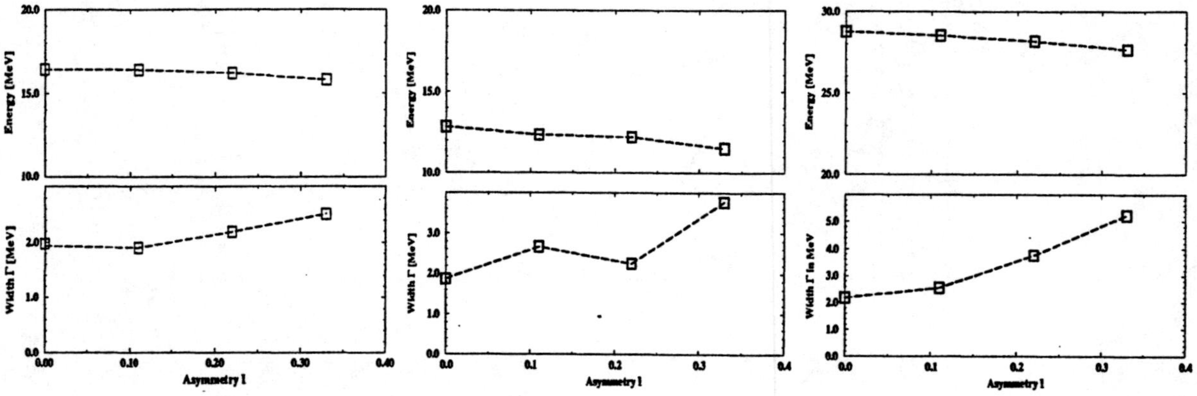

Figure 7: The isospin dependency of the centroid energy and the damping of the isovector dipole (left), isoscalar quadrupole (middle) and isovector quadrupole mode (right).

In the second case we excite protons and neutrons out of phase which excites isovector quadrupole oscillations in figure 9. The isospin energy and the isospin show nice oscillations in the quadrupole multipole moment. There is a slight excitation of an isovector dipole mode too. We would like to remark that due to the test particle width there occurs a mode coupling [56] which can be seen in the slight excitation of isoscalar modes in the kinetic energy. The isospin dependence shown in figure 7 has the same qualitative behaviour as the isoscalar mode, but is stronger in this case.

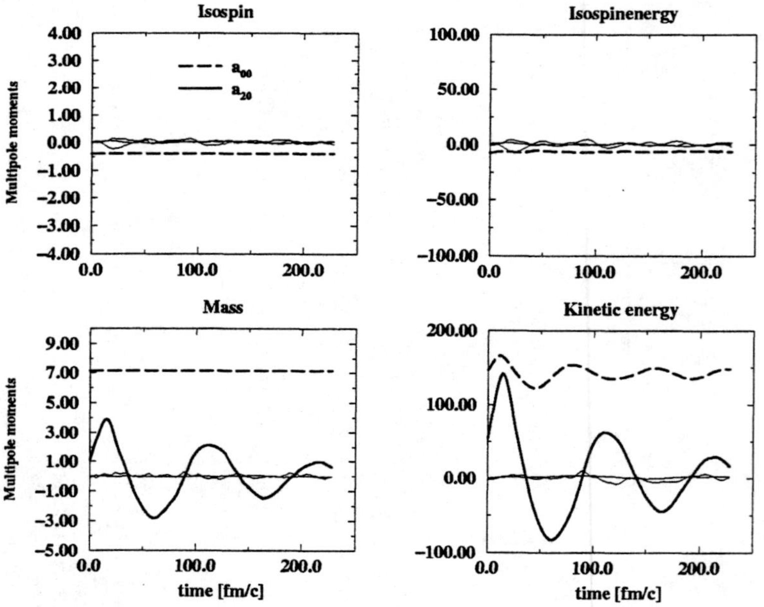

Figure 8: The time evolution of multipole moments for the isoscalar quadrupole excitation of $^{90}_{40}Zr$ analogously to figure 2.

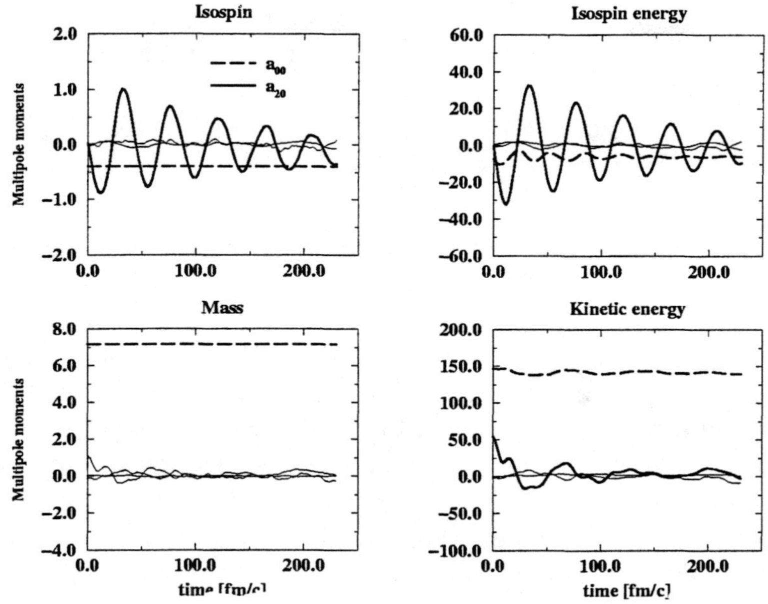

Figure 9: The time evolution of multipole moments for the isovector quadrupole excitation of ${}^{90}_{40}Zr$ analogously to figure 2.

7 Collective modes in linear response

Now we would like to focus on small amplitude oscillations and we will develop the linear response theory. Large amplitude motions and effects beyond linear response are discussed in chapter 16. The simplest microscopic theory which provides basic experimental features and which allows to include collisional correlations is the Fermi-gas(liquid) model including dissipation. We will compare the results from this linear response with the simulation of finite nuclei, and both will be compared with the experimental data.

The principle of linear response is easily explained. When a system which is described by the kinetic equation (1) is disturbed by an external potential $U^{\rm ext}$ it will react and will create a density change $\delta\rho$. The connection of the latter to the external perturbation is called the response function, $\delta\rho = \chi U^{\rm ext}$. Without mean field in (1) we would obtain the polarization Π of the system. The mean field introduces a selfconsistency such that the relation between response function and polarization becomes

$$\chi = \frac{\Pi}{1 - \frac{\delta U}{\delta n}\Pi} \tag{17}$$

where we have assumed that the mean field U is only density dependent. Therefore the potential for the response function is $V = \frac{\delta U}{\delta n}$. The collective mode of a system is now characterized by the condition that the denominator in (17) vanishes

$$\epsilon(\mathbf{q}, \omega) = 1 - V(\mathbf{q})\Pi(\mathbf{q}, \omega) = 0 \tag{18}$$

since then an infinitesimal small external potential can create a finite density oscillation due to the diverging response function. Therefore we will search for complex zeros of the dielectric (or dinuclic) function (DF), $\epsilon(\mathbf{q}, \Omega + i\gamma) = 0$, which provides us with the energy Ω and the damping γ of the collective mode. The strength function is then given by

$$S(\mathbf{q}, \omega) = \frac{1}{\pi}\frac{{\rm Im}\Pi}{(1 - V{\rm Re}\Pi)^2 + (V{\rm Im}\Pi)^2}. \tag{19}$$

The equation of state like the isothermal compressibility can be expressed by

$$\kappa = \frac{1}{n^2}\left(\frac{\partial n}{\partial \mu}\right)_T = \frac{1}{n^2 T}\lim_{q\to 0}\int \frac{d\omega}{\pi}\frac{1}{e^{\omega/T}-1}\mathrm{Im}\chi(q,\omega) \tag{20}$$

and the fluctuations and diffusion coefficients can also be expressed via the response function.

Two lines of theoretical improvements of the response function can be found in recent publications. The first one starts from TDHF equations and considers the response of nuclear matter described by a non time-reversal Skyrme interaction [14, 58, 59]. The other line tries to improve the response by the inclusion of collisional correlations [60, 61, 62, 63, 64] and by considering multicomponent systems [18, 65, 66, 67, 68]. In [64] both lines of improvements have been combined into one expression. We follow this line and derive the response function from a kinetic equation including mean field (Skyrme) and collisional correlations. Therefore we consider interacting matter which can be described by an energy functional (mean field) $\mathcal{E}$ originally introduced by Skyrme [69, 70] and the residual interaction. The latter we condense in a collisional integral additional to the TDHF equation. Then the response to an external perturbation will contain the effect of Skyrme mean field and additionally the effect of the residual interaction.

8 Response function for asymmetric matter

The system now consists of a number of different species (neutrons, protons, etc...) interacting with their own kind and with the others. It is important to consider the interaction between different sorts of particles if we want to include friction between different streams of isospin components, etc. In particular the isospin current may not be conserved in this way. Let us start with a set of coupled quantum kinetic equations[1] for the reduced density operator $\hat{\rho}_a$ of a certain species denoted by the subscript a

$$\partial_t \hat{\rho}_a(t) = i[\hat{\rho}_a, \hat{E}_a(t) + \hat{U}_a^{\mathrm{ext}}] - \sum_b \int_0^t dt' \frac{\hat{\rho}_a(t') - \hat{\rho}_b^{\mathrm{l.e.}}(t')}{\tau_{ab}(t-t')} \tag{22}$$

where $\hat{E} = \hat{P}^2/2m + \hat{U}$ denotes the kinetic as well as mean field energy operator and the external field which is assumed to be a nonlinear function of the density. We approximate the collision integral by a non-Markovian relaxation time which is derived in appendix A. This turned out to be necessary to reproduce damping of zero sound [71, 72]. It accounts for the fact that during a two-particle collision a collective mode can couple to the scattering process. Consequently, the dynamical relaxation time represents the physical content of a hidden three-particle process and is equivalent to memory effects.

Furthermore, we assume relaxation towards a local equilibrium

$$f_a^{\mathrm{l.e.}}(\mathbf{p},\mathbf{R},t) = \sum_q e^{iq\mathbf{R}}\langle p + \frac{q}{2}|\hat{\rho}_a^{\mathrm{l.e.}}|p - \frac{q}{2}\rangle = f_0\left(\frac{\varepsilon_a(\mathbf{p} - \mathbf{Q}_a(\mathbf{R},t)) - \mu_a(\mathbf{R},t)}{T_a(\mathbf{R},t)}\right) \tag{23}$$

[1]The quasiclassical Landau equation (1) follows from gradient expansion as

$$\frac{\partial}{\partial t}f + \partial_{\mathbf{p}}\epsilon\partial_{\mathbf{r}}f - \partial_{\mathbf{r}}\epsilon\partial_{\mathbf{p}}f = -\frac{f - f^{\mathrm{l.e.}}}{\tau} \tag{21}$$

with the Fermi distribution f_0. This local equilibrium will be specified here only by a small deviation of the chemical potential of species a ensuring density conservation [64]. The extension of the method including further conservation laws and specifying also the local current and the temperature can be found in [64].

Linearizing the kinetic equation (22) we obtain the matrix equation for the density deviation $\delta\rho_b$ due to an external perturbation $U^b_{\rm ext}$

$$\sum_b \delta n_b \left\{ \delta_{ab} - \frac{i}{\omega\tau_a + i}\left[\delta_{ab} - \frac{\tau_a}{\tau_b} C_{ab}\right] - \Pi_a(\omega + \frac{i}{\tau_a})\alpha_{ab} \right\} = \Pi_a(\omega + \frac{i}{\tau_a}) U^a_{\rm ext}. \tag{24}$$

The matrix C_{ab} is given by

$$C_{ab} = \sum_c \left\{\frac{1}{\tau}\right\}_{ac} \frac{\Pi_c\left[(\omega + \frac{i}{\tau_a})\frac{m_a}{m_c}\right]}{\Pi_c(0)} \left\{\frac{1}{\tau}\right\}^{-1}_{cb} \tag{25}$$

with $\alpha_{ab} = \frac{\partial}{\partial \rho_b} U_a$ the linearization of the mean field with respect to the deviation of density from equilibrium value caused by the external perturbation $U_{\rm ext}$. The partial polarization function of species a is

$$\Pi_a(\omega) = 2\int \frac{dp}{(2\pi)^3} \frac{f_a(p + \frac{q}{2}) - f_a(p - \frac{q}{2})}{\epsilon_a(p + \frac{q}{2}) - \epsilon_a(p - \frac{q}{2}) - \omega + i0}. \tag{26}$$

The factor 2 in front of the integral accounts for the spin degeneracy according to $\rho_a = 2\int \frac{dp}{(2\pi)^3} f_a(p)$.

Equation (24) represents the complete polarization of the system, because $\delta\rho = \Pi\, U_{\rm ext}$. It represents a matrix equation which is solved easily. The collective modes are given by the zeros of the determinant of the matrix on the left hand side of (24) because these are the poles of the polarization matrix. Since we took into account relaxation processes between all species we are able to cover current - current friction. For a two component system, e.g. neutrons with density ρ_n and protons with density ρ_p, we write explicitly

$$(1 - \Pi^{\rm M}_n \alpha_{nn})(1 - \Pi^{\rm M}_p \alpha_{pp}) - (D_{np} + \Pi^{\rm M}_n \alpha_{np})(D_{pn} + \Pi^{\rm M}_p \alpha_{pn}) = 0 \tag{27}$$

with the generalization of the Mermin polarization function [73] to a multicomponent system

$$\Pi^{\rm M}_a = \frac{\Pi_a(\omega + \frac{i}{\tau_a})}{1 - \frac{i}{\omega\tau_a + i}(1 - C_{aa})}, \tag{28}$$

and the additional coupling due to asymmetry in the system

$$D_{np} = \frac{\tau_n}{\tau_p} \frac{C_{np}}{C_{nn} - i\omega\tau_n}. \tag{29}$$

The D_{pn} are given by interchanging species indices. This term does not appear for symmetric matter. Therefore we call this term the asymmetry coupling term further on. The result (27) represents the generalization of the dispersion relation (18) to a two - component system.

It is illustrative to recover known results for symmetric nuclear matter. This is performed for the case of equal relaxation times $\tau_p = \tau_n = \tau$ and equal deviation from the mean field

$\alpha_1 = \alpha_{nn} = \alpha_{pp}$ and $\alpha_2 = \alpha_{np} = \alpha_{pn}$. Eq. (27) takes then the known Mermin form of dispersion relation [73]

$$1 - (\alpha_1 \pm \alpha_2)\frac{\Pi(\omega + \frac{i}{\tau})}{1 - \frac{i}{\omega\tau + i}\left[1 - \frac{\Pi(\omega + \frac{i}{\tau})}{\Pi(0)}\right]} = 0 \tag{30}$$

with the isovector mode $\alpha_1 - \alpha_2$ and the isoscalar mode $\alpha_1 + \alpha_2$. Please note that for zero temperature the Mermin expression (30) agrees with the response function derived from taking into account multipole expansion in momentum of the disturbed distribution [36].

We have presented a general dispersion relation for the multicomponent system including known special cases. The dispersion relation (27) is similar to the one derived recently in [67, 74] if we neglect the collisional coupling D_{np}. Also a more general polarization function (28) is presented here including collisions within a conserving approximation [60]. In the following we will apply this expression to the damping of giant dipole resonances in symmetric as well as asymmetric nuclear matter.

9 Application to nuclear matter

Before we can solve the dispersion relation (27) we have to specify the wave vector characteristic of the considered mode. This is performed according to the Steinwedel-Jensen [75] model. We assume that the surface remains constant and the density oscillation obeys a wave equation with the boundary condition that the radial velocity vanishes on the spherical surface with radius $R = 1.13A^{1/3}$. This leads to $j_l'(kR) \equiv 0$ with the spherical Bessel function of order $l = 0, 1...$ associated with the monopole, dipole... resonances. From this condition one obtains a connection between the wave vector and the radius of the nuclei or the mass number in the form

$$k_{i,l} = \frac{c_{il}}{1.13A^{1/3}} \tag{31}$$

for the i-th mode of multipolarity l where c_{il} is the i-th zero of $j_l'(c) \equiv 0$.

For the monopole modes as compression modes we have no zero of first order for $l = 0$. For the dipole modes one has in first order $k = 2.08/R$ which describes the giant isovector dipole resonance (IVGDR) while the ISGDR is a spurious mode in first order. This would just mean an unphysical oscillation of center of mass motion. However, in second order $k = 5.94/R$ the isoscalar giant dipole resonance (ISGDR) has been observed recently ([76] and references therein). This can be considered as a density oscillation inside a sphere as we will discuss in chapter 13.

Besides the occurring wave vector we have also to specify the relaxation times which contain the effect of collisions. The dynamic relaxation time has been derived by Sommerfeld expansion [6] in appendix A as

$$\frac{1}{\tau_{ab}^{\rm gas}(\omega)} = \frac{1}{\tau_{ab}(0)}\left[1 + \frac{3}{4}\left(\frac{\omega}{\pi T}\right)^2\right] \qquad \frac{1}{\tau_{ab}^{\rm liq}(\omega)} = \frac{2}{\tau_{ab}(0)}\left[1 + \frac{1}{2}\left(\frac{\omega}{\pi T}\right)^2\right] \tag{32}$$

for a, b neutrons or protons respectively and dependent on whether we use a Fermi liquid or Fermi gas model. The Markovian relaxation time is given in terms of the cross section σ_{ab} between species a and b as $\tau_{ab}^{-1} = \frac{4m}{3\hbar^3}\sigma_{ab}T^2$.

We will use as an illustrative example a slightly different mean field potential than (2) given by [77, 78]

$$U_a = t_0 \left[(1 + \frac{x_0}{2})(\rho_n + \rho_p) - (x_0 + \frac{1}{2})\rho_a \right] + \frac{t_3}{4} \left[(\rho_n + \rho_p)^2 - \rho_a^2 \right] \tag{33}$$

with $a = n, p$ the density of neutrons or protons respectively. The Coulomb interaction leads to an additional contribution for the proton mean-field

$$U_p^C(q) = \frac{4\pi e^2}{q^2} \rho_p(q). \tag{34}$$

The model parameters used here reproduce the Weizsäcker formula

$$\frac{\mathcal{E}}{A} = -a_1 + \frac{a_2}{A^{1/3}} + \frac{a_3 Z^2}{A^{4/3}} + a_4 I^2 \tag{35}$$

with the volume energy $a_1 = 15.68$ MeV, Coulomb energy $a_3 = 0.717$ MeV and the symmetry energy $a_4 = 28.1$ MeV with the asymmetry parameter $I = \frac{n_n - n_p}{n_n + n_p}$.

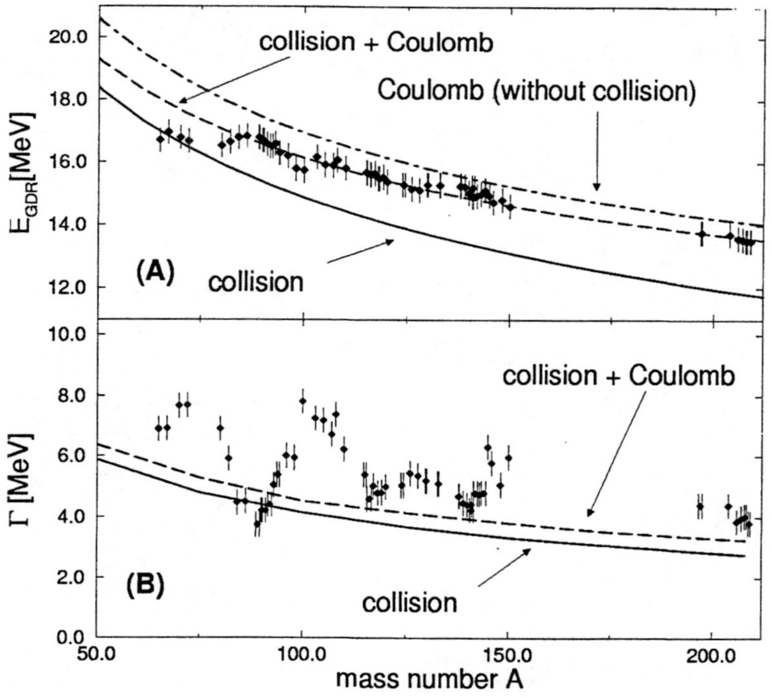

Figure 10: The experimental centroid energies **(A)** and damping rates **(B)** [filled symbols] of the giant dipole resonances vs mass number (data from Ref. [79]) together with different approximations at $T = 0$.

The energy and damping rates are now determined by the zeros of the (Mermin) polarization function (30). First we plot the solution of the dispersion relation (27) for symmetric nuclear matter. In figure 10 we have plotted the real and imaginary (FWHM) part of complex energy for different approximations with relaxation time (collisions) and with and without Coulomb mean field (34). In figure 10 **(A)** we find that the inclusion of Coulomb effects reproduces the experimental shape of the centroid energies at higher mass numbers (dot-dashed line). Taking only collisions into account fails to describe higher mass numbers (solid line). Considering Coulomb together with collisions (dashed line), the centroid energies are reduced towards the data.

The experimental values of the damping rates are also presented versus mass number (figure 10 (**B**)). We recognize that the FWHM is just twice the damping rate $\Gamma = 2\gamma$ which has been recently emphasized [15]. It has to be stressed that the experimental data are accessible by this FWHM. Here we have considered only collisions and have a vanishing Landau damping for the infinite matter model [6].

10 Comparison between simulation and linear response

We have already stressed that the simulation of finite nuclei neglecting collisional correlations lead to a damping of about $\Gamma_{\text{Land}} = 2\text{MeV}$ for the giant dipole resonance at zero temperature. This was attributed to the surface effects of spherical symmetric nuclei. We consider this as the numerical value of the contribution attributed to the wall [36] at zero temperature. For finite temperature the will be considered an additional contribution due to thermally deformed shapes in section 12. Nevertheless it is now interesting to compare the isospin dependences of the simulation without collisions and the linear response including collisions but no surface effects.

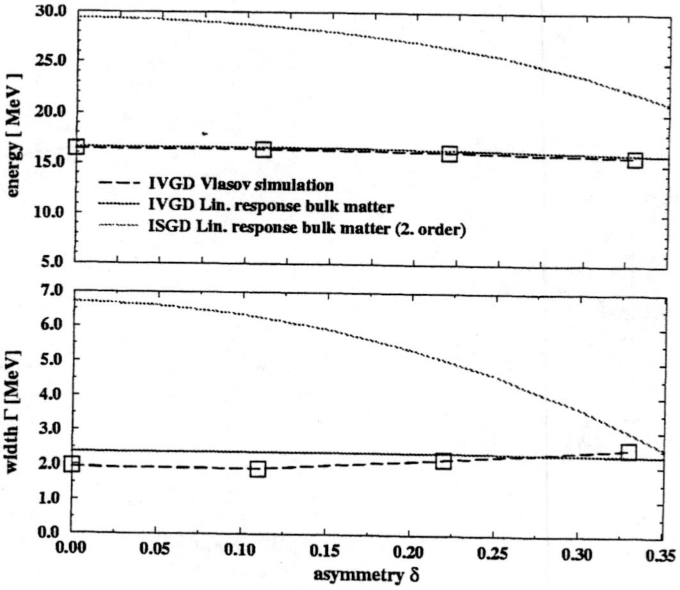

Figure 11: The dependence of isovector dipole resonance energy and damping on the asymmetry obtained from simulations and from linear response for $A = 90$.

In figure 11 we see that the damping is slightly increasing with increasing asymmetry for simulations without collisions while the collisional contributions cause a slight decrease of damping. This decrease is much more pronounced in the second order isoscalar dipole mode. The energy in turn shows the same behaviour for simulations without correlations and for linear response with correlations; it slightly decreases the value.

The same qualitative feature can be recognized for the quadrupole modes in figure (12). However the importance of collisional damping is in this case much larger than the finite size effects.

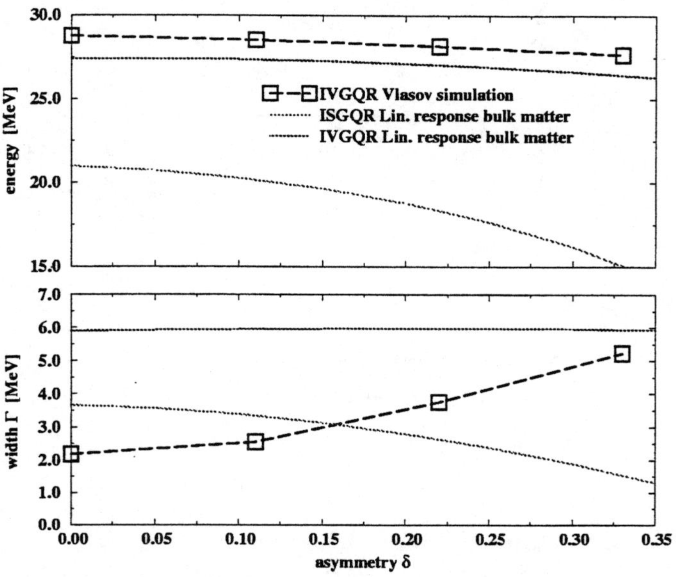

Figure 12: The dependence of isovector and isoscalar quadrupole resonance energy and damping on the asymmetry obtained from simulations and from linear response for $A = 90$.

11 Giant resonances in excited nuclei

We consider now the temperature dependence of the GDR. We have found that the centroid energy only slightly decreases with increasing energy. In Fig. 13 the theoretical damping rates $\Gamma = 2\gamma$ of the IVGDR modes in ^{120}Sn and ^{208}Pb are plotted as a function of temperature together with experimental values. The results of the Fermi gas model (84) and the Fermi liquid model (92) are very close and in good agreement with the data at low temperature. The higher temperature dependence still remains too flat compared to the experiments. The small difference between both models for T=0 vanishes with increasing temperature.

While the figures indicate that the linear response including collisional correlations together with the Steinwedel Jensen model can describe quite well the gross features of giant dipole resonances we should keep in mind that we have neglected here the surface effects. From the simulation results of the second chapter we have seen that there is a damping of about $\Gamma_{\rm Land} = 2$MeV for finite nuclei at zero temperature without collisional correlations. Simply adding now both contributions would overestimate the experimental damping. We would like to point out that the inclusion of only density conservation in the derivation of the response function so far is overestimating the width. In [64] we have shown that the inclusion of momentum conservation reduces this width by approximately 2MeV. Therefore we might anticipate that both the finite size effects together with the collisional contribution lead to the correct value for low temperatures. In figure 14 we see that the temperature

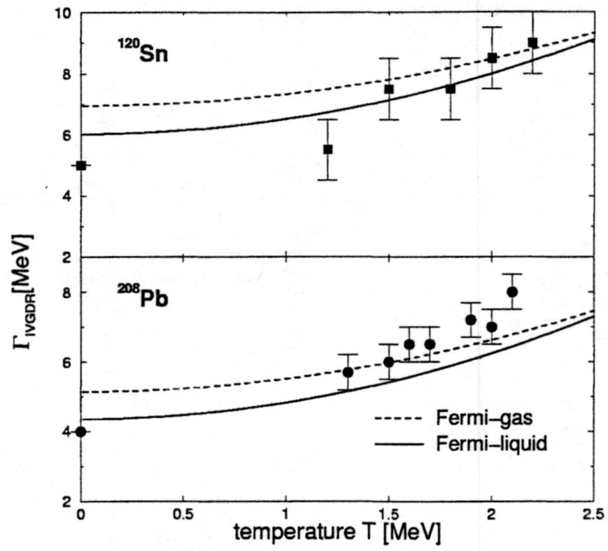

Figure 13: Experimental damping rates vs. temperature of IVGDR for ^{120}Sn and ^{208}Pb (^{120}Sn from Ref. [80] and ^{208}Pb from Ref. [81]) compared with the solution of the dispersion (30) relation $\Gamma = 2\,\gamma$ for the Fermi gas (dashed lines) and the Fermi liquid model (solid lines). Please note that this damping is reduced when the momentum conservation is included like it is done in figure 14.

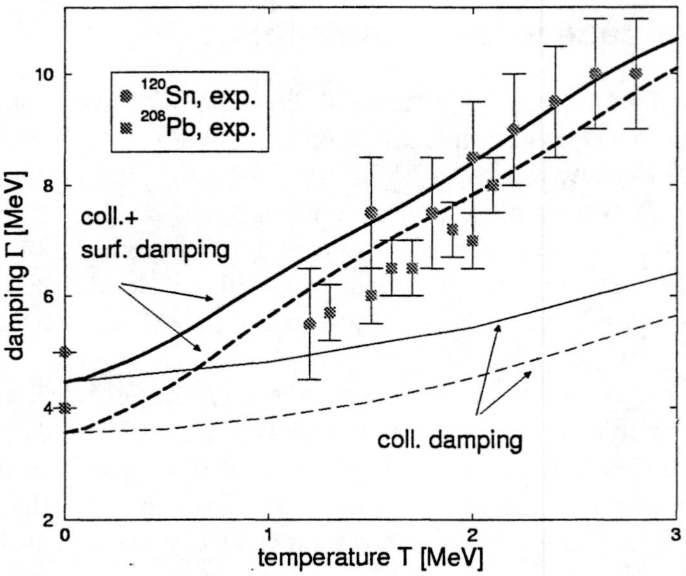

Figure 14: The effective damping consisting of only collisional damping compared with collisional and surface damping [^{120}Sn (solid lines) and ^{208}Pb (dashed lines)] together with the experimental data (filled symbols: Sn from Ref. [80] and Pb from Ref. [81]).

increase is too flat compared to the experimental finding if we consider only collisional damping (thin lines). From simulation results of ground state we have already found a contribution from finite size effects. We might now expect that the discrepancy in temperature behaviour is due to finite size effects. This should be explained and understood within a more simplified model.

12 Nuclear surface contribution - excited nuclei

Besides the $\Gamma_{\rm Land} = 2$MeV of the simulation in the ground state we can have additional damping due to shape deformation if we increase the temperature. In Ref. [82] we have presented a method to include also scattering with the non-spherical nuclear surface. This improves the temperature dependence remarkably (thick lines) in figure 14.

The idea is to consider the surface scattering contribution to the damping determined by the additional Lyapunov exponent due to the surface. It has been shown that such Lyapunov exponents appear in the response function additive to the frequency as imaginary shift provided the Lyapunov exponent is small compared to the product of wave vector times Fermi velocity [82, 83]

$$\Pi_0^{\rm surf}(q,R,\omega) = \Pi_0^{\rm inf}(q,p_f(R),\omega + i\lambda). \tag{36}$$

The Lyapunov exponent by itself is calculated for different deformations of the surface corresponding to different temperatures

$$R_\lambda(\theta) = R_0\,(1 + \alpha_{00} + \alpha_\lambda\, P_\lambda(\cos{(\theta)})) \tag{37}$$

with the nuclear radius $R_0 = 1.13A^{1/3}$fm, and where $\lambda = 2$ corresponds to the quadrupole and $\lambda = 3$ to the octupole deformation [84]. The corresponding mean deformation [2] is linked to the temperature

$$\langle\alpha\rangle = \frac{\int d\alpha|\alpha| \exp{(-E_B(\alpha)/T)}}{\int d\alpha \exp{(-E_B(\alpha)/T)}} \tag{38}$$

where the surface dependent energy $E_B(\alpha)$ is given by the Bethe-Weizsäcker formula [3]

$$E_B(\alpha) = -a_1 + \frac{a_2}{A^{1/3}} + \frac{a_3 Z^2}{A^{4/3}} + a_4 I^2 + a_5 A^{2/3}\frac{S(\alpha)}{S(0)} \tag{39}$$

with the volume energy $a_1 = 15.68$ MeV, Coulomb energy $a_3 = 0.717$ MeV, the symmetry energy $a_4 = 28.1$ MeV and the surface energy $a_5 = 18.56$ MeV. For the quadrupole, S_2, and octupole, S_3, deformations one gets

$$\begin{aligned} \frac{S_2(\alpha)}{S(0)} &= 1 + \frac{2}{5}\alpha^2 + \mathcal{O}(\alpha^3) \\ \frac{S_3(\alpha)}{S(0)} &= 1 + \frac{5}{7}\alpha^2 + \mathcal{O}(\alpha^3). \end{aligned} \tag{40}$$

[2]For small deviations we found identical Lyapunov exponents for prolate $\alpha > 0$ and oblate $\alpha < 0$ deformations and therefore we do not distinguish the sign of α.

[3]Please remember that in principle the Coulomb energy changes with small deformation as well according to the factor [85] $1 - 5(\lambda-1)/(2\lambda+1)^2\alpha_\lambda^2$ while the surface term changes as $1 + (\lambda-1)(\lambda+2)/2/(2\lambda+1)\alpha_\lambda^2$. Only the latter correction is considered since the Coulomb energy deformation would lead to corrections of around 0.3% and are neglected here.

This represents the lowest order expansion in α, however the next term gives already corrections in fractions of percent for the highest deformations considered here. By this way, the statistical model (38) leads to a connection between temperature and deformation as

$$T = c\pi a_5 A^{2/3} < \alpha >^2 \tag{41}$$

where the constant c is given by the coefficient of α^2 in Eq. (40) for the corresponding quadrupole or octupole deformation.

Using Eq. (41) we can translate the Lyapunov exponent λ calculated as a function of deformation into a function of the temperature. In figure 15 the contribution to the damping of IVGDR for ^{120}Sn (circles) and ^{208}Pb (squares) is presented for different shape deformations versus temperature. If we add quadrupole and octupole deformations we come up with a damping curve very similar to [27]. The damping starts at zero and increases rapidly with increasing temperature. We see that the main contribution comes from the quadrupole deformation while the octupole deformation is only sizeable at higher temperature. Let us note that the qualitative difference between Sn and Pb is reproduced by including surface scattering as well as collisional damping.

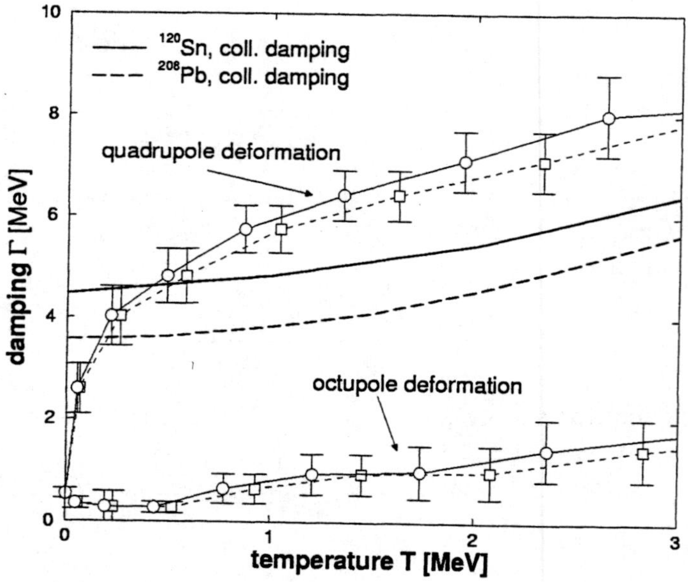

Figure 15: The scaled collisional damping is compared with the damping according to the chaotic scattering from the surface of quadrupole and octupole deformed shapes for ^{120}Sn and ^{208}Pb.

The collisional contribution $\Gamma = 2\gamma$ is plotted as well (Sn: solid line, Pb: dashed line). We recognize that both contributions by themselves, collisional as well as surface scattering, account almost for the same amount required by the experimental values, see figure 15. A proper relative weight between both processes is therefore necessary which will be introduced in the following.

Let us note before that we have considered the surface contribution to the damping for a deformed surface by temperature which therefore vanishes at zero temperature. However, for a spherical surface we got already from simulation a zero temperature contribution of $\Gamma_{\text{Land}} = 2\text{MeV}$. Consequently for the total damping caused by surface both contribute.

So far we have not considered that only particles close to the surface can appreciably contribute to the surface scattering and to this chaotization process, while particles deep inside the nuclei are screened out of this process. Consequently we consider the corresponding collision frequencies as the measure to compare surface collisions with inter-particle collisions. The collision frequency between particles is given by $1/\tau_0$ of (32). The collision frequency of particles with the deformed surface beyond a sphere, ν_{surf}, is given by the product of the density with the surface increase $S(\alpha) - S(0) = c\alpha^2 4\pi R_0^2$ according to Eq. (40) and with the mean velocity in radial direction $v_r = 3/8 v_F$. The result is

$$\nu_{\text{surf}} = 1.5 T n_0 v_F r_0^2 / a_5 \tag{42}$$

where we have used Eq. (41) to replace α. We see that the frequency (42) is independent of the size of the nucleus and linearly dependent on the temperature. We use the ratio of these two frequencies to weight properly the two damping mechanisms, the surface collisional, λ, and the inter-particle collisional, γ, contributions. Consequently the full - width - half - maximum (FWHM) reads

$$\begin{aligned} \Gamma_{\text{FWHM}} &= 2\zeta\gamma + 2(1-\zeta)(\lambda) + 2\gamma_{\text{Land}} \\ &\equiv \Gamma_{\text{coll}} + \Gamma_{\text{surf}} + \Gamma_{\text{Land}}. \end{aligned} \tag{43}$$

With the help of (32) and (42) the weighting factor ζ is given by

$$\zeta(T) = \frac{\frac{1}{\tau_0(T)}}{\frac{1}{\tau_0(T)} + \nu_{\text{surf}}(T)}. \tag{44}$$

One sees that for zero and high temperatures $\zeta = 1$ and due to Eq. (43) only the collisional contributions matter. Since ν_{surf} is linear in the temperature and $1/\tau$ depends quadratically on the temperature, the weighting factor ζ has a minimum at temperatures around $T_c = \frac{\sqrt{3}}{2\pi}\omega$ for the gas model (32) and the surface contributions become important. In the case of the IVGDR this corresponds to a temperature of $T \approx 3.7\text{MeV}$, which is the upper limit of currently achievable experimental temperatures. Therefore we can state that at low and high temperatures the collisional damping is dominant while for temperatures around T_c the surface contribution becomes significant.

In figure 14 we have compared the effective damping according to Eq. (43) with the experimental data. Let us recall that the collisional damping value is reduced here since we include momentum conservation too for the Mermin polarization function. On the other side the surface damping consists now of the deformation contribution by temperature and by the zero temperature contribution obtained by simulation of Vlasov equation. We find a reasonable quantitative agreement combining all these parts. This is illustrated more in detail in Fig. 16 where we have plotted the strength function (19) for ^{120}Sn (LHS) and ^{208}Pb (RHS) within the Fermi gas model (84) (dashed lines) and Fermi liquid model (92) (solid lines) with the normalized data from Ref. [86]. The good overall agreement of the shape evolution with the experiment is again accompanied by only minor differences between the Fermi gas and the Fermi liquid model.

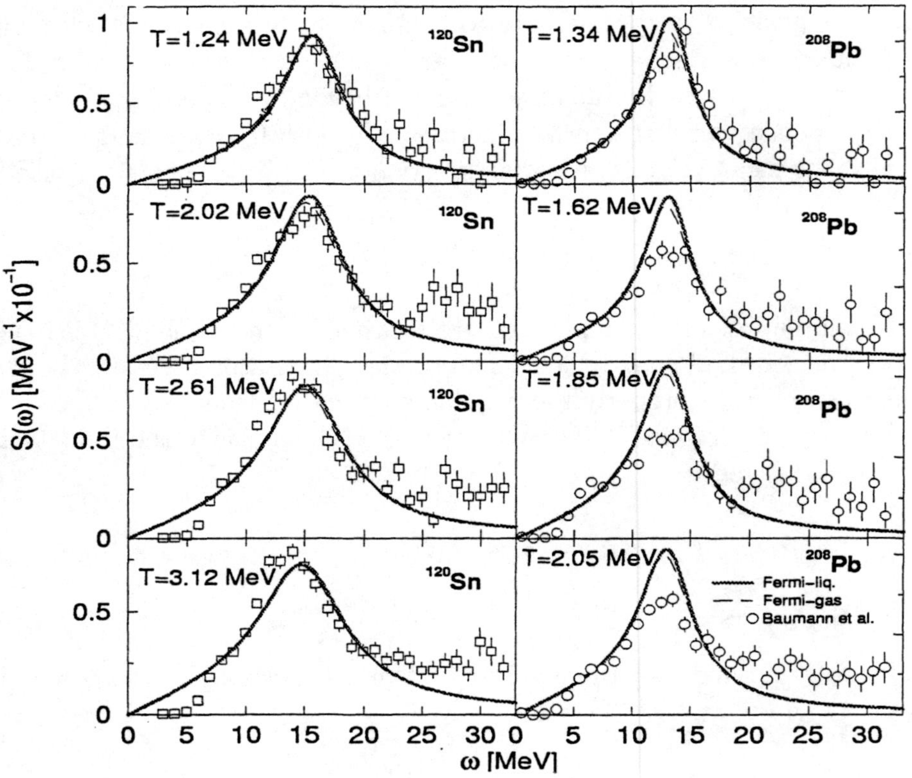

Figure 16: The IVGDR strength function in ^{120}Sn (LHS) and ^{208}Pb (RHS) within the Fermi gas model (dashed lines) and Fermi liquid model (solid lines) at several temperatures compared with normalized data from Ref. [86].

13 Higher order modes - surface modes

The existence of isoscalar giant dipole resonance (ISGDR) in nuclear matter is considered as a spurious mode in most text books since one associates with it a center of mass motion. The more surprising was the experimental justification of a giant resonance carrying the quantum numbers of a isoscalar and dipole mode [87, 76, 88]. Consequently one has to consider higher harmonics as explanation of such a mode [87, 89, 90]. Usually this mode is associated with a squeezing mode analogous to a sound wave [87, 76, 91, 88].

In this chapter we want to discuss the influence of surface effects on the ISGDR compression mode. We will show that even in the frame of the Fermi liquid model such modes can be understood. Moreover we claim that the surface effects are not negligible for reproducing the strength function. While we have already given a phenomenological approach to include surface effects we want to show now a straightforward possibility to include surface effects in the response function. Consequently we give first a short derivation of response function including surface effects. This will result in a new formula in the temperature-dependent extended Thomas-Fermi approximation [92].

The starting point is again the semi-classical Vlasov equation (1). Instead of using now a spatial homogeneous equilibrium as done so far we consider now explicitly the spatial dependence. Provided we know the response to the external potential without self-consistent mean field, which is described by the polarization function Π

$$\delta n(\mathbf{x},t) = \int d\mathbf{x}'\Pi(\mathbf{x},\mathbf{x}',t)U^{\text{ext}}(\mathbf{x}',t), \tag{45}$$

the response including mean field, χ, is given by

$$\chi(\mathbf{x},\mathbf{x}',t) = \Pi(\mathbf{x},\mathbf{x}',t) + \int d\mathbf{x}_1 d\mathbf{x}_2 \Pi(\mathbf{x},\mathbf{x}_1,t)\frac{\delta U^{\text{ind}}(\mathbf{x}_1,t)}{\delta n(\mathbf{x}_2,t)}\chi(\mathbf{x}_2,\mathbf{x}',t). \tag{46}$$

Therefore we concentrate first on the calculation of the polarization function Π and linearize the Vlasov equation (1) according to $f(\mathbf{p},\mathbf{R},t) = f_0(\mathbf{p},\mathbf{R}) + \delta f(\mathbf{p},\mathbf{R},t)$ such that the induced density variation $\delta\rho(\mathbf{R},t) = \int d\mathbf{p}\delta f/(2\pi\hbar)^3$ reads

$$\delta\rho(\mathbf{x},\omega) = \int \frac{d\mathbf{q}}{(2\pi\hbar)^3}e^{i\mathbf{q}\mathbf{x}} \int \frac{d\mathbf{p}d\mathbf{x}'}{(2\pi\hbar)^6}e^{-i\mathbf{q}\mathbf{x}'}\frac{\nabla_p f_0(\mathbf{p},\mathbf{x}')\nabla_{x'}U^{\text{ext}}(\mathbf{x}',t)}{i(\omega - \frac{\mathbf{p}\mathbf{q}}{m})}. \tag{47}$$

Here we have employed the Fourier transform of space and time coordinates of (1) to solve for δf and inverse transform the momentum into the form (47). Comparing (47) with the definition of the polarization function (45) we extract with one partial integration

$$\Pi(\mathbf{x},\mathbf{x}',\omega) = -\nabla_{x'}\int \frac{d\mathbf{p}d\mathbf{q}}{(2\pi\hbar)^6}e^{i\mathbf{q}(\mathbf{x}-\mathbf{x}')}\frac{\nabla_p f_0(\mathbf{p},\mathbf{x}')}{i(\omega - \frac{\mathbf{p}\mathbf{q}}{m})}. \tag{48}$$

With (48) and (46) we have given the polarization and response functions for a finite system.

In the following we are interested in the gradient expansion since we believe that the first order gradient terms will bear the information about surface effects. Therefore we change to the center of mass and difference coordinates $\mathbf{R} = (\mathbf{x}_1 + \mathbf{x}_2)/2$, $\mathbf{r} = \mathbf{x}_1 - \mathbf{x}_2$ and retaining only first order gradients we get from (48) after Fourier transform of $\mathbf{r}$ into $\mathbf{q}$

$$\Pi(\mathbf{R},\mathbf{q}) = -\int \frac{d\mathbf{p}}{(2\pi\hbar)^3}\frac{\mathbf{q}\nabla_p f_0(\mathbf{p},\mathbf{R})}{\omega - \frac{\mathbf{p}\mathbf{q}}{m}} + \frac{i}{2}\nabla_R \int \frac{d\mathbf{p}}{(2\pi\hbar)^3}\left(\frac{\nabla_p f_0(\mathbf{p},\mathbf{R})}{\omega - \frac{\mathbf{p}\mathbf{q}}{m}} - \frac{\mathbf{p}}{m}\frac{\mathbf{q}\cdot\nabla_p f_0(\mathbf{p},\mathbf{R})}{(\omega - \frac{\mathbf{p}\mathbf{q}}{m})^2}\right)$$

$$= -\int \frac{d\mathbf{p}}{(2\pi\hbar)^3}[\mathbf{q} - i\nabla_R(1 + \frac{\omega}{2}\partial_\omega)]\frac{\nabla_p f_0(\frac{p^2}{2m}, \mathbf{R})}{\omega - \frac{\mathbf{pq}}{m}} \tag{49}$$

where in the last equality we have assumed radial momentum dependence of the distribution function f_0. We recognize that besides the usual Lindhard polarization function as the first part of (49) we obtain a second part which is expressed by a gradient in space. The first part corresponds to the Thomas-Fermi result where we have to use the spatial dependence in the distribution functions and the second part represents the extended Thomas-Fermi approximation. So far we did not assume any special form of the distribution function. Therefore the expression (49) is also valid for any high temperature polarization of finite systems.

What remains to be shown is that the response function (46) does not contain additional gradients. This is easily confirmed by two equivalent formulations of (46), $\Pi^{-1}\chi = 1 + V\chi$ and $\chi\Pi^{-1} = 1 + \chi V$, which by adding yield the anticommutator

$$[\Pi^{-1}, \chi]_+ = 2 + [V, \chi]_+. \tag{50}$$

This anticommutator does not contain any gradients up to second order. Therefore we have $[V = \delta U^{\mathrm{ind}}/\delta n]$

$$\chi(\mathbf{R}, \mathbf{q}, \omega) = \frac{\Pi(\mathbf{R}, \mathbf{q}, \omega)}{1 - V(\mathbf{R}, \mathbf{q}, \omega)\Pi(\mathbf{R}, \mathbf{q}, \omega)} + \mathcal{O}(\partial_R^2). \tag{51}$$

Equation (51) and (49) give the response and polarization functions of finite systems in first order gradient approximation.

Now we are ready to derive approximate formulae for spherical nuclei. In this case we can assume $\mathbf{q}||\mathbf{R}$ and we have

$$\Pi(R, \mathbf{q}, \omega) = \Pi^0(R, \mathbf{q}, \omega) - \frac{i}{q}\partial_R\left[1 + \frac{\omega}{2}\partial_\omega\right]\Pi^0(R, \mathbf{q}, \omega) \tag{52}$$

where Π_0 is the usual Lindhard polarization with spatial dependent distributions (chemical potentials, density). We use now further approximations. In the case of giant resonances we are in the regime of small q and $\mathrm{Im}\Pi^0 \sim \omega$ such that

$$\left[1 + \frac{\omega}{2}\partial_\omega\right]\Pi^0 = i\frac{3}{2}\mathrm{Im}\Pi^0 \tag{53}$$

since $\mathrm{Re}\Pi^0 \sim 1 - c^2q^2/\omega^2$. Within the local density approximation we know that the spatial dependence is due to the density $\rho(R) = \rho_0\Theta(R_0 - R)$. Since we have for zero temperature $\mathrm{Im}\Pi^0 \propto p_f(\rho)$ we evaluate

$$\partial_R\mathrm{Im}\Pi^0 = -n_0\delta(R_0 - R)\partial_n\mathrm{Im}\Pi^0 = -\frac{1}{3}\delta(R_0 - R)\mathrm{Im}\Pi^0 \tag{54}$$

where we assumed the density dependence carried only by the Fermi momentum. Now it is straightforward to average (52) over space with the help of (54)

$$\begin{aligned} \Pi(\mathbf{q}, \omega) &= \frac{3}{R_0^3}\int_0^{R_0} dRR^2\Pi(R, \mathbf{q}, \omega) \\ &\approx \Pi^0(\mathbf{q}, \omega) + i\frac{3}{qR_0}\left[1 + \frac{\omega}{2}\partial_\omega\right]\frac{1}{3}\Pi^0(\mathbf{q}, \omega) \\ &= \Pi^0 + \Pi^{\mathrm{surf}}. \end{aligned} \tag{55}$$

Consequently the surface contribution to the polarization function reads finally with (53)

$$\Pi^{\rm surf}(\mathbf{q},\omega) = -\frac{3}{2qR_0}{\rm Im}\Pi^0(\mathbf{q},\omega) \tag{56}$$

which is real. With (55), (56) and (51) we obtain now for the structure function

$$S(q,\omega) = \frac{1}{\pi}\frac{{\rm Im}\Pi^0}{(1-V({\rm Re}\Pi^0+\Pi^{\rm surf}))^2+(V{\rm Im}\Pi^0)^2}. \tag{57}$$

For small q expansion we see that the pole of the structure function becomes renormalized similar to what is known from the Mie mode or surface plasmon mode [93, 94]

$$\omega^2 = \frac{\omega_0^2}{1-V\Pi^{\rm surf}}. \tag{58}$$

After establishing the structure function including surface contribution we specify the model for actual calculations. We choose as mean field parameterization a Skyrme force (33) following Vautherin and Brink [95] which leads to the isoscalar potential

$$V = U_n - U_p = \frac{3t_0}{4} + \frac{3t_3}{8}n_0 \tag{59}$$

with $t_0 = -983.4$ MeV fm^3, $t_3 = 13106$ MeV fm^6, $x_0 = 0.48$ at nuclear saturation density $n_0 = 0.16$ fm^{-3} and the incompressibility of $K = 318$ MeV. Further we employ again the Steinwedel-Jensen model [96] where the basic mode inside a sphere of radius R_0 is given by a wave vector

$$q_{sp} = \frac{\pi}{2R_0}. \tag{60}$$

This would correspond to the first order isovector mode [97]. Since this mode is spurious we have to consider the next higher harmonics [98] which is

$$q_{isgdr} = \frac{\pi}{R_0}. \tag{61}$$

The polarization function with this second order mode contains still contributions from the spurious mode such that we have to subtract this part [89, 90]

$$\Pi^0_{\rm ISGDR}(\omega) = \Pi^0(q_{isgdr},\omega) - \Pi^0(q_{sp},\omega). \tag{62}$$

In figure 17 we have plotted the experimental structure function together with different theoretical estimates according to (62) and (57). The inclusion of surface corrections (dashed lines) shifts the structure function towards the experimental values. The inclusion of collisions (dot-dashed lines), which should be wrong for isoscalar dipole mode due to cancellation of backscattering leads to really worse results. This supports indirectly that the mode is of isoscalar dipole type and surface dominated.

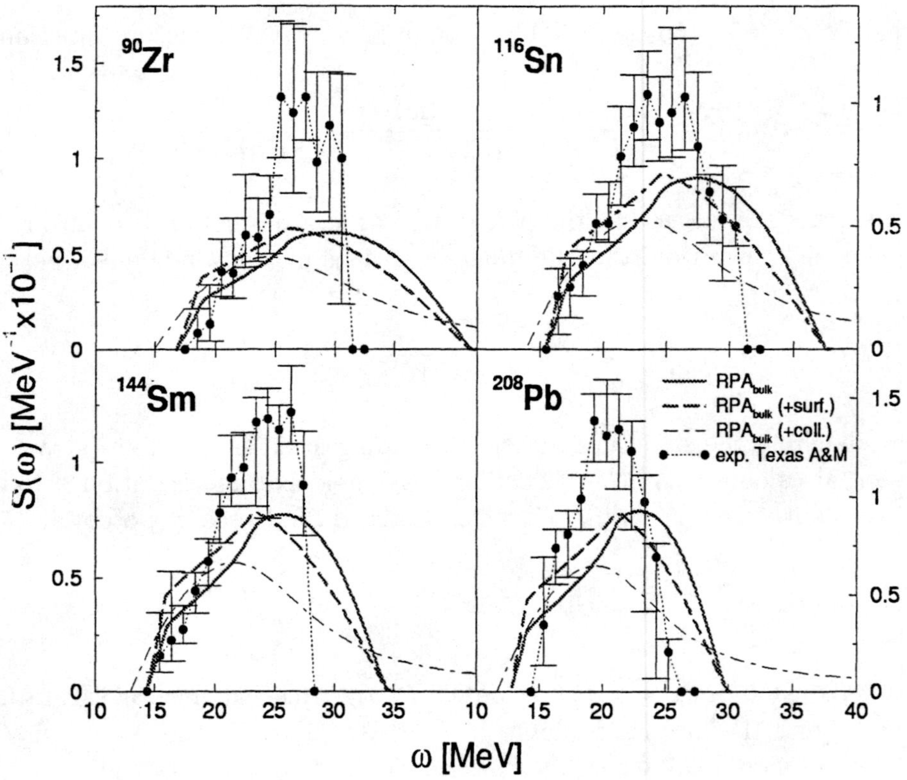

Figure 17: The experimental structure function (T=0) versus theoretical values. The bulk RPA result (solid lines) is compared with the extended Thomas Fermi approximation (surface corrections, dashed lines) and the inclusion of collisions (dot-dashed lines). The latter one should be of less importance due to symmetry of isoscalar mode. The data suggest this case and support surface contributions. Circles: Normalized data from Ref. [99]

14 Simplified model for nuclear matter situation

The resulting wave vectors have very low values compared with the Fermi wave vector in the Steinwedel Jensen model. This allows us to expand the Mermin polarization function (28) with respect to small qv_c/ω ratios and v_c the sound velocity. We obtain

$$\Pi_a^{\mathrm{M}}(\omega) = \frac{\rho_a(\mu_a)}{m_a} \frac{q^2}{\omega(\omega + i/\tau_a)}, \tag{63}$$

$$\rho_a(\mu_a) = 2\,\lambda_a^{-3} f_{3/2}(z_a) \zeta_{corr} \tag{64}$$

where the thermal wave length is $\lambda_a^2 = 2\pi\hbar^3/(m_a T)$, $f_{3/2}$ the standard Fermi integral and $z_a = \mathrm{e}^{\mu_a/T}$ the fugacity. The correction constant $\zeta_{corr} = 1.22$ is introduced to fit the numerical solution of the dispersion relation (18) with the approximative expansion (63). With the help of this expansion the dispersion relation (27) takes the form

$$\begin{aligned} 0 &= \left[\omega\left(\omega + \frac{i}{\tau_n}\right) - c_{nn}^2 q^2\right]\left[\omega\left(\omega + \frac{i}{\tau_p}\right) - c_{pp}^2 q^2\right] \\ &\quad - \left[c_{np}^2 + i\frac{\tilde{c}_{np}^2}{(\omega + i/\tau_n)\tau_p}\right]\left[c_{pn}^2 + i\frac{\tilde{c}_{pn}^2}{(\omega + i/\tau_p)\tau_n}\right] q^4 \end{aligned} \tag{65}$$

with the partial sound velocities c and $\tilde{c}$

$$c_{ab}^2 = \alpha_{ab}\frac{\rho_a(\mu_a)}{m_a}, \qquad \tilde{c}_{ab}^2 = \frac{T}{m_a}\frac{\frac{f_{3/2}(z_a)}{f_{1/2}(z_a)} - \frac{f_{3/2}(z_b)}{f_{1/2}(z_b)}}{\frac{\tau_{ab}}{\tau_{bb}} - \frac{\tau_{aa}}{\tau_{ba}}}. \tag{66}$$

The dispersion relation (27) with the dynamical relaxation times (32) is a polynomial of tenth (sixth) order corresponding to the inclusion of memory (in)dependent relaxation times. While most of these solutions will just lead to parasite solutions ($\mathrm{Re}\,\omega \leq 0$) we will get two coupled modes, i.e. the isoscalar and isovector mode. Furthermore a third mode appears at extreme asymmetries and/or strong collisional coupling which we will describe in the next section.

15 New collective mode

Now we employ the potential (33) and assume different neutron and proton densities. In figure 18 we plot the isoscalar and isovector modes versus temperature for ^{48}Ca with a small asymmetry $\delta = 0.2$ as well as for ^{60}Ca with an asymmetry $\delta = 0.33$. The kinetic energy is linked to a temperature within the Fermi liquid model via Sommerfeld expansion

$$\frac{\mathcal{E}}{A}_{\mathrm{kin}} = \frac{3}{5}\epsilon_f\left[\frac{(1+\delta)^{5/3} + (1-\delta)^{5/3}}{2} + \frac{5}{12}\pi^2\left(\frac{T}{T_f}\right)^2 \frac{(1+\delta)^{1/3} + (1-\delta)^{1/3}}{2}\right]. \tag{67}$$

This connection between temperature and excitation energy is only valid for a continuous Fermi liquid model. For small nuclei, the concept of temperature is questionable. Some improvement can be obtained by the definition of temperature via the logarithmic derivative of the density of states [100]

$$T^{-1} = \frac{1}{\rho}\frac{\partial \rho}{\partial E_{\mathrm{ex}}} = -\frac{5}{4}E_{\mathrm{ex}}^{-1} + \pi(\frac{A}{4\epsilon_f E_{\mathrm{ex}}})^{1/2} \tag{68}$$

which provides $E_{\text{ex}} \approx \frac{1}{4}(E/A)$ in comparison with (67) for small temperatures and small nuclei. We use this temperature to demonstrate possible collective bulk features in an exploratory sense. Of course, the surface energy and shell effects cannot be neglected for realistic calculations of small nuclei.

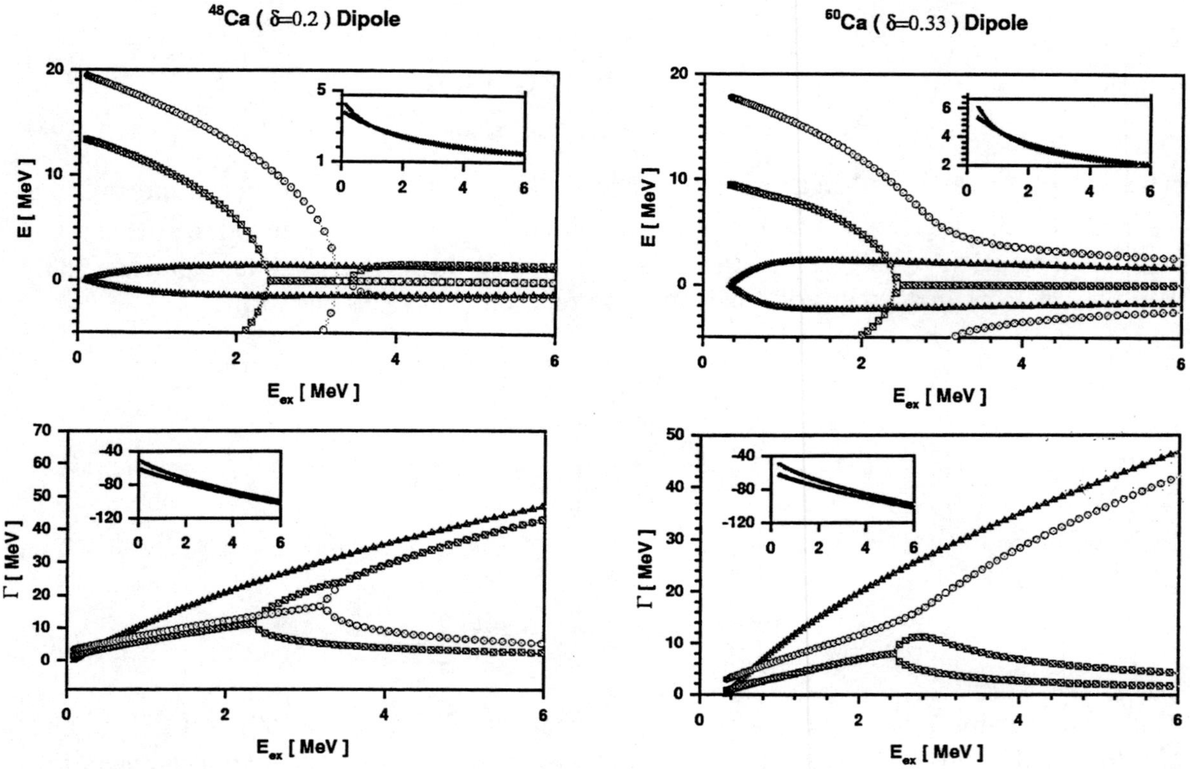

Figure 18: The energy and damping of the IVGDR and ISGDR of ^{48}Ca (left) and ^{60}Ca (right) vs excitation energy. Besides the isovector modes (circles) and the isoscalar modes (squares) a soft third mode appears (triangles).

With increasing temperature the isovector and isoscalar energies decrease and vanish at a certain temperature. At these temperatures the damping becomes twofold because the damping of the spurious mode with negative energy becomes different from the physical mode. We can consider this behavior of damping as a phase transition of isospin demixture. At the same time a very soft mode appears due to collisional coupling which is only present in asymmetric matter [18]. This mode is more pronounced in the next figure 18 for ^{60}Ca with an asymmetry of $\delta = 0.33$.

We see that the isovector mode does not disappear but turns over into a flat decrease with increasing temperature. This behaviour is coupled with the pronounced soft mode. In comparison with the more symmetric ^{48}Ca we see a different behavior of the damping where the isoscalar mode vanishes. Also the isovector mode appears unique and not two-fold. Now a clear transition of damping behavior for the isovector mode is recognizable which can be considered as a transition from zero to first sound damping.

Besides the standard isovector and isoscalar modes we observe a build up of a very soft mode with a centroid energy around 1 MeV. This mode appears due to the collisional coupling $\bar{c}^2_{ab}$ of (66). When we turn off the relaxation times, i.e. the collision integral, this mode is vanishing as well as in symmetric nuclear matter, see discussion after (29). It shows that this mode appears due to collisional coupling of isovector and isoscalar modes. The corresponding damping of the crossed mode is continuously increasing with temperature.

One may argue whether this third mode can really appear in the system. A simple consideration may convince us about the possible existence of such mode. Let us assume a coupled set of two type of harmonic oscillators (neutrons and protons) interacting between the same sort of particles with strength k_n and k_p, respectively and between different sorts with k_{np}. Let us choose for simplicity only two neutrons and two protons. Then we obtain the coupled system of harmonic oscillators with frequencies $\omega_n^2 = k_n/m$, $\omega_p^2 = k_p/m$ and $\omega_{np}^2 = k_{np}/m$. The solution yields three basic modes in the system, i.e. $\omega^2 = 2(\omega_n^2 + \omega_{np}^2), 4\omega_{np}^2, 2(\omega_p^2 + \omega_{np}^2)$. If we neglect the different coupling between neutrons and protons ω_{np} we only obtain two modes analogously to isovector and isoscalar ones. We see that the coupling between neutrons and protons can lead to the appearance of a third mode.

Let us now compare the found new mode with the experimental evidence. There are some hints for a soft mode in ^{11}Be [101]. The authors have observed a low lying structure at around 6 MeV excitation energy with a damping of around 1 MeV which has not been reproduced yet even within refined coupled channel calculations [102]. A standard explanation would give as the origin a weakly-bound single particle neutron orbital. The observed broad structure at 6 MeV might be explained alternatively as the presented new coupled mode. The centroid energy as well as damping width at least seem to suggest this interpretation.

16 Nonlinear effects beyond linear response

While most of the theoretical treatments of oscillations rely on the linear response method or RPA methods, large amplitude oscillations require methods beyond. In particular the question of the appearance of chaos has recently been investigated [44, 45, 46]. The hypothesis was established that the octupole mode is over-damped due to negative curved surface and consequently additional chaotic damping [47, 48, 49]. Here we want to discuss in which conditions one might observe octupole modes at least in Vlasov - simulation of giant resonances. We will consider different initial conditions of isoscalar giant resonances and will demonstrate that the appearance of octupole modes is dependent on the initial configuration which in turn demonstrates the nonlinear behavior beyond linear response.

As a first initial condition we use the ground state distribution of coordinates while the momentum distribution is deformed anisotropically. We have modified the momenta in a way which corresponds to a giant octupole mode. The local densities and currents remain the same as in ground state. Figure 19 shows that at start time $t = 0$ there is a pure giant octupole, which is damped out and a quadrupole resonance develops instead. The monopole and dipole amplitudes which should remain constant document the stability of simulation. In agreement with the already mentioned hypothesis, the octupole mode is over-damped. The figure shows the nonlinear behavior of mode coupling. Within the linear response the damping rate is expected to be independent of initial conditions. We choose now other initial conditions to show that the result is very much dependent on initial conditions. Therefore we split the nucleus in two parts of mass ratio 3:7 in accordance with the symmetry of the octupole oscillation and accelerate both pieces towards each other. Experimentally this might be realized as a central collision of two nuclei with corresponding masses. In figure 20 a clear quadrupole resonance appears and also a smaller octupole resonance can be seen. Both are damped out. Consequently there is no evidence for an over-damped octupole mode

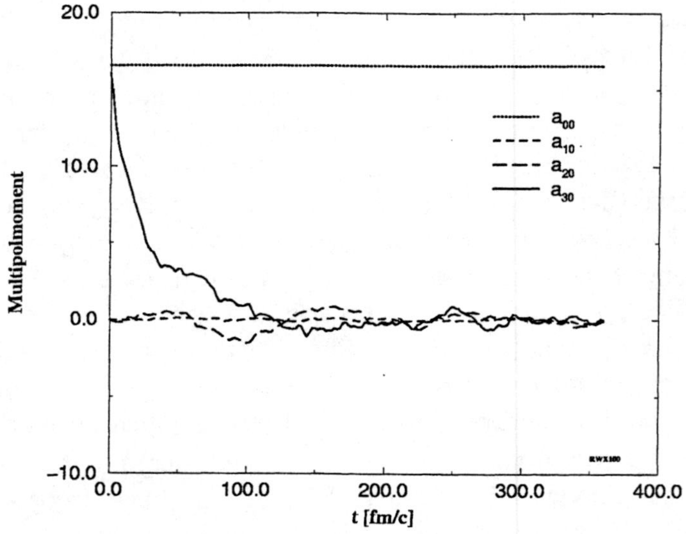

Figure 19: The picture shows the time evolution of multipole moments $\sqrt{\frac{2l+1}{4\pi}}a_{l0}(t), l = 0,1,2,3$ of mass density concerning to normalization, dipole, quadrupole and octupole oscillation. Excitation is due to modification in momentum space only.

in this case.

In order to understand the different initializations, we split the kinetic energy into a thermal part and a collective part according to $\langle \mathbf{p}^2\rangle = \langle(\mathbf{p}-\langle\mathbf{p}\rangle)^2\rangle + \langle\mathbf{p}\rangle^2$. We analyze the time development of the total collective energy in the system

$$E_{coll}(t) = \frac{h^2}{2m}\int d\mathbf{r}\varrho(\mathbf{r})\langle\mathbf{p}\rangle^2(\mathbf{r},t) \tag{69}$$

with the mean current

$$\langle\mathbf{p}\rangle(\mathbf{r},t) = \frac{1}{\varrho(\mathbf{r})}\int\frac{d\mathbf{p}}{(2\pi)^3}\,\mathbf{p}\, f(\mathbf{p},\mathbf{r},t) \tag{70}$$

and the density

$$\varrho(\mathbf{r},t) = \int\frac{d\mathbf{p}}{(2\pi)^3}\, f(\mathbf{p},\mathbf{r},t)\,. \tag{71}$$

In figure 21 the development of collective energy can be seen. There is a background of about 50 MeV due to fixed correlations caused by finite width of pseudo-particles as one can see from the following estimation. Using (23) and (55) in (69) one obtains

$$\begin{aligned} E_{coll} &\approx \frac{1}{\varrho_0}\sum_{i=1}^{AN}\sum_{j=1}^{AN}\frac{1}{N^2}p_ip_j\int dr\, f_S(r-r_i(t),\sigma_r)f_s(r-r_j(t),\sigma_r) \\ &\approx \frac{1}{\varrho_0}\sum_{i=1}^{AN}\sum_{j=1}^{AN}\frac{1}{N^2}p_ip_j f_S(r_i(t)-r_j(t),\sqrt{2}\sigma_r) \\ &\geq \frac{1}{\varrho_0 N}\frac{1}{(\sqrt{2\pi}\sqrt{2}\sigma_r)^3}\sum_{i=1}^{AN}\frac{1}{N}p_i. \end{aligned} \tag{72}$$

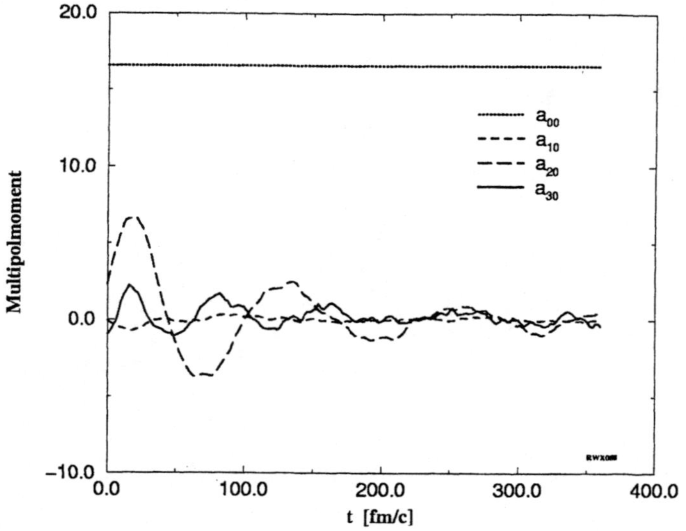

Figure 20: This picture shows the time evolution of multipole moments $\sqrt{\frac{2l+1}{4\pi}}a_{l0}(t), l = 0, 1, 2, 3$ of mass density. Excitation is due to asymmetric splitting (0.3 to 0.7) of the nucleus in two pieces using a plane in coordinate space. Then both pieces are accelerated towards each other.

For simulation parameters of ^{208}Pb, $\rho_0 = 0.162$ fm^{-3}, $N = 75$, $\sigma_r = 0.53$ fm, we obtain a basic collective energy of 54MeV. Using more test-particles would diminish this level.

The solid line corresponding to figure 19 shows no initial collective energy. The exclusive initial excitation in momentum space without correlation in the spatial domain leads to zero initial collective energy. These correlations are forming during time evolution. Of course, there is no center of mass motion, otherwise we would see just the mean streaming velocity.

This situation is changed if we use the second preparation with simple momentum–space–correlations. The long dashed line in figure 21 shows initial collective correlations corresponding to figure 20. Since we can deposit enough collective energy in this case, we observe a clear octupole motion.

In order to compare with [48] we calculate the adiabaticity index η defined in [48] as the ratio of maximum radial surface velocity to the maximum particle speed. A smaller ratio denotes a more adiabatic shape changes in relation to the particle speed. In analogy we define such index as a ratio

$$\eta = \frac{\frac{\partial}{\partial t}\sqrt{< r^2 > (\vartheta)}}{v_F} \tag{73}$$

of the root mean square radius speed in forward direction (opening angle ϑ) and the Fermi velocity. With opening angle 0.4 rad we obtain a maximum $\eta = 0.12$ for figure 19 and $\eta = 0.30$ for figure 20. This shows that we are essentially still in the adiabatic regime described in [48].

The nonlinear behavior described so far already documents that we are in a regime of large amplitude oscillations where linear response fails. The corresponding radius elongation in coordinate space varies about 10 %.

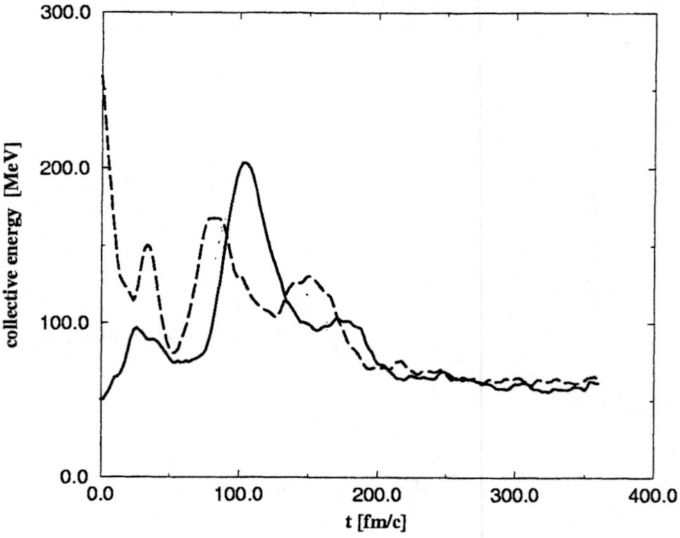

Figure 21: The time evolution of collective energy. The solid line corresponds to the excitation scheme of figure 19 and the long dashed line corresponds to the figure 20, respectively. While the first starts without collective energy, the second one starts with maximal collective energy.

17 Summary

We have investigated the giant resonance oscillations by solving the collision-less Vlasov equation as well as by linear response theory. With the help of numerical solution of the Vlasov equation we could reproduce the mass dependence of energy with one fixed parameter of test-particle width for both giant monopole and giant dipole modes. A multipole analysis was performed which has allowed us to characterize the corresponding excitation. It was shown, that the asymmetry influences the collective behavior. With increasing asymmetry the energy decreases while the damping increases. The damping due to finite systems amount to 2MeV almost constant for all mass numbers which underestimates the experimental values considerably. This motivates to search for additional damping mechanism which was found to be due to the collisional correlations. The linear response can be used in a simplified liquid drop model to describe the giant resonances in asymmetric matter. We find that the collisional contribution as well as surface contributions are both important to reproduce the experimental damping of giant dipole resonances. While for ground state resonances it is sufficient to add the damping of finite size effects from simulation with the collisional contribution from linear response theory, for the temperature behavior we have needed also the deformation of the surface. The combined model between surface and collisional contribution, weighted properly due to their collision frequencies, is able to reproduce the experimental damping curve with temperature as well as the structure function. A higher order mode like the recently measured ISGDR mode has been described within this simple linear response model.

We have observed that due to correlational coupling there can exist a new mode which appears besides isovector and isoscalar modes in asymmetric nuclear matter due to collisions. We suggest that this mode may be possible to observe as a soft collective excitation in asymmetric systems. The transition from zero sound damping to first sound damping behavior should become possible to observe for isovector modes since they do not vanish at this transition temperature like in symmetric matter. At a certain critical temperature the collective isoscalar mode vanishes and the damping becomes two-fold. This can be considered as a phase transition of demixture. On the example of octupole resonances in finite nuclei the dependency on the initial configuration is demonstrated. The appearance of an octupole mode was shown by correlating the spatial and momentum initial excitation. It is possible to excite an octupole mode with sufficient collective energy deposited initially. We suggest that isoscalar giant octupole resonances should be possible to observe in nuclear collisions of mass ratio about 3:7 corresponding to octupole symmetry. This effect illustrates a mode beyond linear response.

18 Acknowledgment

The authors are obliged to M. Vogt who has contributed the calculation of the Lyapunov exponent. M. DiToro is thanked for numerous discussions and the LNS Catania where part of this work was done for hospitality. Valuable hints and comments by J. D. Frankland (GANIL) are also gratefully acknowledged.

A Derivation of non-Markovian relaxation time

Here we shortly sketch the derivation of the damping of a collective mode within a Fermi gas and a Fermi liquid model. We will show that we get the latter one from the Fermi gas model with an additional contribution from the quasi-particles.

We start with the Fermi gas where the dispersion relation between momentum and energy is given by $\epsilon = p^2/2m$ and will show later what has to be changed for a Fermi liquid where ϵ is a solution of the quasiparticle dispersion relation. We will see that the contributions from the quasi-particles alone leads to the Landau formula of zero sound damping [103, 104]

$$\gamma \propto \left[1 + \left(\frac{\Omega}{2\pi T}\right)^2\right]. \tag{74}$$

Our considerations use conveniently the Levinson equation for the reduced density matrix f which is valid at short time processes compared to inverse Fermi energy $\hbar/\epsilon_f$ and which collisional side has the form:

$$I_1(t) = \frac{2g}{\hbar^2} \int_0^\infty d\tau \int \frac{d\mathbf{p}_2 d\mathbf{p}_3 d\mathbf{p}_4}{(2\pi\hbar)^6} |T|^2 \cos\left(\int_t^{t-\tau} \Delta\epsilon(\tau) d\tau/\hbar \right) \delta(\Delta\mathbf{p}) \left(\bar{f}_1 \bar{f}_2 f_3 f_4 - f_1 f_2 \bar{f}_3 \bar{f}_4\right)_{t-\tau} \tag{75}$$

where $\bar{f} = 1 - f$, $\Delta p = p_1 + p_2 - p_3 - p_4$ etc., g is the spin-isospin degeneracy and the transition probability is given by the scattering T-matrix. In case that the quasiparticle energies $\epsilon(t)$ become time independent like in the Fermi gas model, the integral in the *cos* function reduces to the familiar expression $\Delta\epsilon\tau$. We linearize this collision integral with respect to an external disturbance according to

$$f = n + \delta f \tag{76}$$

where n is the equilibrium distribution. Clearly two contributions have to be distinguished, the one from the quasiparticle energy and the one from occupation factors [6]. First we concentrate on the Fermi gas model where we have only the contribution of the occupation factors and will later add the contribution of the quasiparticle energies for Fermi liquid model. We obtain after Fourier transform of the time

$$\delta I_1 = \left\langle \frac{\hbar}{2} \left[\delta_+(\Delta\epsilon + \Omega) + \delta_-(\Delta\epsilon - \Omega)\right] (\delta F_1 + \delta F_2 - \delta F_3 - \delta F_4)(\Omega) \right\rangle. \tag{77}$$

Here we use the abbreviation

$$\langle ... \rangle = \frac{2g}{\hbar^2} \int \frac{d\mathbf{p}_2 d\mathbf{p}_3 d\mathbf{p}_4}{(2\pi\hbar)^6} |T|^2 \delta(\Delta p) ... = \frac{m^3 g}{\hbar^2 (2\pi\hbar)^6} \int d\epsilon_2 d\epsilon_3 d\epsilon_4 \int \frac{d\phi \sin\theta d\theta d\phi_2}{\cos(\theta/2)} |T|^2 ... \tag{78}$$

where the last line appears from standard integration techniques at low temperatures. Further abbreviations are

$$\begin{aligned} \delta_\pm(x) &= \pi\delta(x) \pm i\frac{\mathcal{P}}{x} \approx \pi\delta(x) \\ \delta F_1 &= -\delta f_1 (\bar{n}_2 n_3 n_4 + n_2 \bar{n}_3 \bar{n}_4). \end{aligned} \tag{79}$$

The approximation used in the first line consists in the neglect of the off-shell contribution from memory effects. This is consistent with the used integration technique (78). This terms would lead to divergences which has to be cut off [53].

Neglecting the backscattering terms $\delta F_{2/3/4}$ we obtain from (77) a relaxation time approximations with the relaxation time

$$\frac{1}{\tau(\epsilon_1)} = \frac{3}{4\pi^2 \tau_0} \int_{-\lambda}^{\infty} dx_2 dx_3 dx_4 \left[\delta(\Delta x + \omega) + \delta(\Delta x - \omega)\right] (\bar{n}_2 n_3 n_4 + n_2 \bar{n}_3 \bar{n}_4) \tag{80}$$

with $\omega = \Omega/T$, $x = (\epsilon - \mu)/T$, $\lambda = \mu/T$ and the time

$$\frac{1}{\tau_0} = \frac{2gmT^2}{3\hbar^3} \sigma. \tag{81}$$

Here we have used the definition of cross section $|T|^2 = (4\pi\hbar^2/m)^2 d\sigma/d\Omega$ and have assumed a constant cross section σ. To calculate (80) one needs the standard integrals for large ratios of chemical potentials μ to temperature $\lambda = \mu/T$

$$\int_{-\lambda}^{\infty} dx_2 dx_3 dx_4 n_2 \bar{n}_3 \bar{n}_4 \delta(\Delta x \pm \omega) = \frac{1}{2} \bar{n}_1 (x_1 \pm \omega) \left[\pi^2 + (x_1 \pm \omega)^2\right] \tag{82}$$

to obtain

$$\frac{1}{\tau(\epsilon_1)} = \frac{3}{8\pi^2 \tau_0} \left[2\pi^2 + (x_1 + \omega)^2 + (x_1 - \omega)^2\right]. \tag{83}$$

Further we employ a thermal averaging in order to obtain the mean relaxation time finally[4]

$$\frac{1}{\tau_{\text{gas}}} = \int_{-\lambda}^{\infty} dx_1 n_1 \bar{n}_1 \frac{1}{\tau(\epsilon_1)} = \frac{1}{\tau_0} \left[1 + 3\left(\frac{\omega}{2\pi}\right)^2\right]. \tag{84}$$

[4]One has to use the identities valid up to $o(\exp[-\lambda])$

$$\int_{-\lambda}^{\infty} dx n\bar{n} = 1; \qquad \int_{-\lambda}^{\infty} dx x^2 n\bar{n} = \frac{\pi^2}{3}.$$

If we do not use the thermal averaging but take (83) at the Fermi energy $\epsilon_1 = \epsilon_f$ we will obtain

$$\frac{1}{\tau(\epsilon_f)} = \frac{1}{\tau_0}\left[\frac{3}{4} + 3\left(\frac{\omega}{2\pi}\right)^2\right]. \tag{85}$$

We see that both results disagree with the Landau result of quasiparticle damping (74) by factors of 3 at different places [6, 71, 105, 106]. We have point out that the result at fixed Fermi energy will lead to unphysical results for the Fermi liquid case. Therefore we consider the thermal averaged result as the physical one.

We now turn to the Fermi liquid model and replace the free dispersion $\epsilon = p^2/2m$ by the quasiparticle energy ϵ_p. Than the variation of the collision integral gives an additional term which comes from the time dependence of the quasiparticle energy on the cos-term of (75). We have instead of the sum of two complex conjugate exponentials in (75) an additional contribution from the linearization of the exponential

$$\begin{aligned} \delta \exp\left(i\Delta \int_t^{t-\tau} d\bar{t}\epsilon(\bar{t})\right) &= \mathrm{e}^{-i\Delta\epsilon\tau}\left(1 - i\Delta\int_t^{t-\tau} d\bar{t}[\epsilon(\bar{t}) - \epsilon]\right) - \mathrm{e}^{-i\Delta\epsilon\tau} \\ &= \frac{iT}{\hbar}\mathrm{e}^{-i\Delta\epsilon\tau/\hbar}\Delta\int_t^{t-\tau} d\bar{t}\frac{\delta f(\bar{t})}{n\bar{n}}. \end{aligned} \tag{86}$$

In the last line we have replaced the variation in the quasiparticle energy $\epsilon(t) - \epsilon$ by the variation in the distribution function δf due to the identity [104]

$$\begin{aligned} \delta f(t) &= f(t) - n(\epsilon) = f(t) - n(\epsilon(t)) - [n(\epsilon) - n(\epsilon(t))] \\ &\approx -n'(\epsilon(t) - \epsilon) = \frac{n\bar{n}}{T}[\epsilon(t) - \epsilon] \end{aligned} \tag{87}$$

where we assumed within the quasiparticle picture that $f(t) = n(\epsilon(t))$. This leads now to an additional part in the relaxation time which we write analogously to (77)

$$\frac{1}{\tau_c(\epsilon_1)} = \left\langle \frac{\hbar}{2}\frac{\delta_+(\Delta\epsilon - \Omega) - \delta_+(\Delta\epsilon + \Omega)}{\Omega}\frac{\bar{n}_1\bar{n}_2 n_3 n_4 - n_1 n_2 \bar{n}_3\bar{n}_4}{n_1\bar{n}_1}\right\rangle. \tag{88}$$

Using again (82) we obtain

$$\frac{1}{\tau_c(\epsilon_1)} = \frac{-3}{4\pi^2\tau_0}\left\{\frac{\bar{n}(x_1+\omega)\,[\pi^2 + (x_1+\omega)^2]}{2\omega(\mathrm{e}^{-\omega} - 1)} + [\omega \leftrightarrow -\omega]\right\} \tag{89}$$

and get after thermal averaging (84)[5]

$$\frac{1}{\tau_c} = \frac{1}{\tau_0}\left[1 + \left(\frac{\omega}{2\pi}\right)^2\right]. \tag{90}$$

[5]Here one uses $[o(\exp[-\lambda])]$

$$\int_{-\lambda}^{\infty} dxn = \lambda; \quad \int_{-\lambda}^{\infty} dxxn = -\frac{1}{2}\lambda^2 + \frac{1}{6}\pi^2; \quad \int_{-\lambda}^{\infty} dxx^2 n = \frac{1}{3}\lambda^3$$

Taking instead of thermal averaging the value at Fermi energy ($\epsilon_1 = \epsilon_f$) in (89) we find

$$\frac{1}{\tau_c(\epsilon_f)} = \frac{3(\pi^2+\omega^2)}{2\pi^2\tau_0\omega}\frac{e^\omega - 1}{e^\omega + 1}. \tag{91}$$

Here we like to point out that the Landau result (74) appears in (90) (see also in Ref. [71, 107, 19, 105, 106]).

Adding now (84) and (90) we obtain a final relaxation time for the Fermi liquid model

$$\frac{1}{\tau_{\mathrm{liq}}} = \frac{2}{\tau_0}\left[1+2\left(\frac{\omega}{2\pi}\right)^2\right]. \tag{92}$$

which is the main result in this paper. It contains the typical Landau result of zero sound (74) except the factor 2 in front of the frequencies. Comparing (92) with the Fermi gas model (84) we see that in the limit of vanishing temperature the Fermi liquid value is lower with $\propto 2\Omega^2$ compared to the Fermi gas $\propto 3\Omega^2$. Further for vanishing frequencies (neglect of memory effects) the Fermi liquid model leads to twice the relaxation rate than the Fermi gas model. The coefficient of temperature increase is than twice larger for the Fermi liquid than for the Fermi gas. If we consider the relaxation times at Fermi energy (no thermal averaging) (85) and (91) we find the same results as above in the limit of vanishing frequencies. For vanishing temperature only the Fermi gas (85) coincides with the result of (84) $\propto 3\Omega^2$. Expression (91) goes to zero for $T = 0$ and underlines the necessity to thermal average the value.

References

[1] W. Bothe and W. Gentner, Zeit. f. Physik **106**, 236 (1937).

[2] G. F. Bertsch, P. F. Bortignon, and R. A. Broglia, Rev. of Mod. Phys. **55**, 287 (1983).

[3] J. Bacelar, M. Harakeh, and O. Scholten, Nucl. Phys. A **599**, 1 (1996), proceedings of the Groningen conference on giant resonances.

[4] in *Topical Conference on Giant Resonances Varenna*, No. 3 in *Nucl. Phys. A*, edited by A. Bracco and P. Bortignon (Elsevier, North Holland, 1998), .

[5] in *Selected Topics in Nuclear Collective Excitatons, NUCOLEX99*, Vol. 23 of *Riken Rev.*, edited by N. D. Dang, U. Garg, and S. Yamaji (RIKEN, Hirosawa, Wako, Saitama, 1999), .

[6] U. Fuhrmann, K. Morawetz, and R. Walke, Phys. Rev. C. **58**, 1473 (1998).

[7] F. Catara, E. G. Lanza, M. A. Nagarajan, and A. Vitturi, Nucl. Phys. A **624**, 449 (1996).

[8] S. Kamerdzhiev, Yad. Fiz. **9**, 324 (1969).

[9] H. Sagawa and G. F. Bertsch, Phys. Lett. B **5**, 138 (1984).

[10] V. Abrosimov, M. DiToro, and A. Smerzi, Z. Phys. A **347**, 161 (1994).

[11] V. M. Kolomietz, V. A. Pluiko, and S. Shlomo, Phys. Rev. C **52**, 2480 (1995).

[12] V. Kolomietz, V. Plujko, and S. Shlomo, Phys. Rev. C **54**, 3014 (1996).

[13] V. Kondratyev and M. Di Toro, Phys. Rev. C **53**, 2176 (1996).

[14] E. Hernández, J. Navarro, A.Polls, and J. Ventura, Nucl. Phys. **A597**, 1 (1996).

[15] M. DiToro, V. Kolomietz, and A. Larionov, in *Proceedings of the Dubna Conference on Heavy Ions*, Dubna, 1997 (unpublished).

[16] V. Kolomietz, A. Larionov, and M. DiToro, Nucl. Phys. **A613**, 1 (1997).

[17] V. M. Kolomietz, S. V. Lukyanov, V. A. Plujko, and S. Shlomo, Phys. Rev. C **58**, 198 (1998).

[18] K. Morawetz, R. Walke, and U. Fuhrmann, Phys. Rev. C **57**, R 2813 (1998).

[19] S. Ayik, O. Yilmaz, A. Golkalp, and P. Schuck, Phys. Rev. C **58**, 1594 (1998).

[20] G. Gervais, M. Thoennessen, and W. E. Ormand, Phys. Rev. C, 1998.

[21] S. Shlomo and G. F. Bertsch, Nucl. Phys. **A243**, 507 (1975).

[22] H. Esbensen and G. F. Bertsch, Phys. Rev. C **28**, 355 (1983).

[23] H. Esbensen and G. F. Bertsch, Ann. Phys. **157**, 255 (1984).

[24] P. F. Bortignon, A. Bracco, D. Brink, and R. A. Broglia, Phys. Rev. Lett. **67**, 3360 (1991).

[25] W. E. Ormand, P. F. Bortignon, and R. A. Broglia, Phys. Rev. Lett. **77**, 607 (1996).

[26] S. Kamerdzhiev, J. Speth, and G. Tertychny, Nucl. Phys. **A624**, 328 (1997).

[27] W. E. Ormand, P. F. Bortignon, R. A. Broglia, and A. Bracco, Nucl. Phys **A614**, 217 (1997).

[28] S. Kamerdzhiev, R. J. Liotta, E. Litvinova, and V. Tselyaev, Phys. Rev. C **58**, 172 (1998).

[29] J. Kvasil, N. L. Iudice, V. O. Nesterenko, and M. Kopal, Phys. Rev. C **58**, 209 (1998).

[30] Y. Alhassid and B. Bush, Phys. Rev. Lett. **63**, 2452 (1989).

[31] W. E. Ormand *et al.*, Phys. Rev. Lett. **64**, 2254 (1990).

[32] W. E. Ormand *et al.*, Phys. Rev. Lett. **69**, 2905 (1992).

[33] B. Lauritzen, P. F. Bortignon, R. A. Broglia, and V. G. Zelevinsky, Phys. Rev. Lett. **74**, 5190 (1995).

[34] A. Bracco *et al.*, Phys. Rev. Lett. **74**, 3748 (1995).

[35] S. Leoni, T. Døssing, and B. Herskind, Phys. Rev. Lett. **76**, 4484 (1996).

[36] M. DiToro, V. Kolomietz, and A. Larionov, Phys. Rev. C **59**, 3099 (1999).

[37] D. Lecroix, P. Chomaz, and S. Ayik, Phys. Rev. C **58**, 2154 (1998).

[38] M. Colonna *et al.*, Phys. Lett. B **307**, 293 (1993).

[39] V. Baran *et al.*, Nucl. Phys. **A599**, 29 (1996).

[40] R. Walke and K. Morawetz, Phys. Rev. C. **60**, 17301 (1999).

[41] V. Špička, P. Lipavský, and K. Morawetz, Phys. Lett. A **240**, 160 (1998).

[42] K. Morawetz *et al.*, Phys. Rev. Lett. **82**, 3767 (1999).

[43] D. Brink, A. Dellafiore, and M. DiToro, Nucl. Phys. A **456**, 205 (1986).

[44] W. Bauer, D. McGrew, V. Zeletvinsky, and P. Schuck, Phys. Rev. Lett. **72**, 3771 (1994).

[45] G.F.Burgio, M.Baldo, A.Rapisarda, and P. Schuck, Phys.Rev C **52**, 2475 (1995).

[46] K. Morawetz, Phys. Rev. C **55**, R 1015 (1997).

[47] C. Jarzynski and W. Swiatecki, Nucl. Phys. A **552**, 1 (1993).

[48] J. Blocki, J.-J. Shi, and W. Swiatecki, Nucl. Phys. A **554**, 387 (1993).

[49] J. Blocki, J. Skalski, and W. Swiatecki, Nucl. Phys. A **816**, 1 (1997).

[50] G. F. Bertsch and S. D. Gupta, Phys. Rep. **160**, 189 (1988).

[51] M. Colonna, M. D. Toro, and A. Guarnera, Nucl. Phys. A **589**, 160 (1995).

[52] K. Morawetz, Phys. Lett. A **199**, 241 (1995).

[53] K. Morawetz and H. Koehler, Eur. Phys. J. A **4**, 291 (1999).

[54] C. Wong, Phys. Rev. C **25**, 1460 (1982).

[55] M. Colonna *et al.*, Phys. Rev. C **47**, 1395 (1993).

[56] K. Morawetz and R. Walke, Europ. Phys. J. A (1998), sub. nucl-th/9807045.

[57] J. Blaizot, Phys. Rep. **64**, 171 (1980).

[58] C. Garcia, J. Navarro, V. Nguyen, and L. Salcedo, Ann. Phys. (N.Y.) **214**, 293 (1992).

[59] F. Braghin, D. Vautherin, and A. Abada, Phys. Rev. C **52**, 2504 (1995).

[60] H. Heiselberg, C. J. Pethick, and D. G. Ravenhall, Ann. Phys. **223**, 37 (1993).

[61] N. Mermin, Phys. Rev. B **1**, 2362 (1970).

[62] D. Kiderlen and H. Hofmann, Phys. Lett. B **332**, 8 (1994).

[63] G. Röpke and A. Wierling, Phys. Rev. E **57**, 7075 (1998).

[64] K. Morawetz and U. Fuhrmann, Phys. Rev. E **61**, (2000).

[65] P. Haensel, Nucl. Phys. **A301**, 53 (1978).

[66] E. Hernández, J. Navarro, and A.Polls, Phys. Lett. B **413**, 1 (1997).

[67] M. Colonna, M. D. Toro, , and A. Larionov, Phys. Lett. B **428**, 1 (1998).

[68] F. Braghin, Phys. Lett. B **446**, 1 (1999).

[69] T. H. R. Skyrme, Phil. Mag. **1**, 1043 (1956).

[70] T. H. R. Skyrme, Nucl. Phys. **9**, 615 (1959).

[71] S. Ayik and D. Boilley, Phys. Lett. B **276**, 263 (1992), errata ibd. 284 (1992) 482.

[72] K. Morawetz, M. DiToro, and L. Münchow, Phys. Rev C. **54**, 833 (1996).

[73] N. D. Mermin, Phys. Rev. B **1**, 2362 (1970).

[74] V. Baran and et. al., Nucl. Phys. **A632**, 287 (1998).

[75] H. Steinwedel and J. Jensen, Z. f. Naturforschung **5**, 413 (1950).

[76] B. F. Davis and et. al., Phys. Rev. Lett. **79**, 609 (1997).

[77] D. Vautherin and D. M. Brink, Phys. Rev. C **5**, 626 (1972).

[78] F. Braghin and D. Vautherin, Phys. Lett. B **333**, 289 (1994).

[79] S. Dietrich and B. Berman, Nucl. Data Tabl. **38**, 199 (1988).

[80] E. Ramakrishnan *et al.*, Phys. Rev. Lett. **76**, 2025 (1996).

[81] E. Ramakrishnan *et al.*, Nucl. Phys. **A549**, 49 (1996).

[82] K. Morawetz *et al.*, Phys. Rev. C. **60**, 54601 (1999).

[83] K. Morawetz, Phys. Rev. E. (2000), in press physics/9810046.

[84] P. Ring and P. Schuck, *The Nuclear Many-Body Problem* (Springer-Verlag, New York, 1980).

[85] W. Greiner and J. Maruhn, *Nuclear Models* (Springer-Verlag, Berlin et.al, 1996).

[86] T. Baumann and et. al., Nucl. Phys. **A635**, 428 (1998).

[87] M. Harakeh and A. E. L. Dieperink, Phys. Rev. C **23**, 2329 (1981).

[88] U. Garg, in *Selected Topics in Nuclear Collective Excitatons, NUCOLEX99*, Vol. 23 of *Riken Rev.*, edited by N. D. Dang, U. Garg, and S. Yamaji (RIKEN, Hirosawa, Wako, Saitama, 1999), p. 65.

[89] T. J. Deal, Nucl. Phys. A **217**, 210 (1973).

[90] I. Hamamoto, H. Sagawa, and X. Zhang, Phys. Rev. C. **57**, R1064 (1998).

[91] N. V. Giai and N. Sagawa, Nucl. Phys. A **371**, 1 (1981).

[92] P. Ring and P. Schuck, *The Nuclear Many Body Problem* (Springer, Heidelberg, 1980).

[93] G. F. Bertsch and R. A. Broglia, *Oscillations in finite Quantum Systems* (Cambridge University Press, Cambridge, 1994).

[94] U. Kreibig and M. Vollmer, *Optical Properties of Metal Cluster* (Springer-Verlag, Berlin, 1995).

[95] D. Vautherin and D. Brink, Phys. Rev. C **5**, 626 (1972).

[96] H. Steinwedel and J. Jensen, Z. Naturforsch. **5**, 413 (1950).

[97] F. Braghin and D. Vautherin, Phys. Lett. B **333**, 289 (1994).

[98] K. Morawetz, U. Fuhrmann, and R. Walke, Nucl. Phys. A **649**, 348 (1999).

[99] U. Garg, private communication.

[100] A. Bohr and B. R. Mottelson, *Nuclear Structure* (W. A. Benjamin, Inc., New York, 1969).

[101] F. Cappuzzello and et. al., in *Proceedings of the workshop on Large scale collective motion of atomic nuclei, Brolo 1996*, edited by G. Giardina (World Scientific, Singapore, 1996), p. 88.

[102] H. Esbensen, B. A. Brown, and H. Sagawa, Phys. Rev. C. **51**, 1274 (1995).

[103] A. A. Abrikosov and I. M. Khalatnikov, Rep. Prog. Phys. **22**, 329 (1959).

[104] E. Lifschitz and L. P. Pitaevsky, in *Physical Kinetics*, edited by E. Lifschitz (Akademie Verlag, Berlin, 1981).

[105] V. M. Kolomietz, A. G. Magner, and V. A. Plujko, Z. Phys. A **345**, 131 (1993).

[106] A. G. Magner, V. M. Kolomietz, H. Hofmann, and S. Shlomo, Phys. Rev. C **51**, 2457 (1995).

[107] S. Ayik, M. Belkacem, and A. Bonasera, Phys. Rev. C **51**, 611 (1995).

Pre-equilibrium Hot Giant Dipole Resonance Excitation in N/Z Asymmetric Nuclear Reactions

V.Baran *† Ph.Chomaz ‡ M.Colonna* and M.Di Toro*

Abstract

We study the effect of a direct GDR excitation in intermediate dinuclear systems with very exotic shape and charge distributions formed in charge asymmetric fusion entrance channels. A related enhancement of the GDR γ yield in the evaporation cascade of the fused nucleus is quantitatively analysed. The dynamical origin of such GDR extra strength will show up in a characteristic anisotropy of the dipole γ-emission. Independent information on the damping of the dipole mode in very excited nuclear systems can be extracted. We discuss the beam energy dependence of the effect due to the coupling between the pre-equilibrium dipole mode, the fusion dynamics and other collective dynamical contributions. We derive some optimal entrance channel conditions to get this important pre-equilibrium cooling mechanism.

1 Introduction

The isovector giant dipole resonance represents a well established collective motion of finite nuclei extensively studied since more than fifty years [1]. The dependence on the nuclear structure of the reference state on top of which the collective mode is built has already suggested the use of GDR properties in order to study nuclei far from the ground state. The time structure of the GDR mode actually allows a possibility of using it as a probe of nuclear systems very far from normal conditions, see the nice reviews [2, 3].
We can roughly estimate an oscillation period of $\frac{2\pi\hbar}{E_{GDR}} \simeq 80-100 fm/c$ and a mean life-time $\hbar/\Gamma_{GDR} \simeq 50 fm/c$. Since the spreading width is the largest contribution to the total width Γ_{GDR} we can estimate around $50 fm/c$ also the time needed to build up the collective GDR mode in a Compound Nucleus (C.N.). All these time scales are relatively short and this makes the GDR an ideal probe to study nuclear systems under extreme conditions.
Charge equilibration takes place on time scales of few units in $10^2 fm/c$: in charge asymmetric entrance channels we expect to see a "direct dipole" collective excitation, with special dynamical features, which should give an extra yield to the photon spectrum in the GDR region. The idea is that we can form a fused dinuclear system with the charge not yet equilibrated and therefore with some extra dipole strength of non statistical nature. In this sense the "Dynamical Dipole" is a direct dipole contribution present in dissipative collisions initiated in charge asymmetric entrance channels. This new pre-equilibrium collective mode has been predicted some years ago [4, 5] and recently clearly observed in fusion

*Laboratori Nazionali del Sud, INFN, 44 Via S.Sofia, 95125 Catania, Italy, and Dipartimento di Fisica dell' Universitá di Catania
†NIPNE, Bucharest, Romania
‡GANIL. Caen. France

[6, 7, 8] and deep-inelastic [9, 10] reactions. Some experiments have been performed also at intermediate energies, where other dynamical entrance channel features can be present [11, 12]. The appearance of this prompt dipole effect gives important information on the charge equilibration dynamics in connection to the reaction mechanism. A fully microscopic analysis is performed with quantitative predictions of the GDR photon yield based on a dynamics-statistics coupling model. Independent information on the damping of the dipole mode in very excited nuclear systems can be derived. Moreover the coupling between isovector dipole and large amplitude isoscalar monopole oscillations can be studied, of large interest also for the discussion on the fate of the GDR in hot nuclei. Since the effects is strongly dependent on fusion dynamics we study the beam energy variation of this extra dipole strength and we suggest some optimal conditions for the relative observation.

2 GDR in charge asymmetric fusion: the dynamical dipole

Fusion processes in charge asymmetric entrance channels give a clear example on how emitted GDR photons may carry information on the Compound Nucleus (CN), formed during the reaction dynamics, very far from normal conditions.

In dissipative collisions energy and angular momentum are quickly distributed among all single particle degrees of freedom while charge equilibration takes place on larger time scales [13]. Therefore for charge asymmetric entrance channels at the time of CN formation we can easily have some relic of a pre-equilibrium GDR coming from dipole oscillations in the isospin transfer dynamics [5]. The result will be an enhanced GDR photon emission with special features due to the non-statistical nature of this extra contribution.

An enhancement of the GDR gamma ray yield has been observed experimentally in fusion reaction between heavy ions with different neutron to proton ratios. Flibotte et al. [6], in collisions between ^{40}Ca $(N/Z = 1)$ and ^{100}Mo, $(N/Z = 1.38)$, at $4AMeV$ beam energy, producing a CN at $70MeV$ excitation energy, counted a number of emitted GDR photons over the whole cascade 16% larger than in the case when the same CN was formed with a symmetric N/Z combination. Similar effect has been observed by Cinausero et al.[7]: an increase of 36% resulted when they compared GDR photon yields from a CN populated at 110 MeV excitation energy via the reactions ^{16}O $(N/Z = 1)$ + ^{98}Mo $(N/Z = 1.33)$ and ^{50}Ti $(N/Z = 1.18)$ + ^{64}Ni $(N/Z = 1.28)$ respectively.

In fact, theoretically, an extra-yield of gamma rays was predicted some years ago by Chomaz et al., [4, 5], in a simple model of a GDR phonon gas interacting with the CN. The essential quantity in this approach is the mean number of excited GDR phonons at the time when a CN is formed, $n^{(0)}_{GDR}$. The N/Z asymmetry of colliding heavy ions will trigger in the early stage of the reaction the charge equilibration process. Several experiments and realistic simulations have indicated that the related dipole oscillation has the characteristics of a GDR type quantal collective motion [13], [14], [15], [16], [17], [18]. Expressed in terms of the Brink-Axel hypothesis [19], this means that a GDR may be built not only on equilibrium states, of a warm nucleus for example, but also on non-equilibrium states during the CN formation phase. $n^{(0)}_{GDR}$ represents the number of phonons in this mode at the time of the CN formation, t_{CN}.

Although it was assumed that the quantity $n^{(0)}_{GDR}$ is intimately related to the presence of pre-equilibrium effects, the way it is determined and affected by the projectile target N/Z asymmetry and entrance-channel dynamics has not been investigated yet in detail. From the previous discussion we clearly distinguish three main phases in the entrance channel dynamics: *I* The approaching phase, with the two partners still keeping their own response properties; *II* The dinuclear phase, with relative collective response; *III* The CN formation. We will call $t = 0$ the starting time of the phase *II*, i.e. the onset of the main dissipation mechanisms, including the charge equilibration. In this phase a pre-equilibrium dipole mode is present and damped with a spreading width $\Gamma^{\downarrow} \equiv \hbar\mu(t)$, where the time dependence is expressing a non-equilibrium situation. The correspondent number of phonons will show an exponential decrease:

$$n(t) = n_i e^{-\int_0^t \mu(t)dt} \tag{1}$$

where n_i is the number of phonons at the moment when charge equilibration begins. Fusion experiments seem to indicate that this will happen once the decision for fusion is taken [13]. This means that when the system passes by the strong absorption configuration through the barrier, the neck is large enough to allow for an isospin collective motion. Therefore we can assume that the configuration reached soon after, when the interdistance between the center of mass of the two nuclei is around the sum of their radii, represents the starting point of the collective pre-equilibrium GDR mode. It is important to stress that at this moment we have already reached a quite noticeable density overlap and therefore the considered configuration is not corresponding to the naive picture of the two nuclei as two touching rigid spheres. This will become clear from density contour plots and collective behaviours that we will discuss later in microscopic dynamical simulations.

In order to estimate n_i we consider a harmonic oscillator description for the GDR:

$$H_{GDR}(t) = \frac{\Pi^2(t)}{2M} + \frac{M\omega^2(t)}{2}X^2(t), \tag{2}$$

where $M = \frac{ZN}{(Z+N)}m$ is the reduced mass of the neutron-proton relative motion, $m = 935MeV$ being the nucleon mass, $N = N_1 + N_2$, $(Z = Z_1 + Z_2)$, the total number of neutrons (protons). Here Π denotes the relative momentum:

$$\Pi = \frac{NZ}{A}(\frac{P_p}{Z} - \frac{P_n}{N}) \tag{3}$$

where P_p (P_n) is the center of mass momentum for protons (neutrons), while $X = R_p - R_n$ is the distance between the centers of mass of the two components. The initial number of phonons is defined as:

$$n_i = \frac{H_{GDR}(0)}{\hbar\omega(0)} \approx \frac{1}{2\hbar^2}M(\hbar\omega(0))X^2(0) \tag{4}$$

if we neglect the kinetic part contribution. We expect this approximation not to work at higher incident beam energies, as discussed later. Here $X(0)$ has the following expression:

$$X(0) = \frac{Z_1 Z_2}{NZ}(\frac{N_1}{Z_1} - \frac{N_2}{Z_2})(R_1 + R_2) \tag{5}$$

We can consider for the initial elongated shape a phonon energy [20]:

$$\hbar\omega(0) = \frac{R}{R_1 + R_2}E_{GDR} \tag{6}$$

where $R = r_0 A^{1/3}$, $(r_0 = 1.2)$, is the equilibrium radius of the CN and $E_{GDR} \simeq 78A^{-1/3}$ is the energy of a phonon built on the CN, corresponding to the centroid of the GDR spectrum in medium-heavy nuclei.

In conclusion we can finally express the mean number of phonons at the time when the CN is formed as:

$$\begin{aligned} n^{(0)}_{GDR} &= n(t_{CN}) \approx \frac{M}{2\hbar^2} R(R_1+R_2) \frac{Z_1^2 Z_2^2}{N^2 Z^2} \left(\frac{N_1}{Z_1} - \frac{N_2}{Z_2}\right)^2 E_{GDR} \exp(-\mu_{ave} t_{CN}) \\ &= 1.4 \frac{(A_1^{1/3} + A_2^{1/3})}{A} \frac{Z_1^2 Z_2^2}{NZ} \left(\frac{N_1}{Z_1} - \frac{N_2}{Z_2}\right)^2 \exp(-\mu_{ave} t_{CN}) \end{aligned} \tag{7}$$

where μ_{ave} is an average value of the spreading width:

$$\mu_{ave} = \frac{1}{t_{CN}} \int_0^{t_{CN}} \mu(t) dt \tag{8}$$

We observe that $n^{(0)}_{GDR}$ will depend critically on fusion dynamics (through the time scale for the Compound Nucleus formation) as well as on properties of the spreading width $\mu(t)$.

After the time t_{CN} we have to switch to the approach introduced in [5] for the CN decay phase. The number of GDR phonons is still time-dependent since charge equilibration is still going on, but now we can have some feeding from the CN heat bath with a width $\Gamma_{feed} \equiv \hbar\lambda$. The result is [5] (from now, the origin of time is t_{CN}):

$$n_{GDR}(t) = \frac{\lambda}{\mu}\left[1 - (1 - n^{(0)}_{GDR}\frac{\mu}{\lambda}) \exp(-\mu t)\right] \tag{9}$$

which asymptotically leads to the statistical value:

$$\frac{\lambda}{\mu} = \frac{\rho(E^* - E_{GDR})}{\rho(E^*)} \approx \exp(-E_{GDR}/T) \tag{10}$$

where $\rho(E)$ is the level density, E^* is the CN excitation energy and T the corresponding temperature. From this discussion we see that the quantity Eq.(9) can be directly used as an extra dipole strength in the evaluation of γ-decay in the CN evaporation cascade. From the detailed balance we have a statistical γ emission rate:

$$\begin{aligned} R_\gamma(E_\gamma) &= \frac{\rho(E^* - E_\gamma)}{\rho(E^*)} \frac{\sigma_{abs}(E_\gamma)}{3} \frac{E_\gamma^2}{(\pi\hbar c)^2} \\ &= \frac{\rho(E^* - E_{GDR})}{\rho(E^*)} \left[\frac{\rho(E^* - E_\gamma)}{\rho(E^* - E_{GDR})} \frac{\sigma_{abs}(E_\gamma)}{3} \frac{E_\gamma^2}{(\pi\hbar c)^2}\right] \end{aligned} \tag{11}$$

where $\sigma_{abs}(E_\gamma)$ is the absorption γ-dipole cross section. This naturally leads, in the case of pre-equilibrium contributions, to a time dependent quantity:

$$R_\gamma(E_\gamma, t) = n_{GDR}(t) \left[\frac{\rho(E^* - E_\gamma)}{\rho(E^* - E_{GDR})} \frac{\sigma_{abs}(E_\gamma)}{3} \frac{E_\gamma^2}{(\pi\hbar c)^2}\right] \tag{12}$$

with $n_{GDR}(t)$ given by Eq.(9).

The final yield of emitted photons can be then evaluated using a time-dependent evaporation cascade procedure, see ref.[22]. From Eq.s(12) and (9), once the CN initial excitation energy (temperature) is fixed, the results will be critically dependent on $n^{(0)}_{GDR}$ and $\Gamma^{\downarrow}(T)$.

To show how the method is working and the sensitivity to the above parameters, in Fig.1 we report some results for the systems studied by Flibotte et al [6]. The curves represent the "subtracted" photon spectra (i.e.normalized to a statistical emission without GDR):

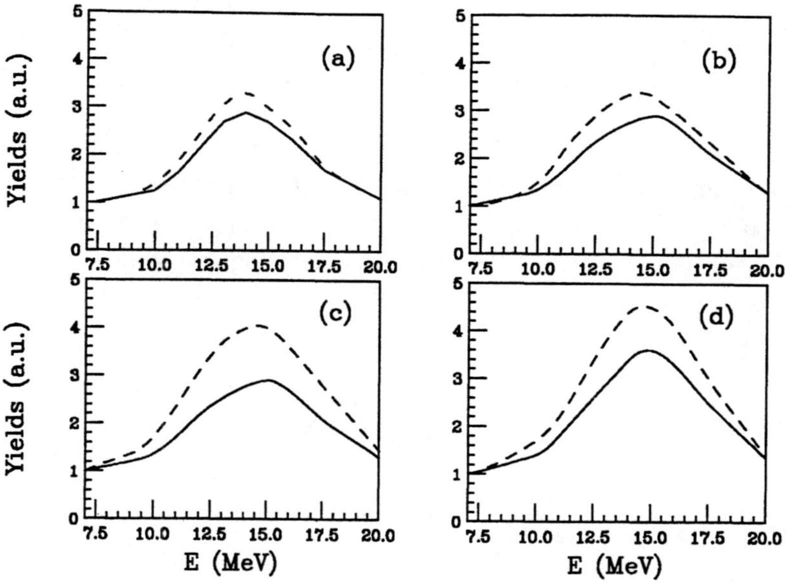

Figure 1 - Subtracted γ spectra from a ^{140}Sm CN formed at $E^* = 71MeV$ in the charge asymmetric ($Ca + Mo$ - dashed lines) and symmetric ($S + Pd$ - solid lines) entrance channel: a) Experiment [6]; b), c), d), simulations (see text) with $(n^{(0)}_{GDR}, \Gamma^{\downarrow})$ respectively equal to $(0.14, 8MeV)$, $(0.6, 8MeV)$, $(0.14, 4.8MeV)$. 10^5 Montecarlo events.

in 1a the data for the two systems, $Ca + Mo$ (charge asymmetric, dashed) and $S + Pd$ (more symmetric, solid), and in 1b, c, d the results obtained from 10^5 runs of a Monte Carlo evaporation cascade [23, 22] with the same initial conditions for the CN and varying only $n^{(0)}_{GDR}$ and $\Gamma^{\downarrow}(T)$.

The best results (Fig.1b) are obtained with the values $n^{(0)}_{GDR} = 0.14$ and $\Gamma^{\downarrow} = 8MeV$. Now while it is reasonable to have a GDR damping width larger than in the ground state of the same compound nucleus Sm^{140} ($4.8MeV$ used in Fig.1c) the choice of $n^{(0)}_{GDR}$ requires a deeper dynamical justification. This is the aim of the next section, where we will also look at the beam energy dependence of this GDR extra strength.

We note however that the pre-equilibrium contribution has some special dynamical features, being associated with a dipole component on the reaction plane, actually along the symmetry axis of the dinuclear system. If we call this as the $z-$component, we expect no dipole strength present on the x, y components at the time of the CN formation. The corresponding number of phonons will also reach asymptotically the statistical value Eq.(10) (times 1/3 on average) but starting from zero at $t = t_{CN}$ [21].

We have then time dependent γ-dipole emission rates different for each component in the intrinsic reference system of the dinucleus:

$$
\begin{aligned}
R_{\gamma,z}(E_\gamma, t) &= \left[\frac{\lambda}{3\mu}(1 - \exp(-\mu t)) + n^{(0)}_{GDR}\exp(-\mu t)\right]\frac{\sigma_{abs}(E_\gamma)}{3}\frac{E_\gamma^2}{(\pi\hbar c)^2} \\
&\equiv W_z(t)\frac{\sigma_{abs}(E_\gamma)}{3}\frac{E_\gamma^2}{(\pi\hbar c)^2} \qquad (13)
\end{aligned}
$$

$$R_{\gamma,x}(E_\gamma,t) = R_{\gamma,y}(E_\gamma,t) = \left[\frac{\lambda}{3\mu}(1-\exp(-\mu t))\right]\frac{\sigma_{abs}(E_\gamma)}{3}\frac{E_\gamma^2}{(\pi\hbar c)^2}$$
$$\equiv W_{x,y}(t)\frac{\sigma_{abs}(E_\gamma)}{3}\frac{E_\gamma^2}{(\pi\hbar c)^2} \tag{14}$$

with $n_{GDR}^{(0)}$ given by Eq.(7).

We remark the physical meaning of the time dependent weights $W_{z,x,y}(t)$ defined before; They represent the average number of dipole phonons on each component at each time step. In Fig.2 we report a typical behaviour of the weights. We have chosen the parameters corresponding to the $Ca + Mo$ system of ref.[6], see before.

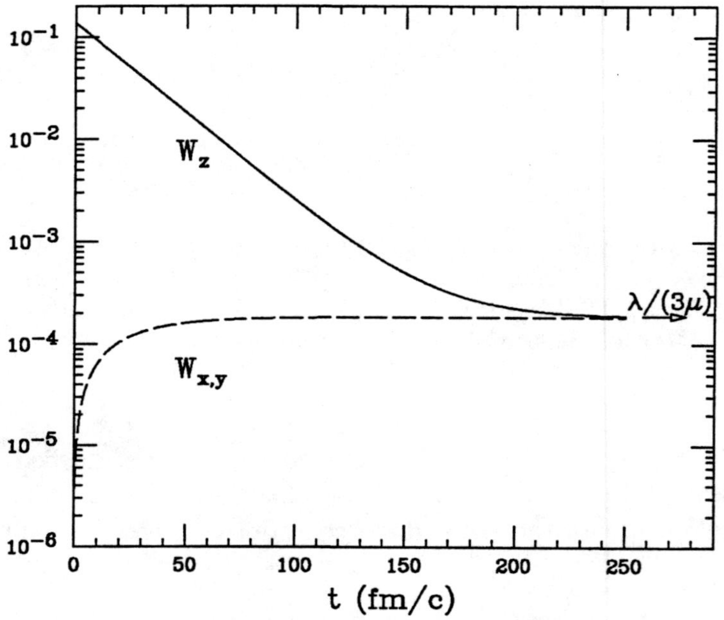

Figure 2 - Time evolution of the number of phonons on the different intrinsic axes for a ^{140}Sm CN formed at $E^* = 71MeV$ in the charge asymmetric $Ca + Mo$ entrance channel. The used parameters are $E_{GDR} = 15MeV$, $T = 2MeV$, $n_{GDR}^{(0)} = 0.14$ and $\Gamma^\downarrow = 8MeV$.

In particular we note the extreme values at $t = 0$ (CN formation time) and asymptotically $t \to \infty$ (full charge equilibration):

$$W_z(0) = n_{GDR}^{(0)} \ , \ W_z(\infty) = \frac{\lambda}{3\mu} \tag{15}$$

$$W_{x,y}(0) = 0 \ , \ W_{x,y}(\infty) = \frac{\lambda}{3\mu} \tag{16}$$

i.e. the GDR phonons are initially only on the $z-$component and finally uniformly distributed on the three axes. This property will have important observable effects on the $\gamma-$angular distributions.

The angular distribution of the emitted photons can be written as [24, 3, 25, 26] :

$$W(\theta, E_\gamma) = W_0(1 + a_2(E_\gamma)P_2(\theta)) \tag{17}$$

where $P_2(\theta)$ is a Legendre Polynomial in the polar angle θ between the direction of the emitted $\gamma-$ray and the beam axis. With our choice of the oscillation axes we have $(x, z)-$stretched and $y-$unstretched transitions [24] and a time-dependent anysotropy parameter:

$$a_2(E_\gamma, t) = \frac{W_y/2 - (W_z + W_x)/4}{W_x + W_y + W_z} \tag{18}$$

with the time dependent weights $W_{z,x,y,}(t)$ given by Eq.s (13,14). The $z-$ extra strength contribution should therefore lead to clear negative values around the GDR energy *also in absence of an intrinsic deformation of the fused nucleus.* We remark that all the previous equations can be easily extended to the use of different Lorentzians for the various axis contributions to σ_{abs}, to take into account the presence of deformations. The dynamical dipole effect will enhance the corresponding characteristic anisotropies in the $\gamma-$ angular correlations.

3 The interplay between fusion dynamics and pre-equilibrium GDR

In this section we present a fully microscopic analysis of the fusion reaction for some systems of interest, to look in particular at the pre-equilibrium GDR population. The calculations are performed in the framework of the BNV transport equation which incorporates in a self-consistent way the mean field and two-body collisions dynamics [27]. The numerical accuracy has been largely improved in order to have a good description also of low energy fusion reactions, see discussion in [26, 18]. For the mean field we have adopted a Skyrme-like parametrization, as described in [18], which well reproduces Nuclear Matter saturation properties (equilibrium density, binding energy, compressibility and symmetry energy). In the collision integral we use in medium reduced nucleon-nucleon cross sections, isospin as well as energy and angle dependent [28].

We have focussed our attention on the fusion reaction induced by ^{16}O $(N/Z = 1)$ on ^{98}Mo $(N/Z = 1.33)$ (see [7]). We have studied in a relatively large range of incident beam energy, $4MeV/n$, $8MeV/n$, $14MeV/n$ and $20MeV/n$, the evolution of the composite system along the fusion path. We will consider results that correspond to an impact parameter $b = 0$ as well representative of the features we are looking for.

In all plots we choose as $t = 0$ the time corresponding to the "touching" configuration. As discussed in the previous section this can be considered the starting time of phase II, when charge equilibration and the relative collective dipole are present in the dinuclear system. In fact this point can be checked from simulations, see later.

At each time step we can evaluate the dinuclear dipole moment in coordinate and momentum space:

$$DR(t) = \frac{NZ}{A} X(t) \tag{19}$$

$$DK(t) = \frac{\Pi(t)}{\hbar} \tag{20}$$

The results are shown in Fig.3. The out of phase behaviour of the two dipoles, clear signature of a collective dinuclear response, is actually starting just after $t = 0$ (touching configuration) practically for all energies. We remark a smaller initial amplitude of the oscillation at the lowest ($4AMeV$) and highest ($20AMeV$) energies, as well as an increase

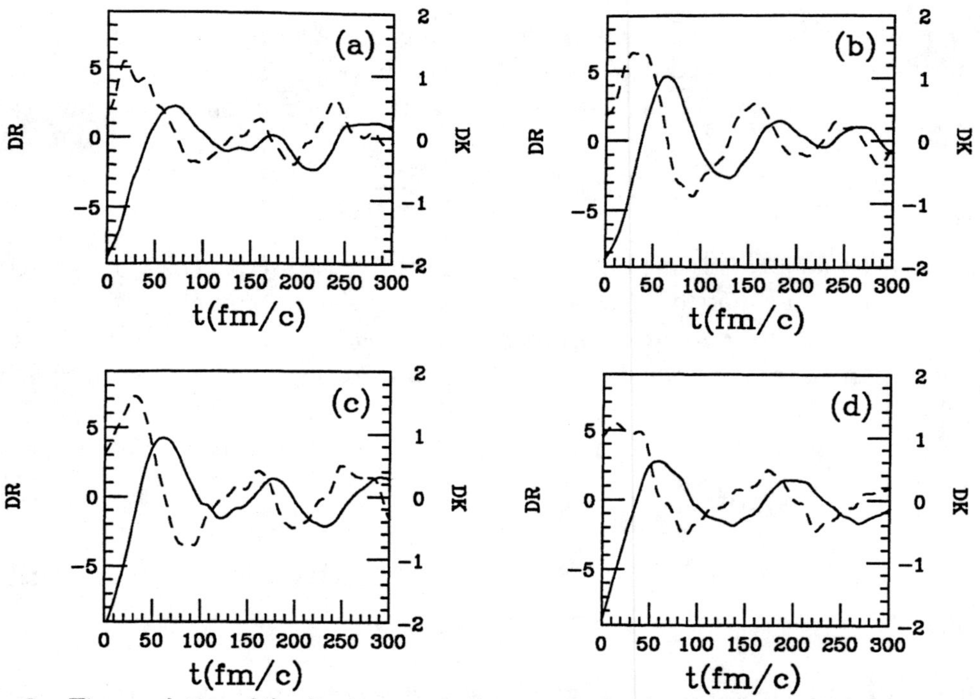

Figure 3 - Time evolution of the dipole moment in $r-space, DR$ (solid lines, left scale) and in $p-space, DK$ (dashed lines, right scale) for the $O+Mo$ system at beam energies: a) $4AMeV$, b) $8AMeV$, c) $14AMeV$, d) $20AMeV$. $t=0$ corresponds to the touching configuration (see text).

of the damping with the beam energy. In conclusion it clearly appears an optimum range of energies for the observation of the dynamical dipole effect, well above the Coulomb threshold (about $4AMeV$ in the $O+Mo$ system) but also well below the Fermi energy domain. This point will be further analysed in the following.

Fig.4 shows phase-space trajectories of the GDR, i.e. the time evolution of the $DK-DR$ correlation. The spiral curves are nicely revealing an out of phase behaviour in presence of some damping. The spiralling trend seems to start at the touching configuration ($t=0$, left point of the curves), maybe with some delay in the lowest energy case. The centre is reached when the CN formation is achieved. The spiral quickly collapses to the central region at high beam energy (Fig.4d) since we have a larger damping of the collective motion, as already seen in Fig.3.

The amplitude reduction at $4AMeV$ and $20AMeV$ is also evident: the nature of the effect is however different in the two cases. At low energy, just above the Coulomb barrier, the neck is formed quite slowly, Fig.5, and so the collective dipole response of the dinuclear system is actually starting sometime after the touching configuration, when some charge equilibration has already taken place. At $20AMeV$ the fusion dynamics is fast but now we can have some prompt particle emission that will reduce the dipole moment of the dinucleus. Just by chance we see the best spiralling behaviour at $8AMeV$, where the experiment has been performed by Cinausero et al. [7] with a very clear evidence of the dynamical dipole enhancement.

Now we can dynamically evaluate the number of pre-equilibrium GDR phonons $n_{GDR}^{(0)}$ present at the time of the CN formation: as shown in the previous section this quantity is essentially ruling the extra emission of GDR γ-rays. At each time step the average number of phonons $n(t)$ can be determined from the simulations just dividing the total

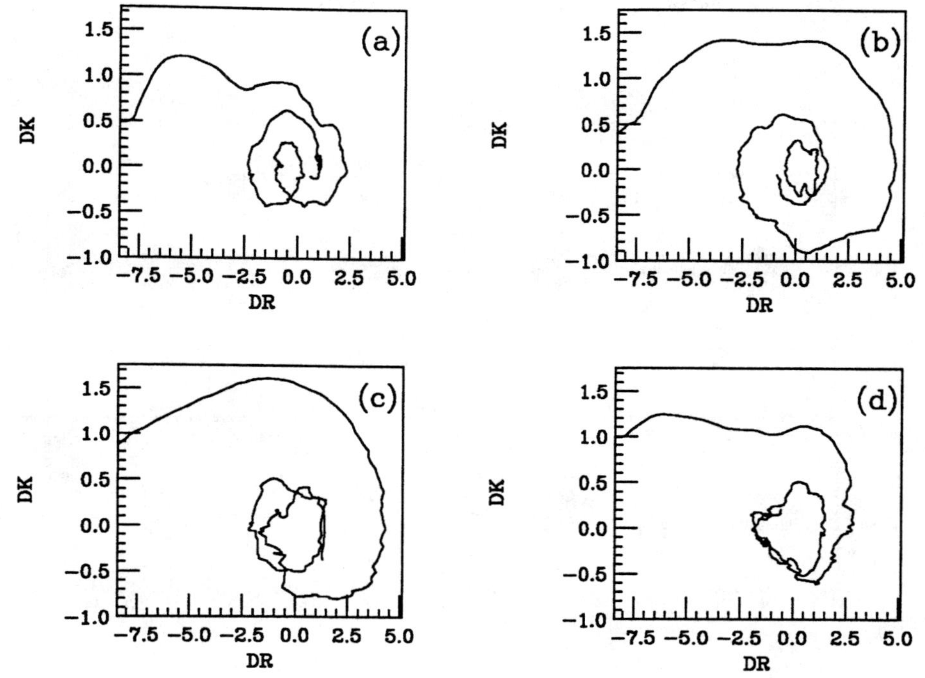

Figure 4 - Phase space trajectory of the entrance channel dipole for the $O + Mo$ reaction. The labels correspond to the same energies of Fig.3.

oscillator energy, as defined by Eq.(2), by the one phonon energy at that instant, $\hbar\omega(t)$, i.e. corresponding to the deformation at the time t. The results are presented in Fig.6. The dashed curve in the pictures represents the analitical prediction of Eq.s(1) and (4) using a constant spreading width value, $\Gamma^{\downarrow} = 5MeV$. We remark a quite good overall agreement at beam energies $8MeV/n$ and $14MeV/n$, although with increasing energy we see in the simulations a higher starting point joint to a faster damping. This is particularly evident at $20MeV/n$ (Fig.6d) where we have to consider, in Eq.(4), a larger value for the spreading width and the kinetic term contribution to the initial number of phonons n_i.

The quantity we are looking for, $n^{(0)}_{GDR}$, is now given by the value of $n(t)$ at the time of the Compound Nucleus formation t_{CN}, Eq.(7). In Fig.7 we plot the time evolution of the mass quadrupole moment in coordinate space at the four incident energies. Generally we remark two distinct behaviours in the evolution: a faster decay until the first minimum is reached, followed by a smoother trend accompanied by small oscillations. During the first stage the conversion of relative motion into heat takes place and therefore, within a short delay needed for a final thermal equilibration, the moment when the first minimum is attained will provide us the Compound Nucleus formation time. Our evaluation is in agreement with the information obtained by looking at the quadrupole in momentum space. We can extract t_{CN} values of 120 fm/c at $4AMeV$, 80 fm/c at $8AMeV$, and 50 fm/c at the higher energies $14AMeV$ and $20AMeV$. From Fig.6 we get the corresponding $n^{(0)}_{GDR} = n(t_{CN})$ parameters: 0.01 at $4AMeV$ going up to 0.07 at $8AMeV$ and 0.08 at $14AMeV$ and finally reducing again, around 0.05, at $20AMeV$.

To test the correctness of our procedure we have performed a complete evaporation cascade calculation using the above extra dipole strength for the system $O+Mo$ at $8AMeV$. The subtracted spectra are shown in Fig.8, from $2 * 10^5$ Montecarlo events, compared with experimental data from ref.[7]. The agreement is quite good.

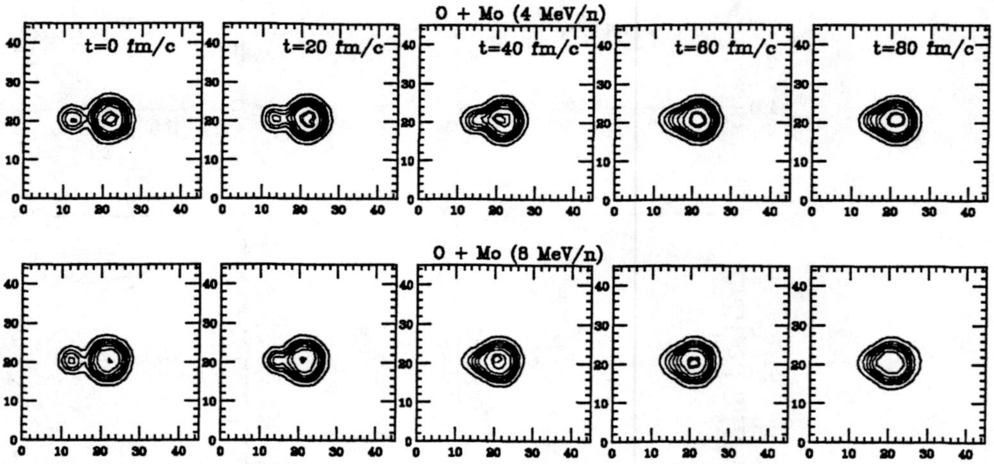

Figure 5 - Density plots of the neck dynamics for the $O + Mo$ system at the two energies $4AMeV$ and $8AMeV$.

Since the time scale of fusion dynamics is playing an important role on the amount of extra dipole strength present at the time of CN formation we expect to see some dependence of the effect also on the mass symmetry in the entrance channel, which is strongly affecting the fusion time, see [22] and ref.s therein. In order to check this point we have studied the system ^{50}Cr $(N/Z = 1.08)$ + ^{64}Ni $(N/Z = 1.28)$ with almost the same charge asymmetry of $O + Mo$ but much more symmetric in mass. We have considered the beam energies 3.5 and $5.5AMeV$ corresponding to the same c.m energies above the Coulomb barrier of the first two energies considered for the $O + Mo$ system. The relative fusion paths are shown in Fig.9, starting from the touching configuration, to be compared with the equivalent Fig.4 of the $O + Mo$. At both energies it is quite evident the longer neck formation and shape equilibration times. As discussed before, we expect a smaller initial amplitude for the collective dipole oscillation (neck dynamics) and a further reduction due to the delay in the CN formation (shape dynamics).

We have finally checked the method also for the ref.[6] system, $Ca + Mo$ at $4AMeV$, where the best results are obtained with a $n_{GDR}^{(0)} = 0.14$, see Fig.1. In Fig.10 we present the time evolution of the relevant quantities, mass quadrupole moment, dipole moments in $r-$ and $p-$ space and average number of phonons. We see a slower fusion process but with a larger initial dipole strength. We have $t_{CN} \simeq 120 fm/c$ and a corresponding $n_{GDR}^{(0)}$ ranging between 0.12 and 0.15, as needed in the fitting procedure of Fig.1.

4 Summary and conclusions

We have shown that important information on the early stage of the fusion path can be obtained studying charge-asymmetric reactions. In fact, in this case it is possible to reveal a "direct" dipole oscillations (*the dynamical dipole*), related to the charge equilibration

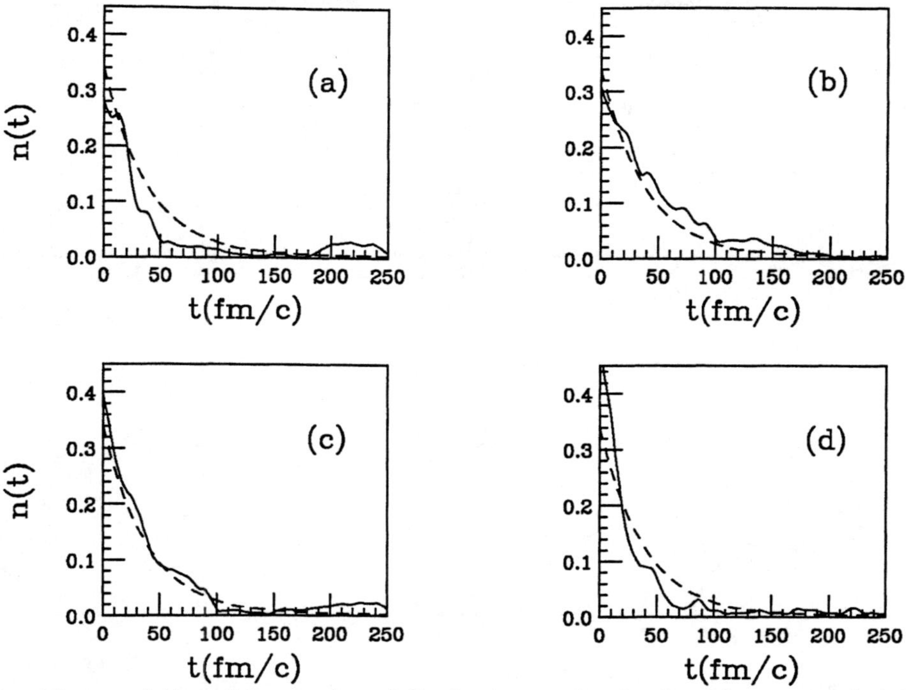

Figure 6 - Time evolution of the number of dipole phonons for the $O + Mo$ reaction (solid lines). The labels correspond to the same energies of Fig.3. The dashed line is the reference curve discussed in the text.

dynamics, which leads to some extra GDR strength in the statistical decay of the fused system. The dynamical nature of such pre-equilibrium contribution, i.e. corresponding to dipole oscillations on the reaction plane, is expected to show up in an anisotropic $\gamma-$emission, also in absence of deformations of the fused system.

In this work we have investigated how N/Z asymmetry and fusion dynamics are affecting the properties of the pre-equilibrium GDR remnants. We have shown the existence of a GDR collective mode, during the formation phase of the compound nucleus, i.e. built on nonequilibrium states. This has allowed us to express the mean number of dipole phonons present in the fused nucleus in terms of the initial isospin asymmetry, CN formation time and GDR spreading width. We can make quite accurate predictions on the optimal choice of the reaction partners, in particular on the interplay between charge and mass asymmetry in the entrance channel.

Particularly interesting is the energy dependence of the effect. We expect to see a *rise and fall* of the dinamical dipole contribution.

At low energies, just above the Coulomb barrier, the effect is strongly reduced for two combined reasons:

- i) A delay in the dinucleus formation (slow neck dynamics) and relative collective response. The pre-equilibrium dipole oscillation of the composite system will have an initial smaller amplitude since some charge equilibration has already taken place.
- ii) A longer Compound Nucleus formation time is decreasing the average number of phonons present in the fused system due to the GDR damping.

At higher energies, close to the Fermi energy domain, the effect is again reduced on average for two main reasons:

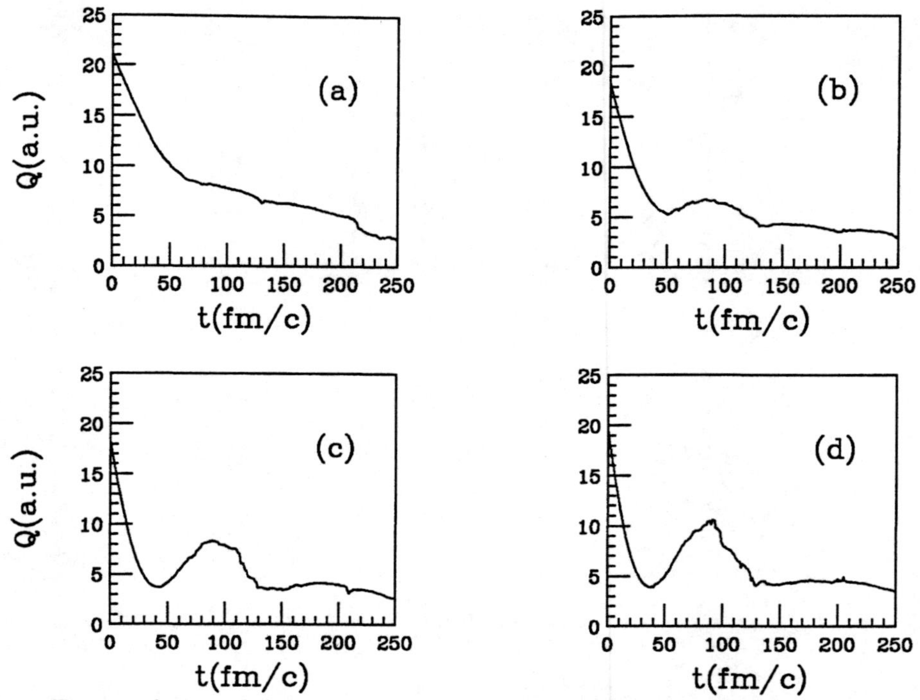

Figure 7 - Time evolution of the mass quadrupole moment for the $O + Mo$ reaction. The labels correspond to the same energies of Fig.3.

- i) We have a relevant pre-equilibrium particle emission (i.e. incomplete fusion events) with a direct reduction of the initial charge asymmetry.
- ii) In the fusion processes we have a large excitation energy deposited in the composite system with a related increase of the GDR spreading width leading to a fast decrease of the number of dipole phonons.

Moreover a new dynamical feature is appearing at energies above $20AMeV$, some large monopole oscillations in the entrance channel dynamics [29, 30, 31]. The dynamical dipole strength is reduced and more fragmented, see the simulations in ref.[31], although still present as confirmed from very recent data [11, 12]. The study of the dynamical dipole effect in the Fermi energy domain, although experimentally quite difficult, would therefore bring new independent information on the spreading width of hot GDR and in particular on the coupling to an expanding collective mode

Finally, for charge asymmetric reactions, the enhanced dipole emission could be an interesting cooling mechanism to favour the fusion of very heavy nuclear systems. From the interplay between charge and shape equilibration time-scales we can also suggest new experiments to study the dipole propagation in excited nuclei. The use of radioactive beams will largely increase the possibility of such observations.

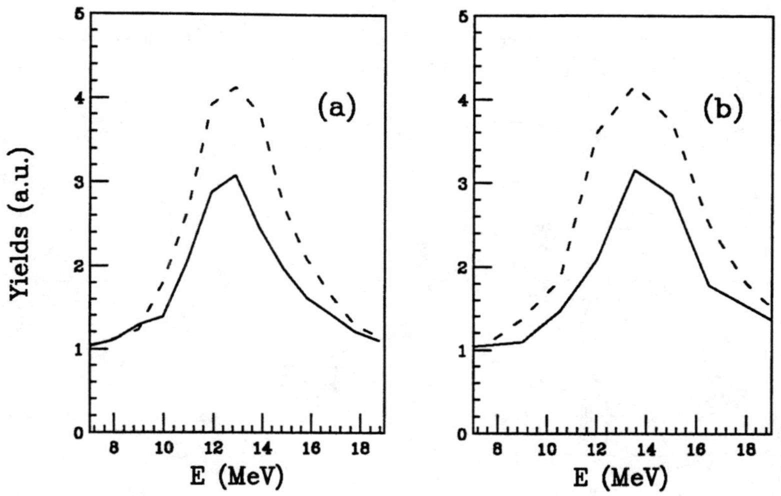

Figure 8 - Subtracted γ spectra from a ^{114}Sn CN formed at $E^* = 108MeV$ in the charge asymmetric ($O+Mo$ - dashed lines) and symmetric ($Ti+Ni$ - solid lines) entrance channel: a) Experiment [7]; b) simulations (see text) with $(n^{(0)}_{GDR}, \Gamma^{\downarrow})$ respectively equal to $(0.07, 7MeV)$. $2 * 10^5$ Montecarlo events.

References

[1] A. Van der Woude, Nucl.Phys. A599 (1996) 393c.

[2] K.A.Snover, Annu.Rev.Nucl.Part.Sci. 36 (1986) 545.

[3] J.J.Gaardhoje, Annu.Rev.Nucl.Part.Sci. 42 (1992) 483.

[4] Ph. Chomaz in "Proceedings of the 6th Franco-Japanese Colloquium", St. Malo, France, 1992, edited by N. Alamanos, S. Fortier and F. Dykstra, Saclay (1993).

[5] Ph.Chomaz, M.Di Toro and A.Smerzi, Nucl.Phys. A563 (1993) 509.

[6] S.Flibotte et al., Phys.Rev.Lett. 77 (1996) 1448.

[7] M.Cinausero et al., Nuovo Cimento A111 (1998) 613.

[8] F.Amorini et al., Phys.Rev. C58 (1998) 987.

[9] M.Trotta et al., RIKEN Review 23 (1999) 96;
M.Sandoli et al., Eur.Phys.J. A6 (1999) 275.

[10] M.Papa et al., Eur.Phys.J. A4 (1999) 69.

[11] F.Amorini, TRASMA exp., Ph.D.Thesis LNS Catania 1998.

[12] S.Tudisco, TRASMA exp., Ph.D.Thesis LNS Catania 1999.

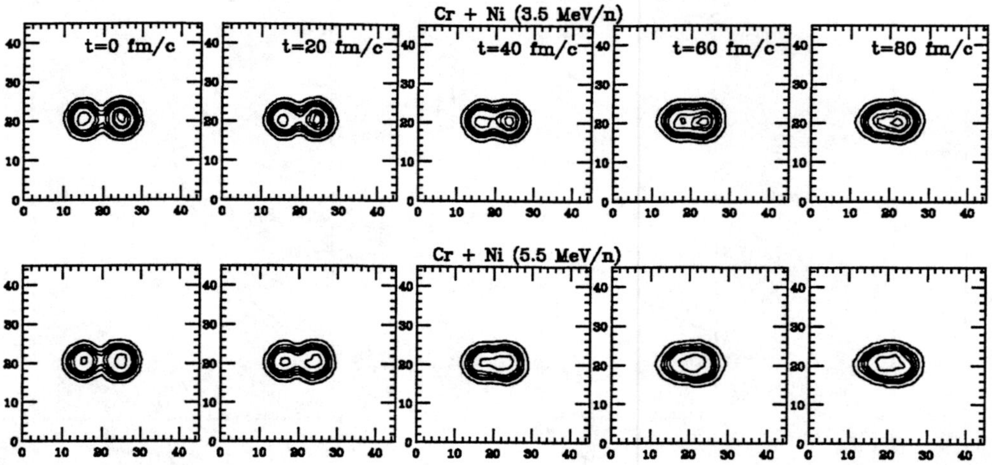

Figure 9 - Density plots of the neck dynamics for the $Cr + Ni$ system at the two energies, $3.5 AMeV$ and $5.5 AMeV$.

[13] C. Gregoire, in "Proceedings of the Winter College on Fundamental Nuclear Physics", vol. 1, Trieste, Italy, 1984, edited by K. Dietrich, M. Di Toro, H. J. Mang, pp.497-643.

[14] H. Hofmann et al., Z. Physik A293 (1979) 229.

[15] P. Bonche, N. Ngo, Phys. Lett. B105 (1981) 17.

[16] E. Suraud, M. Pi, P. Schuck, Nucl. Phys. A482 (1988) 187c.

[17] E. Suraud, M. Pi, P. Schuck, Nucl. Phys. A492 (1989) 294.

[18] V. Baran et al., Nucl. Phys. A600 (1996) 111.

[19] D.M.Brink, D.Phil.Thesis, Oxford 195;
P.Axel, Phys.Rev.126 (1962) 671.

[20] P.Ring and P.Schuck, *The Nuclear Many-Body Problem*, Springer-Verlag, New York 1980.

[21] D.M.Brink, Nucl. Phys. A519 (1990) 3c.

[22] M. Cabibbo, V. Baran, M. Colonna and M. Di Toro, Nucl. Phys. A637 (1998) 374.

[23] M. Cabibbo, *MONTECASCA Code*, Ph.D. Thesis, Univ. of Catania 1998.

[24] P.Thirolf et al., Nucl.Phys.A482 (1988) 93c.

[25] A.Bracco et al., Phys.Rev.Lett. 74 (1995) 3748.

[26] V.Baran et al., Nucl.Phys. A599 (1996) 29c.

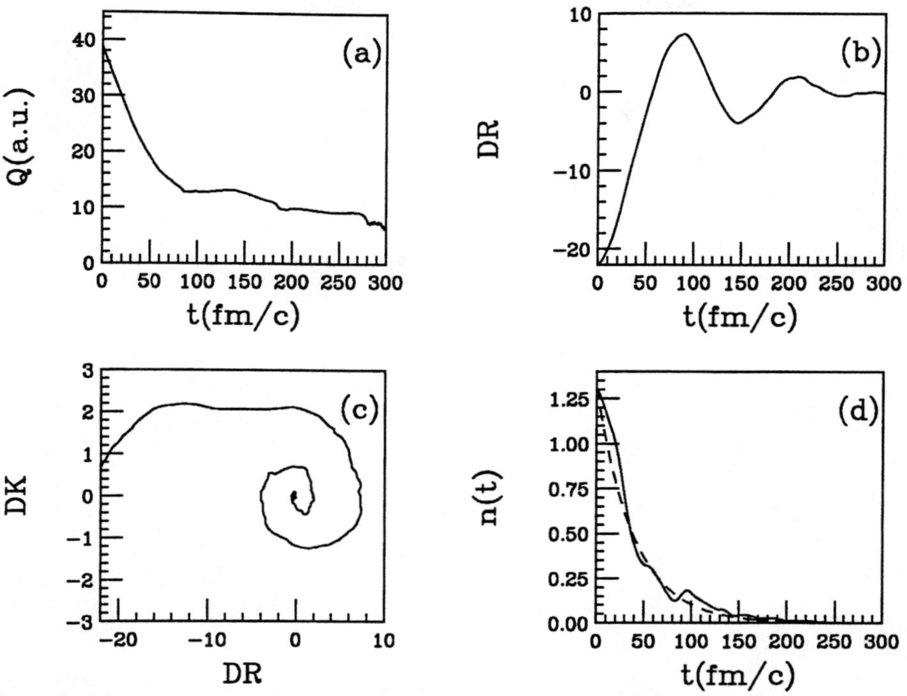

Figure 10 - Time evolution of mass quadrupole moment (a), $r-$dipole moment (b), average number of dipole phonons (d) for the reaction $Ca + Mo$ at $4AMeV$. The panel (c) represents the phase space trajectory of the entrance channel dipole.

[27] A.Bonasera et al., Phys.Lett. B221 (1989) 233;
A.Bonasera et al., Phys.Rep. 243 (1994) 1.

[28] G.Q. Li and R. Machleidt, Phys. Rev. C48 (1993) 1702; Phys. Rev. C49 (1994) 566; G.Q. Li, private communications.

[29] P.Piattelli et al., Nucl.Phys.A649 (1998) 181c.

[30] T.Suomijarvi, RIKEN Symp.on Dynamics in hot nuclei, p.1,1998.

[31] M.DiToro et al., Acta Phys.Polonica B30 (1999) 1331-1352.

Nuclear Transport Models and Effects of Isospin-Dependent Equation of State on Dissipative Heavy-Ion Collisions

M. Colonna* and M. Di Toro *

Abstract

We suggest several ways to study properties of the symmetry term in the nuclear equation of state, EOS, from radioactive beam experiments. After a general discussion on compressibility and saturation density we present various results from dynamical simulations of dissipative heavy ion collisions. The main reaction mechanisms, from fusion to deep-inelastic and fragmentation, appear quite sensitive to the form of the symmetry term, opening some appealing experimental perspectives. From the study of the collective response we predict new features of fragment production, due to the onset of chemical plus mechanical instabilities in a dilute asymmetric nuclear matter. Interesting variances with respect to statistical expectations are also presented in realistic calculations for finite nuclei. Collective flows in heavy ion collisions are shown also to be strongly dependent on the symmetry term of the EOS. A comparison with data leads to a clear evidence for a strong increase of the repulsion of the symmetry term above normal density.

1 Introduction

The availability of exotic (radioactive) beams has driven a large interest on nuclear structure studies of $\beta-$unstable nuclei. It is clear that essential complementary information will come from charge asymmetry effects on non-equilibrium nuclear dynamics. A quantitative analysis of this point, trying to pin down some very sensitive observables in reaction mechanisms, is the aim of the present report.

We will show that it could be possible to extract some unique information on the symmetry term of the nuclear EOS in regions away from normal density. We can put then some experimental constraints on the theoretical predictions used in astrophysical contexts [1], where such information are essential in understanding the lives of neutron stars [2, 3, 4, 5]. Moreover there are quite stimulating predictions on new phases of asymmetric nuclear matter that eventually could be reached during heavy ion reaction dynamics with radioactive beams. The onset of coupled mechanical and chemical instabilities is envisaged [6, 7, 8], that should lead to clear experimental signatures. In particular a quite unexpected isotopic distribution of primary fragments is found, at variance with the predictions of statistical multifragmentation models. We will also show that collective flows and in general the dissipation mechanisms reveal a quite interesting asymmetry dependence.

*Laboratori Nazionali del Sud, INFN, 44 Via S.Sofia, 95125 Catania, Italy, and Dipartimento di Fisica dell' Universitá di Catania

In the last years some first data have appeared on isospin effects in reaction dynamics, with a few theoretical analyses. See the recent reviews [9, 10, 11]. Although the data are mostly of inclusive type and the theoretical studies are not focussing the effect of different charge symmetry terms, it emerges quite clearly a noticeable dependence of the reaction mechanisms on charge asymmetry.

2 Symmetry term effects on compressibility and saturation density

In the nuclear equation of state we have a symmetry term in the form:

$$\frac{E}{A}(\rho, I) = \frac{E}{A}(\rho) + \frac{E_{sym}}{A}(\rho)\, I^2 \tag{1}$$

with $I = (N-Z)/A$ asymmetry parameter, given by kinetic and potential contributions:

$$\frac{E_{sym}}{A}(\rho) = \frac{\epsilon_F(\rho)}{3} + \frac{C(\rho)}{2\rho_0}\,\rho \tag{2}$$

In a Skyrme-like parameterization of effective interactions the function $C(\rho)$ has the form:

$$\frac{C(\rho)}{\rho_0} = -\frac{1}{4}\left[t_0(1+2x_0) + \frac{t_3}{6}(1+2x_3)\,\rho^{\alpha}\right] + \frac{1}{12}\left[t_2(4+5x_2) - 3t_1x_1\right]\left(\frac{3\pi^2}{2}\right)^{2/3}\rho^{2/3} \tag{3}$$

To remark the second term related to isospin effects in the momentum dependence [12].

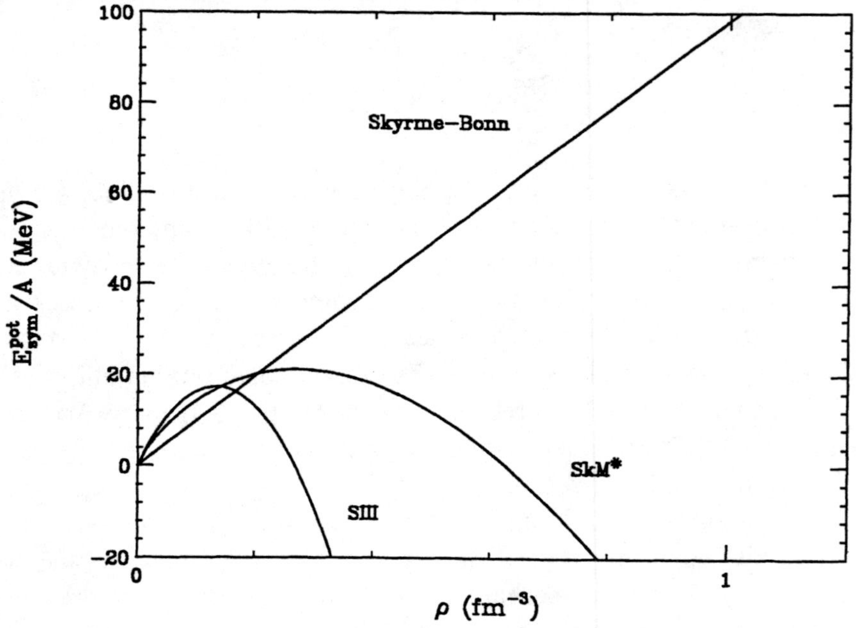

Figure 1 - Density dependence of the potential symmetry term

In Fig.1 we report the density dependence of the potential symmetry contribution, second term of Eq.(2), for three different Skyrme effective interactions, *SIII* (that we will call *asy-soft* case in the following), SKM^* [13] and *Skyrme−Bonn* (*asy-stiff* parameterization), obtained in order to reproduce the EOS of a microscopic nuclear matter calculation with the Bonn A potential [8]. While all curves obviously cross around the normal density ρ_0, quite large differences are present in low density and particularly in high density regions.

We can first mention some simple considerations on asymmetry effects on compressibility and saturation density, observables related respectively to monopole resonances and to the bulk density in a medium heavy nucleus [14]. From a linear expansion around the symmetric value we get for the variation of saturation density a form:

$$\Delta\rho_0(I) = -\frac{9\rho_0^2}{K_{NM}(I=0)} \frac{d}{d\rho} \frac{E_{sym}}{A}(\rho)\Big|_{\rho=\rho_0} I^2 \quad < 0, \tag{4}$$

where $K_{NM}(I=0)$ is the compressibility of symmetric nuclear matter (NM). Eq(4) has an intuitive *geometrical* meaning. Asymmetry brings an extra pressure $P_{sym} = dE_{sym}/d\rho$ that can be compensated just moving to the left the saturation point ($P = 0$) of the quantity $P_{sym}/(dP/d\rho)$ at ρ_0.

For the compressibility shift we have, after some algebra,

$$\Delta K_{NM}(I) = 9\rho_0 \Big[\rho_0 \frac{d^2}{d\rho^2} - 2\frac{d}{d\rho}\Big] \frac{E_{sym}}{A}(\rho)\Big|_{\rho=\rho_0} I^2 \quad < 0 \tag{5}$$

The predictions are quite different also for relatively small asymmetries. E.g., as expected, $SIII$ gives the largest variation for compressibility and the smallest for saturation density [14]. There are therefore good chances of a direct experimental cross-check.

3 Symmetry term effects on Fusion - Deep Inelastic competition

A new code for the solution of microscopic transport equations [15, 16, 17, 18, 19] has been written where asymmetry effects are suitably accounted for and the dynamics of fluctuations is included [20, 21]. This code allows to perform stochastic mean-field calculations. According to one-body theories, the system is still described through the time dependent one-body distribution function, however this function may experience a stochastic evolution in response to the action of a fluctuation source term. Isospin effects on nucleon-nucleon (NN) cross section and Pauli blocking are consistently evaluated.

We use a general Skyrme-like form for neutron and proton mean field potentials:

$$U_q = A\left(\frac{\rho}{\rho_0}\right) + B\left(\frac{\rho}{\rho_0}\right)^{\alpha+1} + C\left(\frac{\rho_i}{\rho_0}\right)\tau_q + \frac{1}{2}\frac{\partial C}{\partial \rho}\frac{\rho_i^2}{\rho_0}, \tag{6}$$

where $\rho \equiv \rho_n + \rho_p$ and $\rho_i \equiv \rho_n - \rho_p$ are respectively isoscalar and isovector densities, and $q = n, p$, $\tau_q = +1$ ($q = n$), -1 ($q = p$).

In order to simplify the analysis of the most sensitive observables to isospin effects we have chosen a Skyrme force with the same $soft$ EOS for symmetric NM ($K = 201MeV$) and with two different choices for the density dependence of the symmetry term, see Fig.1, *asy-stiff* ($Skyrme - Bonn$) and *asy-soft* ($SIII$). In this way we force the symmetric part of the EOS to be exactly the same in order to disentangle dynamical symmetry term effects. The difference on the EOS of asymmetric nuclear matter can be seen in Fig.2 where we show the total energy per nucleon as a function of the density for two values of the asymmetry parameter $I = (N - Z)/A$, $I = 0$ (*symmetric* NM) and $I = 1/3$ ($N = 2Z$).

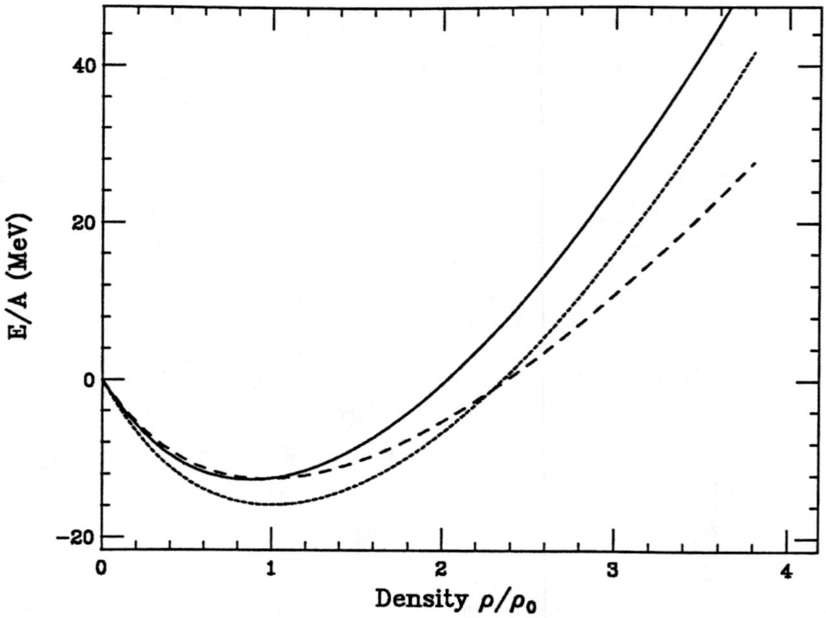

Figure 2 - EOS used in the dynamical simulations. Symmetric case: dotted line. I=1/3 asymmetry: dashed line "asy-soft", full line "asy-stiff"

While the EOS in general softens, see previous section, with asymmetry, we see a clear difference in the stiffness at $I = 1/3$ for the two different choices. Of course the difference is increasing for larger asymmetries. One of the aims of this report is to show that reaction mechanisms at medium energies are sensitive to this effect, although we will certainly not reach high compression regions. The main idea is to get a very asymmetric piece of nuclear matter in the interacting zone just taking advantage of the neutron (proton) skin structure of beta unstable nuclei and considering reaction mechanisms particularly sensitive to a "skin-skin" interaction.

We have started our analysis at relatively low energies to see if the dominant dissipative reaction mechanisms also in that beam energy range can be sensitive to a different density dependence of the symmetry term [20]. In order to see the competition between binary events and heavy residue formation we have chosen an intermediate impact parameter $b \simeq 0.4\ b_{max}$. We have compared the mean trajectories (i.e averaged over many events) of the collisions ^{46}Ar $(N/Z = 1.56)$ + ^{64}Ni $(N/Z = 1.22)$, neutron rich systems, versus ^{46}V + ^{64}Ge (N=Z, proton rich), at 30 $AMeV$ and $b = 4fm$.

In Fig.3 we show, for the neutron rich systems, the corresponding time evolution of density projections on the reaction plane with the two choices of the symmetry term parameterization. The differences are quite evident. In the *asy-soft* case (top panels) we have a stronger interaction between the two partners, larger dissipation of the relative energy and the collision enters a fusion path. The *asy-stiff* case (bottom panels) is more repulsive, we have less dissipation and eventually the dinuclear system breaks on a relative short time scale. So we can claim that just from a measure of the relative rate of fusion events we can get an information on the stiffness of the symmetry term. Moreover for the *asy-soft* case we expect more interaction time available for charge equilibration. This means that also for binary events we will see the final fragments to emerge with more similar N/Z ratios and in

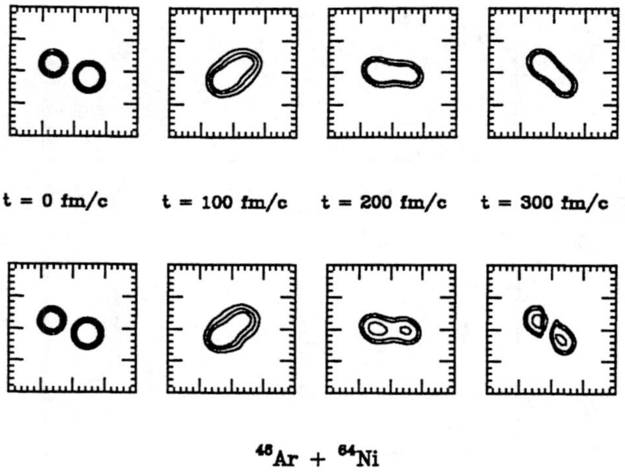

Figure 3 - Simulation of the collision of neutron rich Ar+Ni isotopes at 30 AMeV and b=4fm: Top panels: asy-soft; Bottom panels: asy-stiff.

general with more neutrons emitted since more excited. At variance , for a stiff symmetry term the two final fragments will keep more memory of the entrance channel conditions.

For the isobaric proton rich system the calculations show a quite unexpected new feature. We observe exactly the opposite behaviour with respect to the neutron-rich case. Now the *asy-soft* parameterization leads to a dominant deep-inelastic mechanism. This could be predicted from two combining effects: i) larger Coulomb repulsion; ii) larger number of nucleon-nucleon collisions (more n-p couples, which see larger cross sections) which are reducing the main nucleon-exchange dissipation mechanism. The result observed in the *asy-stiff* case is quite amazing. Since here overall we have $I = 0$ we would not expect any difference, while in fact now the must probable events are of fusion type. A way to understand this result is assuming that the two quite exotic nuclei ^{46}V and ^{64}Ge are developing a proton skin which is overlapping in the interaction zone. The protons in the asy-stiffer case see a stronger repulsive mean field and are more easily emitted. As a consequence the Coulomb repulsion between the two partners decreases and fusion can be easily reached.

At higher beam energies the transition from deep-inelastic to fragmentation has been studied in the case of a quite realistic, charge asymmetric, collision ^{124}Sn + ^{64}Ni at 40 MeV/A, for intermediate impact parameters [22]. As it is shown in Fig.4, at final interaction times the *asy-stiff* case is mostly binary while in the *asy-soft* parameterization neck instabilities develop, together with the possibility to produce fragments in the neck region [23, 24] and in general larger variances in the projectile- and target-like remnants. The isotopic structure of the final products will be also different in the two cases. Like in the lower energy discussion, the *asy-stiff* repulsion will keep a larger N/Z ratio in the PLF and TLF "spectators". In the *asy-soft* case our results seem to indicate a substantial equality in the N/Z ratios of the PLF and TLF remnants and of the fragments produced in the neck region, although the latter should be more neutron rich. A stimulating interpretation could be related to a dynamical effect that tends to restore the charge symmetry in

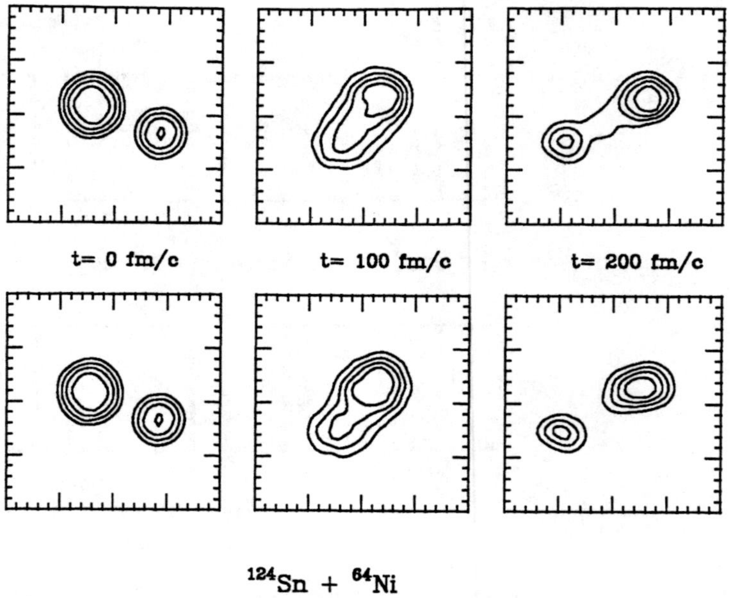

Figure 4 - Simulation of the collision ^{124}Sn + ^{64}Ni at 40 MeV/A, b = 5 fm. Top panels: asy-soft; Bottom panels: asy-stiff.

spinodal instabilities of asymmetric nuclear matter, see ref.s [8, 25]. We would observe an essentially charge symmetric clusterization surrounded by a neutron gas (see the discussion in the next section).

At larger impact parameter binary events of deep-inelastic type are dominant in both cases. Now the important observables are the ones related to the dissipation rate, masses and charge asymmetries of the two final fragments. Also the widths will be sensitive to the symmetry term parameterizations, for the different dissipation as well as for the possibility of some instability contribution.

4 New mechanical plus chemical instabilities in dilute Nuclear Matter

In the previous section we have already seen that the dynamics of fragmentation can be quite sensitive to the symmetry term of the EOS. Now we will focus the attention on more microscopic features of the instabilities in low density asymmetric nuclear matter. Apart the softening of the nuclear EOS it is intuitive to expect some direct isospin effect on the collective response of asymmetric nuclear matter, e.g. a quite strong coupling between isoscalar and isovector modes. Then the instability region of the NM phase diagram will be also affected with observable effects again closely related to the structure of the symmetry term.

The Landau dispersion relation [26] approach can be used to study collective modes in asymmetric nuclear matter just starting from the Vlasov kinetic equations of two fluids, neutrons and protons, coupled through the self-consistent mean field [8].

In the variation of the neutron (proton) mean field both isoscalar and isovector density variations will appear and therefore the linearized Vlasov equations for isoscalar and isovector variations of the distribution function will also be coupled.

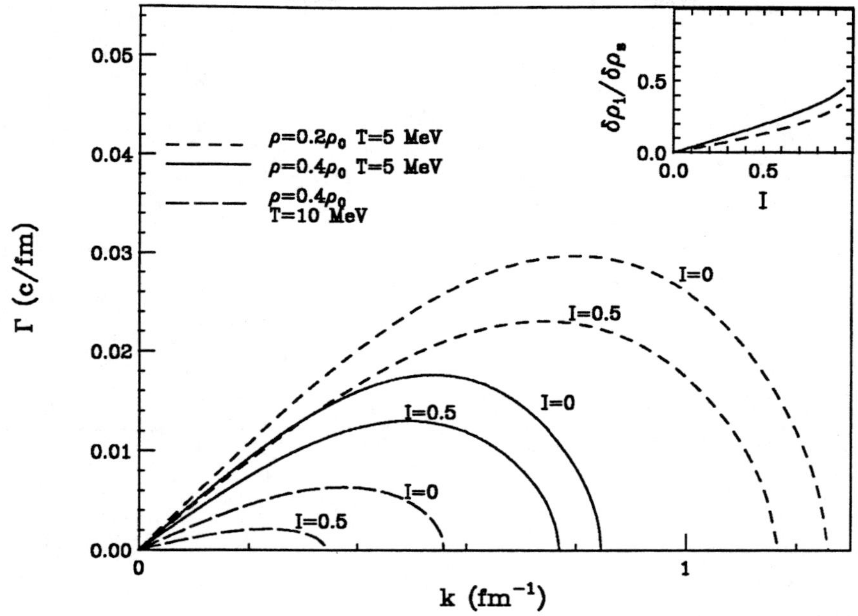

Figure 5 - Instability growth rate for asymmetric nuclear matter (SKM^*force). In the insert we show the isovector to isoscalar component ratio of the perturbation versus I. Solid line: SkM* case; Dashed line: asy-stiff case

We can derive, using the Landau procedure, a system of coupled equations for the k-component of the isovector and isoscalar density variation and then some extended dispersion relations that can be solved in different regions of scalar densities ρ to test various symmetry terms [8].

Here we will focus the attention on microscopic features of the instabilities in low density asymmetric nuclear matter. This "*Spinodal*" region, that is the region where the "isoscalar" modes become unstable, is particularly important since it can be directly related to the fragment formation mechanism [27, 28] and therefore to precise observables in fragmentation reactions with exotic beams. From the generalized Landau dispersion relations we get [8, 28]:

1) The region of low density instabilities is shrinking with increasing asymmetry. In a sense this should be expected from the softening of the EOS. As a consequence the growth time $\tau = 1/\Im(\omega)$ and the most unstable wavelength $\lambda = \frac{2\pi}{k}$ of the instabilities are increasing. Experimentally this effect should show up in a larger size of the primary produced fragments and a reduction of their collective kinetic energy (slower formation, closer to the expansion turning point) (see Fig.5).

2) The coupling between isoscalar and isovector modes leads to a *chemical* contribution in the instability: we see a larger proton component in the collective density oscillation which gives raise to clusterization (see the insert in Fig.5). Therefore we have a collective trend towards charge symmetry restoration in the formed fragments. *This is a pure dynamical effects, on short time scales, due to the charge structure of the instability.* The effect is presented in Fig.6 where we show the time evolution of the asymmetry distribution in various density bins for a piece of unstable asymmetric NM ($I = 0.5$) in a box of size $L = 8fm$ initially at $\rho = 0.4\rho_0$ (periodic boundary conditions [25]).

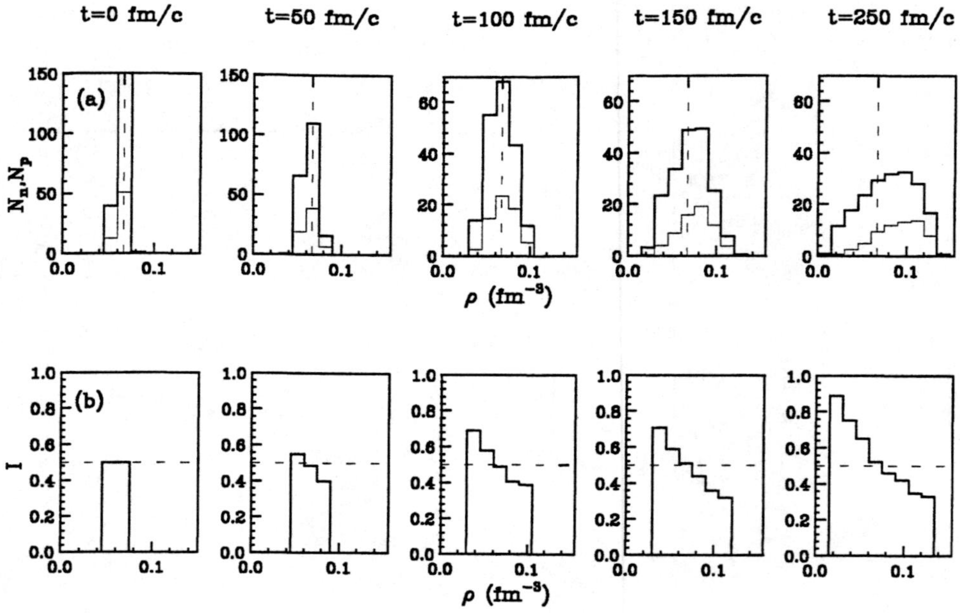

Figure 6 - Time evolution of neutron, proton (top panels) and asymmetry (bottom panels) distributions in various density bins

We see the formation of clusters with a quite reduced asymmetry while the surrounding gas is mostly made of neutrons only. The "$freeze-out$ $time$" will finally fix the proton fraction in the "*gas*" phase, signature of the low density behaviour of the symmetry term [8], in particular of its *slope.*

5 Charge Asymmetric Multifragmentation

In recent years two main models, statistical and dynamical, have been suggested to describe nuclear multifragmentation, i.e. the production of several intermediate mass fragments in heavy ion collisions. In the statistical picture a global equilibrium is assumed in a freeze-out volume and the various break-up channels are selected in a microcanonical way according to the statistical weight of the corresponding partition, given by the channel entropy [29, 30]. In the dynamical model clusters are directly formed during the expansion phase by a collective response of the interacting nuclear matter in low density unstable regions, the spinodal decomposition mechanism [20, 31, 32]. In spite of the quite different physics content the two models seem to lead to very similar predictions for fragmentation events [33, 34, 35].

As a natural consequence of the discussion of the previous section we show that isospin effects are suggesting new observables that should help to disentangle between the two pictures. A detailed analysis can be found in ref. [36]. We show here some results for a neutron excess source, $A = 197$ and $Z = 63$, prepared at the same initial excitation energy.

In the dynamical stochastic mean field (SMF) evolution [31] the temperature of the source nuclei is chosen to be $T = 3$ MeV and a radial flow energy of 3 MeV/nucleon is added in order to prevent the resilience of the initial expansion ($\rho_{init}^{SMF} = 0.5\rho_{eq}$) to the normal density. The total excitation energy per nucleon, which includes thermal, compressional and flow contributions is $E^*/A = 8$ MeV. The time evolution of the expanding source is followed until 150 fm/c. At this time, the fragment formation process due to the density fluctuation growth is already finished, and one can extract various characteristics of hot fragments from the phase space distribution, using a coalescence procedure [31, 37].

In parallel, we have applied the Statistical Multifragmentation Model (SMM) [38] to the same source taking also an overall excitation energy $E^*/A = 8$ MeV and a freeze-out volume $V_{f.o.} = 3V_0$, where V_0 is the volume of a ground state nucleus, consistent with the dynamical estimation at the time of formed fragments.

In Fig. 7 (left part) the fragment charge yields are presented for both models. We see that slopes and absolute values of the charge yields for statistical and dynamical models are very similar for $Z > 5$. This is indeed a good trigger for a meaningful comparison of the fragment properties.

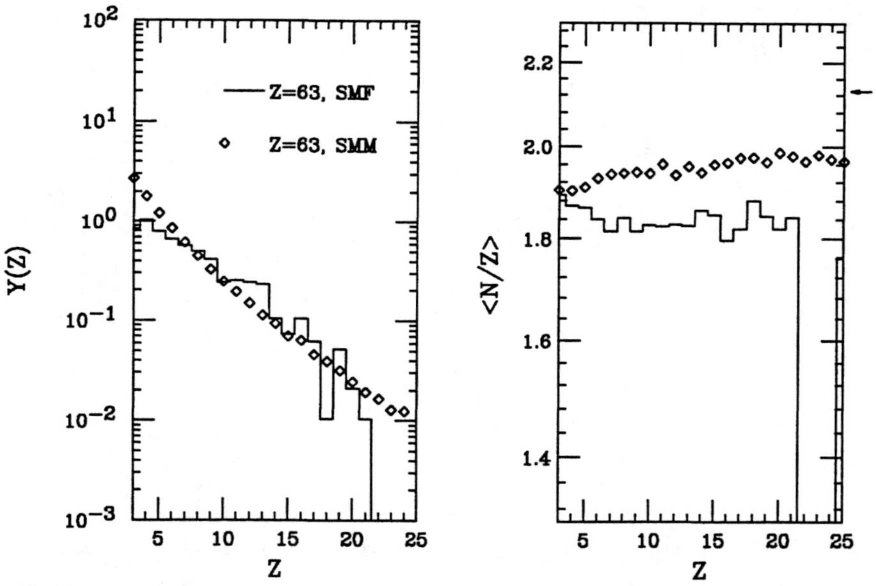

Figure 7 - Primary fragment properties from the excited $A = 197$, $Z = 63$ source: histogram and diamonds – SMF and SMM results respectively. Left part: Charge yields normalized to the number of fragments per event. Right part: Mean neutron-to-proton ratio $\langle N/Z \rangle$ of fragments versus proton number Z. Arrow shows N/Z ratio of the initial source.

It turns out that the primary SMM-fragments are more excited than the SMF-ones. This is a consequence of two main effects: (i) In the dynamical evolution a large fraction of energy is dissipated in the neutron-rich gas and the primary fragments of the liquid phase are less excited; (ii) In the statistical picture an increasing share of surface contributions in light IMFs causes a slight increase of their excitation energy per nucleon.

In Fig. 7 (right part) the average ratio $\langle N/Z \rangle$ of produced hot fragments is shown versus the fragment charge number Z. We observe here a clear effect of the density dependence of the symmetry energy: dynamical calculation (solid histogram) gives a decrease of the N/Z with charge number Z at $3 \leq Z \leq 10$, while the statistical calculation shows the increase of this ratio for primary fragments. Of course this effect will be largely reduced in the detected *cold* fragments. Another interesting difference between SMF and SMM calculations appears if one looks at the width of the N/Z distribution, as a function of the charge Z. As shown in Fig.8, the statistical calculation gives much larger variances than the dynamical one.

This means that, for each Z, a much larger variety of isotopes is produced in the statistical assumption. As a consequence, for instance, the isotopic and isotonic ratios of light particles (an observable that can be measured experimentally) are different in the two models. This is shown in Fig. 9, for initial sources with several N/Z ratios and A = 197.

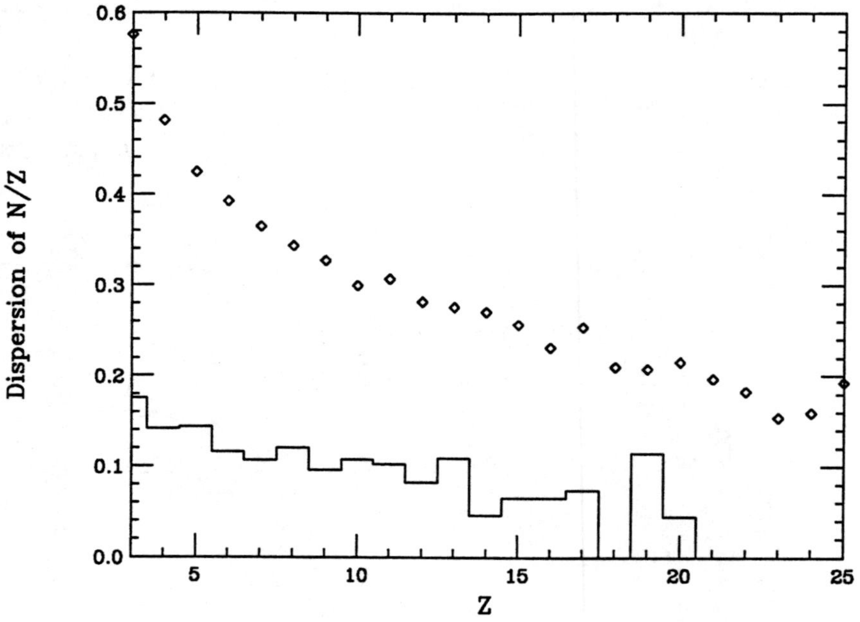

Figure 8 - Dispersion of the N/Z ratio of fragments versus proton number Z: histogram and diamonds – SMF and SMM results respectively.

Before closing this section we would like to discuss the excitation energy dependence of the observed discrepancy in the N/Z distributions of primary fragments predicted by the two models in the very neutron rich case ($Z = 63$). In the "dynamical" picture we expect an enhancement of the effect, i.e. a more pure neutron gas and more symmetric fragments, at lower temperatures, of course inside the spinodal region This is because in the dynamical evolution we have now less energy available to match the increase of symmetry energy needed to form clusters at normal density. In the SMM model we expect results not much dependent on temperature, above some threshold for the opening of multifragmentation channels. At lower temperatures we can actually have a slight increase of the neutron excess in the fragments due to a reduced thermal neutron emission during the multifragment break-up.

Some recent data on charge asymmetry effects on fragment production seem to be in a qualitative agreement with the signatures of the dynamical model, although a more precise analysis must be performed:

1) More neutron rich systems are producing more symmetric and less excited primary intermediate mass fragments [39, 40, 41, 42];

2) Such effects are disappearing for central collisions above 100 $AMeV$ beam energy [42]: in the dynamical picture the expansion phase becomes too fast and the instability growth times cannot be matched;

3) Opposite behaviour is observed for light ions produced in the midrapidity region, that appear more neutron rich than the average N/Z of the composite system [43, 44, 45].

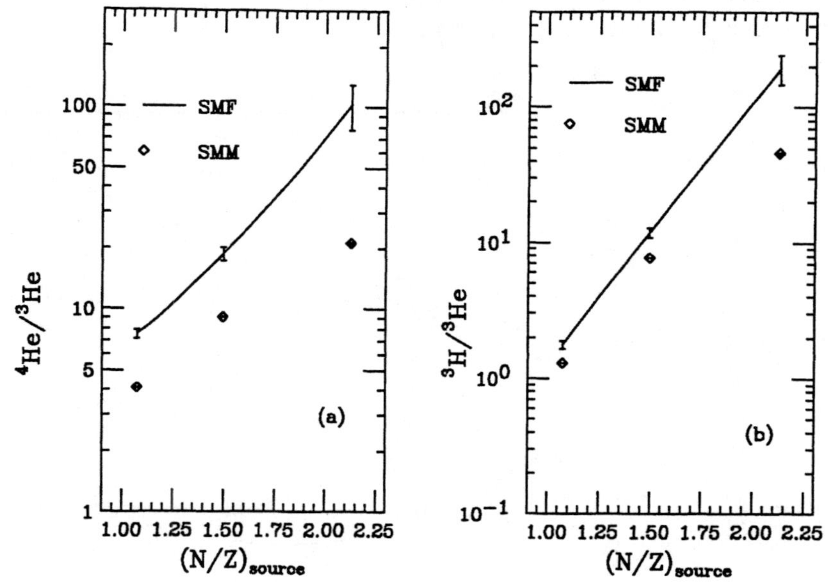

Figure 9 - Comparison of $^4He/^3He$ ratio (left) and $^3H/^3He$ ratio (right) obtained in SMM and SMF models

If all these signatures will be confirmed, from fragmentation studies in medium energy heavy ion collisions we will have a new powerful method to study the symmetry term of the nuclear *EOS* at *subnuclear* densities, of large interest for the structure of unstable nuclei as well as for astrophysics [46].

6 Symmetry term effects on Collective Flows

Collective flows in heavy ion collisions give important information on the dynamical response of excited nuclear matter [47, 48]. Particularly interesting for isospin effects is the directed transverse flow, measured on the reaction plane, which provides a direct signature of the expected transition from mean field to hydro-dynamics in the evolution of dissipative heavy ion collisions at intermediate energies. [48, 9]. Since the transition takes place from a delicate balance between *EOS* properties (mean field and compressibility), Coulomb repulsion and nucleon-nucleon cross sections in the medium, we will see a noticeable isospin dependence [49]. In particular we predict a large sensitivity to the form of the symmetry term in regions above normal density (up to $1.5\rho_0$) which are explored in this energy range (from 40 to 120 $AMeV$).

The symmetry contributions to the mean field, second part of Eq.(6), for the two parameterizations discussed in this report, have a density dependence shown in Fig.10 for a $N = 2Z$ system ($I = 1/3$): in a region just above normal density the field "seen" by neutrons and protons in the two cases is very different.

We expect to see two main direct effects on transverse flows: i) Proton-rich and neutron-rich nuclei will show different proton flows. The difference will be much reduced for a symmetry term of "asy-soft" type. ii) For neutron-rich nuclei proton and neutron flows will be also quite different in the case of a "asy-stiff" symmetry term. The effect will be particularly enhanced for semi-central collisions where we can get a very asymmetric piece of nuclear matter in the interacting zone.

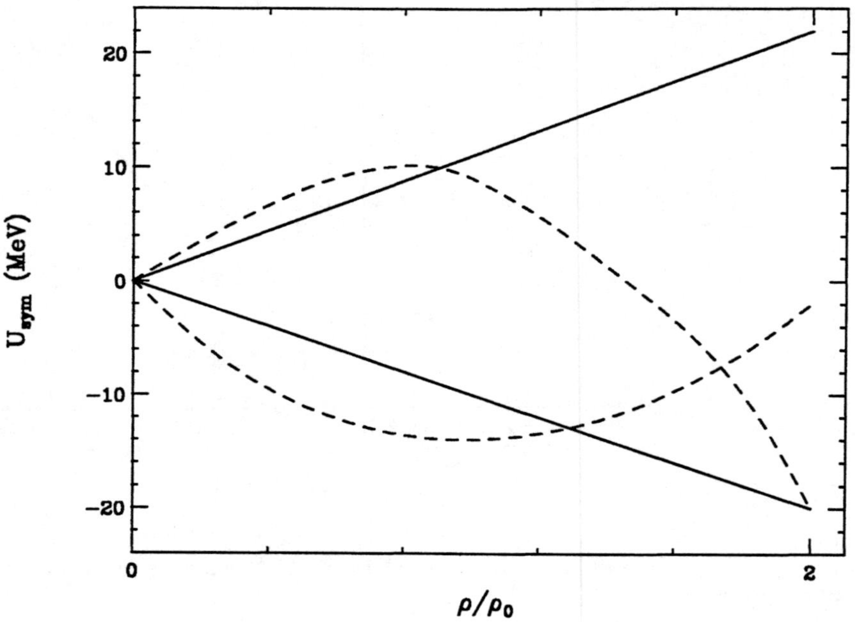

Figure 10 - Symmetry contribution to the mean field at $I = 1/3$ for neutrons and protons: dashed lines "asy-stiff", solid lines "asy-soft"

The transport equations are solved following a test particle evolution on a lattice [19, 50]. Medium effects on NN cross sections are included from the work of ref.[51], based on a Dirac-Brueckner approach with a Bonn NN interaction. In general we have a reduction with respect to the free case, partially due to the effective mass partially to a weakening of the force. In our calculation we use the momentum dependence suggested by Gale-Bertsch-DasGupta (GBD) in a Skyrme-like effective mean field [52]. For a large dynamical range it gives results quite similar to non-local forces of Gogny type [54] that well describe collective flow properties in symmetric systems with The corresponding compressibility value is around $K = 220MeV$. [53, 55, 50].

In order to have a better overall control of this complex calculation and to optimize the parallel structure a new Object-Oriented code has been written in $C{+}{+}$ language, the code $DYN++$ [56].

We have studied the p, n transverse flows for the collisions $^{58}Fe + ^{58}Fe$ $(n - rich)$ and $^{58}Ni + ^{58}Ni$ $(p - rich)$ in the energy range 55 *to* 105 $AMeV$ where nice data are existing for $Z = 1, 2, 3$ flows [9]. For a given energy and reduced impact parameter, $b_{red} = b/b_{max}$, we evaluate the freeze-out momentum distributions of protons and neutrons and we construct the mean transverse component as a function of the parallel velocity (rapidity). In order to eliminate numerical fluctuations we average over twenty events for each fixed initialization. The flow parameter is given by the slope at mid-rapidity.

In Fig.11 we report the energy dependence of proton flows in $Fe - Fe$ and $Ni - Ni$ at fixed $b_{red} = 0.45$ with the two choices of the symmetry term. In the *asy-stiff* case, Fig.11a), we have a larger balance energy, of about $5MeV$, for the neutron rich system $Fe - Fe$. The values are in quite good agreement with the data of ref. [9]. We show also the behaviour

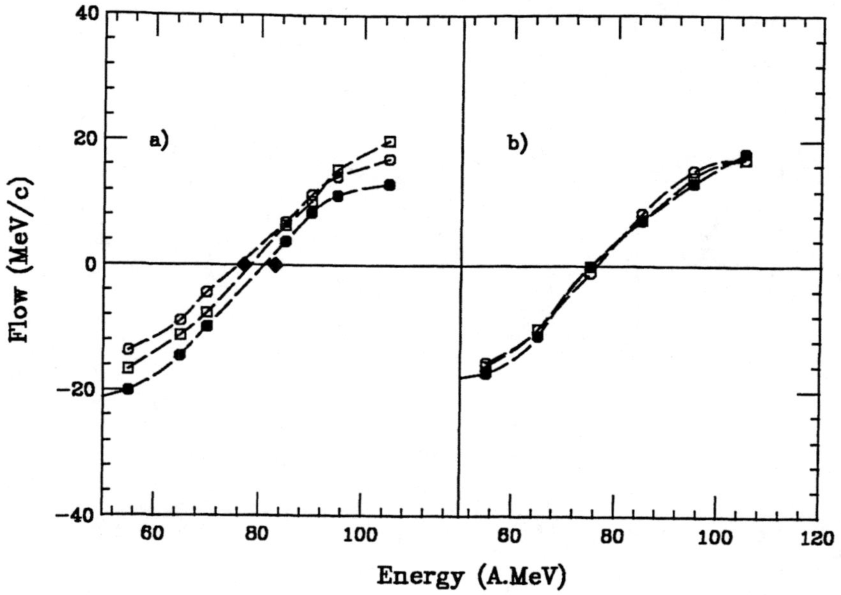

Figure 11 - Energy dependence of flows at $b_{red} = 0.45$: Open circles: Fe-Fe protons; Full circles : Ni-Ni protons; Squares: Fe-Fe neutrons. a) asy-stiff, b) asy-soft.

of neutron flows in the $Fe - Fe$ case. As expected from the more repulsive mean field we have, with respect to protons, a smaller flow below the balance energy and a larger one above. We stress that the collision term is acting in the opposite direction, due to the larger $p - n$ cross section. We remark the increase of the difference at high energy: we are exploring higher densities in the interacting region and so we expect larger mean field differences between neutrons and protons. Everything will disappear in the *asy-soft* choice: in Fig.11b) we report the corresponding results.

A detailed comparison of the flow dependence on centrality of the collision is presented in Fig.12 at a beam energy 55 $AMeV$, i.e. below the balance energy. To remark the asymmetry of the effect with respect to $b_{red} = 0.5$, more pronounced for more peripheral collisions: this is a nice evidence of the larger asymmetries we can reach in a "skin-skin" interaction of neutron rich ions. From this analysis we can draw some quite clear conclusions [57]:

- Collective transverse proton flows for charge asymmetric systems are very sensitive to the density dependence of the symmetry term of the nuclear EOS;
- All present data seem to indicate the need of a quite stiff symmetry term for densities up to $1.5\rho_0$;
- We expect to have a clear difference in neutron and proton flows, in presence of an interacting asymmetric piece of NM. This should be seen in a detailed study of the flows of different isotopes of light ions.

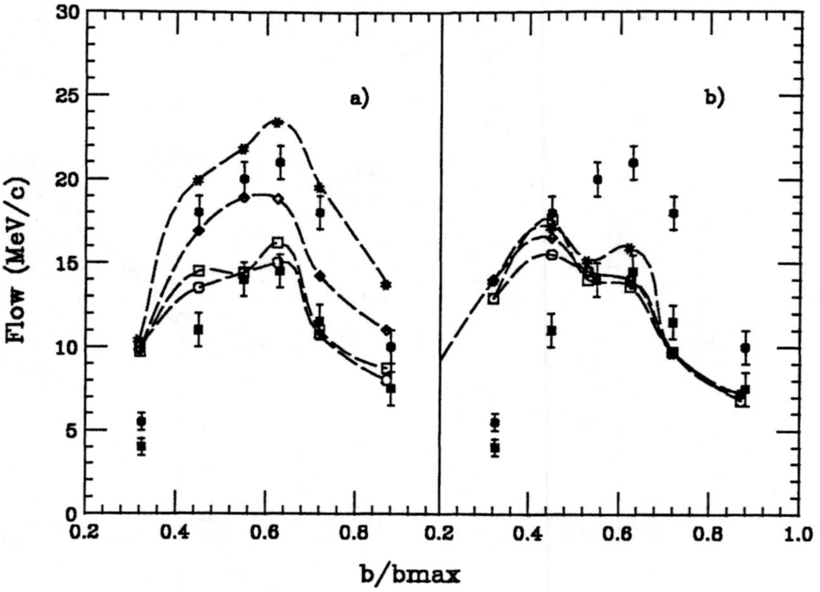

Figure 12 - Centrality dependence of flows at beam energy 55 $AMeV$: Stars: Fe-Fe protons; Open circles: Ni-Ni protons; Diamonds : Fe-Fe neutrons; Squares : Ni-Ni neutrons. Exp. points: Full circles (Fe protons), Full squares (Ni protons). a) asy-stiff, b) asy-soft.

7 Outlook

We have presented many experimental possibilities from physics with radioactive beams to extract direct information on the symmetry term of the nuclear EOS. We have first shown results for compressibility and saturation density. Starting from simulations of reaction dynamics performed with a new transport code where isospin and fluctuation effects are suitably accounted for, we have shown that dissipative heavy ion collisions at medium energies are rather sensitive to the density dependence of the symmetry contribution to the nuclear Equation of State.

For isospin effects on the collective response some simple physical mechanisms are proposed to understand the isoscalar-isovector coupling and the appearance of new instabilities, all directly related to properties of the symmetry term. A new chemical component in the spinodal region is expected, with observable signatures on the isotopic content of the dynamically produced fragments. This effect will be enhanced in the overlapping zone of semi-peripheral collisions (the neck region) where larger shall be the NM asymmetry.

We are now in presence of quite large discrepancies in the predictions of dynamical and statistical multifragmentation models on the disassembly of an excited asymmetric source. In particular *"healthier"* (i.e. more symmetric and less excited) primary intermediate mass fragments are expected in a dynamical mechanism.

We have finally shown that collective transverse flows in heavy ion collisions at medium energies are also strongly sensitive to the density dependence of the symmetry term in the nuclear equation of state. There are some indications towards a stiff behaviour of the symmetry term of the EOS.

We would like to stress that these effects can be clearly seen also with "exotic" but not radioactive beams. Indeed all the flow data analysed here are from stable beams.

A possibility is emerging of obtaining in terrestrial accelerator laboratories important information on the symmetry term of large astrophysical interest. It appears essential to have good charge asymmetric beams available at intermediate energies and to perform more exclusive experiments.

Acknowledgements

We warmly thank V.Baran, G.Fabbri, A.B.Larionov, S.Maccarone and L.Scalone for their substantial contribution to part of the results presented here. We gratefully acknowledge intense and stimulating discussions within the $ISO - DYN$ group, international working group on isospin effects on nuclear dynamics. In particular we would like to mention Bao-An Li, W.Bauer, I.Bombaci, A.Botvina, Ph.Chomaz, F.Gulminelli, I.Hamamoto, H.Horiuchi, U.Lombardo, A.Ono, H.Sagawa, U.Schroeder, L.Sobotka, H.Wolter and S.Yennello.

References

[1] I. Bombaci, T.T.S. Kuo and U. Lombardo, Phys. Reports. 242 (1994) 165 .
I. Bombaci, Phys. Rev. C55 (1997) 1
M.Prakash et al., Phys.Rep.280 (1997) 1

[2] I.M.Irvine, "Neutron Stars" Oxford Univ. Press 1978

[3] D.Pines, R.Tamagaki and S.Tsuruta Ed.s, "Neutron Stars", Addison-Wesley N.Y.1992

[4] C.J.Pethick and D.G.Ravenhall in "The Lives of Neutron Stars" Ed.s M.A.Alpar et al., NATO ASI Series C Vol.450 (1995) pp.59-70

[5] V.A.Urpin, same Proceedings, pp.163-166

[6] H.Mueller and B.D.Serot, Phys.Rev. C52 (1995) 2072

[7] Bao-An Li and C.M.Ko, Nucl.Phys. A618 (1997) 498

[8] M.Colonna, M.Di Toro and A.Larionov, Phys.Lett. B428 (1998) 1,
We remark here that also all relativistic mean field approaches in the presently used approximation scheme are predicting a *stiff-like* simmetry term, proportional to the barion density, coming from the ρ-meson field contribution, see ref.[14].

[9] G.Westfall, Nucl.Phys. A630 (1998) 27c

[10] Bao-An Li, C.M.Ko and W.Bauer, Int.J.Mod.Phys. E7 (1998) 147-229

[11] M.Di Toro et al., Bormio Winter Meeting on Nuclear Physics, Ed.I.Iori, Univ.Milano 1998 p.319,
Progr.Part.Nucl.Phys. 42 (1999) 125-136

[12] E.Chabanat, P.Bonche, P.Haensel, J.Meyer, R.Schaeffer, Nucl.Phys. A627 (1997) 710

[13] H.Krivine, J.Treiner and O.Bohigas, Nucl.Phys. A336 (1980) 155

[14] S.Yoshida, H.Sagawa and N.Takigawa, Phys.Rev. C58 (1998) 2796

[15] G.F.Bertsch and S.Das Gupta, Phys.Rep. 160 (1988)189

[16] Ch.Gregoire et al., Nucl.Phys. A465 (1987) 315

[17] A.Bonasera et al., Phys.Rep. 244 (1994) 1

[18] A.Bonasera, G.F.Burgio and M.DiToro, Phys.Lett. B221 (1989) 233 and A.Bonasera and F.Gulminelli, Phys.Lett. B259 (1991) 399

[19] TWINGO code: A.Guarnera Ph.D.Thesis, Univ.Caen , July 1996

[20] M.Colonna, M.DiToro, G.Fabbri and S.Maccarone, Phys.Rev. C57 (1998) 1410

[21] M.Colonna et al., Nucl.Phys. A642 (1998) 449

[22] M.Colonna et al., *Perspectives in heavy ion physics: 3rd Italy-Japan joint meeting*, Ed.s C.Signorini et al., World.Sci,1999 pp.172-185

[23] M.Colonna, M.Di Toro and A.Guarnera, Nucl.Phys. A589 (1995) 160

[24] J.Toke et al., Phys.Rev.Lett. 77 (1996) 3514 and ref.s therein.

[25] V.Baran, A.Larionov, M.Colonna and M.Di Toro, Nucl.Phys. A632 (1998) 287

[26] G. Baym and C. Pethick, *Landau Fermi-liquid theory* (John Wiley and Sons, New York, 1991)

[27] M.Colonna and Ph.Chomaz, Phys.Rev. C49 (1994) 1908

[28] M. Colonna et al., in "Multifragmentation" see Ref. [29], p.322-327.

[29] J.P. Bondorf in "Multifragmentation" (Hirschegg 99) Ed.s H. Feldmeier, J. Knoll, W. Noerenberg and J. Wambach, GSI 1999, p.11;
I.N. Mishustin, Nucl.Phys. A630 (1998) 111c and ref.s therein.

[30] D.H. Gross, Phys.Rep. 279 (1997) 119 and ref.s therein.

[31] A. Guarnera, M. Colonna and Ph. Chomaz, Phys. Lett. B373 (1996) 267; M. Colonna, Nucl. Phys. A630 (1998) 136c.

[32] M. Colonna, M. Di Toro and A. Guarnera, Nucl. Phys. A580 (1994) 312.

[33] D. Durand, Nucl. Phys. A630 (1998) 52c.

[34] M.F. Rivet et al. (INDRA Collab.), Phys. Lett. B430 (1998) 217 and in "Multifragmentation", see Ref.[1], p.293-305

[35] R. Bougault et al. (INDRA Collab.), in "Multifragmentation", see Ref. [29], p.24-35.

[36] A.B.Larionov, A.S.Botvina, M.Colonna and M.Di Toro, *Multifragmentation of charge asymmetric nuclear systems*, preprint *nucl-th/9906079*, Nucl.Phys. A to appear

[37] M. Colonna, N. Colonna, A. Bonasera and M. Di Toro, Nucl.Phys. A541 (1992) 295.

[38] J.P. Bondorf, A.S. Botvina, A.S. Iljinov, I.N. Mishustin, K. Sneppen, Phys. Rep. 257 (1995) 133.

[39] G.J. Kunde et al., Phys. Rev. Lett. 77 (1996) 2897.

[40] G. Kortemeyer, W. Bauer and G.J. Kunde, Phys. Rev. C55 (1997) 2730.

[41] L.G. Sobotka et al., Phys. Rev. C55 (1997) 2109, and in "Multifragmentation", see Ref. [29], p.72-81.

[42] M.L. Miller et al., Phys. Rev. Lett. 82 (1999) 1399.

[43] J.F. Dempsey et al., Phys. Rev. C54 (1996) 1710.

[44] S. Yennello et al., in "Heavy Ion Physics", Ed.s Yu.Ts. Oganessian and R. Kalpakchieva, World.Sci.1998, p.253-260.

[45] H.S. Xu et al., "Isospin Fractionation in Nuclear Multifragmentation", Preprint MSUCL-1137. Oct 99.

[46] "It is important to put the properties of *neutron* matter at *subnuclear* densities on as firm a footing as possible, not only for astrophysical applications, but also for interpreting terrestrial experiments with coming radioactive beam facilities" C.J. Pethick and D.G. Ravenhall, in "The lives of neutron stars", NATO ASI Series C, Vol.450, 1993.

[47] H.H.Gutbrod, A.M.Poskanzer and H.G.Ritter, Rep.Progr.Phys. 52 (1989) 1267

[48] S.Das Gupta and G.W.Westfall, Physics Today 46 (1993) 34 and ref.s therein.

[49] R.Pak et al., Phys. Rev. Lett. 78 (1997) 1022

[50] V.Greco, A.Guarnera, M.Colonna and M.DiToro, Phys.Rev. C59 (1999) 810

[51] G.Q.Li and R.Machleidt, Phys.Rev. C48 (1993) 1702 and Phys.Rev. C49 (1994) 566

[52] C.Gale, G.F.Bertsch and S.Das Gupta, Phys.Rev. C35 (1987) 1666

[53] M.Prakash, T.T.S.Kuo and S.Das Gupta, Phys.Rev. C37 (1988) 2253

[54] J.Decharge and D.Gogny, Phys. Rev. C21 (1980) 1568 and ref.s therein

[55] V.de La Mota, F.Sebille, M.Farine, B.Remaud and P.Schuck, Phys.Rev. C46 (1992) 677

[56] DYN++ Code: L.Scalone, Master Thesis, Univ.of Catania Dec. 1998

[57] L.Scalone, M.Colonna and M.Di Toro, *Transverse flows in charge asymmetric collisions*, Phys.Lett. B461 (1999) 9

Isospin-dependent Quantum Molecular Dynamics Model and its Applications in Heavy Ion Collisions

Feng-Shou Zhang[1,2,4], Lie-Wen Chen[1,2], Gen-Ming Jin[1,2], Jian-Ye Liu[1,2,4], Zhi-Yuan Zhu[1,3,4]

[1] Center of Theoretical Nuclear Physics, National Laboratory of Heavy Ion Accelerator, Lanzhou 730000, China
[2] Institute of Modern Physics, Academia Sinica, P. O. Box 31, Lanzhou 730000, China
[3] Shanghai Institute of Nuclear Research, Academia Sinica, Shanghai 201800, China
[4] CCAST (World Laboratory) P. O. Box 8730, Beijing 100080, China

Abstract

An isospin-dependent quantum molecular dynamics model to the transient state of nuclear matter with isospin asymmetry, thermal excitation, and compression is introduced. Its applications in heavy ion collisions for studying the isospin effects on nuclear transverse collective flow, on nuclear radial flow, and on nuclear multifragmentation are provided.

1 Introduction

Recent advance in experiments using radioactive nuclear beams with large neutron or proton excess has made it possible create a transient state of nuclear matter with appreciable isospin asymmetry, thermal excitation, and compression[1-12]. It is a challenge for theoretical nuclear physicists to develop some algorithms to describe this transient state. A great deal of theoretical work has been already devoted to the studies of compressed and hot nuclear matter. In the present report we will instead of focus on the role of the isospin degree of freedom. Extensive reviews can be found in Ref. [6].

The dynamics in the formation of the transient state is mainly governed by three components, namely, the mean field, two-body collisions, and Pauli blocking. For an isospin dependent reaction dynamics algorithm it is essential that all the three components should reasonably include isospin degrees of freedom. In addition, it is also important that, in initialization of projectile and target nuclei, the samples of neutrons and protons in phase space should be treated separately since there exists a large difference between neutron and proton density distributions for nuclei far from the β stability line. An isospin dependent quantum molecular model (IQMD) is developed just based on above ideas. It has been shown that the IQMD can be used with large success for studying the effects of isospin in heavy ion collisions at intermediate energies.

In this report, we present an isospin dependent quantum molecular model and its applications in heavy ion collisions for studying the isospin effects. The paper is organized as follows. In Section 2, a description of IQMD is given in detail. In Section 3, we provide some recent calculated results on isospin effects of heavy ion collisions. Conclusions will be given at the end.

Table 1: The parameter sets of Eqs. (2) and (3). The S and H refer to the soft and hard equation of state, the M refers to the inclusion of momentum dependent interaction, and the K refers to the incompressibility.

	K (MeV)	α (MeV)	β (MeV)	γ (MeV)	δ (MeV)	$\varepsilon(\frac{c^2}{GeV^2})$
S	200	−356	303	1.17	–	–
SM	200	−390	320	1.14	1.57	500
H	380	−124	71	2.00	–	–
HM	380	−130	59	2.09	1.57	500

2 Description of the IQMD model

The quantum molecular dynamics model (QMD) model is classical in essence because the time evolution of the system is determined by classical canonical equation of motion, however, many important quantum features are included in this description [2]. The important advantage of the QMD model is that it can explicitly represent the many body state of the system and thus contains correlation effects to all orders. Meanwhile, the QMD model treats static and dynamics on an equal footing. Therefore, the QMD model provides important information about both the collision dynamics and the fragmentation process. Even though the BUU model is quite successful to describe one body observables, yet fails in describing the formation of clusters. In this report, the main ideas in the IQMD model are introduced[13-16]. The following describes briefly the IQMD model from five aspects, i.e., the mean field, two-body collisions, Pauli blocking, initialization, and clusterization.

2.1 Mean field

In the IQMD model, the total interaction potential of the system is given by

$$U^{tot} = U^{dd} + U^{Yuk} + U^{Coul} + U^{sym} + U^{MDI} \tag{1}$$

with U^{dd} the density-dependent (Skyrme) potential, U^{Yuk} the Yukawa (surface) potential, U^{Coul} the Coulomb energy, U^{sym} the symmetry energy term, and U^{MDI} the momentum dependent interaction. The U^{dd} can be written as

$$U^{dd} = \alpha(\frac{\rho}{\rho_0}) + \beta(\frac{\rho}{\rho_0})^\gamma \tag{2}$$

with ρ_0=0.16 fm^{-3}.

For the momentum dependent interaction U^{MDI}, we make use of the real part of the optical potential parametrized as follows

$$U^{MDI} = \delta ln^2[\varepsilon(\mathbf{p_1} - \mathbf{p_2})^2 + 1]\cdot\frac{\rho}{\rho_0}. \tag{3}$$

The parameters of Eqs. (2) and (3) are given in Table 1 from which one can see two kinds of equations of state are commonly used. One is the so-called hard EOS (H, HM) with an incompressibility of K=380 MeV, and the other is the soft EOS (S, SM) with an incompressibility of K=200 MeV. The M refers to the inclusion of the momentum dependent interaction. The U^{Yuk}, U^{Coul}, and U^{sym} have the following forms

$$U^{Yuk} = \frac{1}{2}V_Y \sum_{i \neq j} \frac{1}{r_{ij}} exp(Lm^2)[exp(-mr_{ij})erfc(\sqrt{L}m - r_{ij}/\sqrt{4L})$$

$$-exp(mr_{ij})erfc(\sqrt{L}m + r_{ij}/\sqrt{4L})] \quad (4)$$

$$U^{Coul} = \frac{e^2}{4} \sum_{i \neq j} \frac{1}{r_{ij}} (1 + t_{iz})(1 + t_{jz}) erf(r_{ij}/\sqrt{4L}) \quad (5)$$

$$U^{sym} = \frac{C}{2\rho_0} \sum_{i \neq j} t_{iz} \cdot t_{jz} \cdot \frac{1}{(4\pi L)^{3/2}} \cdot exp[-\frac{(\mathbf{r}_i - \mathbf{r}_j)^2}{4L}] \quad (6)$$

with $V_Y = -0.0024$ GeV, $m = 0.83$, and the L is the so-called Gaussian wavepacket width (here $L = 2.0$ fm^2). The relative distance $r_{ij} = |\mathbf{r}_i - \mathbf{r}_j|$. The t_{iz} is the zth component of the isospin degree of freedom for the ith nucleon, which is equal to 1 and -1 for proton and neutron, respectively. The C is symmetry energy strength. From above equations, one can see that the nuclear mean field is isospin dependent in the IQMD model.

2.2 Two-body collisions

In the IQMD model, two different parametrizations of N-N cross sections may be used optionally. One is the parametrization of Cugnon [17]($\sigma_{Cug.}$) which is isospin independent and the other is the experimental parametrization [18] ($\sigma_{exp.}$) which is isospin dependent. It is shown that the neutron-proton cross section is about three times larger than neutron-neutron or proton-proton cross section for the experimental parametrization at the energies lower than 300 MeV/nucleon.

2.3 Pauli blocking

The method of considering the Pauli blocking effect is as follows. Whenever a collision has occurred, in the phase space we assume that each nucleon occupies a six-dimensional sphere with a volume of $h^3/2$ (considering the spin degree of freedom), and then calculate the phase volume, V, of the scattered nucleons being occupied by the rest nucleons with the same isospin as that of the scattered ones. We then compare $2V/h^3$ with a random number and decide whether the collision is blocked or not. Therefore, the Pauli blocking is isospin-dependent, namely, the Pauli blocking of neutrons and protons is treated separately.

2.4 Initialization

In the IQMD model, the neutron and proton are distinguished from each other in the initialization of projectile and target nuclei. The neutron and proton density distributions for the initial projectile and target nuclei are determined from the Skyrme-Hartree-Fock (SHF) method [19] with parameter set SKM which can give reasonable density distribution for stable and neutron-rich nuclei. Using the neutron and proton density distributions calculated from the SHF method, one can get the radial positions of neutrons and protons in the initial nuclei in terms of Monte Carlo method. Figs. 1 (a) and (b) show the neutron, proton and total density distributions sampled in initial nuclei, ^{58}Fe and ^{58}Ni, respectively. We can see in Fig. 1 that there is a clear neutron skin in neutron-rich nucleus ^{58}Fe. While

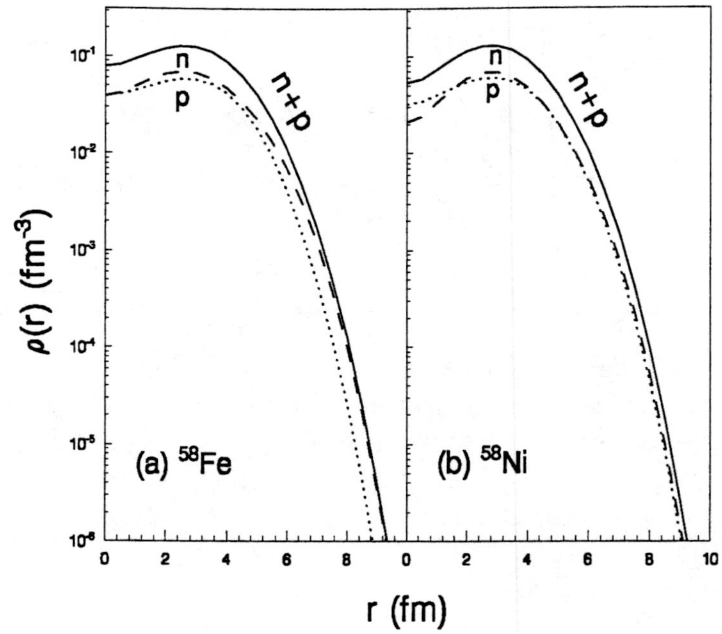

Figure 1: Proton (dot), neutron (dash), and total (solid) density distributions in ^{58}Fe (a) and ^{58}Ni (b).

the total densities $(n+p)$ in ^{58}Fe and ^{58}Ni are almost identical, the total density in ^{58}Fe is more extended than that in ^{58}Ni. These features are in agreement with the results of the nonlinear RMF theory [20]. The momentum distribution of nucleons is generated by means of the local Fermi gas approximation. The local Fermi momentum $p_F^i(\mathbf{r})$ is given by

$$p_F^i(\mathbf{r}) = \hbar(3\pi^2\rho_i(\mathbf{r}))^{1/3}, (i = n, p). \quad (7)$$

2.5 Clusterization

The nuclear clusters are constructed by means of an isospin-dependent modified coalescence model [16, 15], in which firstly particles with relative momenta smaller than P_0 and relative distances smaller than R_0 are coalesced into one cluster (here $R_0 = 3.5$ fm and $P_0 = 300$ MeV/c); then check whether the cluster is or not an isotope existed in the nuclear data sheets, if it is then the cluster can be accepted; finally in order to get rid of nonphysical line-type clusters one should check whether the condition, $R_{rms} \leq 1.14A^{1/3}$, is or not satisfied, if so the cluster is considered as a valid fragment eventually (here, the R_{rms} and A are the root-mean-square radius and mass number of the cluster, respectively). Therefore, the present model to construct clusters is an extended version of the conventional coalescence model which was adopted in Ref. [21, 22, 23] by considering self-consistently the isospin constraint and aggregation procedure. In the calculations, it is indicated that the present calculated results are insensitive to the selection of either $R_0 = 3.5$ fm and $P_0 = 300$ MeV/c or $R_0 = 2.4$ fm and $P_0 = 200$ MeV/c, the latter was used in our previous work[14].

The stability of the propagation of the initialized nuclei has been checked in detail and can last long enough time at least according to the evolution of the average binding energies and root mean square radii of the initialized nuclei.

3 Applications

3.1 Collective flow

The isospin dependence of collective flow has become a very interesting subject of theoretical and experimental studies [24, 25, 26]. One knows that nuclear collective flow is a kind of collective phenomenon found in intermediate and high energy HIC and the study of the dependence of collective flow on the beam energy, mass number and impact parameter, has revealed much interesting physics about the properties and origin of collective flow. Especially, from studying the beam energy dependence it has been found that the transverse collective flow in the reaction plane disappeared at an incident energy, termed the balance energy E_{bal} [27, 28, 29, 30, 31, 32, 33, 34, 35]. Furthermore, detailed theoretical studies using microscopic transport models have shown that both the strength of transverse collective flow and the balance energy can be used to extract information about the nuclear equation of state (EOS) and in-medium nucleon-nucleon (N-N) cross sections [36, 37, 38, 39, 40, 41, 42, 43, 44, 45, 46, 47, 48, 49]. Hence, studying collective flow in reactions induced by radioactive nuclei is very meaningful to explore the isospin-dependent part of the nuclear EOS and isospin-dependent N-N cross sections.

Using the IQMD model, we study the fragment flow and its isospin effects in reactions ^{58}Fe + ^{58}Fe and ^{58}Ni + ^{58}Ni, which have the same mass number but different neutron/proton ratios. The calculated results indicate that the neutron-rich system (^{58}Fe + ^{58}Fe) displays stronger negative transverse collective flow at energy of 55 MeV/nucleon and has a higher balance energy, which could be qualitatively in agreement with the experimental data. Meanwhile, the influence of the symmetry energy and N-N cross sections on the transverse collective flow has been also studied.

3.1.1 Fragment flow

Using the IQMD model with parameter set SM, we calculate the flow parameters of different fragment types in reactions ^{58}Fe + ^{58}Fe and ^{58}Ni + ^{58}Ni at energy of 55 MeV/nucleon. The flow parameter is defined as the slope of the transverse momentum distribution at the center of the reduced center-of-mass (c.m.) rapidity $(y/y_{proj})_{c.m.}$. Figs. 2 (a), (b) and (c) give the calculated flow parameters of fragments with Z=3, 2, and 1, respectively, as a function of the reduced impact parameter b/b_{max} (here b_{max}=8 fm, the calculated values are connected by solid lines in Fig. 2). For each impact parameter, we perform a calculation of 500 events. In the present calculations, it is found that the multiplicity and transverse momenta for different fragment types have basically saturated by the end of 140 fm/c, therefore the calculated results actually correspond to the statistical average value of 3000 "events" which come from the sum of six time points from t=150 to 200 fm/c in each event. All the following calculations adopt this recipe. The errors shown are the statistical errors on the slopes of the linear fits. Meanwhile, the correspondent experimental values are also shown in Fig. 2 (the data points are connected by dotted lines in Fig. 2). The experimental values are extracted at the upper limit of each impact parameter bin [25]. It is shown in Fig. 2 that the calculated results are in agreement with the experimental data in trends. The difference in the magnitude of flow parameter between the two isotopic systems is maximal for semi-central collisions, which is qualitatively in agreement with the previous work [30, 31]. In particular, the calculated results also show the isospin dependence of flow parameter for different fragment types, namely, the neutron-rich system (^{58}Fe + ^{58}Fe) shows stronger negative deflection, which is in agreement with the predictions of BUU model in Ref. [20] where the results are for all nucleons. Moreover, at large impact parameters for fragments with Z=3, we can see that both the calculated results and experimental data have a cross which may be because in the most peripheral collisions the number of N-N collisions is very small and the Coulomb interaction is more important so that the fragments with Z=3 have weaker Coulomb repulse in neutron-rich system. In addition, the mass dependence of the flow parameters shown in Fig. 2 also demonstrates the well known increase in magnitude for heavier fragments [25, 50, 51, 52]. This phenomenon may be because most nucleons are emitted by the hard stochastic collisions and hence the effect of the mean field is largely erased in the nucleon flow. This argument suggests that the flow of composite fragment carries more direct information of the nuclear EOS than the nucleon flow.

From Fig. 2, it is seen that the predictions of IQMD model systematically exhibit stronger values of flow parameter than the experimental data for all three fragment types at all reduced impact parameters. We shall give an explanation for this phenomenon in the following subsection. More importantly here, there is agreement between the data and the IQMD model predictions for the magnitude of the isospin effect, which is displayed in Fig. 3. The open (solid) circles are the difference between the values of flow parameter δF for the data (IQMD model predictions) in the isotopic systems ^{58}Ni + ^{58}Ni and ^{58}Fe + ^{58}Fe at each corresponding reduced impact parameter. The errors shown are statistical. It is shown that the isospin dependence predicted by the IQMD model is in agreement with the experimental data.

In order to see clearly the relation between flow parameter and neutron/proton ratios of the reaction system $(N/Z)_{sys}$, we show in Figs. 4 (a), (b) and (c) the impact-parameter-inclusive values of flow parameter for fragments with Z=3, 2, and 1, respectively, as a function of $(N/Z)_{sys}$ of the systems ^{58}Fe + ^{58}Fe (1.23) and ^{58}Ni + ^{58}Ni (1.07). In the present calculations, we have simply assumed that the number of events is proportional to impact parameter b. Meanwhile, the experimental data are also shown in Fig. 4 (open circles). It is indicated that the calculated results are in good agreement with the experimental data in trends. From above analyses we could conclude that the consideration of the isospin degree of freedom in IQMD model is reasonable.

3.1.2 Disappearance of flow

From studying the beam energy dependence it has been found that the transverse collective flow changes from a negative one to a positive one at an incident energy, i.e., E_{bal}, due to the competition between the attractive nuclear mean field potential and the repulsive N-N collisions [27, 28, 29, 30, 31, 32, 33, 34, 35]. The balance energy has been found to depend sensitively on the mass number, impact parameter and properties of the colliding nuclei, such as the thickness of their surface [54]. Figs. 5 (a) and (b) show the distribution of transverse momentum versus the normalized rapidity $y_{c.m.}/y_{lab}$ for all nucleons at two different energies, 80 and 150 MeV/nucleon, with b = 3 fm for the systems ^{58}Fe + ^{58}Fe and ^{58}Ni + ^{58}Ni, respectively. For both systems, at 80 MeV/nucleon the negative slopes (corresponding to negative scattering angles) are visible whereas at 150 MeV/nucleon the opposite sign slopes (positive scattering angles) are found. The former corresponds to negative flow parameter while the latter corresponds to positive one.

The balance energy E_{bal} is obtained by a linear fit to the energy dependence of flow

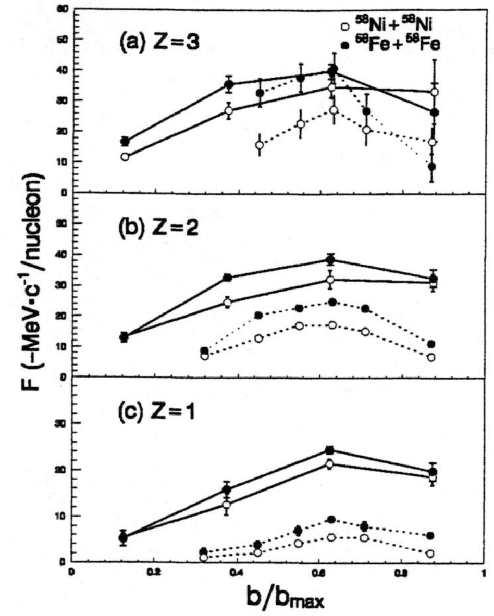

Figure 2: Calculated (data connected by solid lines) and experimental (data connected by dotted lines) values of flow parameter as a function of the reduced impact parameter for three different fragment types, Z=3 (a), 2 (b), and 1 (c), from ^{58}Fe + ^{58}Fe (solid circles) and ^{58}Ni + ^{58}Ni (open circles) collisions at 55 MeV/nucleon.

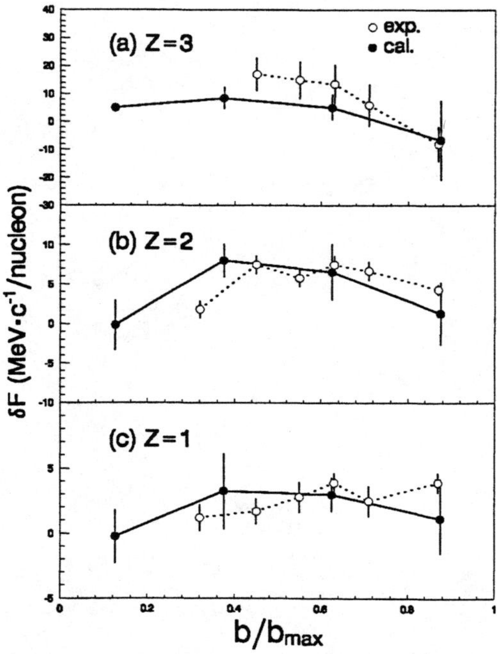

Figure 3: Difference between the values of flow parameter for the data (open circles) and IQMD model predictions (solid circles) for ^{58}Ni + ^{58}Ni and ^{58}Fe + ^{58}Fe collisions at 55 MeV/nucleon versus reduced impact parameter for three different fragment types, Z=3 (a), 2 (b), and 1 (c). The lines included only to guide the eye.

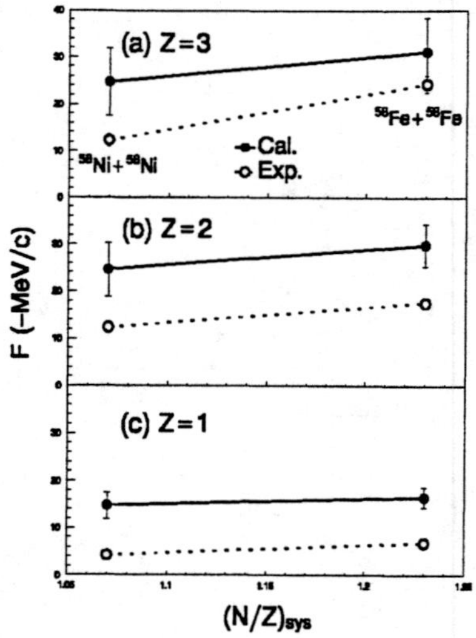

Figure 4: Calculated (solid circles) and experimental (open circles) values of flow parameter as a function of the neutron/proton ratio of the reaction system for different fragment types, Z=3 (a), 2 (b), and 1 (c) from ^{58}Fe + ^{58}Fe and ^{58}Ni + ^{58}Ni at 55 MeV/nucleon. The lines included only to guide the eye.

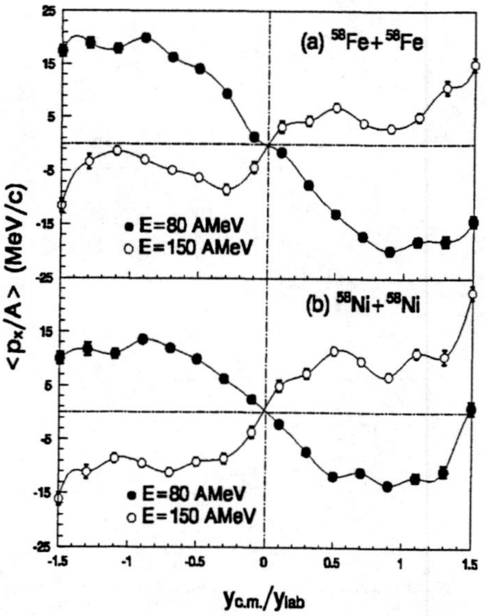

Figure 5: Mean transverse momentum in the reaction plane versus the normalized rapidity for all nucleons from reactions ^{58}Fe + ^{58}Fe (a) and ^{58}Ni + ^{58}Ni (b) at two different incident energies, 80 MeV/nucleon (solid circles) and 150 MeV/nucleon (open circles). The curves are plotted to guide the eye.

parameter at the point where flow parameter passes through zero. Figs. 6 (a), (b), (c) and (d) display the energy dependence of flow parameter at b=2, 3, 4 and 5 fm, respectively, for systems ^{58}Fe + ^{58}Fe (solid circles) and ^{58}Ni + ^{58}Ni (open circles) with potential parameter set SM. It is shown that at different impact parameters the neutron-rich system (^{58}Fe + ^{58}Fe) exhibits systematically smaller flow parameters, which implies a stronger attractive interaction in the reaction of neutron-rich system. This feature is in agreement with the predictions of BUU model [20]. From Fig. 4 in Ref. [20], one can see that the difference between flow parameters in ^{48}Cr + ^{58}Ni and ^{48}Ca + ^{58}Fe (neutron-rich system) decreases as the beam energy increases and finally disappears as the beam energy becomes far above the balance energy. However, this phenomenon is not observed in the present IQMD calculation at the small impact parameters, such as b=2, 3 and 4 fm. It is found in Fig. 6 (a) and (b) that the difference between flow parameters in ^{58}Ni + ^{58}Ni and ^{58}Fe + ^{58}Fe (neutron-rich system) increases with increment of the beam energy. This phenomenon may be because with increment of the beam energy the N-N collisions become dominant and therefore the isospin dependence of N-N cross sections plays a more important role. In fact, in the present calculations it is shown that the difference between the collision numbers per nucleon in ^{58}Ni + ^{58}Ni and ^{58}Fe + ^{58}Fe (neutron-rich system) increases as the beam energy increases. It should be mentioned that this phenomenon is complicated because it is related to many factors, such as the isospin-dependent N-N cross sections, the isospin-dependent Pauli blocking, beam energy, and so on. From Fig. 6 one can extract the balance energy at different impact parameters. The upper window in Fig. 7 displays the calculated E_{bal} with potential parameter sets SM and HM as a function of reduced impact parameter b/b_{max} and the experimental data from Ref. [26] are also included. The points, except for the predictions of IQMD model with SM for ^{58}Fe + ^{58}Fe and the experimental data for ^{58}Fe + ^{58}Fe, have been offset in values of b/b_{max} for clarity. It is indicated that the values of E_{bal} predicted by IQMD model with SM or HM are in agreement with the experimental data in trends. An approximate linear increase of the balance energy with impact parameter is visible in Fig. 7 (upper window), which is in agreement with the previous work [29, 31]. It is also shown that the calculated values of E_{bal} with SM are roughly equal to those with HM, which implies that the balance energy is slightly dependent on incompressibility K and this is in agreement with the results of the BUU model calculations [35, 53, 54, 55].

It is indicated in Fig. 7 (upper window) that the calculated values of E_{bal} with either SM or HM are systematically larger than the experimental data. However, from the lower set of points in Fig. 7 which display the difference of the balance energies δE_{bal} for the data (solid squares), predictions of IQMD model with SM (solid circles), and predictions of IQMD model with HM (solid triangles), respectively, between the isotopic systems ^{58}Ni + ^{58}Ni and ^{58}Fe + ^{58}Fe as a function of the reduced impact parameter, we can see that in error tolerance there is agreement between the data and the IQMD model predictions for the magnitude of the isospin effect. In the lower window of Fig. 7, the values for the predictions of IQMD model with HM have been offset in the horizontal direction for clarity.

In fact, the previous calculated values of flow parameter for different fragment types are also systematically larger than the experimental data. The reason of resulting in these phenomena may be due to the low saturation densities in initial nuclei ^{58}Fe and ^{58}Ni. From Fig. 1 we can see both the saturation densities of ^{58}Fe and ^{58}Ni are about 0.12 fm^{-3} which is smaller than normal saturation density ρ_0, 0.16 fm^{-3}. While the isospin-independent part of the nuclear EOS is attractive at low densities. As a matter of fact, the radial density is related to the Gaussian wavepacket width L. in the IQMD model. Even though the radial positions of the nucleons are reasonable, the radial density will

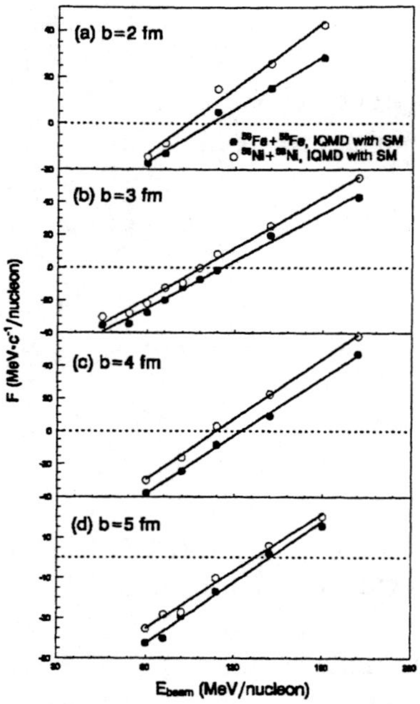

Figure 6: Flow parameter for all nucleons as a function of incident energy at impact parameter b=2 (a), 3 (b), 4 (c), and 5 (d) fm for ^{58}Fe + ^{58}Fe (solid circles) and ^{58}Ni + ^{58}Ni (open circles). The straight lines are the results of linear fits.

Table 2: The flow parameters at different situations(see text) for ^{58}Fe + ^{58}Fe and ^{58}Ni + ^{58}Ni at beam energy of 55 MeV/nucleon and impact parameter b=5 fm.

Reaction systems	C=0 with $\sigma_{Cug.}$	C=0 with $\sigma_{exp.}$	C=32 MeV with $\sigma_{exp.}$
^{58}Fe + ^{58}Fe	−34.8±2.14 MeV	−29.1±1.12 MeV	−27.9±1.24 MeV
^{58}Ni + ^{58}Ni	−33.8±1.96 MeV	−24.5±0.68 MeV	−24.1±1.49 MeV

result in some deviations from the real radial density. Moreover, the larger the L is, the larger the deviations are. Fortunately, this problem does not affect us to study the isospin effects on collective flow. Nevertheless, the previous calculated results are still in qualitative agreement with the experimental data. Particularly, the isospin dependence of the fragment flow and the balance energy is quantitatively in agreement with the experimental data.

3.1.3 Influence of symmetry energy and N-N cross sections on collective flow

Indeed, the isospin dependence of collective flow has been found in BUU model calculations and experiments and it has been explained as a result of the competition among several mechanisms in the reaction dynamics, such as N-N cross sections, symmetry energy, Coulomb energy, the surface properties of the colliding nuclei, and so on. However, the relative importance of these mechanisms is still not clear so far.

Using different symmetry energy strength C and parametrizations of N-N cross sections with potential parameter set SM, we show in Fig. 8 the distribution of average transverse momentum per nucleon for all nucleons versus the normalized rapidity $y_{c.m.}/y_{lab}$ for systems ^{58}Fe + ^{58}Fe (solid circles) and ^{58}Ni + ^{58}Ni (open circles) at energy of 55 MeV/nucleon and

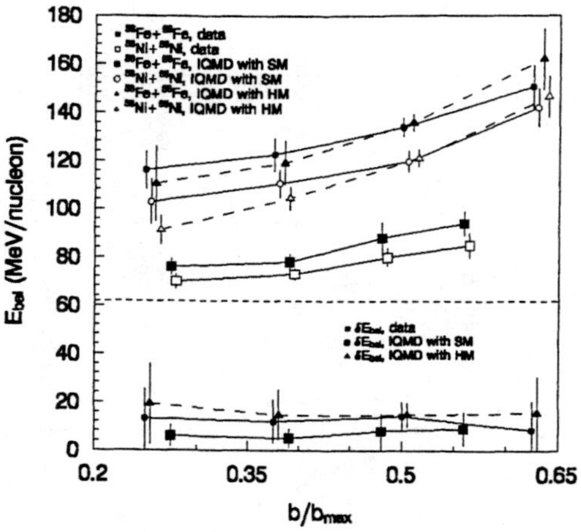

Figure 7: (Upper window) Measured balance energies as a function of reduced impact parameter compared with the predictions of IQMD model with potential parameter sets SM and HM. The solid (open) squares are measured for ^{58}Fe + ^{58}Fe (^{58}Ni + ^{58}Ni). The solid (open) circles are the predictions of IQMD model with SM for ^{58}Fe + ^{58}Fe (^{58}Ni + ^{58}Ni) while the solid (open) triangles are the predictions of IQMD model with HM for ^{58}Fe + ^{58}Fe (^{58}Ni + ^{58}Ni). (Lower window) The difference of the balance energies for the data (solid squares), predictions of IQMD model with SM (solid circles), and predictions of IQMD model with HM (solid triangles), in the isotopic systems ^{58}Ni + ^{58}Ni and ^{58}Fe + ^{58}Fe as a function of reduced impact parameter. The lines included only to guide the eye.

impact parameter b=5 fm. In Fig. 8 (a) we use C=0 (no symmetry energy) and Cugnon's N-N cross sections $\sigma_{Cug.}$. The case of using C=0 and experimental N-N cross sections $\sigma_{exp.}$ is shown in Fig. 8 (b). For the results shown in Fig. 8 (c) we use C=32 MeV and experimental N-N cross sections $\sigma_{exp.}$. Fig. 8 indicates that the neutron-rich system (^{58}Fe + ^{58}Fe) exhibits stronger negative deflection for all above three cases. In order to give a quantitative results, Table 2 shows the calculated values of flow parameter for above three cases. The errors shown are statistical errors of the linear fits. From Table 2 one can see that both the $\sigma_{exp.}$ and symmetry energy make the strength of the flow parameter decrease, which implies that both N-N collisions and symmetry energy are repulsive as one expects [20]. Meanwhile, it is indicated that $\sigma_{exp.}$ has stronger influence in system ^{58}Ni + ^{58}Ni than in system ^{58}Fe + ^{58}Fe, which is easy to understand since N-N collisions result in repulsive flow and this effect is proportional to the number of collisions in the interaction volume. While the number of nucleons in this volume for the two reaction systems is roughly the same and the number of collisions in the reaction of the neutron-rich system is smaller since the neutron-neutron cross section is about a factor of three smaller than the neutron-proton cross section at energy of 55 MeV/nucleon. In addition, it is also shown that symmetry energy has stronger influence in system ^{58}Fe + ^{58}Fe than in system ^{58}Ni + ^{58}Ni, which results from that in ^{58}Fe the difference between neutron and proton densities is larger and thus cause stronger symmetry energy. In the case of C=0 and $\sigma_{Cug.}$, the difference of the flow parameters of the two reactions may mainly come from the influence of the Coulomb potential and the surface properties of ^{58}Fe and ^{58}Ni. Unfortunately, we can not study the influence of Coulomb potential on collective flow in this paper because we can not get stable initial nuclei without Coulomb interaction.

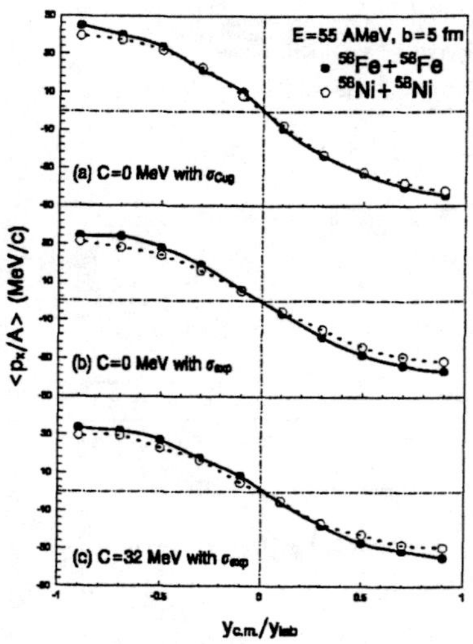

Figure 8: Mean transverse momentum in the reaction plane versus the normalized rapidity for all nucleons from reactions ^{58}Fe + ^{58}Fe (solid circles) and ^{58}Ni + ^{58}Ni (open circles) at 55 MeV/nucleon and b= 5 fm by using different C and parametrizations of N-N cross sections: C=0 with Cugnon's N-N cross sections $\sigma_{Cug.}$ (a), C=0 with experimental N-N cross sections $\sigma_{exp.}$ (b), and C=32 MeV with experimental N-N cross sections $\sigma_{exp.}$. The curves are plotted to guide the eye.

3.2 Radial flow

Central collisions of heavy nuclei are particularly interesting since, for them, stopping, compression, and thermal equilibration are expected to be largest and all the initial kinetic energy is deposited into "one source" located at midrapidity which greatly simplifies the collision dynamics, reduces surface effects, and enhances volume effects[56, 57, 58]. The experimental and theoretical (the BUU and QMD model) studies have shown that the radial flow has the following novel characters: it is proportional to the fragment mass; it is seen to increase with decreasing impact parameter and increasing incident beam energy; it shows no dependence on the nuclear incompressibility; it is sensitive to effective mass m*/m; and it has a rather small system size dependence if quantized in flow energy *per nucleon*. Therefore, the radial flow is very different from the in-plane transverse collective flow.

In order to explore the isospin dependence of the radial flow we study two symmetric, central collisions of systems, ^{124}Ba + ^{124}Ba and ^{124}Sn + ^{124}Sn, which have the same mass number but different neutron/proton ratios of 1.21 and 1.48, respectively. The neutron and proton densities calculated from the nonlinear relativistic mean-field (RMF) theory [59] are used to initialize the ground states of ^{124}Ba and ^{124}Sn in terms of Monte Carlo method. Particularly, in the nonlinear RMF calculations we use the latest force parameters NL3 which is able to provide a very good description not only for the properties of stable nuclei but also for those far from the valley of β stability in all cases considered so far[60].

In symmetric, central collisions, generally, the fragment mean kinetic energy may be decomposed into a random, "thermal" part, a Coulomb repulsive part, and a collective flow

Table 3: Radial flows of reaction systems ^{124}Ba + ^{124}Ba and ^{124}Sn + ^{124}Sn at energies of 100, 150, and 200 MeV/nucleon, and their differences between the two reaction systems Δ.

Beam energies	100 (MeV/nucleon)	150 (MeV/nucleon)	200 (MeV/nucleon)
^{124}Ba + ^{124}Ba	3.06 (MeV/nucleon)	6.90 (MeV/nucleon)	11.94 (MeV/nucleon)
^{124}Sn + ^{124}Sn	2.59 (MeV/nucleon)	4.60 (MeV/nucleon)	6.95 (MeV/nucleon)
Δ	0.47 (MeV/nucleon)	2.30 (MeV/nucleon)	4.99 (MeV/nucleon)

contribution (the so-called radial flow), i.e.,

$$\langle E\rangle = \langle E_{thermal}\rangle + \langle E_{Coul}\rangle + \langle E_{radial}\rangle. \tag{8}$$

As in Refs.[58, 61], the "thermal+Coulomb" part can be calculated in terms of the well-known code WIX[57] which is an extended FREESCO[62] simulation including Coulomb interaction and collective expansion. The radial flows at energies of 100, 150, and 200 MeV/nucleon are illustrated in Figs. 9 (a), (b), and (c), respectively, in terms of the mean kinetic energy per nucleon $\langle E_{kin}/A\rangle$ as a function of the fragment charge Z. The open (solid) circles represent the results of reaction ^{124}Ba + ^{124}Ba (^{124}Sn + ^{124}Sn) and the dotted (solid) curve are the results of "thermal+Coulomb" scenario for ^{124}Ba + ^{124}Ba (^{124}Sn + ^{124}Sn) by modeling a hot single "source". The amount of collective radial flow can be extracted from the difference between the total mean kinetic energy and the mean kinetic energy of "thermal+Coulomb" part of the heavy fragments ($Z \geq 2$). The extracted values of radial flow and their differences between the two reaction systems Δ are displayed in Table 3. It is seen from Table 3 that the neutron-rich system ^{124}Sn + ^{124}Sn shows systematically smaller radial flows. Particularly, the effect is more appreciable as the incident energy increases. It is very clear that the radial flow increases as a function of beam energy for the same reaction system. Figs.9 (d), (e), and (f) display the ratio of the fragment's total kinetic energy E_{kin} to its transverse kinetic energy $\frac{3}{2}E_{trans}$ as a function of the fragment charge Z at different energies. The open (solid) circles represent the results of reaction ^{124}Ba + ^{124}Ba (^{124}Sn + ^{124}Sn). We can see that this ratio is, within a very good approximation, equal to 1, which indicates that the colliding nuclei have completely stopped and the fragments originate from a hot and compressed transient system[63].

The predicted isospin dependence of the radial flow seems to mainly result from the isospin-dependent N-N cross sections. The experimental N-N cross sections adopted in this work for neutron-proton collisions are about three times that for neutron-neutron (proton-proton) collisions in the energy range studied here, which results in weaker relatively N-N collisions in the neutron-rich system and consequently leads to a smaller expansion. As mentioned in[14], the N-N collisions become more and more important with increment of incident energies considered here, which causes the phenomenon observed in Fig. 9, namely, the isospin effect is more appreciable as the incident energy increases. At 150 MeV/nucleon for ^{124}Sn + ^{124}Sn, as previous calculations but set C =0 MeV (dropping the nuclear symmetry energy) and use Cugnon's parametrization of cross sections, $\sigma_{Cug.}$, which is isospin independent, the transverse kinetic energy per nucleon $\langle\frac{3}{2}E_{trans}/A\rangle$ as a function of the fragment charge Z is displayed in Fig. 10 (a) by dotted curve. The dashed curve in Fig. 10 (a) represents the results of setting C =32 MeV and using $\sigma_{Cug.}$. The solid curve corresponds to the case of setting C =32 MeV and using experimental parametrization of cross sections $\sigma_{exp.}$. The results of similar calculations for system ^{124}Ba + ^{124}Ba are displayed in Fig. 10 (b). From Fig. 10, one can see clearly that the mean transverse

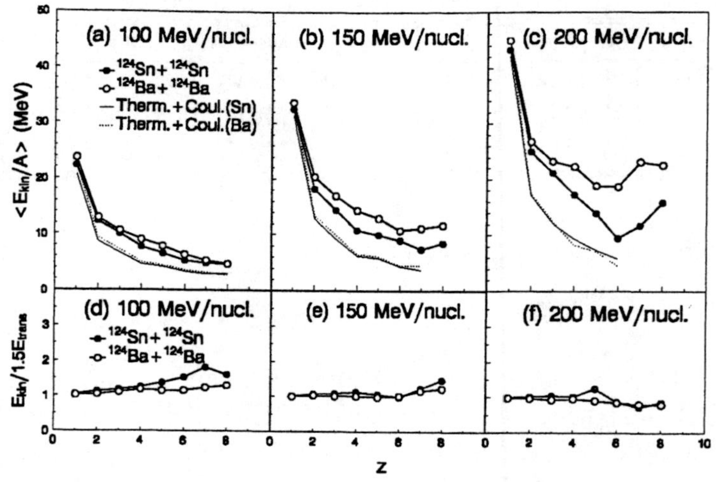

Figure 9: Upper panels: Mean kinetic energy per nucleon $\langle E_{kin}/A\rangle$ as a function of the fragment charge Z at energies of 100 (a), 150 (b), and 200 (c) MeV/nucleon The open (solid) circles represent the results of reaction ^{124}Ba + ^{124}Ba (^{124}Sn + ^{124}Sn) and the dotted (solid) curve are the results of "thermal+Coulomb" scenario for ^{124}Ba + ^{124}Ba (^{124}Sn + ^{124}Sn) by modeling a hot single "source" in terms of code WIX. Lower panels: Same as in upper panels but for $E_{kin}/\frac{3}{2}E_{trans}$.

kinetic energy of heavy fragment is sensitive to the selection of the N-N cross sections and is almost independent of the nuclear symmetry energy. For the same reaction system, i.e., ^{124}Ba + ^{124}Ba or ^{124}Sn + ^{124}Sn, the results using $\sigma_{Cug.}$ show a smaller radial flow energy than that using $\sigma_{exp.}$, which is mainly due to $\sigma_{Cug.}$ being much smaller than $\sigma_{exp.}$ at energy considered here. More importantly here, it is seen that the difference of mean transverse kinetic energy of heavy fragment between the two systems decreases if $\sigma_{Cug.}$ is adopted instead of $\sigma_{exp.}$. The change of N-N cross sections from $\sigma_{exp.}$ to $\sigma_{Cug.}$ has stronger influence in system ^{124}Ba + ^{124}Ba than in system ^{124}Sn + ^{124}Sn, which is easy to understand since the neutron-neutron cross section is about a factor of three smaller than the neutron-proton cross section at energies studied here.

For the experimental N-N cross sections, the difference of cross sections between neutron-proton collisions and neutron-neutron (proton-proton) collisions disappears as the incident energy go up to about 500 MeV/nucleon. In addition, at higher energies the Coulomb interaction becomes not too important. At an incident energy of 600 MeV/nucleon, Fig. 11 displays the mean kinetic energy per nucleon $\langle E_{kin}/A\rangle$ as a function of fragment mass. The open (solid) circles represent the results of reaction ^{124}Ba + ^{124}Ba (^{124}Sn + ^{124}Sn). It is indicated that the difference of mean kinetic energy between ^{124}Ba + ^{124}Ba and ^{124}Sn + ^{124}Sn disappears at 600 MeV/nucleon, which confirms again that the isospin dependence of radial flow is mainly determined by the isospin dependence of N-N cross sections. These features seem to strengthen the intuitive notion that the radial flow is mainly driven by nucleonic collisions or "thermal expansion"[64].

In above calculations, only the free-space N-N cross sections are adopted and the in-medium effect is only simulated by the Pauli blocking. However, the in-medium N-N cross sections and their isospin dependence might be strongly density dependent. The

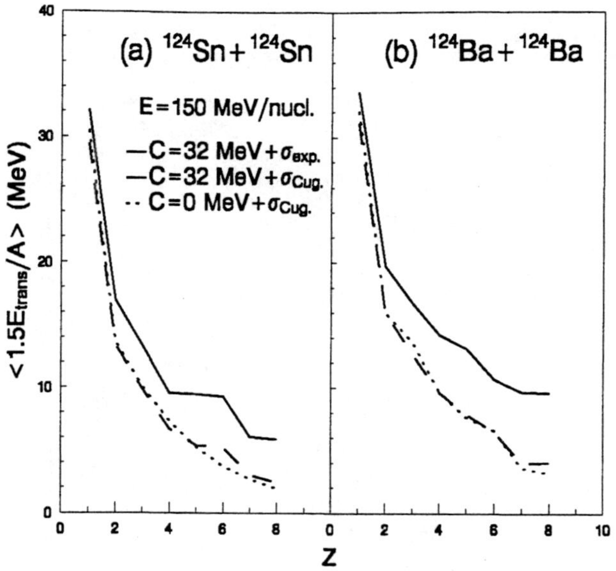

Figure 10: Transverse kinetic energy per nucleon $\langle \frac{3}{2} E_{trans}/A \rangle$ as a function of the fragment charge Z for reaction systems ^{124}Sn + ^{124}Sn (a) and ^{124}Ba + ^{124}Ba (b) in different cases of: C =0 MeV and using Cugnon's parametrization of cross sections $\sigma_{Cug.}$ (dotted curves), C =32 MeV and using $\sigma_{Cug.}$ (dashed curves), and C =32 MeV and using experimental parametrization of cross sections $\sigma_{exp.}$ (solid curves).

isospin dependence of the radial flow may provide a unique opportunity to study the isospin dependent in-medium N-N cross sections by comparing the experimental data with the predictions of theoretical model using the isospin and density dependent in-medium N-N cross sections in the future.

we have shown that there is a strong isospin dependence of the radial flow from the symmetric, central collisions at incident energies of a few hundred MeV per nucleon. It is found that the more neutron-rich system exhibits smaller radial flow, which is shown to result from the isospin dependence of N-N cross sections and be independent of the nuclear symmetry energy. The experimental measurement of the radial flow for reaction systems with different ratios of neutron to proton provides a novel method for determining the isospin dependent in-medium N-N cross sections.

3.3 Nuclear multifragmentation

Heavy-ion collisions (HIC) at intermediate and high energies provide a new way to investigate the properties of nuclear matter at high temperature and high density which is connected with the nuclear equation of state (EOS), multifragmentation, and liquid-gas phase transition[65, 66, 67, 68, 69, 70, 71, 73, 74]. The nuclear multifragmentation, i.e., the production of several intermediate mass fragments (IMF's), particularly attracts a lot of interest since it constitutes a major decay channel of nuclear matter formed in such heavy-ion collisions[75, 77, 76, 78]. Meanwhile, the nuclear multifragmentation offers a unique opportunity to explore the properties of quantum many-body system very far from equilibrium and to discover the universal law of fragmentation, which are theoretically meaningful for investigating the phase transition and critical phenomenon of finite system. A great deal

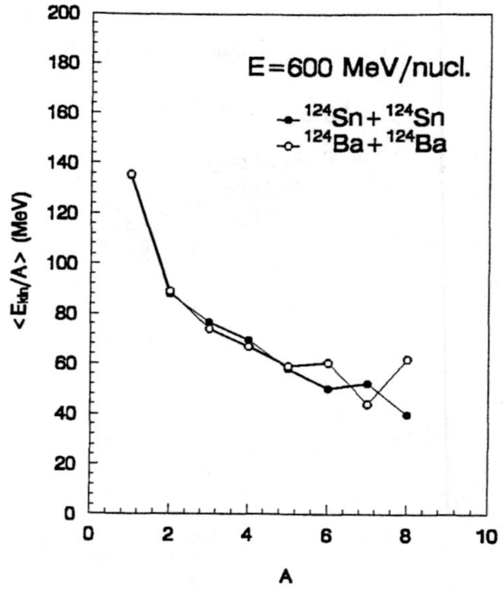

Figure 11: Mean kinetic energy per nucleon $\langle E/A \rangle$ as a function of fragment mass A at energy of 600 MeV/nucleon. The open (solid) circles represent the results of reaction ^{124}Ba + ^{124}Ba (^{124}Sn + ^{124}Sn). The lines are to guide the eye.

of experimental and theoretical work has been devoted to the studies of the mechanism of IMF's emission and the resulting physical pictures mainly focus on the volume instability from the density fluctuations, thermodynamical instability from the temperature fluctuations, and shape instability from the surface fluctuations[66].

More recently, the isospin dependence of nuclear multifragmentation in ^{112}Sn + ^{112}Sn and ^{124}Sn + ^{124}Sn collisions at 40 MeV/nucleon has been observed by the MSU group[79]. This effect was reproduced qualitatively by the expanding evaporating source (EES) model[79 80, 81] whose cooling rates were dependent on the neutron number of the source. An extensive study of these reactions using the percolation model[82, 83] failed to reproduce this effect.

In the present calculations, we simply assume that number of the events is proportional to impact parameter b and b=1, 2, 3, 4, 5, 6, 7, 8, 9 and 10 fm are adopted. The calculation indicates that the number of IMF's has become very small and the multifragmentation mechanism has basically disappeared at b=10 fm. To accumulate the numerical statistics, we select the fragments by summing over the fragments of $t \geq$ 200 fm/c since the charge distributions have been well stable after 200 fm/c.

Figs. 12 (a), (b), and (c) show the average number of IMF's ($\langle N_{IMF} \rangle$: $3 \leq Z \leq 33$) as a function of number of charged particles, N_C, light-charged particles with $Z \leq 2$, N_{LC}, and neutrons, N_N, respectively. The solid (open) circles denote the experimental data for the ^{124}Sn + ^{124}Sn (^{112}Sn + ^{112}Sn) reaction. The solid (dotted) line represents the IQMD model prediction for the ^{124}Sn + ^{124}Sn (^{112}Sn + ^{112}Sn) reaction. From Fig. 12, one can see clearly that the IQMD model predictions are in very good agreement with the experimental data in trends, namely, for the same N_C or N_{LC} the $\langle N_{IMF} \rangle$ for ^{124}Sn + ^{124}Sn is larger than that for ^{112}Sn + ^{112}Sn. The calculated results indicate that the magnitude of the maxima in different correlations has nearly the same values while the experimental data show different

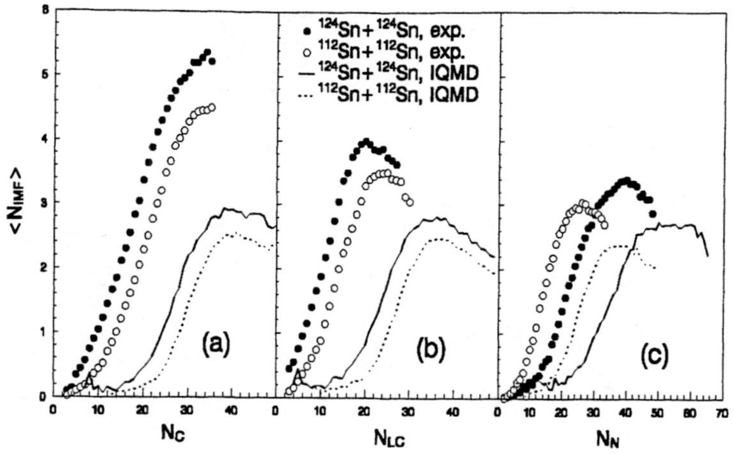

Figure 12: Average number of intermediate mass fragments (IMF's), $\langle N_{IMF} \rangle$, versus number of charged particles, N_C (a), light-charged particles with $Z \leq 2$, N_{LC} (b), and neutrons, N_N (c). The solid (open) circles denote the experimental data for the ^{124}Sn + ^{124}Sn (^{112}Sn + ^{112}Sn) reaction. The solid (dotted) line represents the IQMD model predictions for the ^{124}Sn + ^{124}Sn (^{112}Sn + ^{112}Sn) reaction.

values for the different correlations, which implies that the present IQMD model can not reproduce this feature of the data. More importantly here, there is agreement between the data and the IQMD model predictions for the magnitude of the isospin effect. In addition, the calculated results can not reproduce the maxima of $\langle N_{IMF} \rangle$, which is mainly due to not considering the statistical decay of excited fragments in reaction final state. Meanwhile, we can see that the calculated results can not reproduce the positions of the maxima in $\langle N_{IMF} \rangle$ versus N_C, N_{LC}, or N_N, which is easy to understand since the pre-equilibrium particles are still included and they contribute to N_C, N_{LC}, or N_N. This problem also exists in the calculations of the EES model[79] or percolation model[82]. Therefore, it seems that the pre-equilibrium emission may play a very important role in these correlations.

If the pre-equilibrium particles are excluded, then the results can be displayed in Fig. 13 which is the same as Fig. 12 but in the calculations the pre-equilibrium particles have been excluded. In the present calculations, the pre-equilibrium particle is simply defined as the light fragment with $Z \leq 2$ which satisfies the following conditions: the relative distance between it and any other fragment is larger than 3.5 fm; the average kinetic energy per nucleon of this fragment is larger than 8 MeV in center-of-mass (c.m.) system; it moves far from the center-of-mass in the c.m. system; its relative distance to the center-of-mass is larger than the root-mean-square radius of the reaction system. From the comparison between Fig. 12 and Fig. 13, we can see clearly that the pre-equilibrium emission affects strongly the sorting axis, which results in the calculated results reproducing very well the positions of the maxima in $\langle N_{IMF} \rangle$ versus N_C, N_{LC}, or N_N although the isospin dependence of the horizontal shifts for the different correlations in Fig. 13 become worse than that in Fig. 12 for the peripheral collisions (corresponding to small N_C, N_{LC}, or N_N.). This feature may be a limitation of present determining pre-equilibrium particles due to devoid of experimental information.

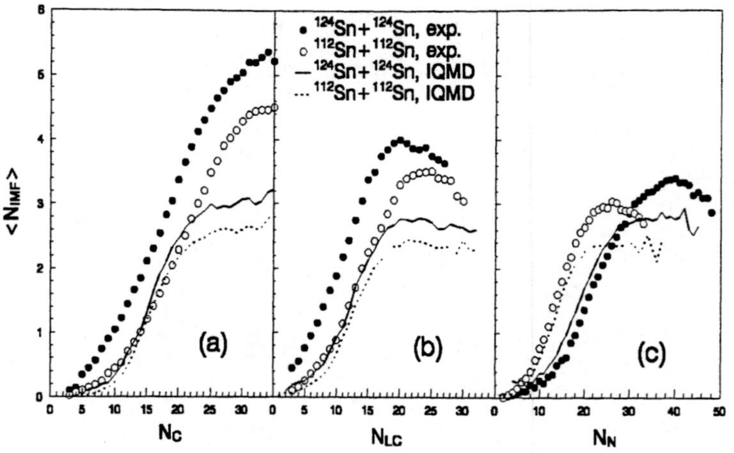

Figure 13: Same as in Fig. 12 but in the calculated results the pre-equilibrium particles have been excluded.

In order to illustrate the influence of the pre-equilibrium emission on the two reactions, at b =0 we display in Figs. 14 (a), (b) and (c) the IQMD model predicted time evolution of number of pre-equilibrium neutrons ($N_N^{pre.}$), light-charged particles ($N_{LC}^{pre.}$) and N/Z of the remnant sources (here the remnant source is defined as the heavy residue after excluding the pre-equilibrium particles), respectively. The solid (dotted) line corresponds to the result of ^{124}Sn + ^{124}Sn (^{112}Sn + ^{112}Sn). From Figs. 14 (a) and (b), one can see clearly that nearly all pre-equilibrium particles are emitted in the time interval from 50 to 100 fm/c for both reactions and the subsequent particle emission is less important. Meanwhile, it is seen that the pre-equilibrium neutron and light-charged particle emissions are very different for the two systems, i.e., the more neutron-rich system ^{124}Sn + ^{124}Sn emits more pre-equilibrium neutrons and fewer pre-equilibrium light-charged particles than the system ^{112}Sn + ^{112}Sn, and the number of neutrons is larger than that of light-charged particles for ^{124}Sn + ^{124}Sn while the opposite result is observed for ^{112}Sn + ^{112}Sn, which may be due to the former existing a thick neutron-skin. This phenomenon is also observed in the analysis of the EES model[79]. As a matter of fact, the more neutron-rich system ^{124}Sn + ^{124}Sn emits more pre-equilibrium neutrons to cool and consequently it remains less energy so as to emit less light-charged particles. In addition, from Fig. 14 (c) we can see that N/Z of the remnant sources have changed from original 1.48 (1.24) to 1.41 (1.23) at 100 fm/c and then 1.49 (1.30) at 400 fm/c, which shows that the pre-equilibrium emission has big influence on the N/Z of the remnant source of ^{124}Sn + ^{124}Sn around 100 fm/c. Therefore, the pre-equilibrium emission affects the N/Z composition of the fragmenting system as expected[82].

It is well known that the total charge contained in fragments with two or more charges, Z_{bound}, is also often measured experimentally and the correlation between $\langle N_{IMF}\rangle$ and Z_{bound} is usually used to investigate the nuclear multifragmentation[67, 68]. It should be noted that the pre-equilibrium light particles have little influence on Z_{bound} since the neutron and hydrogen isotopes do not contribute to the Z_{bound}. The $\langle N_{IMF}\rangle$ as a function of Z_{bound} is shown in Fig. 15. In the left panel of Fig. 15 the effect of pre-equilibrium emission is not considered while the pre-equilibrium particles have been excluded in the right panel of Fig.

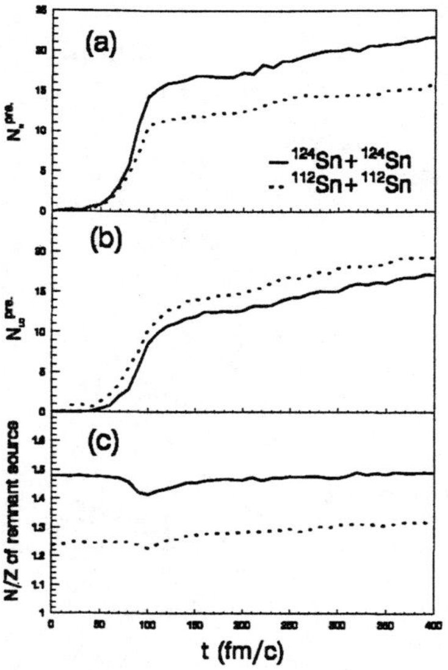

Figure 14: The IQMD model predicted time evolution of number of pre-equilibrium neutrons, $N_N^{pre.}$ (a), light-charged particles, $N_{LC}^{pre.}$ (b), and N/Z of the remnant sources (c) for central collisions of systems ^{124}Sn + ^{124}Sn (solid line) and ^{112}Sn + ^{112}Sn (dotted line).

15. The solid (open) line represents the IQMD model prediction for the ^{124}Sn + ^{124}Sn (^{112}Sn + ^{112}Sn) collisions. From Fig. 15 we can see that there exists very significantly different scalings of $\langle N_{IMF} \rangle$ with Z_{bound} between ^{124}Sn + ^{124}Sn and ^{112}Sn + ^{112}Sn collisions, i.e., for a given value of Z_{bound}, larger $\langle N_{IMF} \rangle$ is observed for ^{124}Sn + ^{124}Sn than ^{112}Sn + ^{112}Sn. Meanwhile, the comparison between the left and right panels of Fig. 15 implies that the pre-equilibrium emission have hardly influence on scalings of $\langle N_{IMF} \rangle$ with Z_{bound} as expected, which indicate that the correlation between $\langle N_{IMF} \rangle$ and Z_{bound} is more convenient than that between $\langle N_{IMF} \rangle$ and N_C, N_{LC}, or N_N for comparison of theoretical calculation and experimental data. Unfortunately, the experimental data involving Z_{bound} is not available.

The experimental data could be reproduced qualitatively within the framework of the IQMD model. In particular, the IQMD model predictions are in quantitative agreement with the experimental data on the magnitude of the isospin effect. The main discrepancy is that the IQMD model can not reproduce the maxima of $\langle N_{IMF} \rangle$ mainly due to not considering the statistical decay of excited fragments in reaction final state. Meanwhile, the IQMD model calculations indicate that the pre-equilibrium emission plays an important role in scaling of the $\langle N_{IMF} \rangle$ with N_C, N_{LC}, or N_N but has hardly influence on scalings of $\langle N_{IMF} \rangle$ with Z_{bound}.

4 Conclusions

The results indicate that the neutron-rich system displays stronger negative deflection and has a higher balance energy, which could be in qualitative agreement with the experimental data. In particular, the magnitude of the isospin effects could be quantitatively in agreement with the experimental data. It is also found that the neutron-rich system exhibits smaller radial flow, which is shown result from the isospin dependence of N-N cross sections and be independent of the nuclear symmetry energy. We suggest that one may extract the isospin dependent in-medium N-N cross sections from the experimental measurements of the radial

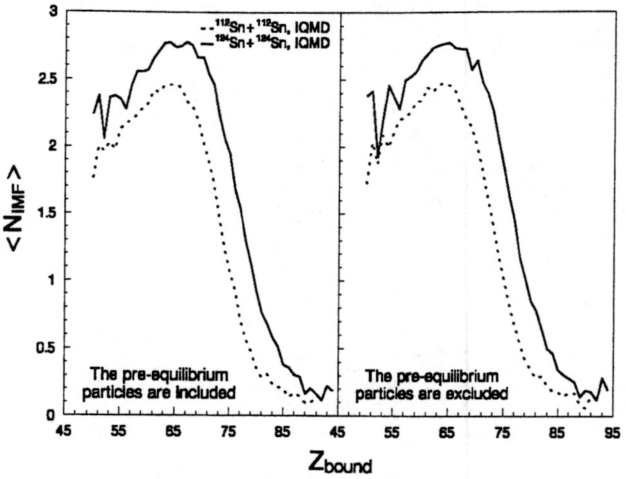

Figure 15: IQMD model predicted average number of intermediate mass fragments (IMF's), $\langle N_{IMF} \rangle$, versus the total charge contained in fragments with two or more charges, Z_{bound}, at two different situations: the effect of pre-equilibrium emission is not considered (left panel) and the pre-equilibrium particles are excluded (right panel). The solid (dotted) line represents results for the ^{124}Sn + ^{124}Sn (^{112}Sn + ^{112}Sn) collisions.

flow for reaction systems with different ratios of neutron to proton. Meanwhile, one may investigate the isospin effects on some other collective flow phenomena, such as the in-plane directed flow, the out-of-plane squeeze-out flow, and the azimuthally symmetry radial flow, which may provide people a better probe of isospin dependent reaction dynamics.

A great deal of experimental and theoretical work has been devoted to the studies of the mechanism of nuclear multifragmentation and the resulting physical pictures mainly focus on the thermodynamical instability from the temperature fluctuations, the mechanical instability (spinodal instability) from the density fluctuations, and shape instability from the surface fluctuations. Up to now, however, very limited knowledge is known about the chemical instability. With the establishment of secondary beam facilities at many laboratories around the world, it is possible to explore the chemical instability of hot and isospin-asymmetric nuclear matter. As a matter of fact, as is shown in the present work, the multifragment production is sensitive to the isospin asymmetry of the reaction systems. More recently, Miller et al. reported the disappearance of the isospin dependence of multifragment production in reactions of ^{58}Fe + ^{58}Fe and ^{58}Ni + ^{58}Ni[84] at higher incident energy. These exciting phenomena may be related to the chemical instability in HIC induced by neutron-rich nuclei and therefore it is a challenge for theoretical and experimental nuclear physicists to investigate the chemical instability in HIC.

Acknowledgements

This work was supported by the National Natural Science Foundation of China under Grant NO. 19609033, 19875068, and 19847002 and the Foundation of the Chinese Academy of Sciences.

References

[1] G. F. Bertsch and S. Das Gupta, Phys. Rep. **160**, 233(1988).

[2] J. Aichelin, Phys. Rep. **202**, 233(1991).

[3] Y. Abe, S. Ayik, P.-G. Reinhard, and E. Suraud, Phys. Rep. **275**, 49(1996).

[4] C. Hartnack, Z. X. Li, L. Neise, G. Peilert, A. Rosenhauer, H. Sorge, J. Aichelin, H. Stocker, and W. Greiner, Nucl. Phys. **A495**, 303c(1989).

[5] W. Cassing and U. Mosel, Ann. Rev. Nucl. Part. Sci. **25**, 235(1990).

[6] B. A. Li, C. M. Ko, and W. Bauer, Int. J. Mod. Phys. **E7**, 147(1998).

[7] M. Colonna, M. Di Toro, G. Fabbri, and S. Maccarone, Phys. Rev. **C57**, 1410(1998).

[8] F. Gulminelli and Ph. Chomaz, Phys. Rev. Lett. **82**, 1402(1999).

[9] F. S. Zhang and L. X. Ge, *Nuclear Multifragmentation*, Science Press, Beijing, 1998.

[10] I. Tanihata, Prog. Part. Nucl. Phys. **35**, 505(1995).

[11] M. V. Zhukov, B. V. Danilin, D. V. Fedorov, J. M. Bang, I. J. Thompson, and J. S. Vaagen, Phys. Rep. **231**, 151(1993).

[12] C. A. Bertulani, L. F. Canto, and M. S. Hussein, Phys. Rep. **226**, 281(1993).

[13] L. W. Chen, L. X. Ge, X. D. Zhang, and F. S. Zhang, J. Phys. **G23**, 211(1997).

[14] L. W. Chen, F. S. Zhang, and G. M. Jin, Phys. Rev. **C58**, 2283(1998).

[15] L. W. Chen, F. S. Zhang, G. M. Jin, and Z. Y.Zhu, Phys. Lett. **B459**, 21(1999).

[16] F. S. Zhang, L. W. Chen, Z. Y. Ming, and Z. Y. Zhu, Phys. Rev. **C60**, 060604(1999).

[17] J. Cugnon, T. Mizutani, and J. Vandermeulen, Nucl. Phys. **A352**, 505(1981).

[18] K. Chen, Z. Fraenkel, G. Friedlander, J. R. Grover, J. M. Miller, and Y. Shimamoto, Phys. Rev. **166**, 949(1968).

[19] P. G. Reinhard, *in* Computational Nuclear Physics 1, editted by K. Langanke, J. A. Maruhn, and S. E. Koonin, Germany, Springer-Verlag, 1991, p. 28-50.

[20] B. A. Li, Z. Z. Ren, C. M. Ko, and S. J. Yennello, Phys. Rev. Lett. **76**, 4492(1996).

[21] H. Kruse, B. V. Jacak, J. J. Molitoris, G. D. Westfall, and H. Stocker, Phys. Rev. **C31**, 1770(1985).

[22] F. S. Zhang and E. Suraud, Phys. Rev. **C51**, 3201(1995).

[23] F. S. Zhang and E. Suraud, Phys. Lett. **B319**, 35(1993).

[24] B. A. Li and S. J. Yennello, Phys. Rev. C **52**, 1746(1995).

[25] R. Pak, W. Benenson, O. Bjarki, J. A. Brown, S. A. Hannuschke, R. A. Lacey, B. A. Li, A. Nadasen, E. Norbeck, P. Pogodin, R. E. Russ, M. Steiner, N. T. B. Stone, A. M. Vander Molen, G. D. Westfall, L. B. Yang, and S. J. Yennello, Phys. Rev. Lett. **78**, 1022(1997).

[26] R. Pak, B. A. Li, W. Benenson, O. Bjarki, J. A. Brown, S. A. Hannuschke, R. A. Lacey, D. J. Magestro, A. Nadasen, E. Norbeck, R. E. Russ, M. Steiner, N. T. B. Stone, A. M. Vander Molen, G. D. Westfall, L. B. Yang, and S. J. Yennello, Phys. Rev. Lett. **78**, 1026(1997).

[27] D. Krofcheck, W. Bauer, G. M. Grawley, C. Djalali, S. Howden, C. A. Ogilvie, A. M. Vander Molen, G. D. Westfall, W. K. Wilson, R. S. Tickle, and C. Gale, Phys. Rev. Lett. **63**, 2028(1989).

[28] C. A. Ogilvie, D. A. Cebra, J. Clayton, P. Danielewicz, S. Howden, J. Karn, A. Nadasen, A. Vander Molen, G. D. Westfall, W. K. Wilson, and J. S. Winfield, Phys. Rev. C **40**, 2592(1989).

[29] S. Soff, S. A. Bass, C. Hartnack, H. Stocker, and W. Greiner, Phys. Rev. **C51**, 3320(1995).

[30] J. P. Sullivan, J. Peter, D. Cussol, G. Bizard, R. Brou, M. Louvel, J. P. Patry, R. Regimbart, J. C. Steckmeyer, B. Tamain, E. Crema, H. Doubre, K. Hagel, G. M. Jin, A. Peghaire, F. Saint-Laurent, Y. Cassagnou, R. Legrain, C. Lebrun, E. Rosato, R. MacGrath, S. C. Jeong, S. M. Lee, Y. Nagashima, T. Nakagawa, M. Ogihara, J. Kasagi, and T. Motobayashi, Phys. Lett. **B249**, 8(1990).

[31] R. Pak, O. Bjarki, S. A. Hannuschke, R. A. Lacey, J. Lauret, W. J. Llope, A. Nadasen, N. T. B. Stone, A. M. Vander Molen, and G. D. Westfall, Phys. Rev. **C54**, 2457(1996).

[32] W. M. Zhang, R. Madey, M. Elaasar, J. Schambach, D. Keane, B. D. Anderson, A. R. Baldwin, J. Cogar, J. W. Watson, G. D. Westfall, G. Krebs, and H. Wieman, Phys. Rev. **C42**, R491(1990).

[33] D. Krofcheck, W. Bauer, G. M. Grawley, S. Howden, C. A. Ogilvie, A. Vander Molen, G. D. Westfall, W. K. Wilson, R. S. Tickle, C. Djalali, and C. Gale, Phys. Rev. **C46**, 1416(1992).

[34] W. K. Wilson, W. Benenson, D. A. Cebra, J. Clayton, S. Howden, J. Karn, T. Li, C. A. Ogilvie, A. Vander Molen, G. D. Westfall, J. S. Winfield, B. Young, and A. Nadasen, Phys. Rev. **C41**, R1881(1990).

[35] G. D. Westfall, W. Bauer, D. Craig, M. Cronqvist, E. Gualtieri, , S. Hannuschke, D. Klakow, T. Li, T. Reposeur, A. M. Vander Molen, W. K. Wilson, J. S. Winfield, J. Yee, S. J. Yennello, R. Lacey, A. Elmaani, J. Lauret, A. Nadasen, and E. Norbeck, Phys. Rev. Lett. **71**, 1986(1993).

[36] C. Gale, G. F. Bertsch, and S. Das Gupta, Phys. Rev. **C35**, 1666(1987).

[37] J. Zhang, S. Das Gupta, and C. Gale, Phys. Rev. **C50**, 1617(1994).

[38] M. B. Tsang, G. F. Bertsch, W. G. Lynch, and M. Tohyama, Phys. Rev. **C40**, 1685(1989).

[39] H. M. Xu, Phys. Rev. Lett. **67**, 2769(1991).

[40] B. A. Li, Phys. Rev. **C48**, 2415(1993).

[41] P. Danielewicz and G. Odyniec, Phys. Lett. **B157**, 146(1985).

[42] J. J. Molitoris and H. Stocker, Phys. Lett. **B162**, 47(1985).

[43] J. J. Molitoris, D. Hahn, and H. Stocker, Nucl. Phys. **A447**, 13c(1986).

[44] G. F. Bertsch, W. G. Lynch, and M. B. Tsang, Phys. Lett. **B189**, 384(1987).

[45] P. Danielewicz, H. Strobele, G. Odyniec, D. Bangert, R. Bock, R. Brockmann, J. W. Harris, H. G. Pugh, W. Rauch, R. E. Renfordt, A. Sandoval, D. Schall, L. S. Schroeder and R. Stock, Phys. Rev. **C38**, 120(1988).

[46] V. de la Mota, F. Sebille, M. Farine, B. Remaud, and P. Schuck, Phys. Rev. **C46**, 677(1992).

[47] Q. Pan and P. Danielewicz, Phys. Rev. Lett. **70**, 2062(1993).

[48] H. B. Zhou, Z. X. Li,Y. Z. Zhuo, and G. J. Mao, Nucl. Phys. **A580**, 627(1994).

[49] G. J. Mao, Z. X. Li, Y. Z. Zhuo, and Z. Q. Yu, Phys. Lett. **B327**, 183(1994).

[50] D. R. Bowman, G. F. Peaslee, R. T. de Souza, N. Carlin, N. K. Gelbke, W. G. Gong, Y. D. Kim, M. A. Lisa, W. G. Lynch, L. Phair, M. B. Tsang, C. Williams, N. Colonna, K. Hanold, M. A. McMahan, G. J. Wozniak, L. G. Moretto, and W. A. Friedman, Phys. Rev. Lett. **67**, 1527(1991).

[51] J. Hubele, P. Kreutz, J. C. Adloff, M. Begemann-Blaich, P. Bouissou, G. Imme, I. Iori, G. J. Kunde, S. Leray, V. Lindenstruth, Z. Liu, U. Lynen, R. J. Meijer, U. Milkau, A. Moroni, W. F. J. Muller, C. Ngo, C. A. Ogilivie, J. Pochodzalla, G. Raciti, G. Rudolf, H. Sann, A. Schuttauf, W. Seidel, L. Stuttge, W. Trautmann, and A. Tucholski, Z. Phys **A340**, 263(1991).

[52] J. Peter, *in* Inter. Symp. on Heavy Ion Phys. and Its Appli., Lanzhou, China, 8-12 Oct., Singapore, World Scientific Press, 1990, p.191.

[53] W. Bauer, Phys. Rev. Lett **61**, 2534(1988).

[54] D. Klakow, G. Welke, and W. Bauer, Phys. Rev. **C48**, 1982(1993).

[55] W. Bauer, C. K. Gelbke, and S. Pratt, Annu. Rev. Nucl. Part. Sci. **42**, 77(1992).

[56] W. Reisdorf and H. G. Ritter, Annu. Rev. Nucl. Part. Sci. **47**, 663(1997).

[57] J. Randrup, Comput. Phys. Commun. **77**, 153(1993).

[58] S. C. Jeong et al., Phys. Rev. Lett. **72**, 3468(1994).

[59] Z. Y. Zhu, W. Q. Shen, Y. H. Cai, and Y. G. Ma, Phys. Lett. **B328**, 1(1994).

[60] G. A. Lalazissis, J. Konig, and P. Ring, Phys. Rev. **C55**, 540(1997).

[61] S. C. Jeong et al., Nucl. Phys. **A604**, 208(1996).

[62] G. Fai and J. Randrup, Comput. Phys. Commun. **42**, 385(1986).

[63] W. Bauer et al., Phys. Rev. **C47**, R1838(1993).

[64] J. Lauret et al., Phys. Rev. **C57**, R1051(1998).

[65] G. Peilert, H. Stocker and W. Greiner, Rep. Prog. Phys. **57**, 533(1994).

[66] L. G. Moretto and G. J. Wozniak, Annu. Rev. Nucl. Sci. **43**, 379(1993).

[67] P. Kreutz, J. C. Adloff, M. Begemann-Blaich, P. Bouissou, J. Hubele, G. Imme, I. Iori, G. J. Kunde, S. Leray, V. Lindenstruth, Z. Liu, U. Lynen, R. J. Meijer, U. Milkau, A. Moroni, W. F. Muller, C. Ngo, C. A. Ogilvie, J. Pochodzalla, G. Raciti, G. Rudolf, H. Sann, A. Schuttauf, W. Seidel, L. Stuttge, W. Trautmann, and A. Tucholski, Nucl. Phys. **A556**, 672(1993).

[68] J. Pochodzalla, S. Aiello, M. Begemann-Blaich, Th. Blaich, D. R. Bowman, R. J. Charity, A. Cosmo, A. Ferrero, C. K. Gelbke, W. C. Hsi, J. Hubele, G. Imme, I. Iori, J. Kempter, P. Kreutz, G. J. Kunde, W. D. Kunze, V. Lindenstruth, M. A. Lisa, W. G. Lynch, U. Lynen, M. Mang, L. G. Moretto, A. Moroni, W. F. Muller, M. Neumann, B. Ocker, C. A. Ogilvie, V. Pappalardo, G. F. Peaslee, G. Raciti, F. Rosenberger, T. Rubehn, H. Sann, R. Scardaoni, A. Schuttauf, W. Seidel, V. Serfling, L. G. Sobotka, J. Stroth, L. Stuttge, W. Trautmann, M. B. Tsang, A. Tucholski, C. Williams, A. Worner, E. Zude, and B. Zwieglinski, Nucl. Phys. **A583**, 553c(1995).

[69] J. Pochodzalla, T. Mohlenkamp, T. Rubehn, A. Schuttauf, A. Worner, E. Zude, M. Begemann-Blaich, Th. Blaich, H. Emling, A. Ferrero, C. Gross, G. Imme, I. Iori, G. J. Kunde, W. D. Kunze, V. Lindenstruth, U. Lynen, A. Moroni, W. F. Muller, B. Ocker, G. Raciti, H. Sann, C. Schwarz, W. Seidel, V. Serfling, J. Stroth, W. Trautmann, A. Trzcinski, A. Tucholski, G. Verde, and B. Zwieglinski, Phys. Rev. Lett. **75**, 1040(1995).

[70] D. H. Gross, Rep. Prog. Phys. **53**, 605(1990).

[71] J. P. Bondorf, A. S. Botvina, I. N. Mishustin, and S. R. Souza, Phys. Rev. Lett. **73**, 628(1994).

[72] X. Campi, Phys. Lett. **B208**, 351(1988).

[73] M. Ploszajczak and A. Tucholski, Phys. Rev. Lett. **65**, 1539(1990).

[74] F. S. Zhang, Z. Phys. **A356**, 163(1996).

[75] D. R. Bowman, G. F. Peaslee, R. T. de Souza, N. Carlin, C. K. Gelbke, W. G. Gong, Y. D. Kim, M. A. Lisa, W. G. Lynch, L. Phair, M. B. Tsang, C. Williams, N. Colonna, K. Hanold, M. A. McMahan, G. J. Wozniak, L. G. Moretto, and W. A. Friedman, Phys. Rev. Lett. **67**, 1527(1991).

[76] K. Hagel, M. Gonin, R. Wada, J. B. Natowitz, B. H. Sa, Y. Lou, M. Gui, D. Utley, G. Nebbia, D. Fabris, G. Prete, J. Ruiz, D. Drain, B. Chambon, B. Cheynis, D. Guinet, X. C. Hu, A. Demeyer, C. Pastor, A. Giorni, A. Lleres, P. Stassi, J. B. Viano, and P. Gonthier, Phys. Rev. Lett. **68**, 2141(1992).

[77] C. A. Ogilvie, J. C. Adloff, M. Begemann-Blaich, P. Bouissou, J. Hubele, G. Imme, I. Iori, P. Kreutz, G. J. Kunde, S. Leray, V. Lindenstruth, Z. Liu, U. Lynen, R. J. Meijer, U. Milkau, W. F. Muller, C. Ngo, J. Pochodzalla, G. Raciti, G. Rudolf, H. Sann, A. Schuttauf, W. Seidel, L. Stuttge, W. Trautmann, and A. Tucholski, Phys. Rev. Lett. **67**, 1214(1991).

[78] M. B. Tsang, W. C. Hsi, W. G. Lynch, D. R. Bowman, C. K. Gelbke, M. A. Lisa, G. F. Peaslee, G. J. Kunde, M. Begemann-Blaich, T. Hofmann, J. Hubele, J. Kempter, P. Kreutz, W. D. Kunze, V. Lindenstruth, U. Lynen, M. Mang, W. F. Muller, M. Neumann, B. Ocker, C. A. Ogilvie, J. Pochodzalla, F. Rosenberger, H. Sann, A. Schuttauf, V. Serfling, J. Stroth, W. Trautmann, A. Tucholski, A. Worner, E. Zude, B. Zwieglinski, S. Aiello, G. Imme, V. Pappalardo, G. Raciti, R. J. Charity, L. G. Sobotka, I. Iori, A. Moroni, R. Scardaoni, A. Ferrero, W. Seidel, Th. Blaich, L. Stuttge, A. Cosmo, W. A. Friedman, and G. Peilert, Phys. Rev. Lett. **71**, 1502(1993).

[79] G. J. Kunde, S. J. Gaff, C. K. Gelbke, T. Glasmacher, M. J. Huang, R. Lemmon, W. G. Lynch, L. Manduci, L. Matin, M. B. Tsang, W. A. Friedman, J. Dempsey, R. J. Charity, L. G. Sobotka, D. K. Agnihotri, B. Djerroud, W. U. Schroder, W. Skulski, J. Toke, and K. Wyrozebski, Phys. Rev. Lett. **77**, 2897(1996).

[80] W. A. Friedman, Phys. Rev. Lett. **60**,2125(1988)

[81] W. A. Friedman, Phys. Rev. **C42**, 667(1990).

[82] G. Kortemeyer, W. Bauer, and G. D. Kunde, Phys. Rev. **C55**, 2730(1997).

[83] W. Bauer, D. R. Dean, U. Mosel, and U. Post, Phys. Lett. **B150**, 53(1985).

[84] M. L. Miller et al., Phys. Rev. Lett. **82**, 1399(1999).

The Use of Coalescence Models to Probe Dynamic Evolution in Nuclear Collisions

J. Cibor[1], A. Bonasera[2], J.B. Natowitz, K. Hagel, R. Wada, M. Murray, T. Keutgen, M. Lunardon[3], N. Marie[4], R. Alfaro[5], W. Shen[6], Z. Majka[7] and P. Staszel[7]

Cyclotron Institute, Texas A & M University, College Station, Texas 77843 USA
[1]Present Address Institute of Nuclear Physics, ul. Radzikowskiego 152, 31-342 Krakow, Poland
[2]Permanent Address INFN, Laboratori Nazionali del Sud, Via S. Sofia 44, I-95123 Catania, Italy
[3]Present Address University Padua, Dipartimento Fisica, Padua, Italy
[4]Present Address Lab. Phys. Corp., 14050 Caen Cedex, France
[5]Present Address University of Indiana, Department of Chemistry, Bloomington, Indiana
[6]Permanent Address Chinese Acad. Sci., Shanghai Inst. Nucl. Res., Shanghai 201800, P.R. China
[7]Permanent Address Jagiellonian Univ. M. Smoluchowski Inst. Phys., PL-30059 Krakow, Poland

Abstract

Light particle emission occurs at essentially every stage of the evolution of composite systems produced in nuclear collisions at intermediate energies. Molecular dynamics calculations which are employed to model such collisions suggest that coalescence model analyses of the light particle emission may be used to probe both the time evolution of these systems and provide information on the degree of thermal, chemical and isospin equilibrium achieved at particular stages of this evolution. In this article the application of coalescence models for this purpose is explored within the framework of a classical molecular dynamics model. The application of techniques suggested by this study is then demonstrated in analyses of light particle emission observed in reactions between 47A MeV projectiles and medium mass targets. For those reactions densities and temperatures of the freeze-out configuration in multi-fragmenting systems are derived.

1 Introduction

In order to understand the properties of highly excited nuclei produced in heavy ion collisions it is very desirable to extract, directly from the experimental data if possible, information on the dynamical and thermodynamical evolution of the interaction region and the extent to which equilibration of various degrees of freedom, thermal, chemical, isospin, etc., is realized. Some time ago the Hanbury-Brown Twiss (HBT) technique, first used to measure the size of stars, was proposed as a means to study the space-time evolution of the collision volume. This technique is now well-established and has been applied in a wide range of studies [1, 2]. Light cluster production in heavy ion collisions also reflects the particle-particle correlations within the interaction region and the detection of a cluster can be viewed as a correlation measurement of its constituent particles in a bound state. Therefore, measurement of emission cross sections of nucleons and clusters of nucleons with suitable application of a coalescence ansatz can offer a complementary means to probe the properties and evolution of the interaction region. Questions of the degree of thermal and chemical equilibration and the isospin asymmetry dependence of the equation of state can be naturally addressed within the framework of coalescence models.

2 Coalescence Models

2.1 Historical Development

The coalescence model was originally introduced by Butler and Pearson [3] to describe deuteron production in proton-nucleon collisions. In their approach the deuteron was formed in a two-step mechanism involving the interactions of pairs of cascade nucleons and a collective mean field acting as a third body, necessary for energy and momentum conservation. A simple phenomenological extension of this model was developed by Schwartzshild and Zupancic [4] for composite particles in nucleon-nucleus collision and Gosset [5] for nucleus-nucleus collisions. Both of these models are based on the assumption that nucleons with relative momenta smaller than a certain value of a coalescence parameter, P_0, representing a radius in momentum space, can coalesce to form a cluster. The kinematical and dynamical correlations between produced clusters are neglected and the multi-particle distribution is a product of single particle distributions. P_0 may depend on cluster type.

The main prediction of this model is a power law relationship between complex clusters and nucleons. The cross section for emission of a light cluster of mass number A, containing Z protons and N neutrons is related to the cross section for emission of protons and neutrons at the same momentum per nucleon:

$$\frac{d^3N_A}{dp^3} = \frac{1}{N!Z!}\left(\frac{2s+1}{2^A}\right)\left(\frac{4\pi}{3}P_0^3\right)^{A-1}\left(\frac{d^3N_p}{dp^3}\right)^Z\left(\frac{d^3N_n}{dp^3}\right)^N \tag{1}$$

where P_0 represents the radius in the momentum space, and s is the cluster spin.

In practice the equation is often modified to take into account the fact that neutron cross sections are usually not measured in experiments, therefore an assumption is made that the neutron and proton momentum distributions are the same (except for a Coulomb shift) and their relative yields are given by the isospin of the combined system. The equation then becomes:

$$\frac{d^3N_A}{dp^3} = R_{np}^N\frac{1}{N!Z!}\left(\frac{2s+1}{2^A}\right)\left(\frac{4\pi}{3}P_0^3\right)^{A-1}\left(\frac{d^3N_p}{dp^3}\right)^A \tag{2}$$

The factor R_{np} is the ratio of neutron and proton numbers, usually taken to be those in the composite system formed from the projectile and target nuclei; *i.e.*, $R_{np} = (N_t + N_p)/(Z_t + Z_p)$.

There have been some comparisons of the source sizes derived from the coalescence formalism and those obtained from interferometry measurements (see [6, 7]). It is important to note, however, that a proper normalization of the cross sections is very important to a quantitative application of Equation (2). The multiplicities of particles employed should be those specifically determined for a particular class of reactions and not averages over a variety of different processes. Normalization using the total reaction cross section is an approximation which has limited earlier works.

Although originally created for high and relativistic energy collisions, the coalescence model has been applied to reactions over a wide range of energies, even to reactions with beam energy as low as 20 A MeV [8]. At lower beam energies, however, certain problems emerge. The most important one, which cannot be neglected, is the Coulomb repulsion between the source and the outgoing particle. To account for this Awes *et al.* [9] presented a Coulomb corrected coalescence model. In the laboratory frame their derived relationship between the differential cross section of the observed cluster and that of the proton is:

$$\frac{d^2N(Z,N,E_A)}{dE_A d\Omega} = R_{np}^N\frac{A^{-1}}{N!Z!}\left(\frac{\frac{4}{3}\pi P_0^3}{[2m^3(E-E_c)]^{\frac{1}{2}}}\right)^{A-1}\left(\frac{d^2N(1,0,E)}{dE d\Omega}\right)^A \tag{3}$$

R_{np}, A, Z, N are the same as in Equation 1, E_c is the Coulomb repulsion per charge unit between the source and outgoing particle and m is the nucleon mass. The double differential multiplicities $d^2N(Z,N,E_A)/dE_A d\Omega$ and $d^2N(1,0,E)/dEd\Omega$ for a cluster A and protons, respectively, should be taken at the energies corresponding to the same surface velocity; *i.e.*, the velocity before Coulomb acceleration. Thus $E_A = AE - NE_c$.

The coalescence model approach is semi-empirical because it relates cross sections to the initial distribution in the momentum space without explaining the processes that lead to this particular distribution. On the other hand the advantage of the coalescence model is its generality. Since it is based on phase space and statistical considerations only it does not introduce any a-priori assumptions about the reaction mechanism.

It should be noted that the results of the coalescence model, as well as of interferometry measurements, can be affected by contributions from statistical evaporation and also by correlations of particles in the emission zone (such as the existence of collective flow, for example). Under suitable circumstances the former effect can be minimized by carefully choosing the kinematic acceptance region. The latter effect presents a more difficult problem. The analytical models do not usually take dynamical correlations between particles in the emitting region into account.

Both correlations and relativistic effects have been accounted for in dynamic coalescence models, which have been recently applied at relativistic energies [10, 11]. In such treatments one can study nucleosynthesis numerically by applying a "coalescence afterburner" to the output of classical microscopic simulations or event generators such as cascade or RQMD. Scheibl *et al.* [12] have recently treated relativistic collisions and noted that in the presence of strong flow, the particles with given momentum are localized in the region of fireball where the flow velocity is close to the particle velocity. Hence only a fraction of the fireball will give particles with a given momentum. In this case the volume accessible by coalescence or HBT measurements is called the homogenity volume.

2.2 Quantitative Size Determinations

Various formulations of the coalescence model have been proposed to establish the relationship between the coalescence parameter P_0 and the size of the interaction volume. Sato and Yazaki [13] have presented a solution to this problem in the frame work of the density matrix formalism. In that model the momentum distributions of emitted particles are represented by the density matrices. The one-body density matrix and the spatial distributions of nucleons in the source are assumed to have Gaussian forms. The many body density matrix is a product of the one body density matrices and thus assumes that correlations between particles in the source may be neglected. An important feature of this approach is that thermal and chemical equilibrium are not required.

Sato and Yazaki showed that the coalescence volume is related to the internal wave function of the composite particle and the spatial distribution of the constituent particles in the emission region.

$$\frac{d^3N_A}{d^3P_A} = R_{np}^N A^{\frac{5}{2}} \frac{2s+1}{2^A} \left(\frac{\hbar}{m_0}\right)^{A-1} \left((1+\beta_T\nu_A)\frac{4\pi\nu_A\nu}{\nu_A+\nu}\right)^{\frac{3}{2}(A-1)} \left(\frac{d^3N_p}{d^3P}\right)^A \tag{4}$$

Here ν and ν_A, respectively, characterize the spatial extent of the emitter and emitted cluster wave functions (assumed to be Gaussian), s is the spin of the cluster and R_{np} is the neutron to proton ratio of the coalescence source. While this model does not assume thermal or chemical equilibrium it does contain a temperature-like parameter,

$$\beta_T = \hbar^2/(2m_0T) \tag{5}$$

which characterizes the momentum distribution of the contributing particles at the time of emission. The numerical values of ν_A are 0.20 fm^{-2} for deuterons, 0.36 fm^{-2} for tritons and ^{3}He and 0.58 fm^{-2} for alpha particles [13]. The equivalent sharp radius of the source is given by $R = \sqrt{5/2\nu}$. This approach was further generalized by Hirenzaki *et al.* [14] to include coalescence of an arbitrary number of constituent particles with different interaction ranges and masses.

The thermal coalescence model was introduced by Mekjian [15]. His model assumes that chemical equilibrium has been reached and the particles are emitted at the freeze-out density where the interactions stop. The momentum radius, P_0, can then be related to the volume of the thermal system at the freeze-out density in the following way:

$$V = \left(\left(\frac{Z!N!A^3}{2^A}\right)(2s+1)e^{\left(\frac{E_0}{T}\right)}\right)^{\frac{1}{(A-1)}} \frac{3h^3}{4\pi P_0^3} \tag{6}$$

Where Z, N and A are the same as in Equations 1 and 2, E_0 is the cluster binding energy and s the spin of the emitted cluster and T is the temperature of the system. If the source is assumed to be spherical its radius is given by

$$R = \left(\frac{3V}{4\pi}\right)^{\frac{1}{3}}.$$

For very high temperatures the factor $e^{E_0/T}$ is approximately equal to 1 and the relation between P_0 and V corresponds simply to what is expected from the phase space density. The inclusion of Planck's constant, h, indicates that the model is quasiclassical in nature.

It should be noted that in this equilibrium model the cluster yields are related to observed nucleon yields, in contrast to non-equilibrium coalescence models where they are related to primary nucleon yields.

3 Coalescence in Model Simulations

Although non-equilibrium coalescence models are based upon a sudden approximation while the thermal coalescence model is based upon an assumption of chemical equilibrium, the very similar formal structures of the various models [13, 14, 15, 16] suggest that they may provide a natural vehicle for following the time evolution of light composite particle emission from the first emission of such particles through freeze-out. To explore this possibility we have turned to molecular dynamics simulations. Such simulations allow us to elucidate the relationship between cluster yields and the properties of the emitting system in a straight-forward fashion.

In this section we report results of a study of the dynamical evolution of a finite system by the use of a classical molecular dynamics (CMD) approach [17, 18]. In this work our goal has not been to reproduce the data coming from nucleus-nucleus and fullerene-fullrene collisions, or other phenomena in finite quantum systems, but rather to provide insight into the information content of the coalescence model applied to an evolving system. Recall that CMD is exact at a classical level; *i.e.*, it contains all the correlations necessary to build finite systems and, at sufficiently high excitation energy, to follow their disassembly. Thus if we follow the dynamics for a long period of time we can obtain final fragment distributions and all of the properties of interest for the final fragments; *i.e.*, their kinetic energies, mass and charge numbers, etc. From the mass distributions we can calculate the corresponding P_0 and deduce the apparent size of the disassembling system. Most importantly we can test the basic ingredients of the models, such as the possible occurrence of chemical equilibrium, as assumed in thermal coalescence models [15].

In model simulations we can also study various features separately. For instance in the next section we consider the expansion of 100 particles in the absence of a Coulomb field. This case is discussed in detail in References [17] and [18]. In the section following that we then repeat the calculation with the Coulomb field turned on. In this way we clarify the role of the Coulomb force, which is of importance since, in many experiments, neutrons are not detected and it is usually assumed that the proton spectra are identical to the neutron spectra apart from some Coulomb shift. We will show that this shift is not the only effect to be considered. Finally in the last subsection we simulate a heavy ion collision and derive methods to analyze the "CMD-data".

3.1 Expansion of an uncharged system

The CMD model was used to simulate the disassembly of 100 nucleons (50 "protons" + 50 "neutrons") without a Coulomb field. In our model the spinless particles interact through Yukawa

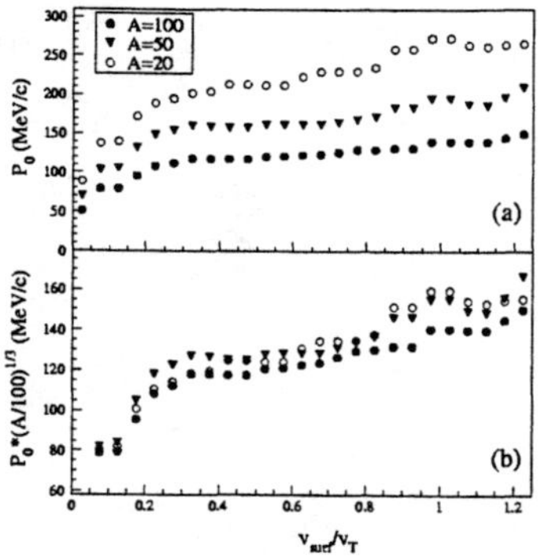

Figure 1: P_0 vs deuteron velocity (in units of a thermal velocity)(top) scaled with the source masses (bottom) at T=5MeV. Results are presented for A=20, 50 and 100.

two-body potentials. The potential between two identical particles is purely repulsive while for non-identical particles the potential is repulsive at short range and attractive at long range. The potentials have a cutoff at 5.4 fm. With this choice it is impossible to make dineutrons or diprotons. The ground state "nuclei" have a deuteron-like cluster structure. This is of course very different from real finite nuclei where, at most, some alpha structure can be found. Also the deuterons have $\sim$ 11 MeV binding energy. In the ground state the system is a solid. More details are given in [17, 18, 19].

To perform a calculation we initially prepare the nucleus in its ground state (a solid) and give to its particles some momenta corresponding to a Maxwell Boltzmann distribution at temperature T. In this way we know the initial source radius and excitation energy. Because of this initial T (or excitation energy E*) the system will expand and if T is high enough, it will undergo fragmentation. The evolution is followed for a long time (1000-2000 fm/c), and the distribution of final fragments, which are then well separated, are constructed. At each T one thousand or more events are generated. From the final fragment distribution we then calculate the temperatures and densities at the fragment formation time using the coalescence models discussed above. We stress that all the formulas employed for the coalescence model are perfectly valid for our classical system apart from some obvious modifications such as setting the spins of the particles equal to zero. All P_0 values determined here are calculated using observed final nucleon yields.

We first studied the dependence of P_0 on the total mass number of the system. We generated 100 events at T=5 MeV for masses A=100, 50 and 20. At T=5 MeV we have previously observed a very rapid collective expansion for such classical systems at this temperature. In this classical system the excitation energy per nucleon is proportional to $\frac{3}{2}T$, while in the nuclear system it is initially proportional to $\frac{T^2}{K}$ where K is in the range of 8-15. As a result, at a particular tempertature, the excitation energy per nucleon in the classical system is significantly higher than that of the nuclear system. In the classical system this 5 MeV temperature is above the critical temperature and the expansion is fairly explosive [17, 18, 19].

In Figure 1 we show (top) P_0 vs. velocity (in units of a thermal velocity $v_T = \sqrt{4T/m}$, and on the bottom, the same data scaled by the total mass (divided by 100) to the $\frac{1}{3}$ power. P_0 decreases as the mass increases. The scaling with the cube root of A is quite good and clearly demonstrates that P_0 gives direct information about the size of the system. This result follows from the Law of Mass Action and is very simple to understand using the coalescence formula. If we simply assume that the yield of a given particle is proportional to the total mass, A, of the emitter, taking the ratio of the yields for two different source sizes A_{T1},A_{T2}, gives:

$$\frac{Y_{T1}(Z,A,E_A)}{Y_{T2}(Z,A,E_A)} = \frac{A_{T1}}{A_{T2}} = (\frac{P_{0T1}}{P_{0T2}})^{A-1} \times (\frac{Y_{T1}(1,0,E_1)}{Y_{T2}(1,0,E_1)})^A = (\frac{P_{0T1}}{P_{0T2}})^{A-1} \times (\frac{A_{T1}}{A_{T2}})^A \tag{7}$$

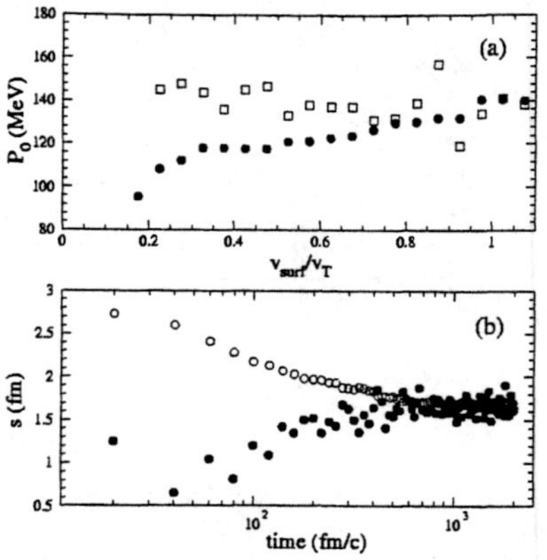

Figure 2: (Top) P_0 for deuterons at T=5 MeV calculated at two different expansion times 100 fm/c(open squares) and 1000 fm/c(filled circles). (Bottom) Average nucleon separation in nascent deuterons as a function of time. $E_d < 30$ MeV - (open circles) $E_d > 30$ MeV (filled circles).

From this equation one easily obtains:

$$P_{0T} \propto A_T^{-1/3}, \tag{8}$$

in agreement with the numerical findings displayed in Figure 1.

It is clear from Figure 1 that the value of P_0 depends on the particle velocity v, and is smaller at lower v. This indicates that, at the time when the less energetic particles are emitted, the system has expanded. Therefore different velocities correspond to different source sizes and times of emission. To further explore the relation between time of emission and velocity of the particle we have repeated the T=5 MeV calculation, stopping the calculation at 100 fm/c.

In Figure 2 (top) the calculated values of P_0 obtained from deuteron yields at 100 fm/c (open squares) are compared with those determined from the earlier calculation which extended to 1000 fm/c (filled circles). Notice that the P_0 values derived from the particles emitted in the very early stages of the reaction are constant (within statistical fluctuations). Also, for high v, P_0 is independent of the time when the calculations are stopped. This confirms that the most energetic particles dominate the early emission. Thus these particles can be used to probe the early stages of the dynamics.

Within the model we can also study the location of the particles in r-space before they are emitted. This gives us useful information regarding the possible establishment of thermal and/or chemical equilibrium. In the bottom of Figure 2 we plot, as a function of time, the average relative separation distances between protons and neutrons which eventually coalesce into a deuteron. In Figure 2 (bottom) the results are displayed for deuterons having kinetic energies less than 30 MeV (open circles) and those with energies larger than 30 MeV (filled circles). Since the initial temperature is 5 MeV particles having energies larger than 30 MeV are quite energetic.

In the figure we see that those very energetic particles have to be initially quite close in r-space in order to form a deuteron. In fact, on average their starting distance is less than 1 fm, which is smaller than the deuteron radius of about 1.5 fm. If two energetic particles are close both in r and p-space, they coalesce and form the deuteron. On the other hand, lower energy deuterons are made of particles that can be located quite far apart in r-space at time t=0. They wander inside the system and it is only at larger time that they find each other to form the deuteron. From this result it is quite clear that a statistical model might be applicable for emission of low energy particles but seems unjustified for particles of high energy. Within the CMD approach the possibility of chemical equilibrium as assumed by Reference [14] becomes more probable for the less energetic particles.

Values of P_0, derived from deuteron yields at various initial temperatures (T=2 (solid circles), 3 (solid inverted triangles), 5 (open circles), 10 (open squares) and 20 MeV (stars)) are plotted in Figure 3. When the velocities are scaled by the thermal velocity the shapes of the curves for

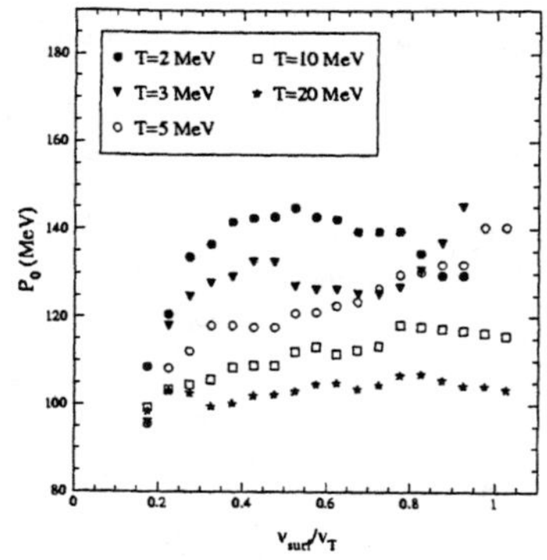

Figure 3: P_0 for the d-clusters vs. normalized velocity at T=2,3,5,10,20 MeV.

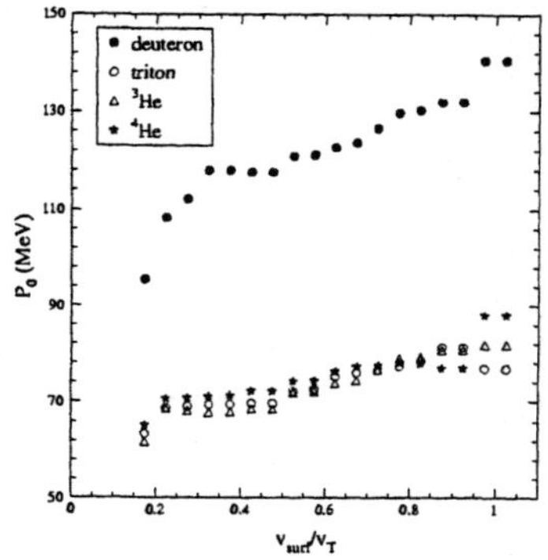

Figure 4: P_0 vs. normalized velocity at T=5 MeV for d,t, ^{3}He and α clusters.

different temperatures show qualitatively similar variations. We note that for high velocity particles P_0 decreases as T increases indicating greater expansion at early times for the higher temperatures.

The models do indicate that P_0 will vary with binding energy and spatial extent of the clusters. In Figure 4 we plot the calculated P_0 at T=5 MeV for deuterons (filled circles), tritons (open circles), ^{3}He (open triangles) and alpha clusters (stars). There is quite a substantial difference between the value derived for deuterons and those for the other clusters reflecting the very strong binding of deuterons in this model.

Finally in Figure 5, we plot the radii derived from P_0 in the two approaches at various temperature. Notice the large difference between the Sato-Yazaki and Mekjian model results. In addition to reflecting the cluster size correction contained in the density matrix model, this difference arises from the different temperature dependencies in the two models. More specifically it arises from the $e^{-BE/T}$ term in the thermal model approach, all the other terms being equivalent in the two formalisms. Since in CMD the clusters are quite strongly bound as compared to real nuclei, the BE term becomes very important, resulting in the large discrepancy at small T. The two approaches give closer results at high T where the BE term can be neglected. The very large values derived from the thermal model at low T argue against the assumption of chemical equilibrium for the case considered.

3.2 Expansion of a Charged System

To delineate the role of the Coulomb force in the coalescence process we have repeated the calculations discussed in the previous section but with the Coulomb field turned on.

In the coalescence model formulated by Awes *et al.* [9] the Coulomb field is taken into account

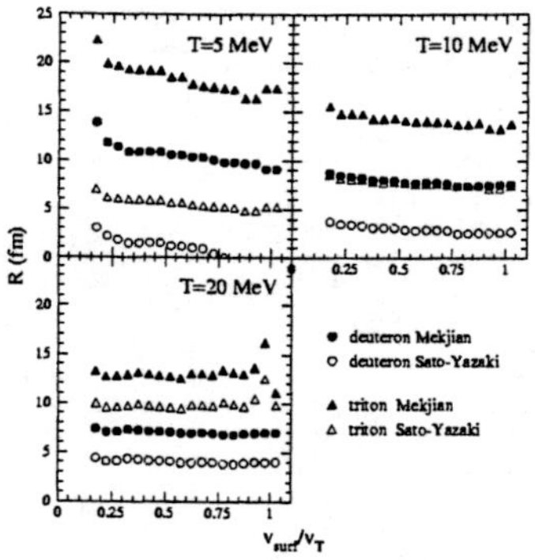

Figure 5: Calculated radii for d and t clusters at T=5,10 and 20 MeV. Circles are for d clusters and triangles for t clusters. Filled symbols for Mekjian model; open symbols for Sato-Yazaki model.

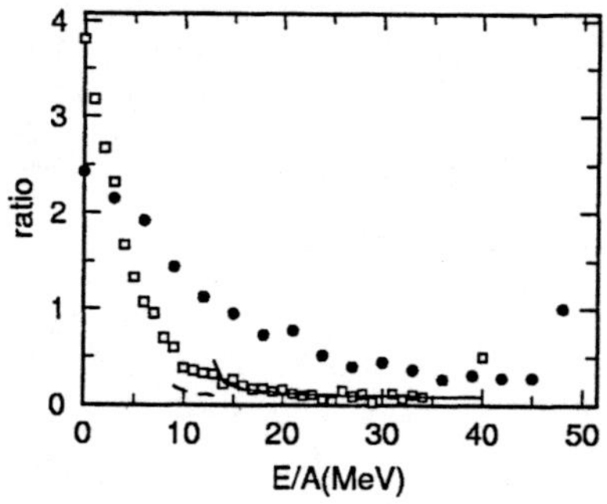

Figure 6: t/^{3}He ratio (filled circles) and n/p ratio (open squares) as a function of energy per nucleon. Lines are calculation with Equation (9) (see text).

by assuming that the proton spectrum is shifted upward by an energy E_c, (which is the Coulomb potential of a sphere of charge Z_T and mass A_T.) relative to the neutron spectrum (which in most experiments is not measured). However, a simple shift of the kinetic energy is not enough to explain the difference between protons and neutrons. Due to the Coulomb charge some protons cannot overcome the barrier and remain trapped in the system. This effect can be estimated in the Weisskopf model as in Reference [20] where the ratio of yields of clusters having the same mass number A, but charges differing by one unit ($Z_1 - Z_2 = 1$) is given by:

$$R_{Z_1/Z_2} = \frac{1 - Z_1 E_c/E_A}{1 - Z_2 E_c/E_A} e^{-E_c/T} \frac{Z_T}{N_T} \tag{9}$$

when the two clusters are detected at the same final kinetic energy $E_A = AE_n = AE_p$ Here E_p is the proton kinetic energy. The last factor in Equation (9) takes into account the neutron to proton ratio in the cluster source.

In Figure 6 we plot the CMD calculated ratio of t/^{3}He (filled circles), and n/p (open squares) at initial T=5 MeV. The corresponding theoretical predictions of Equation (9), are given by the dashed and full lines respectively, with no free parameters adjusted. The agreement is very good for the n/p ratio but not for the heavier clusters. This suggests that some collective effects are becoming important.

Another way to look at the Coulomb effect is to shift the particle velocity by the Coulomb potential and divide the yield by the absorption factor discussed above. This is done in Figure 7 for the various ejectiles at T=5 MeV. Notice that the corrected deuteron yield (filled circles) is the highest yield, consistent with the strong deuteron binding and deuteron cluster-like structure our

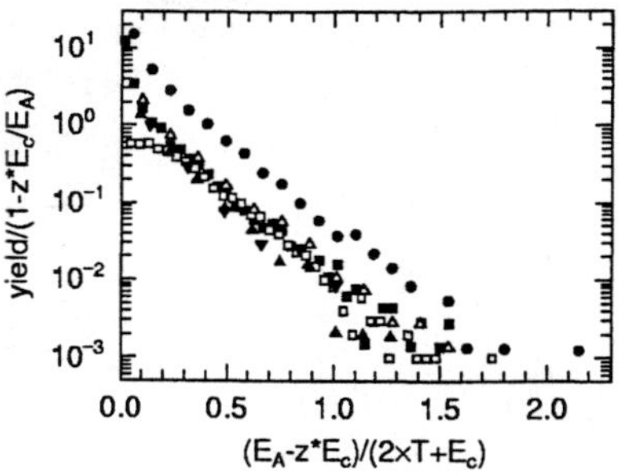

Figure 7: Scaled cluster yield vs. scaled particle velocity for neutrons (open squares), protons (filled squares), deuterons (filled circles), tritons (open triangles) and α-particles (filled inverted triangles).

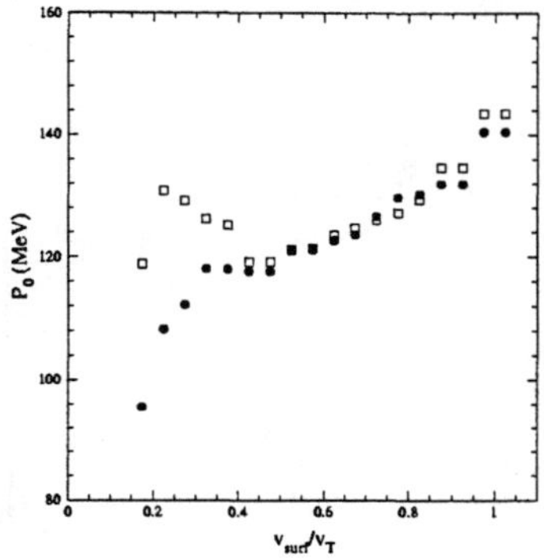

Figure 8: P_0 vs. scaled surface velocity at T=5 MeV. Filled circles-Equation (3); open squares-Equation (11).

"nucleus". The neutron (open squares) and proton yields (filled squares) scale quite well, while a little violation of the scaling is seen for the t (open triangles) and ^{3}He (filled triangles). Notice that the ^{4}He yield (filled inverted triangles) has the same slope as ^{3}He which hints at some collective effect due to the Coulomb force.

From Equation (3) we can write the neutron yield in terms of the proton yield:

$$\frac{d^2N(0,1,E_n)}{dEd\Omega} = R_{n/p}\frac{d^2N(1,0,E_p)}{dEd\Omega}\frac{N_T}{Z_T} \tag{10}$$

Inserting this equation in the coalescence formula we get:

$$\frac{d^2N(Z,N,E_A)}{dE_A d\Omega} = \frac{A^{-1}}{N!Z!}(\frac{4\pi}{3}P_0^3)^{A-1}(\frac{N_T}{Z_T})^N(\frac{e^{-E_c/T}}{1-E_c/E_p})^N/(\sqrt{2m^3E_p})^{A-1}(\frac{d^2N(1,0,E_p)}{dEd\Omega})^A \tag{11}$$

We stress that, at variance with the Awes Equation (9), the energies in Equation (5) are at infinity. In Figure 8, we plot the result for P_0 for deuterons at T=5 MeV (compare to Figure 2). The result obtained with Equation (11) is given by the open squares, while those obtained with Equation (3) are represented by filled circles. There is a decrease of the P_0 at low velocity in the latter case which would translate into a larger source size.

3.3 CMD Simulations of Collisions

Since many macroscopic models assume some form of equilibrium to derive mass yields and energy distributions of particle, the question of whether thermal and/or chemical equilibrium are achieved in a collision is of particular interest. In this section we address heavy ion collisions in the CMD approach and discuss some modifications to the picture of an expanding system which are seen in these collisions.

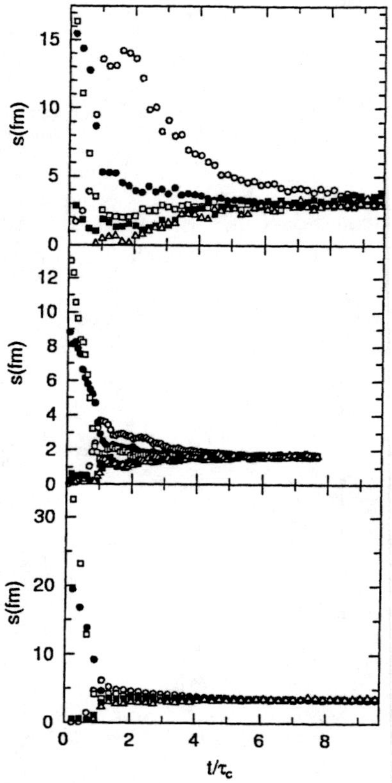

Figure 9: Average nucleon separation for nascent deuterons vs. scaled time (time/time to contact) results are for E/A=20 MeV nucleon (top), E/A=47 MeV nucleon (middle) and E/A=100 MeV nucleon (bottom).

In this context the method discussed in the previous section, following the time evolution of the average separation distances of nucleons which eventually form deuterons, can be equally as well applied to collisions. We discuss CMD simulated Zn+Y collisions at b=1 fm and beam energies E/A=20, 47 and 100 MeV/nucleon. Initially (t=0) the two nuclei were at a relative distance of 22 fm. In Figure 9 we plot, for each incident energy, the average separation distance of nucleons of the nascent deuteron vs. time (in units of a projectile-target crossing time $\tau_c = 2(R_p + R_t)/\sqrt{2(E/A)/m}$). The various symbols correspond to six deuteron kinetic energy bins of width $(E/A)/4$ in the nucleus-nucleus Center of Mass frame starting at E=0 MeV. From the lowest energy bin to the highest energy bin the symbols, in order, are (open circles-first bin, filled circles-second bin, open squares-third bin, filled squares-fourth bin and open triangles-fifth bin).

At the lowest energy, 20 MeV/A, (top) deuterons having kinetic energies per particle less than 15 MeV (first three bins) are seen to be made with contributions both from projectile and target nucleons. For the lowest energy bin the radius becomes constant at a time which is about 800 fm/c. This implies that the constituent particles of the deuteron move in the compound system for a long time before being emitted. The other two cases reach saturation at times of about 400 and 200 fm/c respectively. Assuming that, at the lowest beam energy, a compound system is formed with excitation energy about 4.9 MeV;

(*i.e.*, all the beam energy has been converted into excitation energy), we would expect an average kinetic energy for the deuterons (after Coulomb acceleration) of about 18 MeV. This would imply that deuterons whose kinetic energies are less than 18 MeV can be formed from nucleons coming from either of the colliding nuclei. This is in fact found to be the case; deuterons having an energy

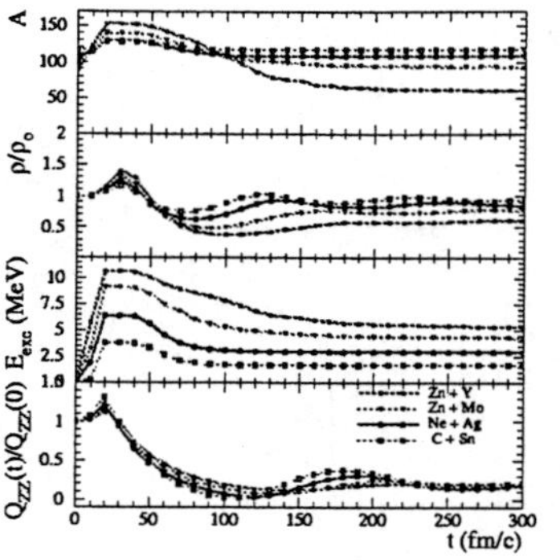

Figure 10: CHIMERA QMD model calculations of the time evolution of the properties of the largest identifiable fragments produced in reactions induced by 47A MeV projectiles. From top to bottom: The mass number, the normalized density, the excitation energy per nucleon and the normalized second moment of the momentum distribution, a measure of the degree of thermal equilibration. The calculations are for an impact parameter range of 0-3 fermis.

less than 30 MeV are initially at average separation distances larger than the sizes of the entrance channel nuclei. The most energetic deuterons, on the other hand, consist of pairs of nucleons which initially reside in the same nucleus. Thus energetic deuterons are pre-equilibrium particles. Chemical and thermal equilibrium might be achieved in this framework but only for excitation energies of 10 MeV/nucleon or less which is roughly the binding energy of the system.

At higher beam energies the deuterons of lowest kinetic energy come either from the projectile or from the target. Apparently no compound system is formed, which implies that the compound system cannot hold the ~11 MeV available energy (middle panel) nor, of course, the 25 MeV available energy (lowest panel). Deuterons whose constituent particles were initially at large distance (*i.e.*, of mixed origin) are now found in the second and third bins only, which implies that in some events the projectile nucleons capture part of the target and vice versa.

With the results of these CMD simulations in mind, we turn, in the next section, to an exploration of the utility of the coalescence model to probe dynamic evolution in intermediate energy collisions.

4 Application to Intermediate Energy Nuclear Collisions

To make a quantitative evaluation of the utility of coalescence model techniques to follow the dynamics of expanding systems we recently used the combined TAMU CsI Ball-Neutron Ball detection system to detect light charged particles, fragments and neutrons emitted in the reactions of ^{12}C + ^{116}Sn, ^{22}Ne + Ag, ^{40}Ar + ^{100}Mo and ^{64}Zn + ^{89}Y, all at 47A MeV projectile energy [21, 22, 23]. In QMD transport model calculations with the code, CHIMERA [24], the particular set of target and projectile combinations used in our experiment are predicted to lead, after pre-equilibrium emission of particles, to excited composite nuclei of very similar mass but having excitation energies and degrees of expansion which increase with increasing projectile mass. Violent events corresponding to the 10% of the reaction cross section having the highest charged particle and neutron multiplicities were selected for detailed coalescence model analyses. For such events multifragment emission was observed to become increasingly more probable as the projectile mass increases. Greater detail on these measurements is provided in References [22] and [23].

4.1 Correlation Between Kinetic Energy and Time

We have carried out analyses of nucleon and light cluster emission as a function of particle velocity, basing our approach on coalescence model techniques [9, 13, 14, 15, 16]. The possible utility of applying the coalescence approach as a function of ejectile velocity to follow the dynamics of an

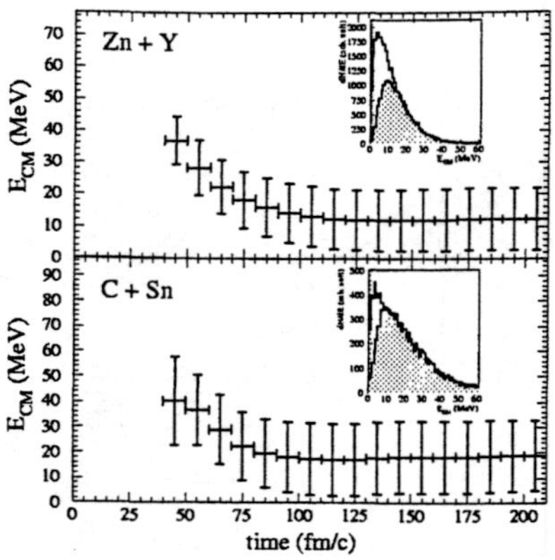

Figure 11: CHIMERA QMD model calculations of the time evolution of the average kinetic energies of protons emitted in the 0-3 fermi impact parameter collisions of $^{64}Zn+^{89}Y$ (top) and $^{12}C+^{116}Sn$ (bottom) at 47A MeV. The inserts show the spectral distributions at 80 and 120 fm/c.

expanding system is already suggested by the discussion of the classical molecular dynamics calculations for expanding systems, presented in the previous section and is reinforced by results of simulations of the reactions we have studied, carried out with the QMD model code, CHIMERA [24] (see Figure 10). In the calculations for 47A MeV projectiles, the first light particles are emitted at ~50 fm/c after contact. These and subsequent pre-equilibrium particles remove significant amounts of both mass and energy from the expanding composite nuclei. For each reaction the hot composite reaches its minimum average density close to 100 fm/c. At that point thermal equilibrium appears to be established. Over the time span from first emission to thermalization there is a monotonic decrease of the average kinetic energies of the emitted particles (Figure 11). A strong correlation between energy and emission time has, in fact, been clearly demonstrated in reactions induced by ^{40}Ar projectile energies of 25 [25] and 34 MeV/u [26] where light particle correlation measurements were employed to determine the mean times for emission of hydrogen ejectiles as a function of particle velocity. Thus both theoretical models and experiments suggest that the relationship between emission time and ejectile kinetic energy may be exploited to follow the time evolution of the system.

4.2 Preliminary System Characterization - Emission Sources

A common technique to characterize light particle emission in this energy range has been to fit the observed spectra assuming contributions from three sources, a projectile-like (PLF) source, an intermediate velocity (NN) source and a target-like (TLF) source [27, 28]. Such a source fit was employed to estimate the multiplicities and energy emission at each stage of the reaction. We emphasize that the event selection is on the most violent and presumably more central collisions. In the fitting process, which assumes isotopic emission and a Maxwellian spectral shape in the particular source frame considered, accounting for forward emitted particles with projectile-like velocities requires the PLF source. We consider these particles to be of pre-equilibrium emission origin and not to be evaporated from a fragment. From the multiplicities of emitted species associated with each source the masses and excitation energies of the hot nuclei which remain after the early (PLF and NN) emission were determined. The mass numbers, obtained by subtracting the mass removed by projectile-source and intermediate source particles from the total entrance channel mass are ~110. The excitation energies, determined using calorimetric techniques to evaluate the excitation of the TLF source, increased with projectile mass from 2.6 MeV/u for $^{12}C+$ ^{116}Sn to 6.9 MeV/u for ^{64}Zn + ^{89}Y. While the source fits establish a qualitative or semi-quantitative picture of ejectile sources and system evolution following the time evolution of a continuously evolving system in more detail than has previously been attempted requires a more sophisticated approach.

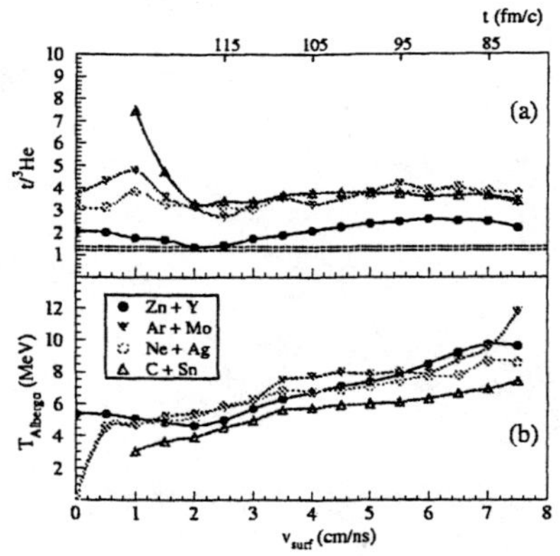

Figure 12: Experimental ratios of ^{3}H to ^{3}He emission yields (top) and double isotope yield ratio temperatures (bottom) as a function of Coulomb-corrected surface velocity. Data below ~4 cm/ns may have residual contributions from statistical evaporation. The horizontal bar in the top portion indicates the range of composite nucleus N/Z values for the systems studied. Time scales derived from CHIMERA QMD model calculations are indicated at the top of the figure.

4.3 Determination of the Coalescence Parameter, P_0

As indicated in the preceding sections the coalescence model analysis is based upon a determination of the coalescence parameter, P_0. Because the goal was to derive information on the time evolution of the emitting system, our analysis was not limited to determining average P_0 values appropriate to the higher energy portions of the particle spectra, as is common in previous work. Instead, for d, t, ^{3}He and ^{4}He, P_0 was calculated as a function of V_{surf}, the velocity of the particle at the nuclear surface prior to Coulomb acceleration, using the Coulomb corrected coalescence model formalism of Awes *et al.*, [9] Equation (3).

To minimize contributions from secondary decays our analysis was

4.4 N/Z Ratios

In Equation (3), N_p, N_t, Z_p and Z_t enter the equation because measurements which include neutron information are relatively rare. It has typically been assumed, in most coalescence model analyses, that the neutron energy spectra are identical in shape to the Coulomb corrected proton spectra and that the neutron yields are simply N/Z times the proton yields, where N/Z is the neutron to proton ratio in the composite system. In this work, also, the neutron spectra are not measured. However, since within the framework of the coalescence model the yield ratios of two isotopes which differ by one neutron are essentially determined by the effective N/Z ratio in the coalescence volume. We have used values derived directly from the observed triton to ^{3}He yield ratio to determine the N/Z ratio used in this analysis. This use of this "effective" N/Z ratio is a self-consistent approach but may mean that some actual differences in neutron and proton spectra are absorbed into this ratio (see discussion regarding Equation 9). In Figure 12 (top), observed values of the t/^{3}He ratio are presented as a function of V_{surf}. These ratios are significantly higher than the N/Z ratios in the composite systems. We understand these to be the ratios of "free nucleons" [30] which participate in the coalescence. This is consistent with results obtained by Albergo *et al.* [30] who often deduce significant free neutron excesses based on integrated yields observed in a variety of early intermediate energy experiments.

Figure 13 shows the results of an analysis of the resultant ejectile spectra (Experiment - (PLF and TLF)) for all four systems studied. As seen in Figure 13, at high V_{surf} the four systems have similar limiting values of P_0 for a given cluster species, supporting the idea of a similar emission mechanism and a similar source size at the time of emission of the higher energy particles in the different reactions studied. The observed decreases of P_0 with decreasing surface velocity indicate changes in the emitting system.

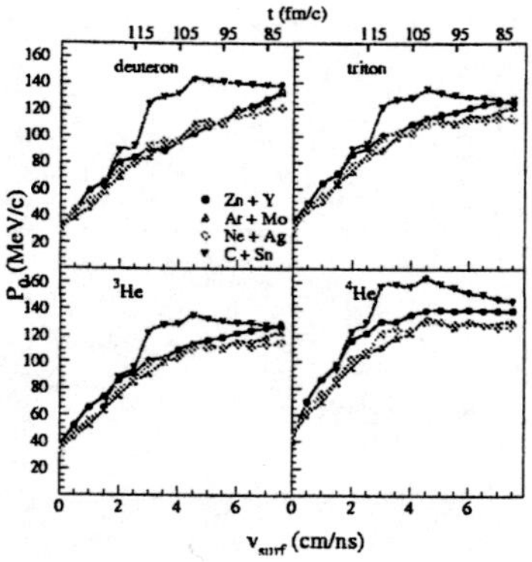

Figure 13: Corrected values of the coalescence parameter, P_0, as a function of surface velocity for p, d, t, ^{3}He and ^{4}He clusters emitted in the four reactions studied. The contributions of the ejectile spectra attributed to the TLF source have been subtracted from the experimental spectra at Lab=38-52 degrees. Time scales derived from CHIMERA QMD model calculations are indicated at the top of the figure.

The time scale presented at the top of Figures 12-15 is derived from the QMD model and indicates that a value of V_{surf} of 4 cm/ns is expected to sample the system near 105 fm/c. Below 4 cm/ns residual contributions from late stage evaporative decay of the target-like source or secondary decay from light fragments [31], not removed by the fitting process, may still contribute.

4.5 Size Evolution of the Emitting System

For each model, deriving the size of the system from P_0 also requires knowledge of the temperature. To characterize the temperature at a particular emission time we have employed double isotope yield ratios. For a system at chemical and thermal equilibrium at a suitably low density, Albergo *et al.* [30] have shown that the temperature of the emitting system can be derived directly from the first chance emission double isotope yield ratios of two adjacent isotopes of two different elements. In a more recent work by Kolomiets *et al.* [32], essentially the same result is derived when only thermal equilibrium is assumed. If the particle energies are well correlated with emission time, and secondary emission contributions contribute primarily at the lower energies, derivations of double isotope yield ratio temperatures as a function of particle energy may be relatively uncontaminated by secondary emission processes, except at the lower energies. On the other hand, it should be clearly noted the apparent temperature derived for the earliest stage, while indicative of the particle momentum distribution at that emission time, is not the temperature since the dynamic transport calculations indicate that the condition of thermal equilibrium is established only after some particle emission occurs.

4.6 Temperature

We have derived the double isotope yield ratio temperature, T_{HHe}, as a function of, V_{surf} from the yields of d, t, ^{3}He and ^{4}He particles, again corrected by subtracting the contributions associated with the TLF. The derived temperatures, presented as a function of V_{surf} in the bottom of Figure 12, increase slowly with projectile mass and decrease with decreasing V_{surf}. The QMD calculations suggest that the system equilibrates rapidly but global thermal equilibrium is not completely established when the first particles are emitted. Thus the temperatures at high V_{surf} should be considered only as approximations.

At the lowest values of V_{surf}, values of T_{HHe} in the 4-5 MeV range, similar to those spectral integrated values seen at comparable excited energies in other experiments [33, 34, 35] are observed. Indeed the T_{HHe} values observed in this lower energy region are very similar to those calculated using the sequential evaporation code GEMINI [36]. Since spectra at these lower velocities may still

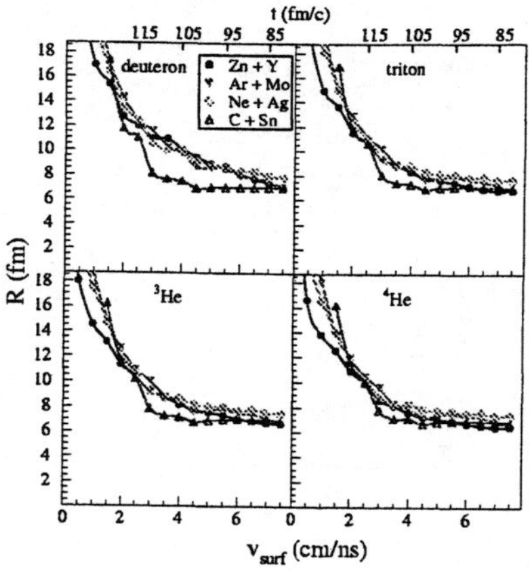

Figure 14: Equivalent sharp radii derived from the Mekjian model. Radii for assumed spherical sources are presented as a function of surface velocity for d, t, ^{3}He and ^{4}He clusters. The values of P_0 employed are those of Figure 14. Time scales derived from CHIMERA QMD model calculations are indicated at the top of the figure.

contain a contribution from late stage evaporation we do not attempt to extract emission system sizes for $V_{surf} < 4$ cm/ns.

4.7 System Sizes

To extract nuclear size information from the P_0 and T determinations, both the density matrix formalism of Sato and Yazaki [13] and the thermal model formalism of Mekjian [15] were employed. Recall that in the Sato-Yazaki model (Equation 4) formulation of the coalescence model, P_0 depends both on the size of the emitter and on that of the emitted cluster. For this evaluation the temperature characterizing the emission spectrum was set equal to the instantaneous Albergo temperature at the corresponding surface velocity.

In the thermal coalescence model of Mekjian [15] thermal and chemical equilibrium are assumed and, as indicated in Section II.B., the model leads to a direct relationship between the volume in momentum space and the radius of the emitting system (Equation 6). Note that the Mekjian model and the model proposed by Albergo *et al.* to derive double isotope ratio temperatures [30] are, in fact, equivalent and the assumed validity of this model incorporating chemical equilibrium is implicit in all recent works which use double isotope yield ratios to determine temperatures [33, 34, 35].

For both models, we present in Figures 14 and 15, equivalent sharp cut-off radii for assumed spherical sources as derived from the P_0 values presented in Figure 13. For each model, the sizes derived from the highest velocity particles show little dependence on projectile type and only for the lower energy ejectiles is a difference seen. The absolute values of the radii are larger for deuterons and smaller for alpha particles, possibly reflecting the very different binding and spatial extent of these clusters [13]. Such differences have been indicated in previous coalescence model studies [9, 13, 15, 16]. That the differences persist in the density matrix model which attempts to take the cluster size into account is interesting and, if it does not result from simplifying assumptions implicit in the model, may imply some different freeze-out densities required for survival of different cluster species.

The size parameters are clearly different, reflecting differences in the models. The S-Y model equivalent sharp radii are smaller than the radii of normal density nuclei in this mass range. This is true even if no correction is made for cluster size and apparently reflects the particular analytical formulation of the Sato-Yazaki model [13]. The values of thermal model equivalent sharp radii at the highest values of V_{surf} are slightly larger than those expected for the composite nuclei at normal density. At V_{surf} = 7 cm/ns they correspond approximately to R=1.3$A^{\frac{1}{3}}$ where A is the total entrance channel mass.

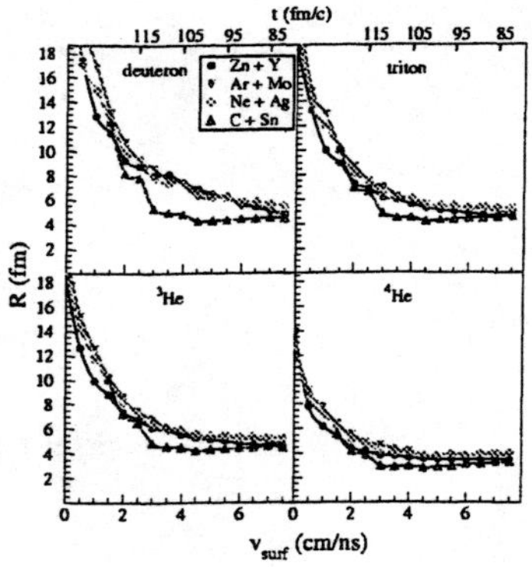

Figure 15: Equivalent sharp radii derived from the Sato-Yazaki model. Radii for assumed spherical sources are presented as a function of surface velocity for d, t, ^{3}He and ^{4}He clusters. The values of P_0 employed are those of Figure 14. Time scales derived from CHIMERA QMD model calculations are indicated at the top of the figure.

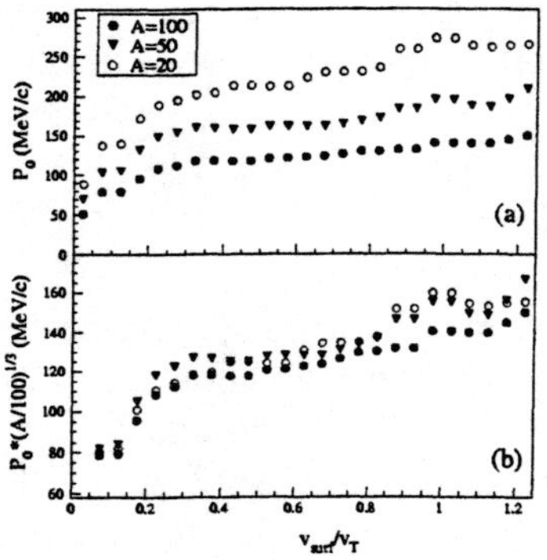

Figure 16: P_0 values, radii and relative radii at V_{surf}=4 and 7 cm/ns derived from the coalescence model analysis. In all four systems studied, Part (a) P_0 values derived from triton data; Part (b) radii derived from triton data; Part (c) the change in relative triton and ^{3}He radii observed in the Mekjian model; Part (d) the change in relative triton and ^{3}He radii observed in the Sato-Yazaki model. Dashed lines show the average changes for all four clusters.

Although subject to experimental uncertainties the thermal model results suggest that, at the time the early emitted particles leave the system, the size of the emitting nucleus is above normal. This is in qualitative agreement with the predictions of the CHIMERA QMD [24] calculations for the early evolution of the system. As seen in Figure 10, in the calculation the densities of the largest masses identifiable pass through normal density at a time ~50 fm/c, as the very first particles are emitted and are less than normal when subsequent particles leave.

For ^{12}C induced reactions the derived radii for the different particles indicate very little change of size during the particle emission phase. Progressively larger increases of the radii during particle emission are indicated for reactions with the heavier projectiles. This is indicated for triton emission in Figure 16. The observed variation of P_0 with system at V_{surf} = 7 and 4 cm/ns, depicted in the top of the figure, leads to the variation of the radii shown in the center of the figure.

While the absolute values of the derived radii are different for the two models, for a given ejectile, the ratio $R(V_{surf} = 4\text{ cm/ns})/R(V_{surf} = 7\text{ cm/ns})$, a measure of the relative radius increase, is found to increase with projectile mass in a very similar fashion in the two models. This is shown in the bottom of Figure 16 for t and ^{3}He emission. Although the relative variation observed is somewhat larger for deuteron emission and somewhat lower for alpha emission, the relative variations in radius

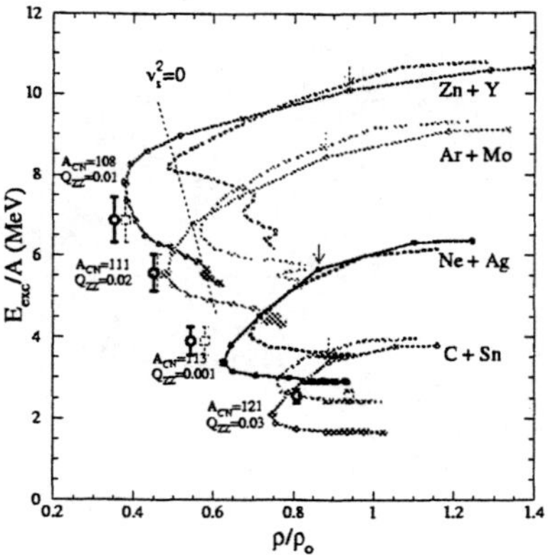

Figure 17: Excitation energy-density values at freeze-out. The symbols represent the excitation energy-density values (open circles-Mekjian model, open squares-Sato-Yazaki model) derived from coalescence model analyses of the light cluster emission. They are compared to CHIMERA QMD model trajectories in the excitation energy per nucleon-normalized density plane calculated for central collisions in the four different systems studied. Calculations for a soft, K=200 MeV, equation-of-state are represented by solid lines. Calculations for a hard, K=380 MeV, equation-of-state are represented by thick dashed lines. The trajectories start at the time of maximum density. The small dots mark time increments of 10 fm/c. Arrows indicate the time of first emission of particles (near 50 fm/c after contact). Both times and Q_{zz} values are indicated at the minimum calculated densities (large solid dots). To the left of the dashed line, ${V_s}^2 = 0$, is the spinodal region.

as a function of V_{surf}, averaged over the four different ejectiles, are quite similar to those observed for t and ^{3}He emission. This is illustrated by the dashed lines in the bottom of Figure 16. Since the derived absolute radii can be subject to systematic uncertainties both in the measurements and in the model assumptions, we choose, in the following section, to derive densities at freeze-out from the relative changes averaged over the four particles.

4.8 Densities

To determine the average density associated with the system when V_{surf} = 4 cm/ns. We first assumed that the highest velocity particles are emitted from an object of mass equal to the sum of the masses of the target and projectile nuclei and of density equal to 0.90 P_0 (see Figure 10). We then used the relative values of the radii derived from the t, ^{3}He and ^{4}He data at 4 and 7 cm/ns and took the mass of the primary emitter at V_{surf} = 4cm/ns to be the mass remaining in the target-like source. From the thermal model we find that the average densities sampled at V_{surf} = 4cm/ns are 0.81 ρ_0, 0.54 ρ_0, 0.45 ρ_0 and 0.36 ρ_0 for the ^{12}C, ^{22}Ne, ^{40}Ar and ^{64}Zn induced reactions, respectively, with uncertainties of $\pm$ 20% of these values. The corresponding values obtained using the Sato-Yazaki model are 0.94 ρ_0, 0.58 ρ_0, 0.45 ρ_0 and 0.38 ρ_0.

Our energy-density results are presented in Figure 17. There we note that they are in very reasonable agreement with results of the CHIMERA QMD calculation when an equation of state with K=200 MeV is employed. For this soft equation of state the calculations indicate entry into the spinodal region to the left of the dashed line, ${V_s}^2 = 0$. A harder equation-of-state with K=380 MeV results in less expansion and poorer agreement with the experimental results. Refined measurements and coalescence analyses, including determinations of the impact parameter dependence of the cluster production might lead to a more precise determination of the incompressibility.

4.9 Caloric Curve

We present in Figure 18 the double isotope yield ratio temperatures at 4.0 cm/ns plotted against excitation energy. The results indicate a nearly flat caloric curve with T~7MeV at excitation energies

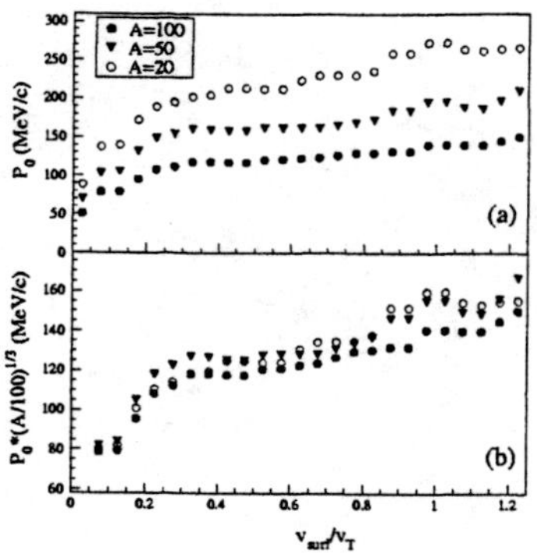

Figure 18: Caloric curve for nuclei with $A\sim110$. Double isotope yield ratio temperatures derived in the present work are combined with results reported previously by Wada *et al.* obtained with a different technique [37]. Dashed lines indicate trends of a Fermi gas model calculation with two different choices of level density parameter.

from 3.5 to 7 MeV per nucleon. At the higher excitation energies, the double isotope yield ratio temperatures near 7 MeV indicated for the expanded low density systems isolated here are consistent with the limit suggested in our earlier work on the caloric curve for $A\sim125$ nuclei in which we found a temperature of 6.8 ± 0.5 MeV at 4.3 MeV/u excitation energy [37]. The general shape of the caloric curve in Figure 18 can then be understood as reflecting first, at lower excitations, primarily the washing out of shell effects and collectivity [38, 39] and later, at higher energies, the expansion of the system.

Both classical and quantum molecular dynamics calculations lead to caloric curves similar to that observed here [40, 41]. If a temperature limit of thermally equilibrated nuclei is reached, these calculations suggest that the system clusters and the nucleons with high kinetic energies stream out of the expanding system, creating a natural limit to the momentum distribution and to the excitation energy of the remaining nucleus over a wide transitional region [40, 41]. This is also suggested by the saturation of the Lyapunov experiment in CMD studies [17]. The evolution of the volume of the system is very important in determining the caloric curve [40, 41, 42]. Observed differences in caloric curves extracted from different reaction systems may reflect the particular dynamic evolution of the system being studied and great care must be taken to understand this dynamics.

5 Summary and Conclusions

The work discussed here indicates that coalescence model studies of light cluster emission can indeed be used in the intermediate energy regime to follow the dynamic evolution of excited systems. Information on the space-time evolution of the system complementary to that contained in HBT measurements can be obtained in a relatively simple manner. It would be interesting to make a direct comparison of the two techniques for some well-chosen cases. It would also be interesting to extend the application of the present coalescence techniques to the study of IMF emission in detailed experiments including both Z and A identification of the IMF. By clearly establishing the relative importance of different mechanisms (and different time-scales) of IMF emission it should be possible to establish the degree to which pre-existing correlations are preserved, reflecting transparency in the collisions [43, 44, 45]. This would provide clearer evidence of the degree of thermal and chemical equilibration attained and may prove to be a powerful tool to probe the isospin asymmetry dependence of the equation of state and the extent to which different species can be said to originate from either gaseous or liquid phases which are present.

6 Acknowledgements

This work was supported by The Robert A. Welch Foundation, the United States Department of Energy (Grant No. DE-FE05-86ER40256) and the Polish Scientific Research Committee (Grant No. 2 PO3B 103 12).

References

[1] W. Bauer *et al.*, Ann. Rev. Nucl. and Part. Sci. **42**, 77 (1992).

[2] D. Ardouin, Int. J. Mod. Phys. E **6**, 391 (1997).

[3] S.T. Butler and C.A. Pearson, Phys. Rev. **129**, 836 (1963).

[4] A. Schwartzschild and C. Zupancic, Phys. Rev. **129**, 854 (1963).

[5] J. Gosset *et al.*, Phys. Rev. C **16**, 629 (1977).

[6] M.C. Lemaire *et al.*, Phys. Lett. **85B**, 38 (1979).

[7] J. Barrette *et al.*, Phys. Rev. C **50**, 1077 (1994).

[8] T.C. Awes *et al.*, Phys. Rev. C **25**, 2361 (1982).

[9] T.C. Awes *et al.*, Phys. Rev. C **24**, 89 (1981).

[10] J.L. Nagle *et al.*, Phys. Rev. Lett. **73**, 1219 (1994).

[11] D.E. Kahana *et al.*, Phys. Rev. C **54**, 338 (1996).

[12] R. Scheibl and U. Heinz, Phys. Rev. C **59**, 1585 (1999).

[13] H. Sato and K. Yazaki, Phys. Lett. **B98**, 153 (1981).

[14] S. Hirenzaki *et al.*, Phys. Rev. C **48**, 48 (1993).

[15] A.Z. Mekjian, Phys. Rev. C **17**, 1051 (1978); Phys. Rev. Lett. **38**, 640 (1977); Phys. Lett **B89**, 177 (1980).

[16] L.P. Csernai and J.I. Kapusta, Phys. Rep. **131**, 223 (1986).

[17] A. Bonasera, Phys.World **12**, 20 (1999); V. Latora, *et al.*, Phys. Rev. Lett. **73** 1765 (1994); P. Finocchiaro, *et al.*, Nucl. Phys. **A600**, 236 (1996); A. Bonasera, *et al.*, Phys. Rev. Lett. **75**, 3434 (1995).

[18] C.O. Dorso *et al.*, Phys. Rev. C **60**, 34606 (1999).

[19] R.J. Lenk *et al.*, Phys. Rev. C **42**, 372 (1990).

[20] A. Bonasera and G.F. Bertsch, Phys. Lett. **B195**, 521 (1987).

[21] J. Cibor *et al.*, *Proc. of the Workshop on Nuclear Matter in Different Phases and Transitions*, Les Houches, France; Ed. J-P. Blaizot, X. Campi and M. Ploszajczak, p. 479 (1998).

[22] J. Cibor *et al.*, Phys. Lett. B **473**, 29 (2000).

[23] K. Hagel *et al.*, Submitted to Phys. Rev. C, (March 2000).

[24] J. Lukasik and Z. Majka, Acta. Phys. Pol. **24B**, 1959 (1993).

[25] Z.Y. He *et al.*, Phys. Rev. C **57**, 1824 (1998).

[26] C.J. Gelderloos *et al.*, Phys. Rev. C **52**, R2834 (1995).

[27] D.H.E. Gross Phys. Rep. **279**, 119 (1997).

[28] J. Bondorf *et al.*, Nucl. Phys. **A433**, 321 (1985).

[29] W.A. Friedman, Phys. Rev. C **42**, 667 (1990).

[30] S. Albergo *et al.*, Nuovo Cimento A **89**, 1 (1985).

[31] N. Marie and the INDRA Collaboration, Phys. Rev. C **58**, 256 (1998).

[32] A. Kolomiets *et al.*, Phys. Rev. C **55**, 1376 (1997).

[33] J. Pochodzalla *et al.*, Phys. Rev. Lett. **75**, 1040 (1995).

[34] J.A. Hauger *et al.*, (EOS Collaboration), Phys. Rev. C **57**, 764 (1998).

[35] Y.G. Ma *et al.*, Phys. Lett B **390**, 41 (1997).

[36] R.J. Charity *et al.*, Nucl. Phys. **A483**, 371 (1988).

[37] R. Wada *et al.*, Phys. Rev. C **39**, 497 (1989).

[38] S. Shlomo and J.B. Natowitz, Phys. Rev. C **44**, 2878 (1991).

[39] J.N. De *et al.*, Phys. Rev. C **57**, 1398 (1998).

[40] A. Strachan and C.O. Dorso, Phys. Rev. C **58**, R632 (1998); *ibid*, C **59**, 285 (1999).

[41] Y. Sugawa and H. Horiuchi, Phys. Rev. C **60**, 607 (1999).

[42] L.G. Moretto *et al.*, Phys. Rev. Lett. **76**, 2822 (1996).

[43] E. Colin *et al.*, Phys. Rev. C **57**, R1032 (1998).

[44] M. Cavinato *et al.*, Z. Phys. **A347**, 237 (1994).

[45] B.A. Li and S.J. Yennello, Phys. Rev. C **52**, R1746 (1995) (See also contribution to this volume).

Dynamics of Heavy-Ion Collisions at Fermi Energies: Challenges and Opportunities

W. Udo Schröder* and Jan Tõke *

Abstract

The dynamics of heavy-ion reactions at Fermi energies is dominated by a dissipative mechanism modified by the concurrent emission of non-statistical nucleons, light particles, and nuclear clusters. Experimental observables are available to monitor the relaxation processes driving the evolution of an interacting nuclear system towards equilibrium. Isospin degrees of freedom provide interesting new access to fundamental information on the reaction mechanism and the effective in-medium nucleonic interactions.

1 Introduction

It is well known that heavy-ion reactions at bombarding energies near the interaction barrier are "fueled" by the stochastic exchange of nucleons between the reaction partners [Sch84]. In such dissipative collisions, the interacting projectile and target nuclei exchange independent nucleons, giving rise to one-body transport phenomena occurring on a time scale which is fast compared to that on which the collective degrees of freedom change (separation of time scales). At near-barrier energies, the concept of an underlying stochastic nucleon-exchange mechanism explains qualitatively multi-dimensional experimental correlations between the net loss in kinetic energy of the massive reaction partners, their individual intrinsic ("thermal") excitation energies, their gains and losses in mass, charge, and spin. As illustrated by numerous examples at low energies [Sch84], the progress of a heavy-ion reaction can be mapped in terms of a number of experimental indicators ("clocks", "thermometers", "impact-parameter filters"). These indicators include the average fragment kinetic-energy loss, the average multiplicity of neutrons or of light particles evaporated sequentially from the hot primary projectile-like (PLF) and target-like (TLF) reaction fragments, the mass-to-charge (A/Z) asymmetry of PLF and TLF, among others. Experimental results [Sch84, Pla88] have clearly demonstrated that in general equilibrium is not reached in dissipative heavy-ion collisions at energies close to the barrier, except possibly for the most central collisions and the longest interaction times, or when the system is initially already close to such equilibrium. It seems unlikely that such equilibrium is reached for higher bombarding energies.

Heavy-ion reactions at Fermi bombarding energies [$\epsilon = E/A \approx (20 - 100)$ MeV] are expected to be much more complex than those near the barrier. Of practical concern is that the energy available in the entrance channel equals several times what is required for the complete disassembly of a system into its constituents, opening up an enormous number of reaction and decay channels. Under such conditions, it is experimentally difficult to reconstruct the interesting primary reaction events (prior to evaporative decay). On a conceptual level, at Fermi energies, there is no separation of the respective time scales for intrinsic and collective degrees of freedom. Macroscopic motion occurs on a time scale comparable to estimated internal relaxation times ($t_{rel} \sim 10^{-22}s \approx 40$ fm/c). This concurrence presents difficulties for the interpretation of reaction phenomena, for example, in terms of effective reaction products and the forces acting on their collective motion, or an interpretation in terms of statistical thermodynamics.

On the other hand, and precisely because the time scales for intrinsic nucleonic motion and collective variables merge at Fermi energies, the mean field cannot adjust adiabatically to the rapidly

*Department of Chemistry, University of Rochester, Rochester, New York 14627, USA

approaching or receding projectile and target nuclei. This situation may lead to sudden compression or dilution of nuclear matter, resisted by the density-and isospin-dependent in-medium interactions. In this fashion, diabatic nucleus-nucleus interactions arise, which can influence heavy-ion collision dynamics in a measurable way, and differently for neutrons and protons. Here, the reaction dynamics become dependent on the projectile-target A/Z asymmetry. There are iso-scalar and iso-vector parts of the nucleus-nucleus liquid-drop and proximity interactions [Pom80] implying different effects for different projectile-target A/Z asymmetries [Sob95, Col98]. Another, related consequence of these isospin-dependent effects is that the properties of the reaction products, e.g., their differential branching ratios for different A/Z ratios, emission angles, and energies, depend on the dynamics of a collision. This behavior of interacting nuclei establishes a connection between the isospin-dependent nuclear equation of state and observable reaction dynamics and opens up an interesting new access to a fundamental property of nuclear matter.

In the following, first examples of experimental data showing a dependence of dynamical reaction variables on projectile and target "isospin" will be illustrated and discussed, mostly in terms of the qualitative physics. Here and in the following, the term "isospin" is used in its colloquial meaning, referring to the third component of the corresponding isospin, i.e., the A/Z or N/Z ratios, making no claim as to a conservation of isospin quantum numbers in heavy-ion reactions. Available experimental experience demonstrates that the reaction environment is more complicated than often assumed and that a host of simultaneously measured experimental parameters and sophisticated analytical procedures are needed to deduce a meaningful reaction scenario. Therefore, first the need for a consistent view of the reaction environment at bombarding energies in the Fermi energy domain is explained in Section II, followed in Section III by a discussion of newly observed isospin-dependent trends and some of the concepts useful for their interpretation. In particular, model predictions for the relevant isospin-dependent dynamics of heavy-ion reactions are confronted with experimental data.

2 Empirical Reaction Environment at Fermi Energies

2.1 General Reaction Dynamics

As has been stated in the Introduction, heavy-ion reactions at bombarding energies near the interaction barrier are driven by the stochastic exchange of nucleons between the reaction partners [Sch84]. In such dissipative collisions, projectile and target nuclei are believed to develop a matter bridge between them, a "neck," and to form a transient di-nuclear system. Through this neck interface, the interacting fragments exchange independent nucleons (single particles), giving rise to one-body transport phenomena occurring on a time scale which is fast compared to that on which the collective degrees of freedom change (separation of time scales). This exchange process dissipates kinetic energy of relative motion, as the dinuclear complex rotates ("orbits") about its center of gravity. The dinuclear system is mechanically unstable and holds together only temporarily by the action of adhesive liquid-drop and proximity forces, which are counterbalanced by repulsive Coulomb and centrifugal forces. For all but light reaction systems and a narrow range of impact parameters associated with central collisions, the repulsive forces are sufficiently strong to drive the di-nuclear system apart again, leading to projectile-like and target-like fragments, PLF and TLF, respectively. The primary fragments are typically highly excited and decay subsequently by evaporation or fission-like processes. Both, approach and re-separation are retarded by one-body frictional forces mediated by nucleon exchange. The experimentally observed reaction features are well reproduced, often quantitatively, by semi-classical nucleon-exchange models (NEM) developed by Randrup and collaborators [Ran76-82] and others [Bor90, Fel85], even though some of the simplifying assumptions made in these models appear difficult to justify.

In any case, at near-barrier energies, the concept of an underlying nucleon-exchange mechanism explains qualitatively multi-dimensional experimental correlations between the net loss in kinetic energy of relative motion of the massive reaction partners, their individual intrinsic ("thermal") excitation energies, their gains and losses in mass, charge, and spin. Based on such reaction models, it is plausible that the energy loss is stochastically (anti-) correlated with impact parameter and associated interaction time. This means that all variables are characterized by stochastic probability functions, as they are determined by a stochastic nucleon exchange mechanism. Therefore, as demonstrated by numerous examples [Sch84], the progress of a heavy-ion reaction at low energies can be followed in terms of experimental indicators such as the average fragment kinetic-energy loss, the average multiplicity of evaporated light particles, and the mass-to-charge (A/Z) asymmetry of reaction products.

As already stated, heavy-ion reactions at Fermi bombarding energies are much more difficult to analyze and to understand than those at near-barrier energies. For example, at these energies it is conceivable that upon impact the system breaks up directly into many light particles and clusters. Even if such catastrophic events are unlikely to occur, massive remnants of the primary PLFs and TLFs, or their collective energies, may not be identifiable any longer. It is then not obvious how to reconstruct important reaction parameters such as the amount of initial kinetic energy dissipated into intrinsic heat. On a conceptual level, at Fermi energies there is no separation of time scales between the motion along intrinsic and collective degrees of freedom. This concurrence presents difficulties for the interpretation of reaction phenomena in terms of effective system trajectories, reaction products, and forces acting on the collective motion, or even in terms of an interpretation in terms of statistical thermodynamics. In fact, classical concepts such as that of an average system trajectory may not be applicable any longer.

A meaningful experimental access to heavy-ion reactions at Fermi energies has been gained through so-called 4π experiments. In such measurements, all important reaction observables are sampled with a broad dynamic range and throughout the entire solid angle (in the corresponding center-of-mass reference frame). Typically, such coverage is achieved for light and intermediate-mass charged reaction products. Experimental detection includes less frequently also the emitted neutrons, and the massive PLFs, TLFs, and other heavy residues (HR), in addition to the lighter charged products. So far in a few cases only has it been possible to reconstruct primary reaction fragments. Approximately, this task was accomplished in the reaction ^{209}Bi+^{136}Xe both at bombarding energies per nucleon of ϵ = E/A = 28 [Bal93, 95, Lot93, Tok94-96] and 55 MeV [Sku97, 98]. Much of the overall reaction dynamics prevailing at Fermi energies can be gleaned from these two reactions. Other heavy and intermediate reaction systems such as ^{197}Au+^{86}Kr [Sku96, Dje96] and ^{112}Sn+^{40,48}Ca [Agn97, 98] at ϵ = 35 MeV provide consistent and supplementary information.

In Fig.1, experimental results are displayed in form of two-dimensional contour diagrams, for the angle-energy correlations (Wilczyński Plots) for the above heavy-ion reactions and the reaction ^{112}Sn+^{40}Ca [Agn97, 98], all measured at Fermi-type bombarding energies. All but the latter reaction have been measured with 4π coverage. In this figure, the observed yield of PLFs is plotted *vs.* laboratory kinetic energy and angle. For all reactions, one observes consistent trends in the yields, which in each case form a ridge of elastically scattered projectiles at high (elastic-type) energies, extending up to the respective grazing angle [Sch84]. Starting from the grazing angle, the correlation ridge of reaction events indicates PLF emission at successively smaller (more forward) angles, as energy is dissipated. Eventually, the forward-going ridge escapes the angular acceptance of the detection system. However, the picture appears consistent with the trend continuing, until PLFs are emitted on the other ("negative") side of the beam. For large enough angles, PLFs emitted at negative angles are again accepted by the detector, leading to the appearance of a backward-going ridge at lower energies, the reflection of the negative-angle ridge. In the plots included in Fig. 1, no correction has been made to the data for the decay of the primary PLFs. This is the main reason for the lack of definition of the yield ridges. Considering this circumstance, these data are reminiscent of the process of "dissipative orbiting" of a dinuclear complex formed in low-energy heavy-ion reactions. The curves superimposed on the data in Fig. 1 represent calculations with the nucleon exchange model (NEM) mentioned earlier. The calculations shown have been corrected on average for the number of particles evaporated sequentially from the primary hot PLFs. The resulting agreement between data and calculations, while not quantitatively excellent, nevertheless supports a generally dissipative character of the reactions studied in the Fermi energy domain. From this type of dissipative reaction, in a first step, two massive reaction fragments emerge, which preserve memory of projectile and target nucleus, in several respects. This feature is recovered for

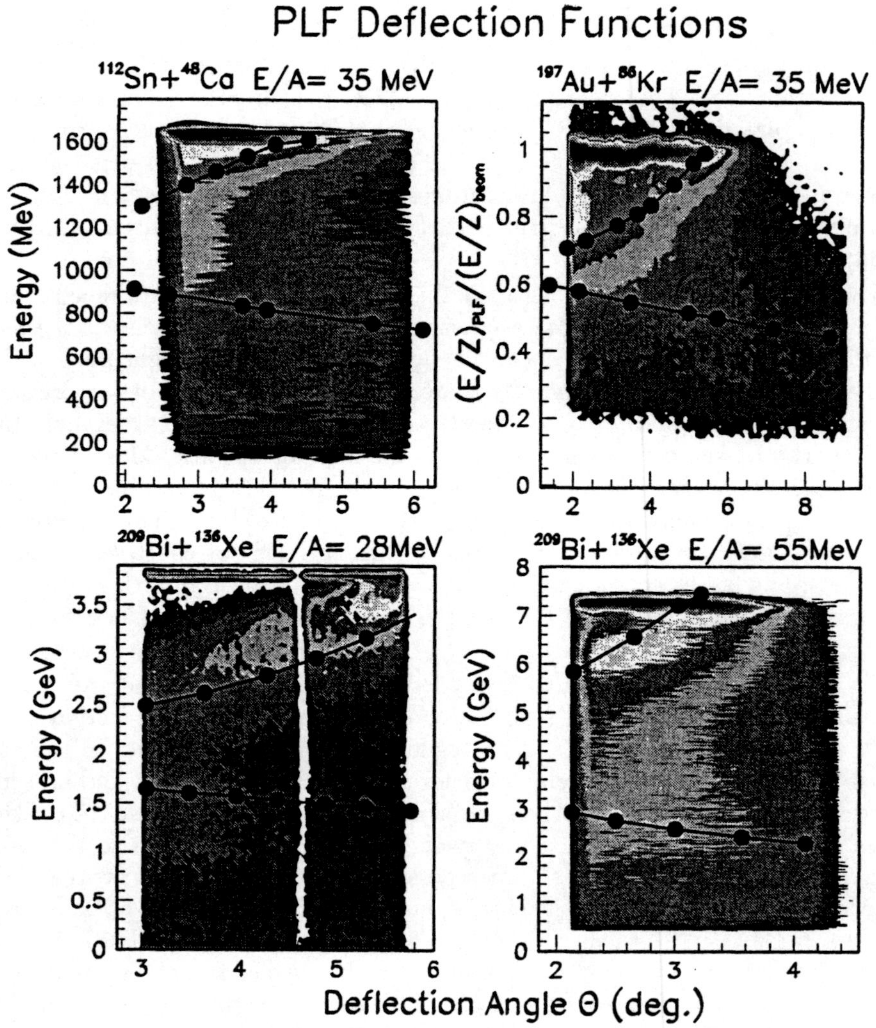

Figure 1: Two-dimensional contour diagrams of the yield of charged products *vs.* lab energy and angle, for the reactions ^{112}Sn+^{48}Ca (top left), ^{197}Au+^{86}Kr (top right), and ^{209}Bi+^{136}Xe (bottom left and right), for the indicated bombarding energies. The solid dots and lines represent predictions by the NEM, corrected for sequential evaporation. (From [Agn98])

the two ^{209}Bi+^{136}Xe reactions, for which a partial reconstruction of the primary PLF has been possible, adding the average of the combined charge of light charged particles evaporated from the primary PLF to the measured charge of the PLF evaporation residue. Figure 2 displays the so-called diffusion plot for the reaction ^{209}Bi+^{136}Xe at $\epsilon = 28$ MeV. Here, the angle-integrated yield of reconstructed primary PLFs is plotted *vs.* the PLF atomic number (Z) and multiplicity (m_n) of evaporated neutrons. The multiplicity of neutrons is a measure of the dissipated kinetic energy. In this plot, the ordinate reflects essentially the number of neutrons evaporated from the PLF, measuring the internal excitation energy of the primary PLF. One observes that, for small and intermediate amounts of dissipated energy, the yield is centered at the atomic number ($Z = 54$) of the projectile. The distribution broadens considerably, as energy is dissipated into heat of the reaction products. The degree of broadening is quantitatively consistent with expectations based on the NEM, given the sizeable uncertainties inherent in the average reconstruction procedure. A discrepancy definitively exists for the higher neutron multiplicities, where an average shift of the PLF Z distribution to smaller atomic numbers is observed. This effect is understood to be due to the emission of intermediate-mass fragments (IMF) from the intermediate dinuclear system, which were not assignable to either PLF or TLF, as explained below. It is not due to a diffusive "drift" of the Z-distribution reflecting the underlying potential [Sku98] The particles evaporated from PLF or TLF

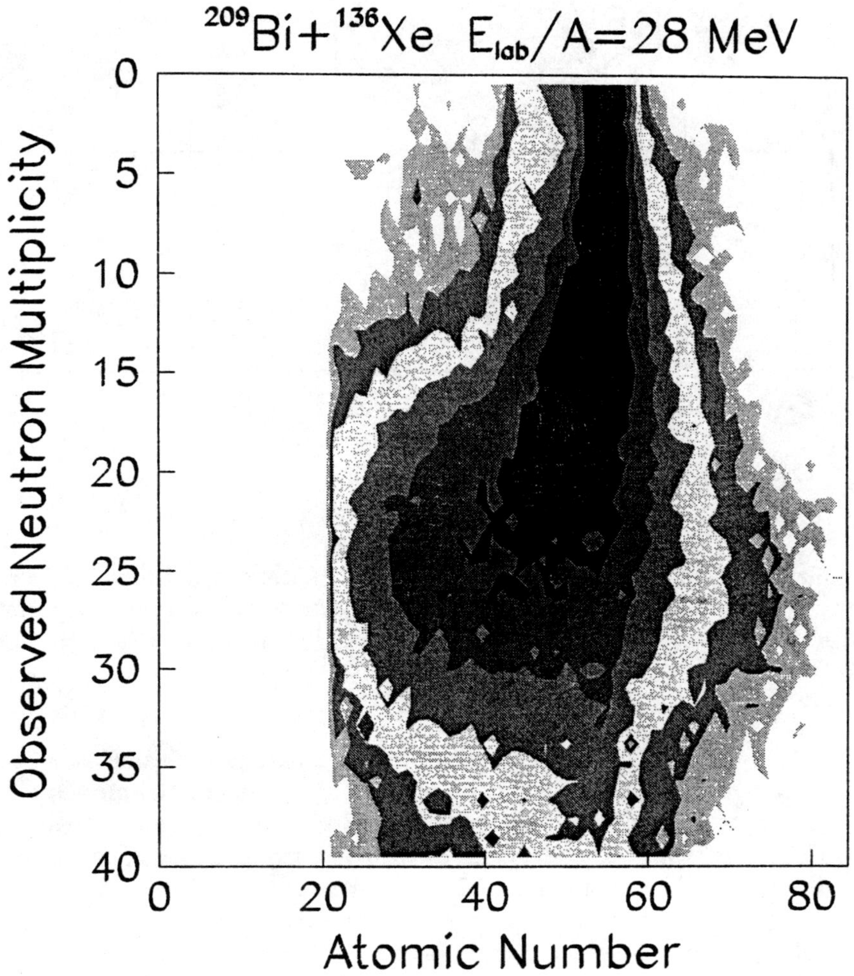

Figure 2: Contour plot of the correlation between PLF atomic number and the multiplicity of neutrons for the ^{209}Bi+^{136}Xe reaction. The PLF Z has been corrected for evaporated light charged particles. (From [Bal95])

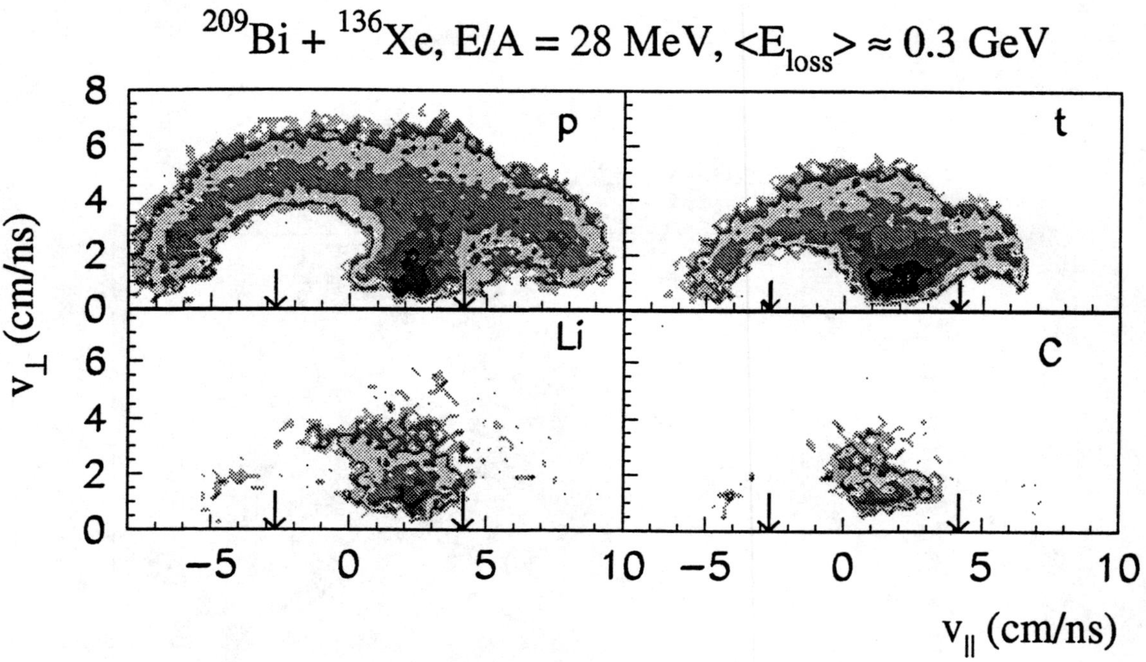

Figure 3: Galilei-invariant cross section of protons (p), tritons (t), lithium (Li), and carbon (C) fragments emitted from the ^{209}Bi+^{136}Xe reaction. Contours of constant yields are plotted vs parallel and transverse particle velocity. The arrows indicate the velocities of PLF and TLF, respectively (From [Tok96]).

have been identified based on their characteristic emission patterns. Examples of such patterns are displayed in Fig. 3, showing the Galilei-invariant plots of the differential cross section for protons, tritons, Li, and C fragments, emitted in the reaction ^{209}Bi+^{136}Xe for peripheral collisions with energy losses around $E_{loss} = 0.3$ GeV. The yield is presented as a two-dimensional contour diagram of the yield plotted *vs.* the particle velocities parallel and perpendicular to the beam. The origin represents the c.m. system. The geometry consistently seen in the invariant cross section diagrams, for the light charged particles (p,d,t, and α-particles) and essentially all energy losses, is that of two intersecting semi-circles, one centered at the velocity of the PLF (arrows on right) and one at the velocity of the TLF reaction partner (arrows on left). The radius of each semi-circle corresponds to the most probable (Coulomb plus thermal) energy of the emitted particle. Each semi-circle is centered at the velocity of the emitter. These "Coulomb circles" are indicative of an evaporation of light charged particles (LCP) from primary hot PLF and TLF, occurring after significant acceleration in the mutual Coulomb field has taken place already. Owing to these characteristic emission patterns, the origin of the evaporated particles can be assessed, not event by event but at least on average. The emission patterns, shown in the two bottom panels of Fig.3, for intermediate-mass (IMF) reaction products Li and C are distinctly different from those of the LCPs. They exhibit no Coulomb circles but are produced on average at rest in the cm system.

2.2 Non-Statistical Emission of Particles and Clusters

The data in the above three figures (Figs. 1 to 3) demonstrate the dominantly dissipative character of the reaction mechanism. However, in addition to the similarities between heavy-ion reaction mechanisms at low and Fermi energies, there are also important differences. As was noted in an early experiment [Que93] on the system ^{197}Au+^{208}Pb at $\epsilon = 29$ MeV, measuring neutrons along with LCPs, there is a limit to the amount of entrance-channel kinetic energy which can be dissipated, converted into intrinsic heat, and finally emitted again in form of nucleons and light particles. For heavy systems and at bombarding energies around ϵ = 30 MeV, typically only (50-60)% of the available energy is converted into heat. In particular in central collisions, the remaining energy can be consumed in the emission of energetic non-statistical light particles and IMFs.

Non-statistical particle emission has been studied extensively for fusion-type reactions [Hil88]. As sample illustration of non-statistical particle emission, experimental neutron energy spectra observed [Agn97, 98] for semi-peripheral $^{112}Sn+^{48}Ca$ collisions at $\epsilon = 35$ MeV and three selected (out of 12 available) angles are depicted in Fig. 4, together with theoretical fits. These spectra are very similar to those measured [Que93], both for the heavy system $^{197}Au+^{208}Pb$ reaction at 29 MeV and for light system $^{12}C+^{35}Cl$ at $\epsilon = 43$ MeV [Lar 98]. In order to understand the data in either of these cases, a minimum of three separate moving neutron emitters has to be assumed. The spectra at backward angles (e.g., at $\Theta_{lab} = 120^o$ for $^{112}Sn+^{48}Ca$) are dominated by evaporation from a target-like reaction fragment, i.e., have a Maxwell-Boltzmann ("moving source") form, in the laboratory given by

$$\frac{dm}{d\Omega dE} \propto \sqrt{E}e^{\frac{E-2\sqrt{\epsilon E}\cos\Theta+\epsilon}{T}} \tag{1}$$

for temperature T. Here, ϵ is the emitter energy per nucleon and Θ the particle detection angle. Transformed to forward angles (e.g., $\Theta_{lab} = 15^o$), this component is found in the data at very low laboratory energies. The remainder of this forward-angle spectrum is due to evaporation from a fast-moving PLF. The spectrum is well represented already by the two evaporative (PLF and TLF) components. However, at observation angles sideways to the beam, e.g., at $\Theta_{lab} = 55^o$ in Fig. 4, these two evaporation components are insufficient to explain the exponential high-energy tail of the neutron spectrum. This region is kinematically essentially inaccessible to evaporation from PLF or TLF, and the neutron spectrum is much harder than the evaporation spectra observed for the same PLF/TLF events both at more forward or more backward angles. Clearly, a third, non-statistical neutron component is needed. The energy-angle emission pattern (12 angles) of this non-statistical component can be described well by a hypothetical third source, moving with about half the beam velocity and emitting neutrons isotropically in its own rest frame. This frame of reference is approximately identical with the NN cm frame. As suggested by fusion-like reaction studies, the corresponding non-statistical neutron energy spectrum is very similar to a Maxwell-Boltzmann distribution (cf. Equ. 1), in the rest frame of the emitter given by

$$\frac{dN}{dE} \propto \sqrt{E}e^{\frac{E}{E_o}} \tag{2}$$

even though the emission is non-thermal. For the case shown in Fig. 4, the inverse exponential slope parameter is large, of the order of $E_o \approx 10$ MeV. In comparison, the evaporation spectra are characteristic of a nuclear temperature of only $T = 3.3$ MeV in the respective emitter frames. This discrepancy is due to the fact that non-statistical emission is initiated in collisions between a few nucleons only, much before thermalization, i.e., distribution of the kinetic energy over all degrees of freedom, is completed. The obvious, two-component nature of the neutron spectra exhibited in Fig. 4 for neutrons emitted in peripheral collisions disappears for higher bombarding energies and/or more central collisions. Here the evaporation temperatures are not so different anymore from the inverse logarithmic slopes E0 characterizing the spectra of non-statistical particles. Similar observations are made for the spectra of protons measured for reactions such as $^{112}Sn+^{40,48}Ca$. Because of the Coulomb barrier for emission, the proton energy spectra exhibit a higher effective energy threshold than that for neutrons, thus suppressing an important part of the proton evaporative spectra. On the other hand, because of their shift to high energies, non-statistical proton spectra are not significantly affected by Coulomb barriers. Non-statistical nucleon emission is weak for peripheral collisions, but its multiplicity increases with decreasing impact parameter. To put the importance of non-statistical nucleon emission in perspective, for $^{112}Sn+^{40,48}Ca$, this non-equilibrium process may consume up to 25% of the available energy.

IMF emission presents another modification of the dissipative reaction mechanism. Like emission of non-statistical light particles, this process occurs infrequently in peripheral collisions but becomes important for collisions that are more central. Although, the IMF data in the lower two panels of Fig. 3, taken by themselves, suggest a single source emitting at intermediate velocity, there

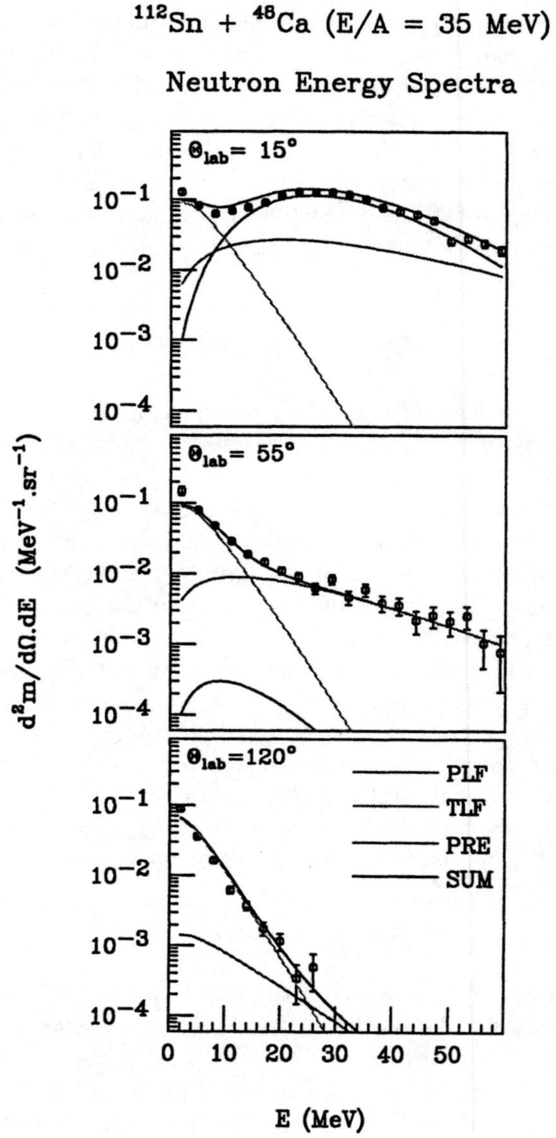

Figure 4: Neutron energy spectra for three lab angles for the reaction $^{112}Sn+^{48}Ca$ at $\epsilon = 35$ MeV. Solid lines indicate moving- source fits. (From [Agn98])

is no doubt as to the presence of two massive fragments (PLF and TLF) from the same events. These projectile and target remnants survive violent central nuclear collision. Apparently, even at Fermi energies, nucleon exchanges between the reaction partners, as well as the frictional effects caused by this transport process, remain important. This behavior appears to be universal for the Fermi energy domain. For example, in the reaction ^{197}Au+^{86}Kr at 35 MeV, correlated PLF/TLF pairs, as well as TLF evaporation residues, have been measured [Sku96, Dje96] for energy losses corresponding to a range of peripheral to central collisions. Even for lighter reaction systems and higher energies, such as ^{12}C+^{35}Cl at $\epsilon = 43$ MeV, and for events in which the entire system has vaporized, event reconstruction reveals [Lar 98] that PLF and TLF survive the first dissipative reaction step. The general conclusions on the reaction mechanism, drawn already from studies of the reactions ^{197}Au+^{208}Pb [Sch92, Que93]and ^{209}Bi+^{136}Xe [Lot93, Sch92, 94, Bal95] have now been confirmed up to relatively high energies. Recently, the reaction ^{58}Ni+,^{36}Ar has been studied [Bor99] at $\epsilon = 95$ MeV, reporting the reconstruction properties of vaporized PLFs, i.e., for very high excitation energies or energy losses. Uncertainties as to the competition between dissipative and fusion-like mechanisms persist only on the level of the innermost 1% of the impact parameter range.

Experimentally, it is not yet clear at what reaction stage intermediate-mass fragments are emitted, e.g., whether at the beginning or at the end of the dissipative interaction between projectile-like and target-like fragments. Certainly, because of substantial masses and energies, IMF emission influences significantly the balance of mass, charge, and energy available to the rest of the reaction products. As an example, a reduction of the average PLF atomic number by IMF emission is seen in Fig. 3 for large neutron multiplicities, i.e., for central collisions and several IMFs. Properties of the IMFs emitted in the reaction ^{209}Bi+^{136}Xe at $\epsilon = 28$ MeV are shown in Fig. 5. Here, the atomic-number (Z) distributions for IMFs emitted in the reaction ^{209}Bi+^{136}Xe at $\epsilon = 28$ MeV are plotted on the left, for various numbers of IMFs measured in coincidence, i.e., different IMF multiplicities (m_{IMF}). These Z distributions have a generally exponential character. It is an unexpected fact that the shapes of the Z distributions are very similar and independent of the IMF multiplicity. The average charge of an IMF is $< Z_{IMF} >\approx 8$, corresponding to oxygen fragments. On the r.h.s. of this figure, the corresponding transverse-energy spectra are shown for the same events. These latter spectra look again very much like thermal Maxwell-Boltzmann distributions (see Equ. 2), albeit with very large inverse slope parameters ($E_o \approx 30$ MeV) and average IMF energies of $< E_{IMF} >\approx 60$ MeV. For the same events, the statistical neutron and LCP spectra show temperature parameters of only a few MeV ($T \leq 5.5$ MeV).

The IMF energy spectra are also seen to be independent of the IMF multiplicity. The resulting reducibility of the total kinetic energy spectrum of several IMFs, in terms of a fundamental single-particle energy spectrum, has given rise to a statistical interpretation [Mor97] of IMF emission from a nuclear system in equilibrium with an infinite heat bath at temperature T. While this schematic approach can account for some of the combinatorial aspects of multi-IMF emission, the model is inconsistent with fundamental properties of the reaction mechanism in general and with the IMF emission process in particular. For an equilibrated (micro-canonical) nuclear source, emission of IMFs would compete with that of other particles, such as neutrons, protons, and other LCPs, dividing the same fixed total excitation energy according to the available numbers of microstates. The joint neutron/LCP multiplicity distributions $P(m_n, m_{LCP})$ plotted in Fig. 6 for different values of m_{IMF} demonstrate that such statistical competition does in fact not exist. Apart from the distributions for a transitional region ($0 \leq m_{IMF} \leq 2$), the distributions are relatively well defined, with only small fluctuations, and approximately independent of m_{IMF}. This implies that the thermal energy of the system "saturates" at energies per nucleon of $\epsilon \approx (4-5)$ MeV, i.e., remains constant, while hundreds of MeV of entrance-channel kinetic energy are still available, and are eventually expended in, IMF emission from the system. A trend consistent with thermal IMF emission would have consisted in a significant shift away from the origin and a considerable broadening of the joint multiplicity distributions with increasing IMF multiplicity. However, there is no mystery as to where the energy comes from which is carried away by multiple IMFs: With increasing m_{IMF}, the velocity of the PLF remnant is seen to decrease [Tok96].

Certainly, the hypothetical IMF-emitting source is not in thermal equilibrium with the PLF and TLF in the same reaction event. Kinetic energy of relative motion is channeled more or less directly

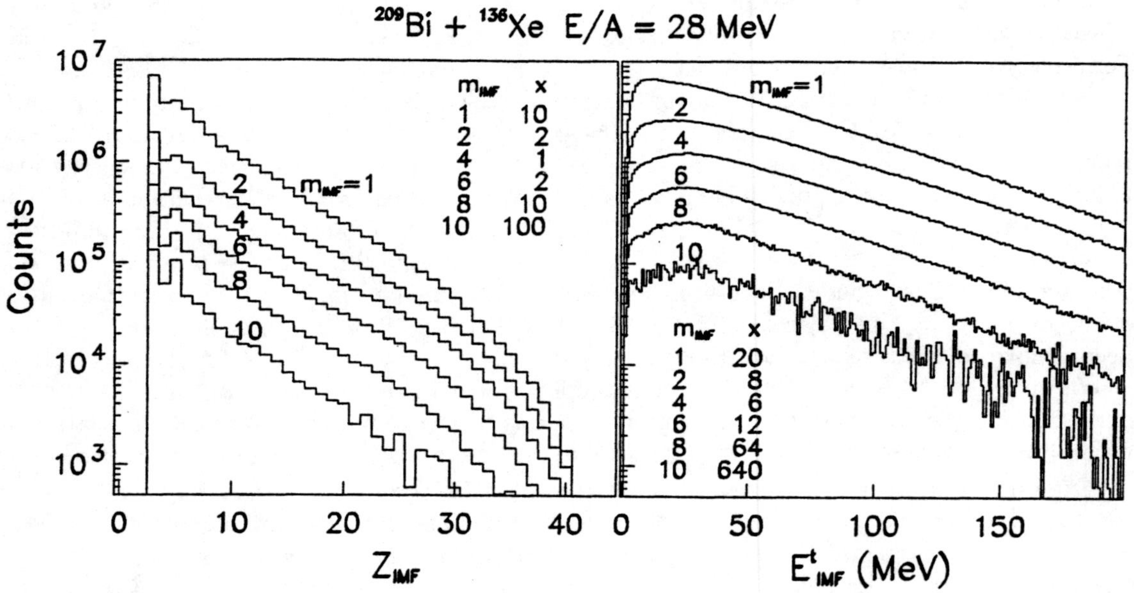

Figure 5: Atomic-number (left) and transverse-energy distributions for intermediate-mass fragments emitted in the ^{209}Bi+^{136}Xe reaction. The distributions are sorted for different IMF multiplicities and displaced along the ordinate (From [Tok98]).

into dynamical emission of IMFs and other particles. The non-statistical processes of nucleon and IMF emission appear to have similarities but are not obviously reducible to one another. Both types of process lead to an emission of relatively hard Maxwell-Boltzmann-type distributions with large slope parameters and show some isotropy in a system of reference kinematically between those of PLF and TLF and close to the cm system. A possible scenario attributes both, nucleon and IMF emission, to slightly different stages of the collision impact process. Fast nucleons could be emitted in the first encounters of nucleons from the projectile with those of the target. Presumably, a breakup of the projectile into IMFs requires significant overlap, mixing, and compression of the matter density distributions of projectile and target. Thus, the question could be decided by experiments studying both types of non-statistical emission as functions of projectile-target A/Z asymmetry, impact parameter, and bombarding energy. First experiments [Agn97] of this kind have yielded indeed information about the time evolution of these processes, encouraging further, and more specific investigations.

2.3 The Charge Density Asymmetry Degree of Freedom

As stated previously, the charge density, or A/Z, asymmetry of reaction products has been found to be an effective indicator of the degree of (chemical) equilibrium reached in a heavy-ion reaction. At near-barrier energies, interacting heavy-ion systems evolve slowly from an initial state towards equilibrium [Sch84]. Chemical equilibrium can be defined in terms of a local minimum in the driving potential energy surface (PES) or by the bulk A/Z ratio of the composite system. More rigorously, the time-dependent free energy F defines an evolution of the chemical equilibrium along a system trajectory. Practically, however, only the asymptotic potentials, the binding, and the interaction energies of representative nuclei with well-defined shapes are accessible. Therefore, experiments have often been evaluated only with respect to some static PES or its derivatives, the chemical potentials. To devise a more model-independent method to explore chemical equilibration processes represents one of the long-standing challenges in heavy-ion studies. This applies in particular to the fluctuations in these variables, which contain independent information on the reaction and its

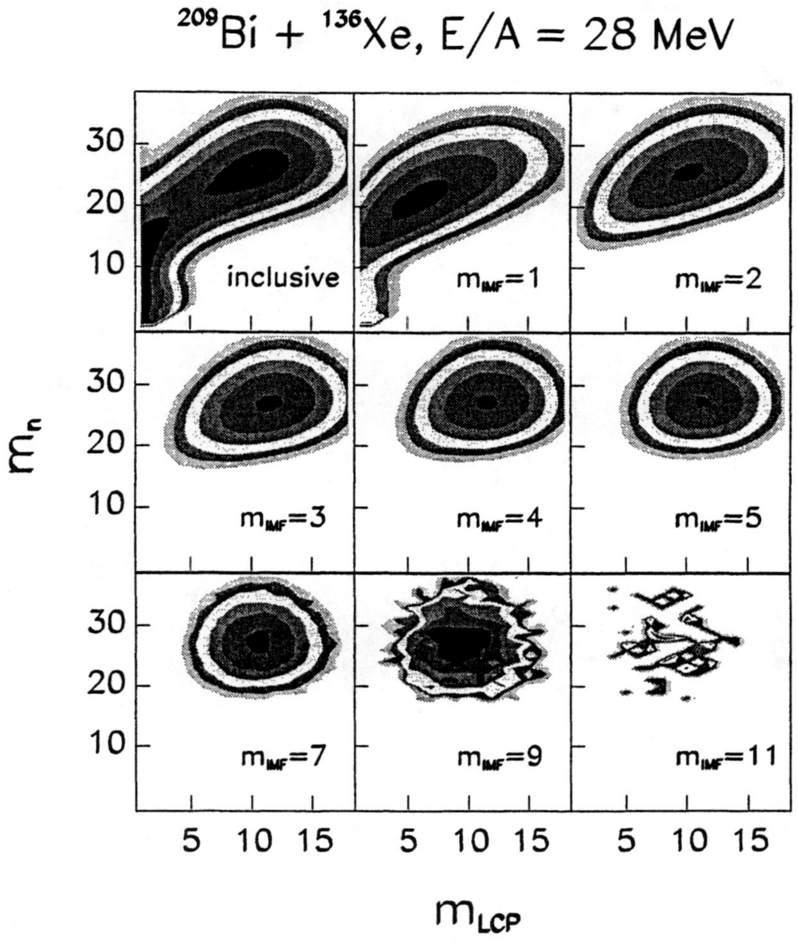

Figure 6: Contour plots of the joint multiplicity distributions $P(m_n, m_{LCP})$ for different IMF multiplicities. The upper left distribution represents the unconditional multiplicity distribution.

products. In the absence of such methods, ambiguities in the interpretation of experimental data caused by such over-simplifications should be kept in mind. The approximations made in such analysis are probably best justified for peripheral collisions, where distortions due to nuclear overlap are smallest.

Experimental results [Sch84, Pla88] have clearly demonstrated that, in general, equilibrium (in the above operational sense) is not reached in dissipative heavy-ion collisions at bombarding energies close to the barrier, except possibly for the most central collisions, the longest interaction times, or when the system is initially already close to such equilibrium. It is interesting that, at these lower energies, the rapidity $d < Z_{PLF} > /dE_{Loss}$ of the drift of the overall PLF-Z distribution clearly correlates with the gradient dV/dZ of the static $PES(V)$ for nucleon exchange. Data are shown in Fig. 7, as obtained [Des88] for reactions involving a ^{238}U target and several medium-weight projectiles at the same bombarding energy of $\epsilon = 8.5$ MeV. The correlation suggests that the propensity of ^{40}Ca to transfer one of its protons to the ^{238}U target is related to the initially steep gradient of the PES in that direction. Possibly because of the large curvature (d^2V/dZ^2) of the PES, one observes net (average) transfers between projectile and target of never more than a few nucleons. It should be noted that otherwise successful NEM models have never been able to describe consistently the details of the average mass and charge drifts of the fragment distributions at near-barrier energies [Sch84].

Because of the merging interaction and relaxation time scales in the Fermi energy domain, one

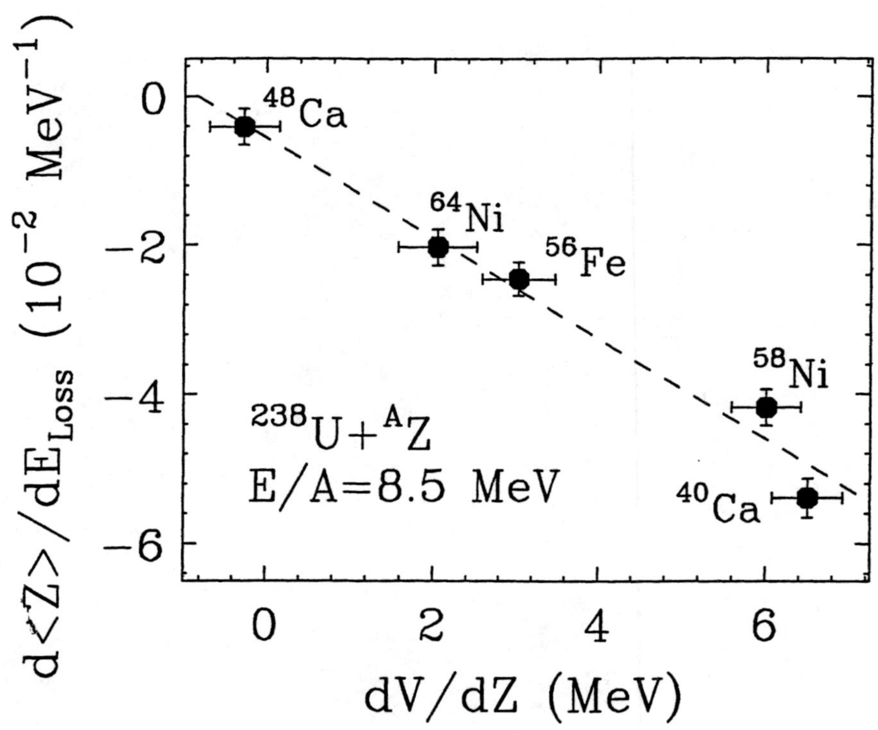

Figure 7: Rate of initial drift in average PLF-Z value *vs.* potential gradient. (From [Des88]).

does not intuitively expect energetic collisions to drive a system closer to chemical equilibrium than at lower energies. In addition, because of the diminishing role of potential energies and Q-values for nucleon transfer at these energies, chemical equilibrium should be defined somewhat differently, i.e., more in terms of thermodynamic properties of free projectile and target nucleon gases. However, detailed experiments in this energy region studying A/Z equilibration in dissipative reactions [Agn97,98] or fusion-like reactions [Yen94, Joh96,97] are rare. Experimental systematics similar to those of Fig. 7, or theoretical model predictions, are largely unavailable for these higher energies.

Interesting first information about A/Z relaxation, and a host of information about other "isospin" effects in the Fermi energy domain, has been reported by Agnihotri et al. [Agn97, 98], who have measured PLFs in coincidence with neutrons, LCPs, and IMFs emitted from the two reactions ^{112}Sn+^{40}Ca and ^{112}Sn+^{48}Ca at $\epsilon = 35$ MeV. These two systems have been chosen because of their different initial conditions with respect to the local minima of the corresponding static potential energy surfaces. The *PES* for fragmentations of the composite system ^{112}Sn+^{40}Ca is depicted in Fig. 8. It is defined as,

$$V(Z_1, N_1; Z, N) = V_{LD}(Z_1, N_1) + V_{LD}(Z - Z_1, N - N_1) + V_{Coul}(Z_1, Z_2) + V_{Nucl}(A_1, A_2) \quad (3)$$

Here, the different terms represent the liquid-drop binding energies of two nuclei, the Coulomb and the nuclear interactions, respectively. The potential is arbitrarily normalized. As shown in this figure, the initial configuration ("injection point") for the ^{112}Sn+^{40}Ca reaction is associated with a point on the steep wall of the *PES*, some 8-10 MeV above the local minimum. In contrast, the injection point for the ^{112}Sn+^{48}Ca reaction is very close to the minimum in the corresponding *PES*. Based on the low-energy systematics depicted in Fig. 7, one would expect very different neutron and proton pickup *vs.* stripping tendencies of the two projectiles ^{40}Ca and ^{48}Ca. Only the reaction ^{112}Sn+^{40}Ca is expected to show an evolution towards chemical equilibrium at all and provide thus

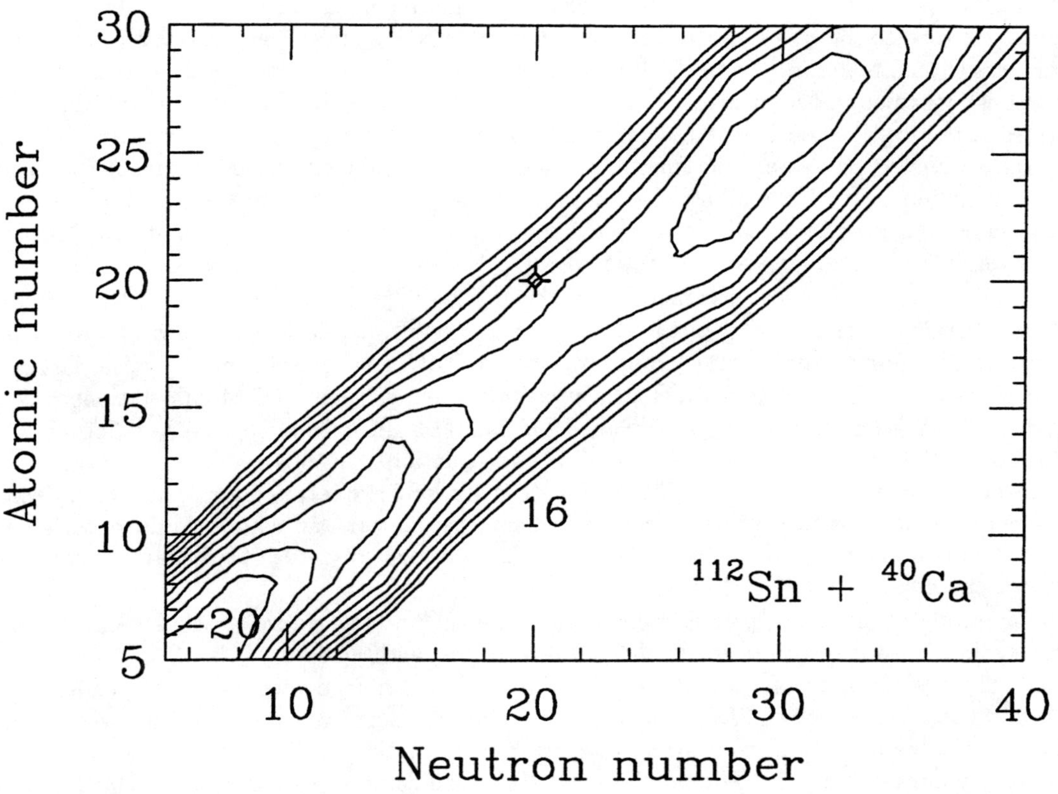

Figure 8: Static liquid-drop (plus shell corrections) potential energy surface for fragmentations of the composite system ^{152}Yb. The spacing between the contour lines is 4 MeV and the "injection point" fragmentation ^{112}Sn+^{40}Ca is indicated by a cross.

information about the degree to which this equilibrium is reached in a heavy-ion reaction at Fermi energies. The experiment sampled almost the entire reaction cross section. The measured cross sections are consistent with an exclusively dissipative reaction mechanism. However, the data allow for a 10% contribution of fusion-like events, which were not detected in the experiment.

Some of the neutron spectra measured [Agn98] in coincidence with PLFs for the reaction ^{112}Sn + ^{48}Ca have been shown already in Fig. 4. The solid lines included in that figure indicate the contributions assigned to the PLF, TLF, and non-statistical (PRE) sources. Similar spectra were obtained for the associated reaction ^{112}Sn+^{40}Ca, as well as for protons and LCPs in both cases. From the information contained in the measured post-evaporative properties of PLFs and the multiplicities and energies of the evaporated light particles, mostly neutrons, the properties of the primary fragments were reconstructed on average, as functions of energy loss or reconstructed fragment excitation energy. Corrections were made allowing for mass and energy loss due to pre-equilibrium neutron and proton emission, processes to be discussed further below. Iterative evaporation calculations were performed with the statistical-model code GEMINI [Cha88].

An important observation was made from the ^{112}Sn+^{40,48}Ca data with respect to thermal equilibrium, pertains to the sharing of the dissipated energy between the reaction partners. The data indicate that thermal equilibrium (equipartition) has not been reached, even for the largest excitation energies ($< E^* > \approx 400$ MeV) measured in the experiment. NEM calculations predict an almost equal sharing of this energy between PLF and TLF, in spite of their very different numbers of nucleons. This is a more extreme disequilibrium situation than observed in the data, which follow a gradual evolution towards equilibrium, with increasing excitation energy, i.e., with decreasing impact parameter. For the events accessible in the measurements, an equilibrium excitation energy distribution is never reached. In this sense, the two reactions ^{112}Sn+^{40,48}Ca behave similarly at low and high energies.

Relaxation of the A/Z degree of freedom in the two ^{112}Sn+^{40,48}Ca reactions is illustrated in Fig. 9. Here, the neutron-to-proton multiplicity ratio (circles with error bars) is plotted *vs.* excitation energy. This ratio of multiplicities of evaporated particles, combined with measured properties of the secondary (post-evaporative) PLF, represents an observable sensitive to the A/Z ratio of the primary PLF. The sizable errors reflect overall systematic uncertainties in the method, e.g., pertaining to level density parameters, etc. The abscissa scale in Fig. 9 can be thought of as an effective impact-parameter or time scale. The curves included in this figure represent calculations assuming different primary A/Z (or N/Z) ratios and are meant to give a conservative estimate of the sensitivity of the data. The label "Equilibrium" refers to the respective local *PES* minima. It is obvious from the different behavior shown by the reactions ^{112}Sn+^{48}Ca and ^{112}Sn+^{40}Ca that the charge density asymmetry is a dynamical variable, evolving with impact parameter and interaction time. The large multiplicity ratios M_n/M_p for ^{48}Ca at low excitations could be taken to reflect simply the large neutron excess of the projectile and, perhaps to a lesser extent, the efficiency of the Coulomb barrier in hindering the emission of protons. In such picture, higher excitation energies would simply reduce the Coulomb effect for proton emission. However, a similar hypothetical scenario for the reaction ^{112}Sn+^{40}Ca would predict a trend opposite to that actually observed. One concludes that the observed different evolution of the neutron/proton multiplicity ratios must reflect differences in the constitutions of the emitting PLFs, which change with dissipated energy or impact parameter.

From the experimental data, it appears that the mass-to-charge ratio of the ^{48}Ca-induced reaction remains fixed, at A/Z = 2.4, solely because the initial projectile/target fragmentation is already at the minimum of the *PES*. This behavior could have been mistaken for a fast A/Z equilibration of the system. On the other hand, PLFs from the ^{40}Ca-induced reaction, where the injection point is removed from the local *PES* minimum, show an obvious evolution of their average A/Z ratio with increasing excitation energy (or decreasing impact parameter). The change, an increase from the initial value of $(A/Z)_o = 2$ to $(A/Z)_\infty = 2.15$, is significant. Within the present experimental accuracy, one has to conclude that chemical equilibrium can possibly be reached in semi-central ^{112}Sn+^{40,48}Ca collisions at $\epsilon = 35$ MeV. However, a more accurate determination of cross sections and primary fragment A/Z ratios would be desirable, in order to improve the present, somewhat broad constraints on reaction models. Nevertheless, the above measurements suggest already that the mass-to-charge (A/Z) or neutron-to-proton (N/Z) ratios of the reaction products represent important

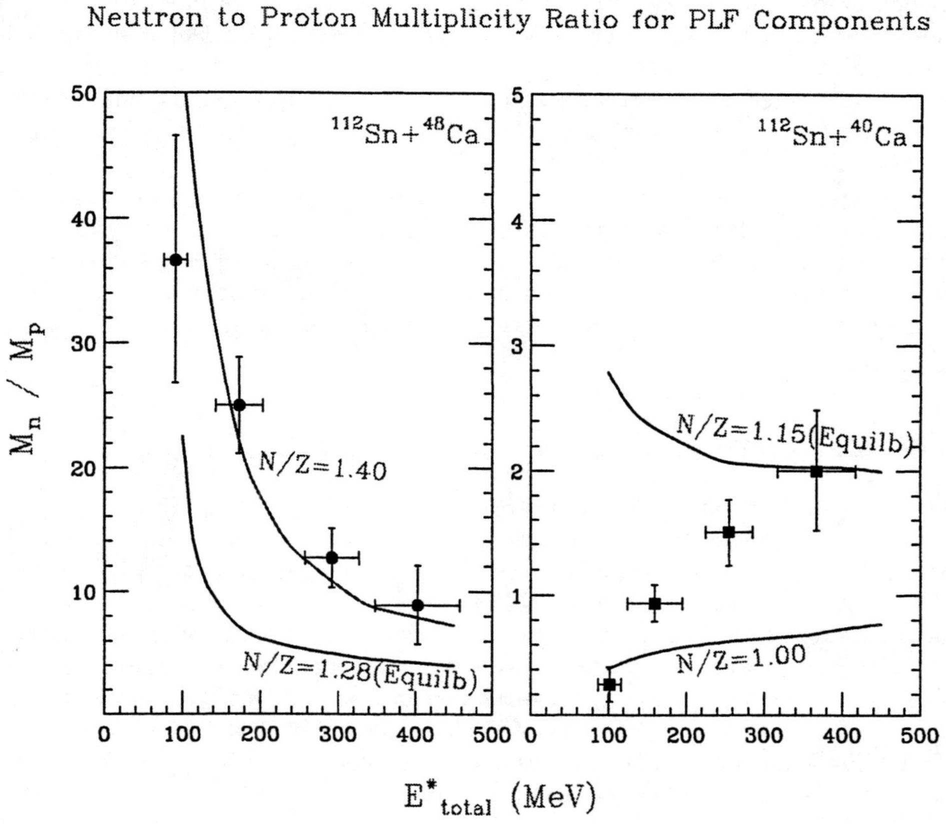

Figure 9: Neutron-to-proton multiplicity ratio vs. total excitation energy, for the two reactions $^{112}Sn+^{48}Ca$ (left) and $^{112}Sn+^{48}Ca$ (right) at $\epsilon = 35$ MeV.

variables providing independent information needed for an interpretation of the reaction dynamics in terms of non-equilibrium transport processes. There are indications [Ram00] that these variables remain powerful indicators of reaction dynamics at even higher energies.

Average A/Z ratios can be determined in a relative straightforward fashion from the isotopic distributions of light particles and clusters emitted in heavy-ion reactions. However, taken as isolated aspects, these observables may be ambiguous. For example, a neutron-rich flow of particles emitted from an equilibrated composite system could indicate nothing more interesting than the size of the available phase space, whereas in a direct breakup scenario, this fact may reflect initial-state correlations or incomplete mixing of projectile and target constituents. However, taken together with simultaneously measured other observables, the isotopic distributions of the final reaction products can play a crucial role in establishing the reaction mechanism, e.g., for its similarity to thermal (statistical) dynamics.

In this context, it should be pointed out that a determination of only the first moments of the multivariate probability distribution for system observables, a task that already exceeds the capacity of many contemporary experimental setups, provides an incomplete picture of heavy-ion reaction mechanisms. For example, statistical processes characterized by certain correlations between different moments of the probability distribution, e.g., as governed by a fluctuation-dissipation theorem [Sch84].

3 Isospin-Physics: The Next New Thing

In addition to its role as an indicator of proximity to chemical equilibrium, the A/Z degree of freedom may provide genuinely new information on effective nucleon-nucleon interaction forces in heavy-ion reactions at Fermi energies. Because here the time scales for intrinsic nucleonic motion and collective variables merge, the mean field cannot adjust adiabatically to the collective motion. The resulting sudden compression or dilution of nuclear matter, mainly in the overlap region, is resisted by diabatic, density-and momentum-dependent in-medium interactions. These regions of non-normal nuclear matter density (ρ) can develop mechanical and chemical instabilities, which influence reaction dynamics and the decay of the intermediate nuclear system.

In a heavy-ion collision, to first order, the average (mean-field) energy (ϵ) of a nucleon in the nuclear medium changes by amounts that have to be provided by the relative motion of projectile and target nuclei. In this fashion, diabatic nucleus-nucleus interactions arise, which can influence the heavy-ion reaction dynamics in a measurable way. This effect has the interesting consequence that, in principle, information on the nuclear equation of state (EOS) may be obtained from *dynamical* observables. Furthermore, because of the nuclear symmetry energy, the interaction energy is different for neutrons and protons. Therefore, there are iso-scalar and iso-vector parts of the nucleus-nucleus liquid-drop and proximity interactions [Pom80] leading to effective interactions depending on the asymmetry in projectile-target A/Z values or isospins, I=(N-Z)/A. It is customary also to consider the relative neutron excess degree of freedom,

$$\delta = (\rho_n - \rho_p)/\rho \tag{4}$$

and to express the isospin-dependent nuclear equation of state $\epsilon(\rho, \delta)$ in terms of this variable.

Different models for isospin-dependent interactions are reviewed in other sections of this book. For example, a model by Myers and Swiatecki [Mye98] considers a semi-classical Thomas-Fermi approach to the motion of neutrons and protons in finite nuclei, interacting via a short-range effective interaction of the Yukawa type. In this model, the isospin-dependent single-particle energy is given by a series expansion in the relative density (ρ/ρ_o),

$$\epsilon(\rho, \delta) = \epsilon_F[a(\delta)(\frac{\rho}{\rho_o})^{2/3} - b(\rho)\frac{\rho}{\rho_o} + c(\delta)(\frac{\rho}{\rho_o})^{5/3}] \tag{5}$$

where ϵ_F is the Fermi energy and a, b, and c are functions of the (local) neutron excess δ. This relation represents the equation of state (EOS) in the model, containing kinetic and (mean-field) potential energy parts. In other approaches, effective Skyrme forces are used to describe these interactions. The mean-field interaction potential applicable to heavy-ion collisions has the typical form of an expansion in the relative density (ρ/ρ_o),

$$V^{n(p)}(\rho,\delta) = a(\delta)\frac{\rho}{\rho_o} + b(\delta)(\frac{\rho}{\rho_o})^{\sigma} + V^{p}_{Coul} + V^{n(p)}_{asym}(\rho,\delta) \tag{6}$$

In this expression [Li98], the coefficients a and b are functions of the (local) neutron excess δ, and σ is an exponent characteristic of the Skyrme type used for the effective force. The Coulomb interaction acts only on the protons in the system. The last term in Equ. (6) is the symmetry term, containing the dependence of the nuclear asymmetry energy on the matter density and the neutron excess. These are poorly known physical variables, which are of fundamental interest for several fields of science, for example for cosmology.

Of interest for heavy-ion reactions are also the chemical potentials, since they determine the net exchange of neutrons and protons between the reaction partners, an experimental observable. For moderately high nuclear temperatures, $T \geq 4$ MeV, these potentials can be expressed as

$$\mu_{n(p)} \approx V^{n(p)} + Tf(\rho_{n(p)}, T) \tag{7}$$

in terms of the mean field and a density- and temperature-dependent function f. In this fashion, a connection is drawn between the isospin-dependent equation of state (the mean nucleus-nucleus interaction) and the dynamic driving forces influencing nucleon exchange between the reaction partners in a heavy-ion collision. As a result, one expects the essentially binary reaction dynamics of heavy-ion reactions in the Fermi energy domain to be sensitive to these fundamental in-medium interactions revealed, for example, by the nature of PLF deflection functions, the relative importance of fusion as compared to the dissipative reaction cross section. Furthermore, these interactions determine the moments of the probability distributions for the transfer of neutrons and protons between projectile and target, e.g., average and (co-)variances. Additional information is carried by fast particles promptly emitted during the nuclear interaction.

In the absence of residual nucleon-nucleon interactions, the above diabatic interactions are essentially conservative, returning compressional energy back into kinetic energy of collective expansion. However, in general, higher-order in-medium NN scattering exists, giving rise to dissipative nucleus-nucleus interaction forces. These latter dissipative forces are very different from the one-body, wall-and window-type [Sch84] dissipation prominent at near-barrier bombarding energies and should be detectable in heavy-ion reactions at Fermi energies. In addition to causing dissipative retardation of collective nucleus-nucleus motion, isospin-dependent in-medium NN scattering effects the prompt emission of scattered nucleons into the continuum. While multiple NN scattering quickly leads to a distribution of the initial kinetic energy onto many intrinsic degrees of freedom, at every stage of such an equilibration cascade, fast nucleons can be emitted from the interacting system as "non-statistical" (pre-equilibrium) particles. Again, the expected effects are different for neutrons and protons. Thus, the multiplicities, angular and energy distributions of these non-equilibrium particles contain independent information on the isospin-dependent in-medium nucleonic interactions. Together with the mean-field influences on the reaction dynamics, the isospin-dependent dissipation and particle-emission processes should fit into an internally consistent picture of the reaction mechanism.

Quantitative comparisons to data rely on an intermediate layer of model descriptions, namely the modeling (propagation) of the essential reaction dynamics. A number of very different theory "platforms" proposed in the literature have been utilized for such comparisons, from relatively simple trajectory type calculations, e.g., based on the nucleon exchange model NEM referred to earlier, to more detailed microscopic, multi-dimensional transport models. The latter models include various realizations of Boltzmann-Ühling-Ulenbeck-type (BUU, etc.) [Ber88, Bau93] approaches, classical or quantal molecular-dynamics models (QMD, AMD, FMD, etc.) [Aic91, Ono92, Fel90, 97]. In addition, models have been proposed [Bon85, Gro93] for the statistical decay of hot and rather diluted nuclear systems. To date, most work is based on NEM or BUU simulation calculations to collisions between heavy ions at Fermi energies.

As an example, calculations of PLF deflection functions with a BUU model [Sob95] for the ^{209}Bi+^{136}Xe reaction are shown in Fig. 10. In this figure, the open and filled squares represent a stiff and a soft EOS, respectively, for symmetric nuclear matter. The diamonds are the results of a calculation assuming an isospin-soft EOS. The deflection functions are meant to illustrate expected sensitivities of the data, but no attempt has been made as yet to adjust parameters to fit experimental particular angle-energy data. Arguably, such a parameter adjustment should only be done to describe the entire set of observables simultaneously.

The trends illustrated in Fig. 10 are intuitively expected: The isospin-dependent interaction (open diamonds) is the most repulsive one, not leading to orbiting but to a bouncing-off behavior. Consequently, for this case, the cross section for fusion is smallest, which is best consistent with experiment, where there is no positive evidence for fusion at all [Bal93, 95]. All calculations predict a monotonic relation between impact parameter and PLF kinetic energy (E/A) or energy loss. Experimentally, this latter behavior is confirmed qualitatively. However, experimentally a damping exceeding 60% of the initial kinetic energy, i.e., a slowing-down further to full damping ($E/A \approx 10$ MeV), is unlikely for this system. Furthermore, orbiting is observed for the ^{209}Bi+^{136}Xe reaction from bombarding energies of E/A = 28 MeV to E/A = 55 MeV [see Fig. 1]. Therefore, the so-called balance energy must be significantly higher than 55 MeV per nucleon.

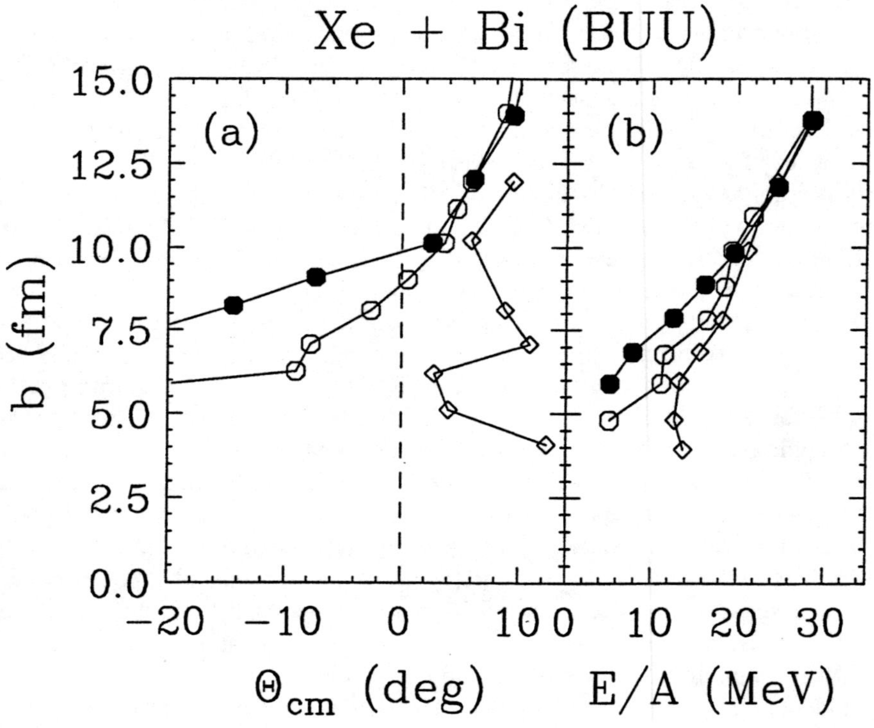

Figure 10: al correlations of angle/impact parameter (a) and angle/energy per nucleon (b), predicted by BUU for symmetric matter (squares) and an isospin-soft EOS (diamonds). (From [Sob94])

In other BUU-type isospin-dependent calculations [Col98] including density fluctuations in the nuclear interior, the different effects of isospin-dependent interactions on isobaric reaction pairs are emphasized. For example, for the n-rich light and more symmetric system 64Ni+46Ar at 30 MeV per nucleon, an isospin-soft EOS results in more attraction between projectile and target, and there is more dissipation, than for the case of an isospin-stiff EOS. In the former case, the system may fuse at a certain impact parameter, while the isospin-stiff EOS leads to more repulsion and binary dissipative collisions. For the N=Z isobaric analog reaction ^{64}Ge+^{46}V, the behavior is inverted. A test of such interesting predictions has to await future experiments with exotic, secondary beams with a range of projectile A/Z ratios.

Fragmentation dynamics is expected to provide additional testing grounds for the exploration of isospin dependent interactions. Figure 11 provides a two-dimensional (x-z plane) overview over the time sequence of a simulated ^{112}Sn+^{40}Ca collision (left column) and of a ^{112}Sn+^{48}Ca collision (right column). The z direction is that of the beam. In both cases, an impact parameter of b = 7 fm was chosen for the calculations, corresponding to a semi-peripheral collision. The nuclear EOS was chosen to be soft in the symmetry energy degree of freedom, as selected by the scaling function F_2 [Li98]. can be seen from the right column in this figure corresponding to the neutron-rich

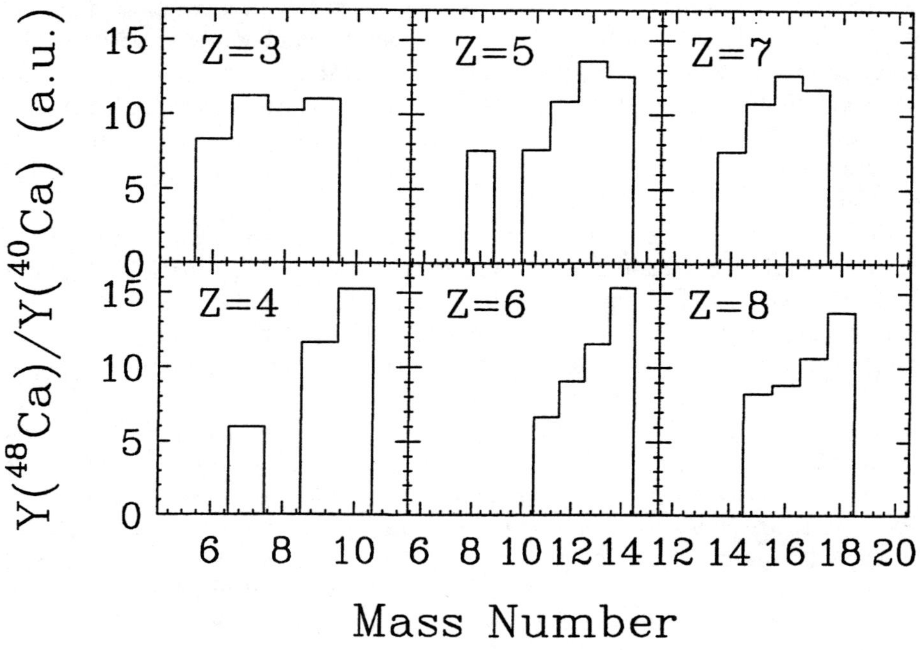

Figure 11: Two-dimensional time sequence of BUU simulations of ^{112}Sn+^{40}Ca (left) and ^{112}Sn+^{48}Ca (right) reactions at ϵ= 35 MeV and b = 7 fm. An iso-soft EOS has been used.

system ^{112}Sn+^{48}Ca, projectile and target fuse and intermix their nucleons significantly during the first (100-200) fm/c of the interaction time. Later, the system stretches and forms a neck. This neck subsequently breaks and leaves one or two intermediate-mass clusters behind, in addition to a number of energetic light particles. A very different situation is predicted for the neutron-poor reaction system ^{112}Sn+^{40}Ca illustrated on the left of Fig. 11. Here, none or only very small nuclear clusters are produced in a similar collision. Such an inhibition of IMF emission for neutron-rich systems appears to be at variance with experiment [Dem96]. It should also be noted that the above BUU predictions for ^{112}Sn+^{40,48}Ca depend on the stiffness or softness of the EOS employed and on impact parameter. For example, calculations done with a stiff EOS predict no IMF production at all. Predicted IMF emission probabilities depend significantly also on the number of test particles used in the calculations. In fact, IMF cluster emission is predicted to occur only in a narrow region of impact parameters around $b = 7$ fm, while experimentally, such clusters are observed for a large range of impact parameters. This is an striking puzzle, an inconsistency between model and experiment that demands closer scrutiny. In addition, it is a known fact that, the more test particles are used in the BUU calculations, the smoother the reaction features become, including a disappearance of the cluster emission process. However, one can argue that, in reality, density fluctuations would be present, having a similar effect as the artificial fluctuations introduced by poor sampling statistics in the calculations. This observation calls for a simultaneous measurement of observables that measure fluctuations independently, such as the widths of fragment A and Z distributions.

The above BUU calculations illustrate a dynamical process of nuclear cluster production in heavy-ion reactions, which would fit the general kinematics of IMF emission discussed previously (see Fig. 3). The calculations suggest origin and structure for an "intermediate, mid-rapidity source," revealed in experiments. In the simulations, IMFs are formed from both projectile and target matter and appear relatively late in a collision, when the isospin asymmetry has largely relaxed. The neck matter is pulled out of both PLF and TLF, when the intermediate complex breaks apart. The average velocities of such IMFs are expected to be intermediate between those of PLF and TLF and probably somewhat closer to the velocity of the TLF or that of the cm system. In addition, the fact that IMF emission occurs late and after considerable intermixing of all nucleons in the model, the IMF clusters should reflect the A/Z ratio of the composite system, characteristic for overall isospin equilibration. The model makes definite predictions about the isospin relaxation as a function of impact parameter and bombarding energy, which can be subjected to equally unambiguous experimental test.

Unfortunately, experiments are extremely rare, where IMFs have been identified with respect to A and Z, and none is known yet that would allow one to perform a detailed, multi-facetted comparison to reaction models. Some summary information is available for the ^{112}Sn+^{40,48}Ca reactions introduced already above, as well as for "incomplete fusion" reactions induced by 32A-MeV ^{14}N projectiles on Sn and Au targets [Mur95].

In the former, ^{112}Sn+^{40,48}Ca experiments, IMFs were detected in addition to PLFs and light particles. In a range of forward and sideways angles, the intermediate "neck-like" source is emphasized. It is kinematically not likely that the IMFs considered here have been evaporated statistically from excited primary PLF or TLF reaction fragments. However, because of poor statistics, the inclusive isotopic yields are averages over energy loss or impact parameter, emphasizing semi-central collisions.

Ratios $Y(^{48}\text{Ca})/Y(^{40}\text{Ca})$ of isotopic IMF yields from the corresponding intermediate, neck-like source are available for the elements lithium through oxygen, as depicted in Fig. 12. It is obvious from these yield distributions that in the reaction ^{112}Sn+^{48}Ca, more neutron-rich, as well as fewer neutron-poor clusters are emitted than in the reaction induced by the neutron poorer ^{40}Ca projectile. The enhancement is significant, as the yield ratios vary by up to factor of 3, in particular, when considering the fact that the N/Z ratios for the two composites differ by only 10%. The variation in yield is even larger than the difference in the N/Z ratios of the projectiles ^{40}Ca and ^{48}Ca, which amounts to only 40%. Certainly, the results displayed in Fig. 12 are not indicative of overall, or even local, isospin equilibration. For these two reactions, it is known that the PLF/TLF isospin asymmetry relaxes with increasing excitation energy and decreasing impact parameter. As illustrated by Fig. 9, the PLF N/Z ratio either stays constant (for the ^{112}Sn+^{48}Ca reaction) or increases (for the ^{112}Sn+^{40}Ca reaction) with increasing energy dissipation. Therefore, with increasing energy dissipation, the PLF N/Z ratios of the two reactions approach each other. Large IMF emission probabilities sample presumably more dominantly the central, rather than peripheral reactions and should reflect the nearly equilibrated charge density in the overlap zone between PLF and TLF. If the BUU scenario discussed above were realistic, one would expect more mixing of PLF and TLF matter to have occurred in the formation of a neck than reflected in the IMF isotopic yield ratios of Fig.12.

A simple and convincing explanation of the isotopic IMF yields discussed above is currently not available. The BUU scenario implies several inconsistencies with the data, both concerning the range of impact parameters leading to IMF emission and the variation of isotopic IMF yields. If the data shown in Fig. 12 can be confirmed, they could indicate that significant spatial charge polarization occurs within projectile and target. This process would occur necessarily early on in a collision approach phase. It would be followed by an early IMF emission preserving this polarization. Such a process could perhaps also show up in the abundance and other properties of the light particles emitted non-statistically on a fast time scale.

The diagrams of Fig. 11 picturing the progress of ^{112}Sn+^{40,48}Ca collisions in BUU simulation also illustrate properties of the non-statistical particle emission process. In fact, the calculations predict that the emission of such nucleons and other light particles occurs on a "mesoscopic" timescale, during the collision but not confined to the shortest times. Although the emission process commences in the approach phase, at about 50 fm/c, the bulk of the fast particles is emitted throughout the collision. The first of each of the 4 frames in Fig. 11 represents an early stage (t = 50 fm/c) of the collision. At this time, significant matter overlap between projectile and target has already developed, and one expects substantial mixing of projectile and target nucleons to have occurred.

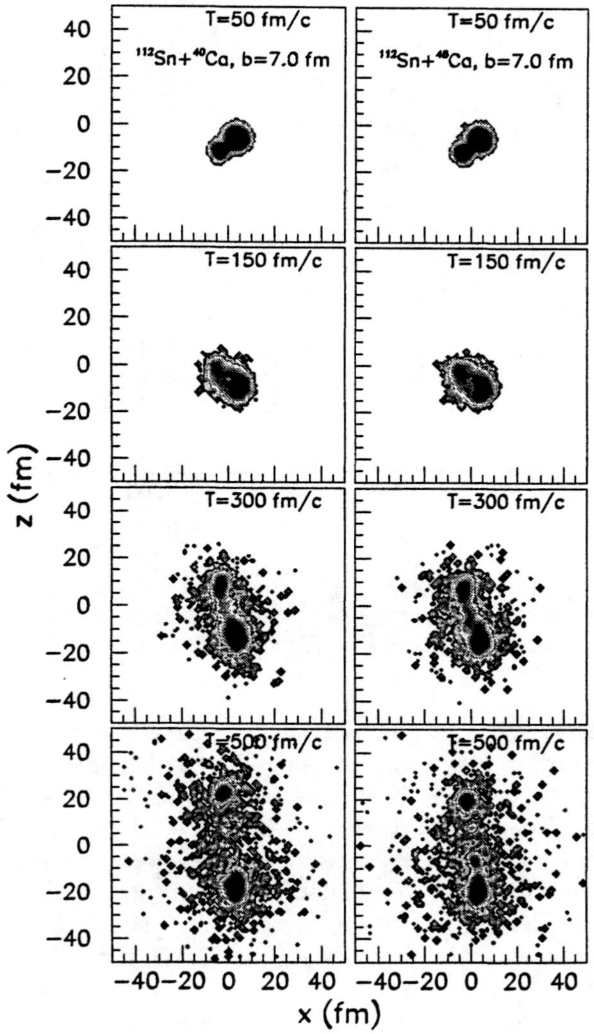

Figure 12: Ratio of yields of IMF isotopes in the reactions ^{112}Sn+^{48}Ca and ^{112}Sn+^{40}Ca.

However, there is very little emission of fast particles. Only for times larger than a few hundred fm/c are there many fast (test) particles visible in the plots. Hence, in the model, particle emission takes place during the heavy-ion collision, but it occurs dominantly at its later, and not in its early stages. Since many particles are emitted late, one might also expect a relatively soft energy spectrum for these particles.

Figure 13 puts this observation on a quantitative basis. Here, the predicted ratio of neutron-to-proton multiplicities is plotted *vs.* elapsed interaction time, for an impact parameter of b =7 fm. Two different parameterizations of the symmetry energy V_{asym} are compared in the two panels. The parameterization on the left ($asym = 1$) is more repulsive for neutrons and more attractive for protons than the parameterization depicted on the right ($asym = 3$). The latter form has actually a negative curvature for neutrons. (These symmetry potentials correspond to the scaling functions F_2 and F_3, respectively, in [Li98]).

For both parameterizations shown in Fig. 13, the neutron-to-proton ratio is initially larger for the neutron-rich projectile ^{48}Ca than for ^{40}Ca. At later times, the ratio drops precipitously and takes on an asymptotically constant value. The time dependencies of the multiplicity ratios are very different for the two reactions and show characteristic sensitivity to the symmetry energy. The dependence on the projectile N/Z ratio is particularly obvious for the parameterization on the right.

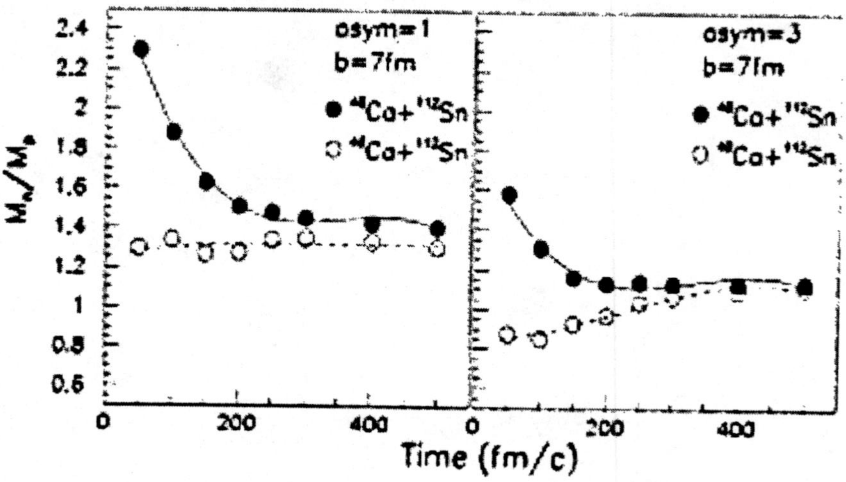

Figure 13: BUU simulations of the multiplicity ratios of neutrons and protons emitted promptly in ^{112}Sn+^{40,48}Ca reactions. Isospin-soft (left) and stiff (right) EOS assumed in BUU.

In the latter case, the emission of fast neutrons builds up in time.

The examples of the two isospin parameterizations in Fig. 13 demonstrate a notable sensitivity of the asymptotic nucleon multiplicity ratios to the effective asymmetry potentials (and the associated EOS). However, appreciating that an average correlation between the energy of a non-statistical particle and its time of emission may exist, one can realize an even greater sensitivity to this symmetry energy. Typically, in an exciton model [Bla75] of non-statistical (pre-equilibrium) particle emission, early emission times lead to higher particle kinetic energies. This behavior is expected also for other nuclear models emphasizing nucleonic interactions such as QMD, but it is not necessarily the case in BUU-type approaches. In BUU transport models, fast, non-statistical nucleon emission can occur in at least two ways. The first, presumably less important mechanism is that of in-medium nucleon-nucleon scattering, which bears similarities to an exciton model. The second and more powerful mechanism has a collective origin: Nucleons can be ejected from the strong mean-field distortions caused by compressional effects generated in a heavy-ion collision.

If non-statistical nucleon emission is due to collective compressional effects, the particle spectra will be harder for delayed than for early emission, because of the time needed to produce a significant matter overlap between projectile and target. This effect should be strong for the parameterization of the symmetry energy illustrated on the *r.h.s.* of Fig. 13, since for this case, substantial compressional energy is built up, before neutrons are finally ejected from the mean field. This feature seems to be reflected in the energy spectra of non-statistical neutrons predicted by isospin-dependent BUU simulations. In this figure, the neutron yields are plotted *vs.* energy for two emission angles. Obviously, there is a very strong dependence of the particle spectra on the isospin-dependent EOS. If the strength of the effect is confirmed, the particle energy spectra could present the most sensitive indicators of the strength of the isospin-dependent mean field. The energy spectra on the right (for $asym = 3$), are much harder than typically observed experimentally (cf. Fig. 4). Even the spectra for the parameterization illustrated on the left are somewhat harder than observed. In addition to the shape of the energy spectra of the fast particles emitted, their angular distributions carry information about the emission process. Anisotropies in the angular distributions may result from various sources, including the velocity of the effective emitting source, the anisotropy of the nucleon-nucleon scattering process, as well as final-state interactions leading to attenuation and multiple scattering in the nuclear medium. All these effects need to be treated consistently within a single model, in order to derive meaningful information even on selected aspects of the effective nucleonic interactions, e.g., the EOS. However, it is encouraging already that detailed BUU calculations, e.g., by Li et al. [Li98], predict a net sensitivity of the kinetic-energy spectra of fast particles to the isospin-dependent EOS.

The validity of these general ideas can be subjected to first tests, utilizing the data on non-statistical nucleon emission in the same two reactions $^{112}Sn+^{40,48}Ca$ discussed above. Some instructive data on neutron and proton spectra from these reactions are depicted in Fig. 15. Here, for a detection angle of $\Theta = 60^o$, spectra are shown, arranged from top to bottom according to increasing energy loss, $< E^* > \approx 100$ MeV $\rightarrow$ $< E^* > \approx 400$ MeV, in the reactions. In the same direction, the collision impact parameter should decrease. The left column in Fig. 15 depicts neutron (circles) and proton (squares) data for $^{112}Sn+^{40}Ca$, while the right column shows the same for $^{112}Sn+^{48}Ca$. The neutron spectra show the two spectral components discussed already in the context of Fig. 4. At the present detection angle of $\Theta = 60^o$, the spectra are a superposition of a low-energy evaporative component, associated with the TLF, and a high-energy, non-statistical component. Because of Coulomb barrier and detection thresholds for protons, the low-energy evaporative component is suppressed and less obvious in the proton spectra. Within present experimental accuracy, the angular distribution of the non-statistical particles is isotropic in the cm system of the nucleon-nucleon system. Hence, one can fit these spectra at measured angles and determine the overall multiplicities, M_n and M_p, by interpolation and extrapolation. To no great surprise, the neutron multiplicities are generally larger than proton multiplicities. However, there is a definite evolution with increasing energy loss of the relative yields of non-statistical neutrons and protons. In addition, there is a marked difference in the behavior of the two reaction systems.

As discussed previously (see Fig. 9), for the reaction $^{112}Sn+^{40}Ca$, with increasing energy loss an evolution of the PLF/TLF isospin asymmetry degree of freedom towards equilibrium has been observed. Presumably, this evolution is associated with an increasing overlap of PLF and TLF matter distributions and the ensuing mixing of the nucleons of the two fragments. In agreement with this picture, proceeding from top to bottom, one observes a decrease of the non-statistical multiplicity ratio $R = (M_n/M_p)|_{neq}$ from $R \approx 2$ to $R \approx 1$.

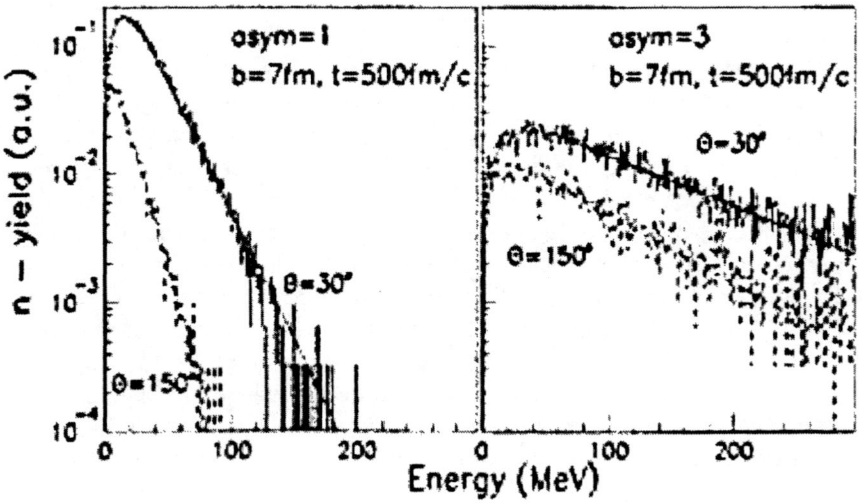

Figure 14: Lab energy spectra of promptly emitted neutrons in the reaction $^{112}Sn+^{48}Ca$, as predicted with BUU with the same parameterizations as used for Fig. 13.

The situation is very different for the reaction illustrated on the *r.h.s.* of Fig. 15 for the neutron-rich system $^{112}Sn+^{48}Ca$. First, the multiplicity ratio R for the non-statistical particles is very much larger than for the neutron-poorer system. With increasing energy loss, one observes the same type of evolution as for the neutron-poor system. The decrease in R is very steep, which requires a steep asymmetry potential, i.e., significant chemical-potential gradients. However, the multiplicity ratio never decreases below $R \approx 2.5$, although the bulk N/Z ratios of the two ^{112}Sn+^{40,48}Ca systems are identical within 10%. This behavior indicates that even semi-central ^{112}Sn+^{48}Ca collisions do not lead to chemical equilibrium characterized by significant mixing of projectile and target nucleons, *before* non-statistical particles are emitted. The same conditions must hold for the neutron-poor system where, however, the corresponding evidence is hidden.

These observations impose tight constraints on the various timescales for fusion and isospin equilibration. Apparently, fast nucleons are emitted from non-equilibrated and very neutron-rich subsystems. As a likely scenario, prompt emission of particles from the surface regions of the interacting nuclei during the early phases of a collision comes to mind. The experimental observations may also imply a more important role of in-medium nucleon-nucleon scattering than previously thought. This is a trend expected from the large average values and n-p asymmetries for the in-medium cross sections in low-density regions, regions that are also neutron rich. Alternatively, one also has to consider the possibility of breakup of PLF and TLF on impact. This scenario would be challenged by other experimental data, such as deflection functions, should it require a stiff EOS with a stiff asymmetry term. Such a situation can be modeled and studied more appropriately in molecular-dynamics simulations.

The few, but rather clear-cut, experimental data for non-statistical nucleon emission discussed above show some generic similarities with the emission of IMFs in heavy-ion reactions at Fermi energies. Such a similarity is apparent also in other data [Mur95], where both, fast particles and energetic clusters (Fig. 5) have been measured. Both types of particles appear to be emitted from a hypothetical third source traveling at some speed intermediate between those of PLF and TLF. In BUU, such a source is modeled as the decay of a neck between the main fragments. Experimentally, this source is not in thermal, mechanical, or chemical equilibrium with the massive PLF and TLF nuclei. It appears to be neutron rich, or the underlying process that this source simulates seems to emit preferentially neutron-rich particle and cluster species. For fast nucleons, this experimental fact has been demonstrated above. The emission of light particles and clusters such as ^{3,4,6}He and Li ions has been studied [Dem96] in the reactions ^{112,124}Sn+^{124,136}Xe at $\epsilon = 55$ MeV. It was concluded from this study that particles at intermediate velocities are substantially more neutron rich than those evaporated from the heavy, Xe-like PLFs. This is consistent with the scenario for IMF and fast particle emission.

In view of this experience, it seems currently unlikely that isospin equilibrium is reached in the Fermi energy domain, where reactions still lead to massive remnants of projectile and target, even in central collisions [Lot93, Bal95, Lar98]. In fact, a dominantly non-statistical, non-equilibrium nature of reactions at Fermi bombarding energies is suggested by an observation of incomplete damping [Que93] of the kinetic energy available in the entrance channel and by the dynamic emission patterns of intermediate-mass fragments [Tok95-96] in such reactions. A basically dissipative reaction environment is reflected also in the display of memory of the entrance-channel A/Z asymmetry observed [Tok94-96, Agni98,99, Dem96] for different types of products from the reactions ^{209}Bi+^{136}Xe, ^{112}Sn+^{40,48}Ca, and ^{120}Sn +^{136}Xe. The lack of chemical equilibrium suggested at the present state of research would fit well into this scenario. Such a scenario is fundamentally dynamical, following a multi-dimensional path toward relaxation, which can be followed experimentally. Thus, the challenge to understand the motion along the isospin degrees of freedom also provides new experimental opportunities to monitor the time evolution of the microscopic reaction mechanism.

4 Conclusions

Accepting the dominance of dynamics in nuclear collisions at Fermi energies and a substantial absence of final nuclear configurations in mutual statistical equilibrium, the pursuit of fundamental scientific goals in nuclear mechanistic studies at Fermi energies continues to present a formidable task. These goals include gathering information on the effective isospin-dependent in-medium nucleonic interactions, i.e., the mean-field equation of state (EOS) of nuclear matter, the residual interactions leading to in-medium nucleon-nucleon scattering, and nucleonic correlations responsible for nucleonic aggregation and properties of intermediate-mass nuclear clusters.

Conventional strategies aiming at information on the nuclear EOS have consisted in searching for critical behavior of nuclear systems produced in central heavy-ion collisions, such as critical temperatures and cluster distributions reflecting a nuclear liquid-gas phase transition. While it may be premature to abandon altogether the hope of observing this statistical nuclear phase transition in heavy-ion reactions, a new access to the fundamental information on the nuclear EOS and in-medium effects has been found, provided by new observations associated with the isospin degrees of freedom.

To date, there have been very few experiments exploring these new degrees of freedom. Available data are not sufficient to develop a coherent reaction scenario with these new observables. However, it is clear that the initial projectile-target isospin (A/Z) asymmetry influences the reaction dynamics, as reflected consistently in the isotopic distributions of the various types of final products.

On the other hand, non-equilibrium emission of light particles and clusters are illustrative examples demonstrating interesting new isospin-dependent effects. For example, it has been shown that relative abundances and energy spectra of non-statistical (pre-equilibrium) nucleons and clusters contain information about the nuclear EOS, in particular on its dependence on the nuclear asymmetry energy. However, there is no consistent reaction scenario as yet that would explain consistently all observations. Some of the observed dependencies are expected in certain reaction models, others are not. Exploiting the additional isospin degree of freedom will help to develop such an internally consistent, holistic reaction scenario explaining simultaneously a host of reaction features, which also incorporates important features of the nuclear equation of state. While there is no detailed roadmap for future research available that could guarantee success in this venture, new experimental and theoretical discoveries are pointing to new horizons.

Much of the experimental work on the reactions ^{112}Sn+^{40,48}Ca reported here has been done by Dr. D.K. Agnihotri, whose contributions we gratefully acknowledge. One of us (WUS) has benefited from discussions with B.A. Li and M. DiToro, whose interest in our experiments has been encouraging.

This work has been supported by the United States Department of Energy Grant No. DE/FG02-88ER40414.

5 References

[Agn97] D.K. Agnihotri, S.P. Baldwin, B. Djerroud, B. Lott, B.M. Quednau, W. Skulski, J. Tõke, W.U. Schröder, and R.T. de Souza, Advances in Nuclear Dynamics (A.C. Mignerey, Edt.) 3, 67 (1997);

[Agn98] D.K. Agnihotri , University of Rochester Ph.D. Thesis, and D.K. Agnihotri et al., to be published.

[Aic91] J. Aichelin, Phys. Rep. 202, 233(1991).

[Bal93] S.P. Baldwin, B. Lott, B.M. Szabo, B.M. Quednau, W.U. Schröder, J. Tõke, L.G. Sobotka, J. Barreto, R.J. Charity, L. Gallamore, D.G. Sarantites, D.W. Stracener, R.T. de Souza, Advances in Nuclear Dynamics, B.Back, W.Bauer, J. Harris (Editors), World Scientific (Singapore 1993), p.36.

[Bal95] S.P. Baldwin, B. Lott, B.M. Szabo, B.M. Quednau, W.U. Schröder, J. Tõke, L.G. Sobotka, J. Barreto, R.J. Charity, L. Gallamore, D.G. Sarantites, D.W. Stracener, R.T. de Souza, Phys. Rev. Lett. 74, 1299 (1995).

[Bau93] W. Bauer, Prog. Part. Nucl. Phys. 30, 45 (1993).

[Ber88] G.F. Bertsch and S. DasGupta, Phys. Rep. 160, 189(1988), and references cited.

[Bla75] M.B. Blann, Ann. Rev. Nucl. Sci. 25, 123 (1975).

[Bon85] J.P. Bondorf et al., Nucl. Phys. A443, 321 (1985); ibid A444, 460 (1985).

[Bor90] B. Borderie. H.F. Rivet, and Tassan-Got, Ann. Phys. Fr. 15, 287 (1990).

[Bor99] B. Borderie et al. Eur. Phys. J. A 6, 197 (1999).

[Cha88] R.J. Charity, GEMINI code, available via anonymous ftp from wunmr.wustl.edu/pub/gemini.

[Col98] M. Colonna, M. Di Toro, G. Fabbri, and S. Maccarone, Phys. Rev. C57, 1410 (1998).

[Dem96] J.F. Dempsey, R.J. Charity, L.G. Sobotka, G.J. Kunde, S. Gaff, C.K. Gelbke, T. Glasmacher, M.J. Huang, R. Lemon, W.G. Lynch, L. Manduci, L. Martin, M.B. Tsang, D. K. Agnihotri, B. Djerroud, W. U. Schröder, W. Skulski, J. Tõke, and W.A. Friedman, Phys. Rev. C54, (1996), 1710.

[Des88] R. T. de Souza, J. R. Huizenga, and W. U. Schröder, Physical Review C 37 1901 (1988).

[Dje96] B. Djerroud, W. Skulski, D.K. Agnihotri, S.P. Baldwin, W.U. Schröder, J. Tõke, L.G. Sobotka, R.J. Charity, J. Dempsey, D.G. Sarantites, B. Lott, W. Loveland, K. Aleklett, Advances in Nuclear Dynamics, W.Bauer and G.D. Westfall (Editors), Plenum Press (New York and London) 2, 333 (1996).

[Fel85] H. Feldmeier and H. Spangenberger, Nucl. Phys. A 435, 229 (1985).

[Fel90] H. Feldmeier, Nucl. Phys. A515, 147 (1990).

[Fel97] H. Feldmeier and J. Schnack, Prog. Part. Nucl. Phys. 39, 393(1997).

[Hil88] Hilscher, Proc. of Specialists's Meeting on Preequilibrium Nuclear Reactions OECD, Semmering, Austria, (B. Strohmaier, Editor), p. 245 (1988).

[Gro93] D.H.E. Gross et al., Prog. Part. Nucl. Phys. 30, 155 (1993), and references cited therein.

[Joh96] H. Johnston et al., Phys. Lett. B 371, 186 (1996).

[Joh97] H. Johnston et al., Phys. Rev. C 56, 1972 (1997).

[Lar98] Y. Larochelle et al., Phys. Rev. C57, R1027 (1998).

[Li98] Bao-An Li, C. M. Ko, and W. Bauer, Int. Jour. Mod. Phys. 7, 147 (1998).

[Lot93] B. Lott, S.P. Baldwin, B.M. Szabo, B.M. Quednau, W.U. Schröder, J. Tõke, L.G. Sobotka, J. Barreto, R.J. Charity, L. Gallamore, D.G. Sarantites, D.W. Stracener, R.T. de Souza, Advances in Nuclear Dynamics, B.Back, W.Bauer, J. Harris (Editors), World Scientific (Singapore 1993), p.159.

[Mor97] L.G. Moretto and G. Wozniak, Phys. Rep. 287, 249 (1997), and references cited.

[Mur95] Yu. Murin et al., Phys. Rev. C51, 2794 (1995).

[Mye98] W.D. Myers and W.J. Swiatecki, Phys.Rev. C57, 3020 (1998).

[Ono92] A. Ono. H. Horiuchi, T. Maruyama, and A. Ohnishi, Phys. Rev. Lett. 68, 2898 (1992); Prog. Theor. Phys. 87, 1185(1992).

[Pla88] R. Planeta, S. H. Zhou, K. Kwiatkowski, W. G. Wilson, V. E. Viola, H. Breuer, D. Benton, F. Khazaie, R. J. McDonald, A. C. Mignerey, A. Weston-Dawkes, R. T. de Souza, J. R. Huizenga, and W. U. Schrder, Physical Review C 38 195 (1988).

[Que93] B.M. Quednau et al., Phys. Lett. 309B, 10 (1993).

[Pom80] K. Pomorski and K. Dietrich, Z. Physik A295, 355 (1980).

[Ram00] F. Rami et al., Phys. Rev. Lett. 84, 1120 (2000).

[Ran76] J. Randrup, Nucl. Phys. A259, 253 (1976).

[Ran 78] J. Randrup, Nucl. Phys. A307, 319 (1978).

[Ran 79] J. Randrup, Nucl. Phys. A327, 490 (1979).

[Ran82] J. Randrup, Nucl. Phys. A383, 468 (1982).

[Sch84] W. U. Schröder and J.R. Huizenga, in Treatise on Heavy-Ion Science, (D.A. Bromley, Editor), Plenum Press (New York and London,1984), Vol.2, pp.113-726; and references cited therein.

[Sku96] W. Skulski, B. Djerroud, D.K. Agnihotri, S.P. Baldwin, W.U. Schröder, J. Tõke, X. Zhao, L.G. Sobotka, J. Barreto, R.J. Charity, L. Gallamore, D.G. Sarantites, B. Lott, W. Loveland, K. Aleklett, Physical Review C53, R2594 (1996).

[Sku97] W. Skulski, J. Dempsey, B. Djerroud, D.K. Agnihotri, S.P. Baldwin, J. Tõke, W.U. Schröder, L.G. Sobotka, R.J. Charity, B. Lott, W. Loveland, K. Aleklett, Advances in Nuclear Dynamics (A.C. Mignerey, Editor) 3, 75 (1997).

[Sku98] W. Skulski, J. Dempsey, D.K. Agnihotri, B. Djerroud, J. Tõke, K. Wyrozebski, W. U. Schröder, R.J. Charity, L.G. Sobotka, G.J. Kunde, S.Gaff, C.K. Gelbke, T. Glasmacher, M.J. Huang, R. Lemon, W.G. Lynch, L. Manduci, L. Martin, and M.B. Tsang, Report UR-NSRL 98-435, 1998, to be published.

[Sob95] L.G. Sobotka, Physical Review C51, R2686 (1995).

[Tok94] J. Tõke, D.K. Agnihotri, S.P. Baldwin, B. Djerroud, B. Lott, B.M. Quednau, W. Skulski, W.U. Schröder, L.G. Sobotka, R.J. Charity, D.G. Sarantites, and R.T. de Souza, Nucl. Phys. A583, 519 (1994).

[Tok95] J. Tõke, D.K. Agnihotri, S.P. Baldwin, B. Djerroud, B. Lott, B.M. Quednau, W. Skulski, W.U. Schröder, L.G. Sobotka, R.J. Charity, D.G. Sarantites, and R.T. de Souza, Phys. Rev. Lett. 75, 2920 (1995).
[Tok96] J. Tõke, D.K. Agnihotri, S.P. Baldwin, B. Djerroud, B. Lott, B.M. Quednau, W. Skulski, W.U. Schröder, L.G. Sobotka, R.J. Charity, D.G. Sarantites, and R.T. de Souza, Phys. Rev. Lett. 77, 3514 (1996).
[Yen94] S. J. Yennello et al., Phys. Lett. B 321, 15 (1994).

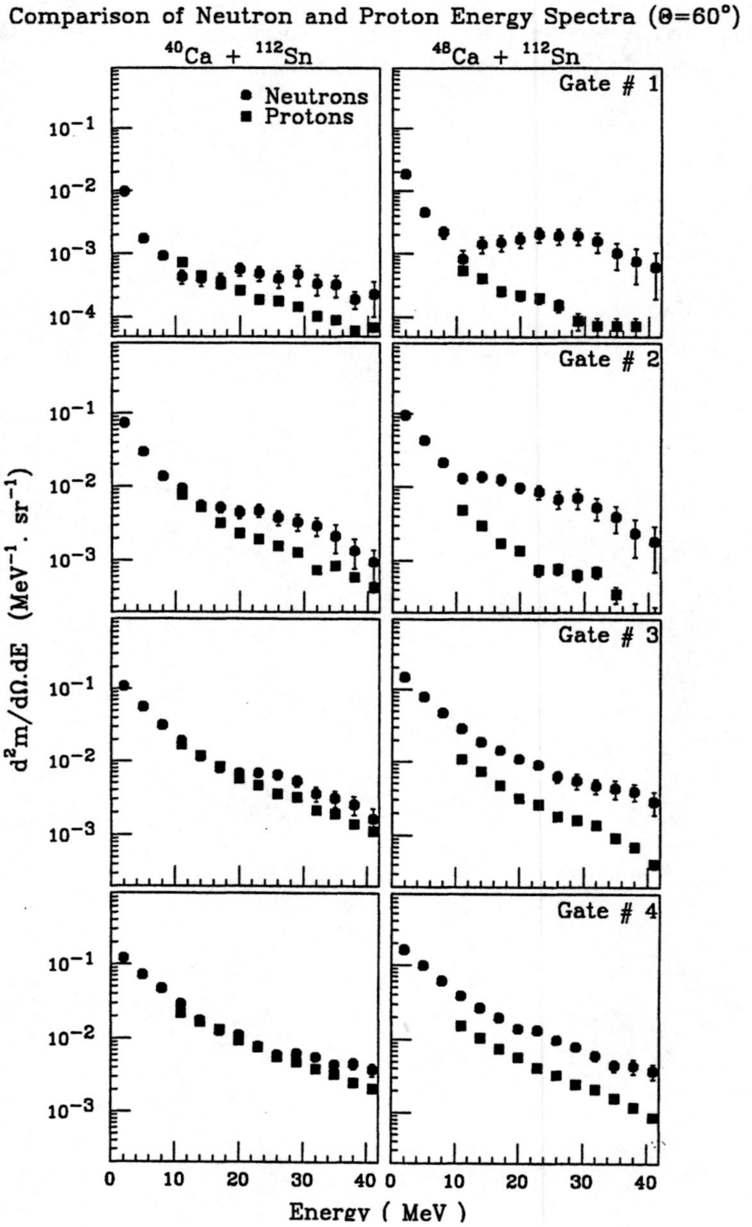

Figure 15: ^{112}Sn+40, ^{48}Ca neutron and proton spectra measured at a lab angle of $\Theta = 60^o$. The spectra are arranged from top to bottom according to increasing energy loss. (From [Agn98])

Isospin and Clusterization at Mid-rapidity

L.G. Sobotka*, R.J. Charity*, and J.F. Dempseym *

Abstract

Data from several intermediate energy experiments clearly show that light fragments from mid-central to peripheral collisions in the mid-rapidity region are enriched in neutrons relative to fragments found in the projectile rapidity region. Furthermore, it is also shown that the free n/p ratio far exceeds the system's n/p ratio and that this ratio is larger at mid-rapidity than other rapidity regions. While a neutron enrichment in the forming "neck" region is the equilibrium expectation, the observations are consistent with the expectation of extensive α-clustering (and the resulting amplification of the neutron excess of the remaining material) and perhaps also a more straight forward statistical kinetic interpretation of fragment production. The latter scenario is not investigated in the present work.

1 Introduction

Much of the recent effort in low and moderate energy nuclear physics has been directed towards making and studying reactions involving nuclei with extreme n/p ratios. Needless to say, understanding extremely asymmetric nuclear material has been a long standing objective of basic nuclear science; an objective with direct astrophysical importance. Peripheral reactions between heavy nuclei (we believe) offer the best opportunity to observe the most interesting isospin effect. This effect is "isospin fractionation" which is mediated by the existence of regions of matter, in equilibrium, with substantially different densities. It is important to understand that there must be both a condensed phase (higher density) and a gas phase (lower density) to observe this fractionation effect. The need for a condensed phase (liquid in this case) is why peripheral collisions, with surviving target and projectile remnants are of most interest. While the possibility that the high density phase is a distributed one comprised of many intermediate mass fragments, in principle allows central collisions to address the isospin fractionation issue, the connection to nuclear matter is far less transparent. In this contribution we wish to highlight a few recent results which relate to the possibility of making meaningful observations pertaining to phase equilibration of the two component nuclear system. These results are:

1. Peripheral reactions of heavy nuclei at intermediate energies do produce fragments which tend to be neutron-rich in the intermediate rapidity or "neck" region as compared to the fragments produced in the projectile and target rapidity regions.

2. Free neutrons are not only quite abundant relative to protons at mid-rapidity but more so at mid-rapidity than in the rapidity regions associated with the target and projectile remnants.

*Washington University Departments of Chemistry and Physics St. Louis, MO 63130, USA

3. Equilibrium logic predicts that the dropping density of the "neck" region will create a neutron-rich intermediate rapidity zone if the bulk matter is enriched in neutrons relative to the azeotropic composition.

4. Equilibrium considerations also indicate that common isospin observables are strongly influenced by clusterization in the low density neck region.

5. Dynamical BUU simulations *with clusterization logic* suggest that the time scale for isospin equilibration is incommensurate with the contact time of intermediate HI reactions while that for substantial clusterization is not. This suggests that (α-particle) clusterization will influence observables that one would like to relate to isospin fractionation.

An issue not addressed in this contribution, but which deserves study, is that of whether a statistical kinetic treatment, one in which chemical equilibrium is never achieved, can provide a natural explanation of n/p fractionation observables.

2 Some data

Some years ago we presented data indicating that light intermediate mass fragments found at intermediate rapidity tend to be more neutron-rich than those found at the rapidity of the projectile[1]. Figure 1 shows an isotopic ratio (a,b,c) and an approximately isobaric ratio (d,e,f) as functions of both parallel velocity and projectile remnant charge, for our study of the reactions $^{124,136}Xe+^{112,124}Sn$ at 55 MeV/A. Due to the fact that a projectile remnant is required, mid-central to peripheral collisions are selected in this data set. All four systems and both types of ratios (isobaric and isotopic) show the trend mentioned above. The trend is enhanced (larger enhancement of the neutron rich species) when the overall system is neutron rich. It also seems that this trend is rather ubiquitous in that it is observed for all gates on the charge of the projectile-like fragment.

An additional point, which is dealt with more directly in the chapters by Li and Yennello and by Schröder and Tõke, is the lack of isospin equilibration in reactions of the type studied here. This is most clearly seen for the lower Z_{plf} bins, (frames a and b) of Figure 1 when the "crossed" systems with the same total n/p ratio (dashed and dot-dashed lines) are compared. The projectile clearly manifests itself in the emissions in the sense that neutron-rich fragment emission is enhanced when the projectile is neutron-rich. This observation holds not only in the projectile velocity region ($V_{plf}^{cm} = 5.0\ cm/ns$) but also to some extent back to just above the intermediate velocity region. This indicates that the projectile and target nucleons are not well mixed in these collisions. While this study does not probe the most central collisions, the lack of equilibration does seem to hold for the bulk of the reaction cross section (as most of the cross section produces projectile fragments.) For a further discussion of this topic, we direct the reader to the other chapters referenced above.

The tendency to find more neutron-rich light fragments in the intermediate-rapidity region than in the projectile-rapidity region, was first noticed for H isotopes in our study of the $^{136}Xe+^{207}Bi$ system at 28 MeV/A [2]. These results are summarized in Table I.

The fraction of the hydrogen isotopes emitted at mid-rapidity increases in going from p to d to t. The latter yield is up to 1/4 of the total t emissions. This is about equal to the fraction of α−particles from this source but less than the values for the heavier ions. The trend of the increasing fraction of hydrogen isotopes at mid-rapidity in going from p to d

to t, is one of the most striking features of the invariant cross section maps of intermediate energy collisions[2].

The data listed for the $^{136}Xe + ^{207}Bi$ in Table I is for a bin in total kinetic energy loss of about 300 MeV. The analysis of the beautiful data collected with the INDRA detector not only supports the relative n-enrichment at mid-rapidity but also illustrates that the prominence of $t's$ at mid-rapidity occurs at least as much in peripheral collisions as it does for mid-central collisions. Table II shows the multiplicities and percentages for $p's$, $d's$, and $t's$ for two bins in transverse energy. These two bins correspond to average impact parameters of about 1) 0.9 and 5) 0.3 of $b_{\max}$. (The multiplicities and percentages given in this table result from an average of the values obtained from two different extraction procedures, see [3] for details.)

TABLE I. Relative yields for the system $^{136}Xe + ^{207}Bi$

Particle→	p	d	t	α	Li
$< M_{plf} >$	0.28	0.08	0.07	0.28	0.01
$< M_{tlf} >$	0.35	0.16	0.15	0.48	0.02
$< M_{neck} >$	0.02	0.02	0.08	0.33	0.06
% neck	**4**	**9**	**26**	**30**	**68**

TABLE II. Relative yields for the system $^{129}Xe + Sn$. The charged particle data are shown for two transverse energy bins from an experiment at $E/A = 50$ MeV[3], while the neutron data are for an experiment at $E/A = 40$ MeV.

Bin[3]→	1	1	1	5	5	5	-
Particle →	p	d	t	p	d	t	n
$< M_{neck} >$	0.14	0.17	0.15	0.18	0.40	0.45	-
% neck	**12**	**48**	**70**	**5**	**20**	**35**	**44**

What is clearly missing from the data presented above, and absent in the literature to date, is information on the (free) neutrons. One of the difficulties of course is that such data cannot be obtained along with 4π charged particle information due to the need to avoid excess material on which the neutrons can scatter. We have recently collected neutron data for both $^{129}Xe + ^{120}Sn$ and $^{129}Xe + Au$ reactions at E/A=40 MeV. In this experiment, excess material was minimized and 15 discrete BC501 detectors, with charged-particle veto paddles, were used to collect the time-of-flight data. The projectile-like fragment (PLF) was detected in an annular Si-CsI(Tl) array spanning angles just beyond the grazing angle. The charge of this fragment (Z_{plf}) was used as a crude impact parameter selector [4]. A neutron invariant cross section map is shown in Figure 2 for the $^{129}Xe + ^{120}Sn$ system. These data are well explained with a three source model (fit with only the velocities fixed) in which the multiplicity breakdown PLF : TLF : MID-(rapidity) is 27% : 29% : 44%. The mid-rapidity percentage is 4 times the proton percentage in the most peripheral collisions (from the INDRA data) and 9 times the proton percentage for the mid-central collisions (as determined from gates on the transverse energy.) While it is difficult to estimate the impact parameter from the Z_{plf} gate, our estimate is in the range of 0.3-0.5 b_{max}. Regardless of this uncertainty, it is quite clear that the mid-rapidity free n/p ratio: 1) exceeds the total n/p ratio by a large factor and 2) is larger at mid-rapidity than in the target or projectile remnant rapidity regions.

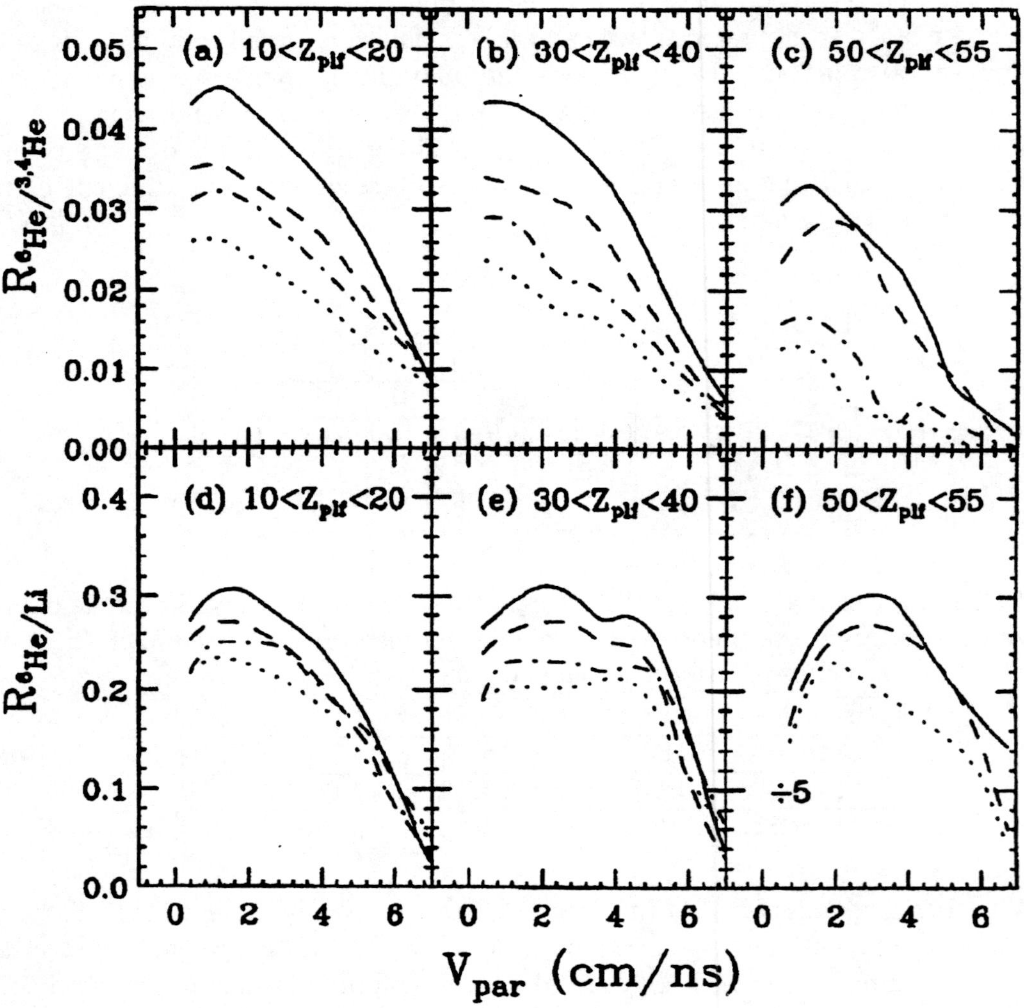

Figure 1: Isotopic and "Isobaric" ratios as a function of parallel velocity for the reactions $^{124,136}Xe +^{112,124} Sn$ at 55 MeV/A. The abscissa values are those for the center-of-mass system. The projectile + target combinations are represented by; 136 + 124 (solid), 136 + 112 (dashed), 124 + 124 (dot-dash), and 124+ 112 (dotted). The data points have been suppressed for clarity.

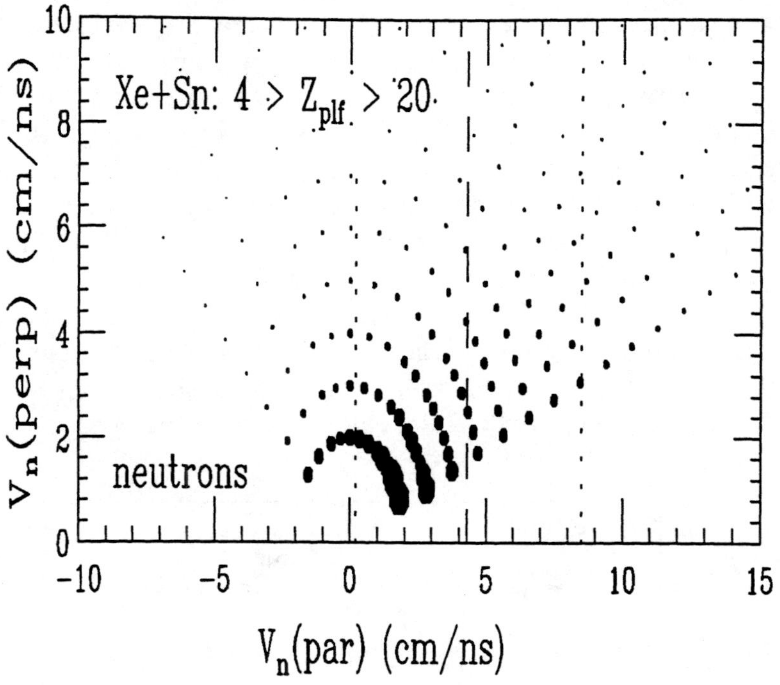

Figure 2: Invariant cross section plot of neutrons emitted in the reaction E/A= 40 MeV $^{129}Xe+^{120}Sn$. The abscissa offset is that for the laboratory frame and the center-of-mass (dashed) and projectile and target like (dotted) source velocities are indicated.

3 Relevant models

In order to gain insight into the possible explanations for the apparent n-richness of the mid-rapidty region, three types of model calculations are of value. These are: 1) equilibrium isospin fractionation (with density) for both asymmetric (uncharged) nuclear matter and asymmetric charged matter drops, 2) equilibrium clustering and 3) dynamical BUU simulations with (as best as one can do at present) inclusion of both isospin fractionation and clusterization effects. Ultimately, each of these calculations must be coupled with statistical kinetic evaporation models to treat sequential decays and thus predict the fragments one is likely to detect. We must admit at this point that it is the last feature, statistical decay (which does not require chemical equilibrium) which might very well dictate all observables. Thus it is possible that concrete evidence of isospin fractionation effects cannot be obtained, even if chemical equilibrium is established, in the primary reaction. We will not deal with this issue here but rather, what one might expect before the statistical kinetic (sequential) decay mechanisms have been played out.

3.1 Equilibrium fractionation

The benchmark equilibrium calculations for nuclear matter (NM) were done by H. Müller and B. Serot (using a relativistic mean field model). This work is presented in Chapter 4 of this book, yet it is so important to understand this work that we summarize the relevant features here. While symmetric matter has a critical temperature of about 16 MeV, T_c for asymmetric matter drops with asymmetry. As is commonly the case (in chemical reactions), phase separation is accompanied by component fractionation. The physics behind this effect is the fact that the symmetry energy decreases with decreasing density ($\rho^{2/3}$ for the kinetic contribution) and therefore free energy minimization will favor putting the component in excess in the low density phase. Several points are worth stressing. First, for (uncharged) NM an azeotrope exists at symmetry. This mix can be taken through the phase transition without a fractionation, i.e. as if it were a single component fluid. Second, the overall picture is symmetric with respect to symmetry, however the region with more $p's$ than $n's$ is of less physical significance. Third, this fractionation implies that if equilibrium is attained in a system with both high and low density phases, and an overall neutron excess exists, then the low density phase should become enriched in neutrons. One can imagine that the projectile+target regions versus the overlap (or emerging surface) region in a mid-central or peripheral collision could be such a case. Finally, these calculations beg the question of how these results would be changed if finite charged systems were considered.

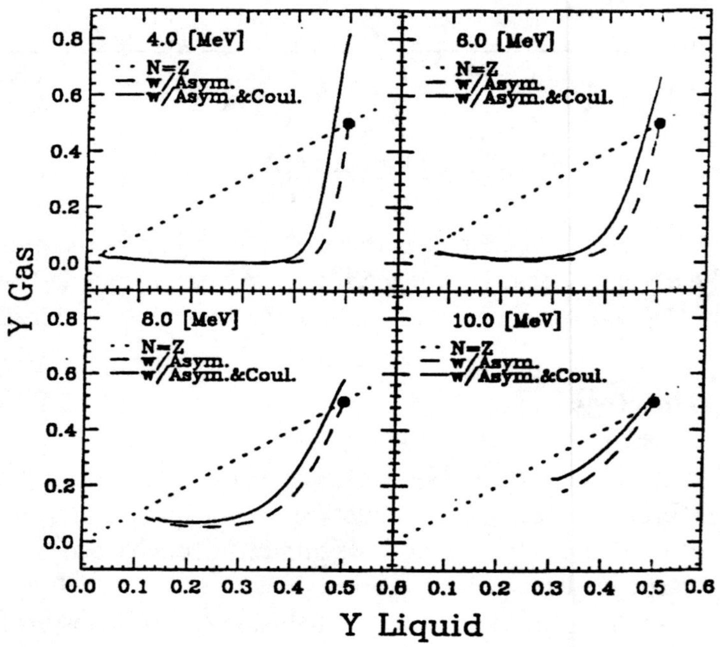

Figure 3: The proton fraction (Y) in the low density phase as a function of the proton density in the high density phase, for the nuclear matter drop (charged) calculations of [4]. The results for four different temperatures are shown each with and without the Coulomb effect.

While we cannot answer this question quantitatively, we can present some calculations which point to the likely influence of charge. We have performed both two component van der Waals and Skyrme model calculations in which we have included the effect of charge. In brief, we investigated three different forms for the potential part of the symmetry energy and the effect of charge was included in an ad-hoc fashion by lifting the free energy by the inclusions of a term with a $\rho^{1/3}$ dependence and with a magnitude, at ρ_o, corresponding to the Coulomb energy of a Xe nucleus. (Full details can be found in [4].) Some of the results of the Skyrme model calculations are shown in Figure 3 which displays the proton fraction (Y) in the low and high density phases for several temperatures. Without the Coulomb term, an azeotrope exists at symmetry (same proton fraction in each phase) and otherwise (when there is an excess of n's) the gas phase is strongly depleted in protons.

This fractionation decreases with increasing temperature. These are all results found in the work by H. Müller and B. Serot[5]. The inclusion of charge decreases the proton depletion in the gas phase and shifts the azeotrope to a proton fraction closer to $\beta-$stability. While we do not believe these calculations provide quantitative benchmarks, we do believe that the inclusion of charge will have these two effects, namely, that the fractionation will exist, albeit weakened relative to uncharged matter, and that the azeotropic composition will move in the direction of $\beta-$stability. (It reasonable that the equilibrium expectation would in fact fall on $\beta-$stability.) More serious calculations, like those presented elsewhere in this book, are needed to pin down the magnitude of component fractionation in finite systems. Before moving on to some dynamical calculations, we must consider another equilibrium effect.

As the density drops in the "neck" region (as the target and projectile remnants recede from one another) one will enter the state where α-matter is more stable than normal nuclear (nucleon) matter. Again if the expansion is adiabatic, the matter will cluster. This point, central to all sub-saturation nuclear physics, was made by Randrup and Konnin almost 20 years ago from the perspective of how clusterization influences isotopic and isobaric ratios [6]. A microcanonical version of the relevant equilibrium calculation in shown in Figure 4. This figure displays the results of a calculation for 20 nucleons with an overall proton fraction of $Y = 0.4$. All clusters are considered with A $<$ 5 and the fragment volume (needed for the free volume portion of phase space) is calculated from the radial sizes taken from electron scattering. Both the free n/p and $t/^3He$ ratios increase well above the bulk n/p ratio (1.5). These ratios sky-rocket with decreasing energy per nucleon. Due to free volume effects and the small relative size of the α-particle, the effect is stronger at high density. Needless to say, the trends seen in this figure are driven by α-clustering which results in amplifying the n-enrichment of the remaining material. The lesson to be learned from this calculation is that the case for fractionation cannot be made by looking at n/p or $t/^3He$, or any other small collection of isotope ratios, but rather from the relevant sums so that the total n and p content can be determined. (Cognizance of the physics implies that this should be done for regions of rapidity which can be associated with low and high density regions of physical space.)

Figures 3 and 4 highlight the conundrum concerning the extent to which certain isospin observables reflect the clusterization process rather than the overall n/p ratio of the material. As is often the case, the issue might be decided by the relevant time scales, in other words the time scale of isospin equilibration versus that for α-clustering. In our attempt to address this problem we have performed BUU calculations which contain cluster formation. Danielewicz and Bertsch have explicitly included the degrees of freedom for d, t, and ^3He clusters in a BUU code [7]. Results for such calculations are shown in Figure 5 for one of the intermediate energy systems our group has studied [8]. These calculations show the total density profile (a,b,c) and that for the *free* n's (d,e,f) and *free* p's (g,h,i) at three different times (t = 0, 50, and 100 fm/c). The final time step shows a great excess of free neutrons relative to protons in the "neck" region. An observation which, we remind the reader, was also made for experimental data.

Which one of the effects discussed above is this due to? This question is answered in the three frames on the right hand side (j,k,l). Here the total (open symbols) and free (closed symbols) n/p ratios are presented as a function of the distance (D) from the projection, onto the separation axis, of the coordinate position to the system center-of-mass. Overall, the center of the neck region ($D = 0$) is only slightly enriched in neutrons (reflecting some isospin equilibration) however the free neutrons far exceed the free protons. The difference is

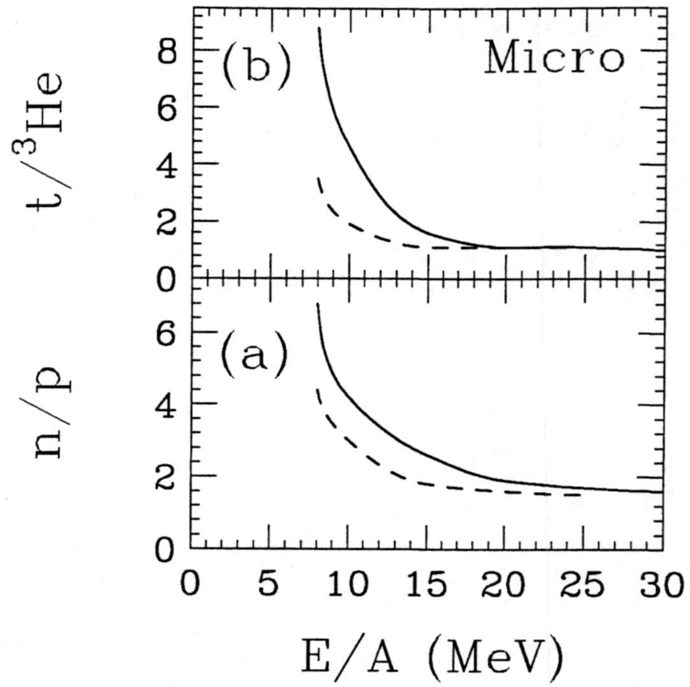

Figure 4: Free n/p (a) and $t/^3He$ (b) ratios from microcanonical cluster calculation involving 12 neutrons and 8 protons and considering all fragments with $A < 5$. Two densities are shown: $\rho = \rho_o$ (solid line) and $\rho = \rho_o/2$ (dashed line).

due to the fact the central region is highly clustered with an overall symmetric participation in the clustering. In other words, the clustering has resulted in an amplification of a modest n-excess in the overlap region. The results for such calculations for three impact parameters are shown in Figure 6. Here the free (solid) and total (dashed) n/p ratios as shown as a function of time. The amplification is clearly seen, an effect which is more dramatic for the more peripheral collisions. The root of this effect is shown on the right where the free n/p ratios are shown as a function of the fraction of nucleons bound in clusters for three different total n/p ratios.

This suggests that (approximately) symmetric clustering is the explanation of the mid-rapidity n-enhancement. However as sequential decay has not been considered, such a conclusion is premature. What is clear is that it is important to include α-particle clustering in dynamical simulations as clustering in general and α-particles in particular, can strongly influence the n/p balance of the remaining material.

4 Summary

In intermediate energy heavy-ion collisions, it is peripheral collisions which provide the best opportunity of observing interesting isospin equilibration effects. This is due to the contrast in density between the mid-rapidity region and the regions associated with the target and projectile remnants. The former region has an ever decreasing density while the latter remains near saturation. These two (density) regions provide a stage on which one might observe isospin equilibration effects of the type studied by Müller and Serot [5]. The study of "neck" or ternary fission may also offer the same possibility, except in this case

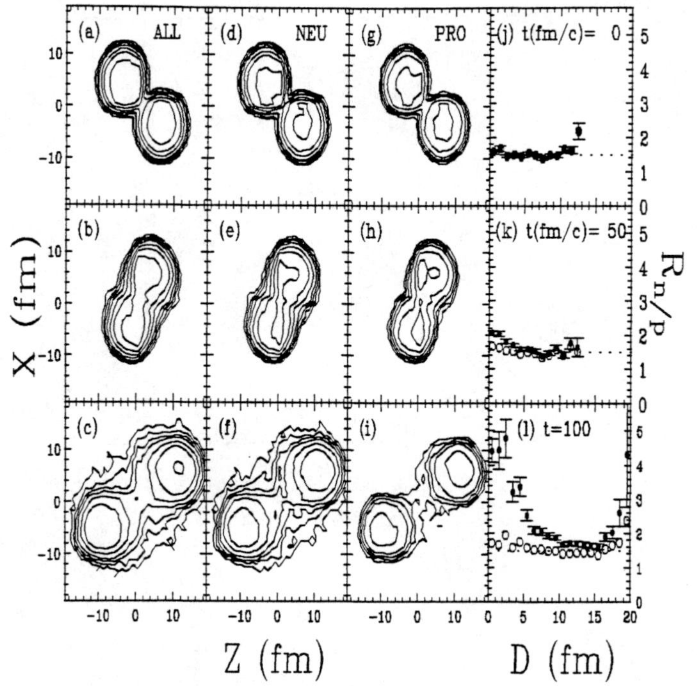

Figure 5: Boltzmann-Uehling-Uhlenbeck simulation of a collision between ^{132}Xe and ^{124}Sn at E/A = 55 MeV and an impact parameter of 8.8 fm. The different panels show projections (in the plane defined by the projectile momentum (Z-axis) and the impact parameter (X-axis)) of the total (a,b,c), neutron (d,e,f) and proton (g,h,i) densities. The nucleons bound in light clusters are excluded from the contour plots (d-i). Sections (j-l) display the neutron-to-proton ratio $R_{n/p}$ as a function of the magnitude D of the projection of the space coordinates on the projectile-target separation axis, excluding (solid) and including (open) the nucleons bound in small clusters. D = 0 corresponds to the center of mass position.

the observables are associated with very rare events rather than common ones. However, the dynamics for the formation of $\alpha-$clusters is clearly important in that the propensity for α-particle formation strongly influences isospin observables of the remaining matter. In this regard dynamical work building on the study of the energetics of "quarteting" by Röpke et al. [9] is vital for progress in the study of isospin dynamics. It is quite likely, and some evidence suggests, that the reactions are too fast to observe substantial effects of equilibrium isospin fractionation (with density) while sufficient for the formation of small clusters and the requisite alteration of the isospin observables.

As a final note we would like to add a word of caution. It is possible that phenomenon resulting from equilibration may be little different from statistical kinetic reaction scenarios. The only difference might lie in the difference between the equilibrium azoetropic composition, which should be close to $\beta-$stability, and the statistical kinetic dividing line, which is the Evaporation Attactor defined (and discussed in this book) by Charity [10].

This work was supported by the Director, Office of High Energy and Nuclear Physics, Nuclear Physics division of the U.S. Department of Energy under Contract No. DE-FG02-87ER-40316.

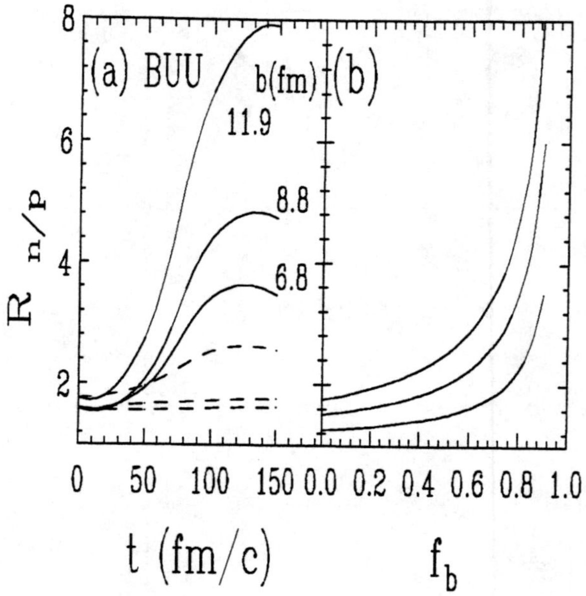

Figure 6: The free (solid) and total (dashed) n/p ratios are shown on the left for three impact parameters as a function of time. On the right, the free n/p ratios as a function of the charge fraction bound in symmetric clusters for total n/p ratios of 1.25, 1.50 and 1.75 (bottom to top) are shown.

References

[1] J.F. Dempsey, R. J. Charity, L.G. Sobotka, G.J. Kunde, S. Gaff, C.K. Gelbke, T. Glasmacher, M. J. Huang, R.C. Lemmon, W.G. Lynch, L. Manduci, L. Martin, M.B. Tsang, D. K. Agnihotri, B. Djerroud, W.U. Schröder, W. Skulski, J. Tõke, and W.A. Friedman, Phys. Rev. C**54**, 1710(1996).

[2] J. Tõke, B. Lott, S.P. Baldwin, B.M. Quednau, W.U.Schröder, L.G. Sobotka, J. Barreto, R.J. Charity, D.G. Sarantites, D.W. Stracener, and R.T. de Souza, Phys. Rev. Lett. **75**, 2920(1995).

[3] J. Łukasik et al.(INDRA collaboration), Phys. Rev. C**55**, 1906 (1997).

[4] J.F. Dempsey, Thesis, Washington University (1997).

[5] H. Müller and B.D. Serot, Phys. Rev. C**52**, 2072(1995), see contribution by these authors in the present book.

[6] J. Randrup and S. Koonin, Nucl. Phys. **A356**, 223(1981).

[7] P. Danielewicz, Phys Rev. C**51**, 716(1995).

[8] L.G. Sobotka, J.F. Dempsey, R.J. Charity, and P. Danielewicz, Phys. Rev. C**55**, 2109(1997).

[9] G. Röpke, A. Schnell, P. Schuck, and P. Nozières, Phys. Rev. Lett. **80**, 3177 (1998).

[10] R.J. Charity, Phys. Rev. C**58**, (1998) 1073.

The Evaporation Attractor Line

R. J. Charity *

Excited primary fragments produced in nuclear reactions undergo sequential light–particle evaporation. Even if a primary fragment is produced with an unusual proton or neutron excess, the secondary fragment or evaporation residue remaining after the evaporation cascade maybe less unusual as the action of evaporation moves its location in the chart of nuclides towards a line called the evaporation attractor line[1]. This line lies on the neutron deficient side of the line of β–stability and only for heavy system does it enter a region of extremely exotic nuclei. The existence of the evaporation attractor line is easy to understand, neutron-rich primary fragments evaporate more neutrons than protons, on average, and become less neutron-rich. On the other hand, proton-rich fragments evaporate more protons and become less proton-rich. Fragments on the evaporator attractor line have similar neutron and proton decay rates and hence, on average, their neutron richness does not change, rather the decaying systems follow the attractor line down in mass.

For a more formal definition of the evaporation attractor line, let us also consider the evaporation of fragments other than protons and neutrons and consider the relative rates of emission of all neutrons and protons, whether as free nucleons or in a cluster. For example including deuterons and alpha particles, this ratio is obtained from the appropriate decay widths as

$$\frac{\Gamma_N}{\Gamma_Z} = \frac{\Gamma_n + \Gamma_d + 2\Gamma_\alpha}{\Gamma_p + \Gamma_d + 2\Gamma_\alpha} \tag{1}$$

and is related to the average direction, in the chart of nuclides, a fragment moves due to evaporation. The location of the evaporation attractor line will now be defined by

$$\frac{dN}{dZ} = \left\langle \frac{\Gamma_N}{\Gamma_Z} \right\rangle \tag{2}$$

which enforces the condition that a fragment on the attractor line, decays, on average, along this line. The reason why this definition is useful is that the value of the ratio $\frac{\Gamma_N}{\Gamma_Z}$ at the attractor line is generally not very sensitive to the excitation energy. As an example, values of this ratio calculated with the GEMINI code[2] for Zn isotopes are plotted in Fig. 1 for various excitation energies. Only for the one isotope, located on the attractor line, do all the curves intersect.

The slope of the attractor line, given by the value of $\frac{\Gamma_N}{\Gamma_Z}$ at this point, is close to unity so similar numbers of protons and neutrons are lost, on average, to evaporation. The definition of the attractor line is in terms of a local average of $\frac{\Gamma_N}{\Gamma_Z}$ as shell effects are responsible for some local fluctuations in this ratio. If we were to ignore the Coulomb barriers hindering charged particle emission, the attractor line would coincide with the line of β–stability where the separation energies for proton and neutron removal are similar. However, Coulomb barriers suppress charged particle emission and this pushes the attractor line towards the

*Department of Chemistry, Washington University, St. Louis, Missouri, 63130, USA.

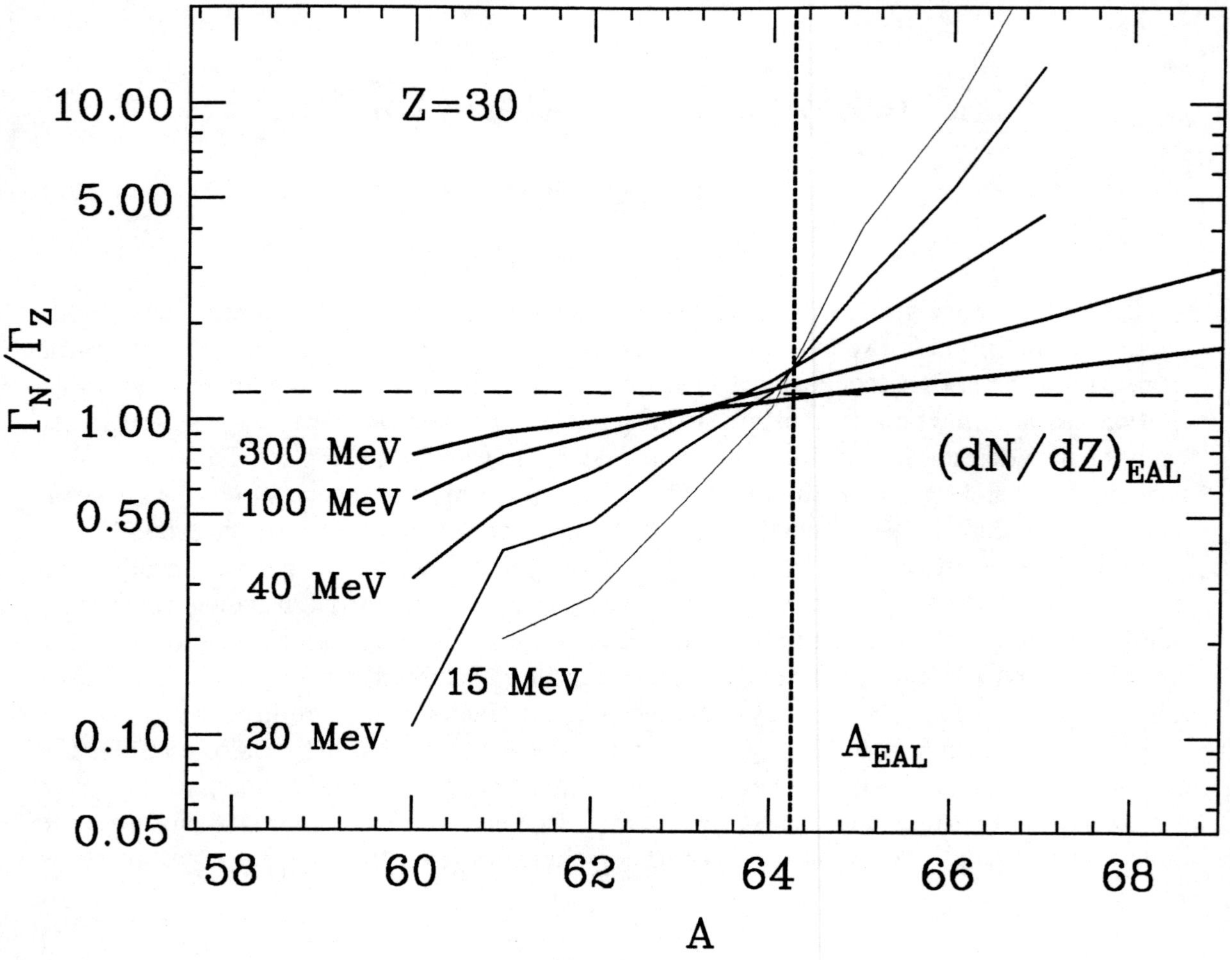

Figure 1: The ratio Γ_N/Γ_Z predicted by the statistical model calculations for Zn isotopes as a function of mass number. Results are shown for the indicated excitation energies. The position and slope of the attractor line are indicated by the dotted and dashed curves, respectively.

neutron-deficient direction as shown in Fig. 2. For very light fragments, where Coulomb barriers are small, the attractor line and the line of β–stability merge. For heavier fragments as the Coulomb barriers increase, the attractor line moves further from the line of β–stability and at $Z \sim 82$ it crossed the limits of known isotopes, although for such heavy systems fission is also competing with evaporation and may limit our ability to produce such exotic fragments (see later).

To aid in locating the attractor line, the following approximations are useful;

$$Z = 0.909N - 1.12 \times 10^{-3}N^2, \tag{3}$$
$$Z = 0.479A - 2.04 \times 10^{-4}A^2, \tag{4}$$
$$N = 1.072Z + 2.32 \times 10^{-3}Z^2, \tag{5}$$
$$N = 0.514A + 2.45 \times 10^{-4}A^2. \tag{6}$$

They are all accurate to less than a nucleon.

Although the ratio $\frac{\Gamma_N}{\Gamma_Z}$ is independent of excitation energy at the attractor line, this is

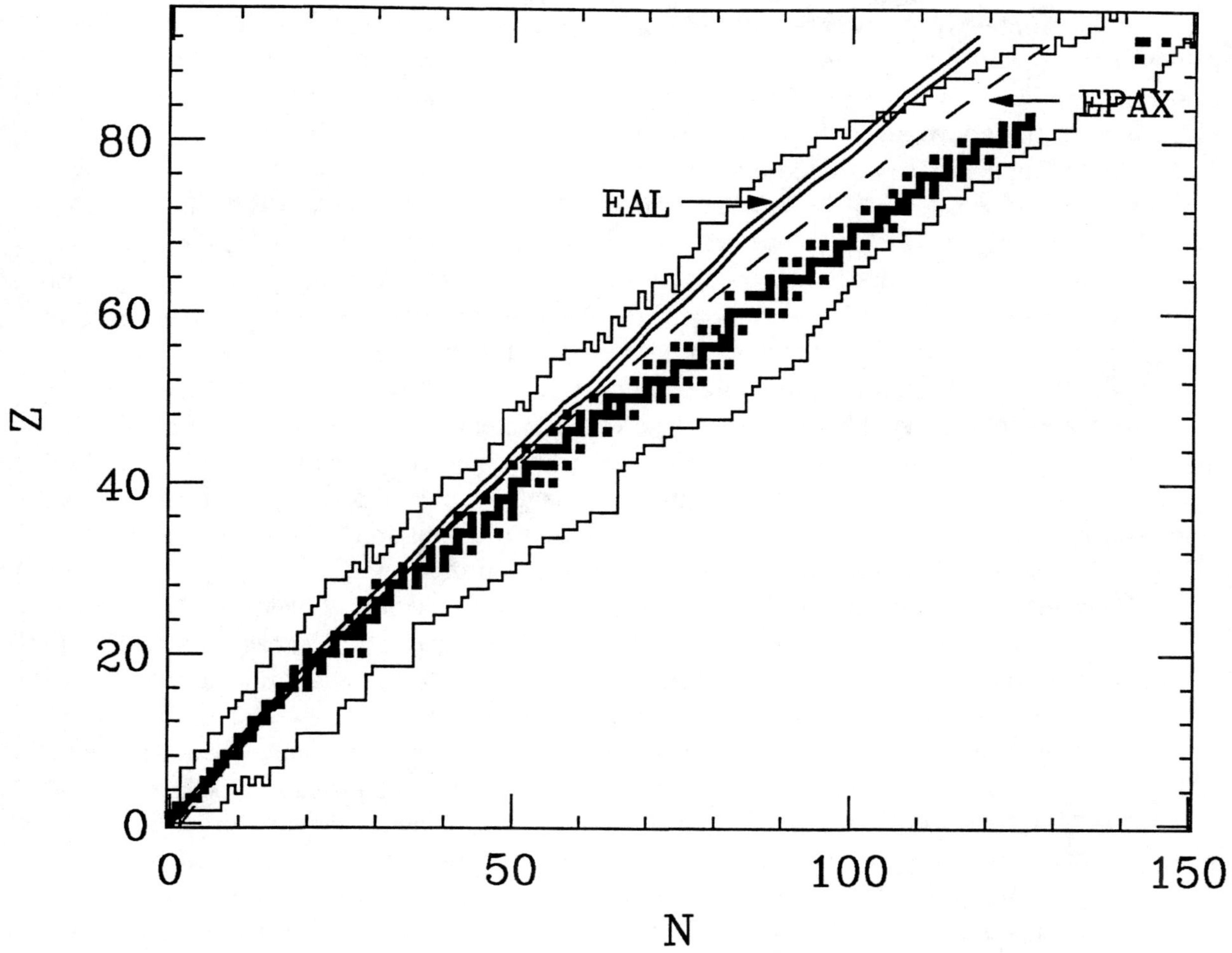

Figure 2: Location of the evaporation attractor line (EAL) with respect to the stable isotopes (solid squares) and the limits of known isotopes. For comparison, the location of the EPAX fits[5] to fragmentation products is indicated by the dashed line.

not true as we move away from it (Fig. 1). Proton and neutron emission rates are only similar when the difference of their "energy cost", the sum of their separation energy and Coulomb barrier, are of the order of the nuclear temperature or less. Thus relative rates of proton and neutron emission change more rapidly with distance from the attractor line when the temperature and excitation energy are low. Depending on which side of the attractor line a fragment is on, either neutron or proton emission quickly dominate. As the excitation energy and temperature increases, the change in relative emission rates with distance from the attractor becomes smaller, and vanishes in the limit that the temperature approaches infinity. Not only are differences between the energy costs for neutron and proton removal less important at large temperatures, but alpha particle and other cluster emissions become more common. Clusters, on average, remove similar numbers of protons and neutrons and thus the net effect of evaporation is a loss of mass with little change in the neutron or proton excess. One can say that at high excitation energies, evaporation, on average, preserves a memory of the initial neutron or proton excess of the primary fragments. Only when a fragment cools down does the attractor line become "attractive". Substantial movement towards the attractor line only occurs at low excitation energies and thus there are only a limited number of evaporation steps possible before this energy is exhausted. Hence, movement towards the attractor is also limited. As a consequence, primary fragments located too far from the attractor will evaporate towards, but never reach, this line. As an example of the above ideas, Fig. 3 shows the predicted location in the chart of nuclides of secondary fragments from the decay of a very neutron–rich Gd isotope (A=164) and a very proton–rich Gd isotope (A=137) at low (100 MeV) and high (400 MeV) excitation energies.

The predictions were obtained with the GEMINI code[2] including the evaporation of neutrons and the ground and excited states ($E^* < 5$ MeV) of all H, He, Li, and Be isotopes. Separation energies were obtained from the experimental and predicted masses tabulated in Ref. [3]. The value of other statistical model parameters can be found in Ref. [4]. The predicted location of the secondary fragments are indicated by the contours (linear scale) in Fig. 3. The average trajectory a system follows from its initial position (solid point) to its final location is indicated by the solid (100 MeV) and dashed (400 MeV) arrows. For the neutron–rich ^{164}Gd fragment, located a significant distance from the attractor line, decay is almost exclusively by neutron emission for the calculation with 100 MeV of excitation energy. In contrast, the very proton–rich ^{137}Gd fragment is not extremely proton rich compared to the attractor line and, even at 100 MeV of excitation energy, some neutron emission occurred. At the higher excitation energy, for both the neutron-rich and proton-rich systems, the initial direction of the evaporation trajectories are similar, almost parallel to the attractor line illustrating that the attractor line is not very alluring to these hot fragments. For the hot ^{164}Gd primary fragment, the evaporation trajectory changes direction after about half the evaporative mass loss has occurred. The final low–temperature decay phase is again dominated by neutron emission moving the location of the decaying fragments towards the attractor line. For the hot ^{137}Gd fragment, the change in trajectory is less dramatic as it is location is closer to the attractor line.

Figure 4 gives a general overview of the final position of evaporation residues in the chart of nuclides as a function of initial compound nucleus excitation energy and neutron–proton asymmetry. A selection of compound nuclei which are very neutron–rich, very neutron–deficient and intermediate examples were calculated with the GEMINI code. The large solid points indicate the position of the selected compound nuclei in the chart of nuclides. The curves extending from these points are the loci of the predicted, average N and Z of the evaporation residues as the excitation energy is increased. The small points indicate the average position of the residues as the excitation energy is increased in increments of 100 MeV. These curves should not be confused with the average trajectories of the decaying systems which, as shown in Fig. 3, can be quite different at high excitation energies.

These curves clearly show that at large excitation energies, all evaporation residues are found in the vicinity of the evaporation attractor line indicated by the dashed curve. Generally, there is an initial rapid movement towards the attractor when the excitation energy is of the order of 100–200 MeV. Once there is enough excitation energy to bring the residues close to the attractor, increasing the excitation energy only causes the average residue position to approach the attractor asymptotically, following it down to lower masses.

The very neutron–rich systems, as noted early, never reach the evaporation attractor line. For the selected neutron–rich compound nuclei shown in Fig. 4, one sees that the final separation from the attractor increases as the mass of the system increases. This is partially a result of the fact that the heavier neutron–rich systems were selected further from the attractor. However in addition to this selection bias, the increasing separation from the attractor can also be attributed to the shape of the β–valley of stability which is broader for larger masses and hence the "forces" driving the decaying systems towards the attractor are not as large.

For $Z \gtrsim 50$, even primary fragments located on the line of β–stability are sufficiently distant from the EAL that their secondary fragments never reach the evaporation attractor. To illustrate this, in Fig. 5 the loci of average positions of secondary fragments are shown for selected primary fragments located on the line of β–stability. Rather than EAL, the location of the secondary fragments approach and subsequently decay, on average, down the dotted line in Fig. 5 which can be approximated by the equations:

$$Z = 0.933N - 1.65 \times 10^{-3}N^2, \quad (7)$$

$$Z = 0.487A - 3.01 \times 10^{-4}A^2, \quad (8)$$

$$N = 1.018Z + 3.71 \times 10^{-3}Z^2, \quad (9)$$

$$N = 0.512A + 3.01 \times 10^{-4}A^2. \quad (10)$$

More neutron–rich or proton–rich primary fragments will of course approach other lines. However with increasing proton richness, these lines move towards and merge with the evaporation attractor line.

The influence of the spin of the primary fragment is not very important in terms of the approach to the attractor line. As an example in Fig. 6, the predicted average locations of secondary fragments from the decay of ^{118}Sn systems with spins of zero and 60 $\hbar$ are plotted as the solid and dashed curves, respectively. The separation between these lines is exceeding small, less than half a nucleon. The most important effect of spin is to modify the total evaporated mass. At low excitation energies, increasing the spin reduces the thermal excitation energy and the evaporated mass is reduced. At much higher excitation energies, spin promotes cluster emission which are more efficient at removing mass and the evaporate mass is increased, however, the magnitude of this effect is not large.

The line described by Eqs. 7 to 10, giving the average location of highly excited β–stable primary fragments, lies very close to the EPAX parameterization of Sümmerer *et al.*[5] which has been fit to the average location of the fragmentation products produced in reactions of GeV protons on target nuclei located close to the line of β–stability. The EPAX parameterization also works very well for projectile fragmentation products[5, 6, 7], and thus all these fragmentation products can be understood as evaporation residues associated with highly excited primary fragments located close to the β–valley of stability. This interpretation is further strengthened when the width of the residue distributions are compared to the EPAX fits. In Fig 7, the predicted distribution of $Z - \langle Z(A) \rangle$ (histogram), where $\langle Z(A) \rangle$ is given by Eq. 8, is plotted for products from ^{118}Sn primary fragments with excitation energies between 200 and 400 MeV. The EPAX results obtained from fits to experimental data are indicated by the curve. This curve is well reproduced by the statistical model calculations, including the width and skewness of the distribution. The small displacement in average location between the EPAX formula and the dotted line in Fig. 5 (Eqs. 7-10), may indicate that the primary fragmentation products are actually somewhat more neutron rich compared to the valley of stability. However, modifications to statistical model parameters at high excitation energies, for instance decreasing the Coulomb barrier for charged particle emission or decreasing the level density parameter, can bring the predicted average location of residues in agreement with the EPAX result.

So far we have concentrated mostly on the average location of the secondary fragments, Fig. 3 also gives examples of the predicted distributions. At high and moderate excitation energies, the distributions are typified by a long ridge orientated parallel to the attractor line. The length of the ridge, which is approximately the width of the mass distributions $\sigma(A)$, increases as the excitation energy increases. The width of the ridge, which we will gauge by the standard deviation $\sigma(Z - Z_{EAL})$, is a measure of the range of neutron excesses sampled by the fragment distribution. The width has little excitation energy dependence and is smaller for distributions which are located on the attractor line. As an example in Fig. 8, the predicted width and length of the ridge are shown as a function of excitation energy for the ^{118}Sn system at spins of zero and 60 $\hbar$. As for the case of the mean positions discussed previously, the effect of spin is not very important at moderate to high excitation energies. The width of the ridge $\sigma(Z - Z_{EAL})$ quickly saturates as the excitation energy increases whereas the length $\sigma(A)$ continues to increase. High excitation energies are therefore not any more useful than moderate values for producing fragments with unusual isospins as even the tails of the residue distribution do not extent to more exotic regions.

The mass dependence of width $\sigma(Z-Z_{EAL})$ along the attractor line is well approximated by the smooth function

$$\sigma(Z - Z_{EAL}) = \frac{0.73}{(1 + \exp[-A/33.4])} - 0.02. \tag{11}$$

As already noted, for systems that do not reach the attractor line, the width of the ridge is larger than this value. In Fig. 2, the attractor line is indicated by two curves representing $Z_{EAL} + \sigma(Z - Z_{EAL})$ and $Z_{EAL} - \sigma(Z - Z_{EAL})$. It is quite clear from this figure that the width of the attractor line is quite narrow over all masses and saturates at 0.71 for heavy fragments.

Although the products from fragmentation reactions do not reach the attractor line for $Z \gtrsim 50$, this region is accessible with fusion reactions which form compound nuclei that are typically proton–rich compared to β–valley of stability. As an example, the reaction products produced at zero degrees in the E/A=14.2 and 20.6 MeV, ^{96}Mo + ^{58}Ni reactions were selected with a Wien filter with narrow velocity gates around 3.3 and 4.0 cm/ns, respectively[8]. From the systematics of momentum transfer, it is estimated that the excitation energy of the compound nuclei are approximately 2.7 and 4.1 MeV per nucleon, respectively, sufficiently large that the evaporation residues are predicted to lie on the EAL. The distributions of detected products in the chart of nuclides are shown in Fig. 9. The filled circles gives the average N for each Z and the error bars indicated the width (1σ) of the distributions. The results for the lower energy reaction follow our expectation for a secondary fragment distribution, a long-narrow ridge orientated along the evaporation attractor line. The average mass of these residues is consistent with statistical model calculations using the assumed excitation energy[8]. For the higher bombarding energy, the lightest detected elements are located nearer to the line of β–stability, with the distribution approaching the attractor line for the heavy elements. These lighter fragments have mass below that of the projectile and hence are probably not from fusion reactions, but are projectile-like fragments associated with lower excitation energies. As the action of evaporation can never cause the average decay trajectory to cross the attractor line, but only approach it, the parents of these fragments must be closer to the line of β–stability. For the heavier fragments, their location is again consistent with significant evaporation.

For $Z > 82$, the evaporation attractor line crosses the limits of known isotopes (It is still inside the proton drip line). The exact location of the attractor line is this region is not precisely known, as its depends on separation energies which have not been measured. Its exact location, at the moment, is thus dependent on which predicted mass tables are used. If fission competition is not overpowering, then it would seem reasonable that "hot" fusion reactions can be used to extend the limits of known isotopes in this region out to the evaporation attractor line. In fact, many of the most neutron–deficient isotopes in this region have been produced in hot fusion reactions. For example, ^{206}Ra was first detected with the E/A ~9.5 MeV ^{19}F+^{197}Au reaction (10n channel) [9] and ^{213}Th was detected in the E/A ~9 MeV ^{16}O+^{208}Pb reaction (9n channel). In both of these reactions, the measured complete fusion cross section are surprisingly large (~tens of mb) and do not have a strong bombarding energy dependence[11]. Enhanced residue cross sections (compared to statistical model estimates) have also been measured for ^{110}Pd[12] and ^{40}Ar[13] induced reactions producing compound nuclei in the same region of the chart of nuclides. Thus for this $82 \lesssim Z \lesssim 90$ region at least, fission competition would not seem to be a limiting factor in extending the region of known isotopes to the attractor line. As a example, one can propose the reaction E/A=10 MeV ^{24}Mg+^{194}Pt to produce, by the 12n channel, ^{206}Th (proton bound[3]) which is located on the attractor line and accounts for 8% of the residue yield in the statistical model calculations using masses from Ref. [3]. Based on experimental evaporation residue excitation functions, one might expect a yield of the order of 1–10μb. Other unknown isotopes with similar or greater yields would also be produced, many of these involving channels containing light charged particles.

In conclusion, it has been shown that the process of evaporation moves the position of a fragment in the chart of nuclides towards a line which has been called the evaporation attractor line (EAL). Provided the primary fragments have sufficient excitation energy ($E^*/A <$ 1.3 MeV) and are not located too far form the EAL, the evaporation residues on average will be located on the evaporation attractor line and the fragments will have lost memory of their initial neutron–richness or neutron–deficiency.

The direction in N–Z space followed by an evaporating fragments depend on its excitation energy. At the lowest excitation energies, proton or neutron emissions move the decaying systems directly towards the evaporation attractor line. At larger excitations energies, the evaporation of nucleons and clusters removes similar number of neutrons and protons from the decaying systems and the initial trajectories followed by the decaying system are almost parallel to the evaporation attractor line. As the movement towards the EAL occurs only at the end of the decay sequence, it was found that for very neutron–rich primary fragments, the average position of the evaporation residues never reaches the EAL even for large excitation energies. The location of detected secondary fragments in the chart of nuclides relative to evaporation attractor line can give some information as to the initial proton–neutron composition and/or the initial excitation energy of the fragments.

Heavy fragmentation products are, on average, more neutron rich that the attractor line as expected if the primary fragments are located close to the line of β–stability. However, fusion reactions can produce heavy residues which are more proton rich that fragmentation products. For $Z > 82$, the attractor line crosses the limits of known isotopes and there seem no reason why "hot" fusion reaction cannot extend these limits out to the attractor line.

This work was supported by the Director, Office of High Energy and Nuclear Physics, Nuclear Physics Division of the U.S. Department of Energy under Contract No. DE–FG02–87ER–40316.

References

[1] R.J. Charity, Phys. Rev. C **58**, 1073 (1998).

[2] R.J. Charity, Computer code GEMINI (unpublished), This code is available via anonymous FTP from wunmr.wustl.edu in directory /pub/gemini.

[3] P. Möller, J.R. Nix, W.D. Myers, and W.J. Swiatecki, At. Data Nucl. Data Tables **59**, 185 (1995).

[4] R.J. Charity, M. Korolija, D.G. Sarantites, and L.G. Sobotka, Phys. Rev. C **56**, 873 (1997).

[5] K. Sümmerer, W. Brüchle, D.J. Morrissey, M. Schädel, B. Szweryn, and Yang Weifan, Phys. Rev. C **42**, 2546 (1990).

[6] M. Weber, C. Donzaud, J.P. Dufour, H. Geissel, A. Grewe, D. Guillemaud–Mueller, H. Keller, M. Lewitowicz, A. Magel, A.C. Mueller, G. Münzenberg, F. Nickel, M. Pfützner, A. Piechaczek, M. Pravikoff, E. Roeckel, K. Rykaczewski, M.G. Saint-Laurent, I. Schall, C. Stephan, K. Sümmerer, L. Tassan–Got, D.J. Vieira, and B. Voss, Z. Phys. A **343**, 67 (1992).

[7] J. Reinhold, J. Friese, H.-J. Körner, R. Schneider, K. Zeitelhack, H. Geissel, A. Magel, G. Münzenberg, and K. Sümmerer, Phys. Rev. C **58**, 247 (1998).

[8] L.G. Sobotka, R.J. Charity, J.F. Dempsey, H. Scheit, and T. Glasmacher, unpublished.

[9] K. Valli, W. Treytl, and E. Hyde, Phys. Rev. **161**, 1284 (1967).

[10] K. Valli and E. Hyde, Phys. Rev. **176**, 1377 (1968).

[11] K.T. Brinkmann, A.L. Caraley, B.J. Fineman, N. Gan, J. Velkovska, and R.L. McGrath, Phys Rev. C **50**, 309 (1994).

[12] W. Morawek, A. Ackermann, T. Brohm, H.–G. Clerk, U. Gollerthan, E. Hanelt, M. Horz, W. Schwab, B. Voss, K.–H. Schmitt, and F.P. Hessberger, Z. Phys. A **341**, 75 (1991).

[13] F.P. Hessberger, V. Ninov, and D. Ackermann, Z. Phys. A **343**, 301 (1992).

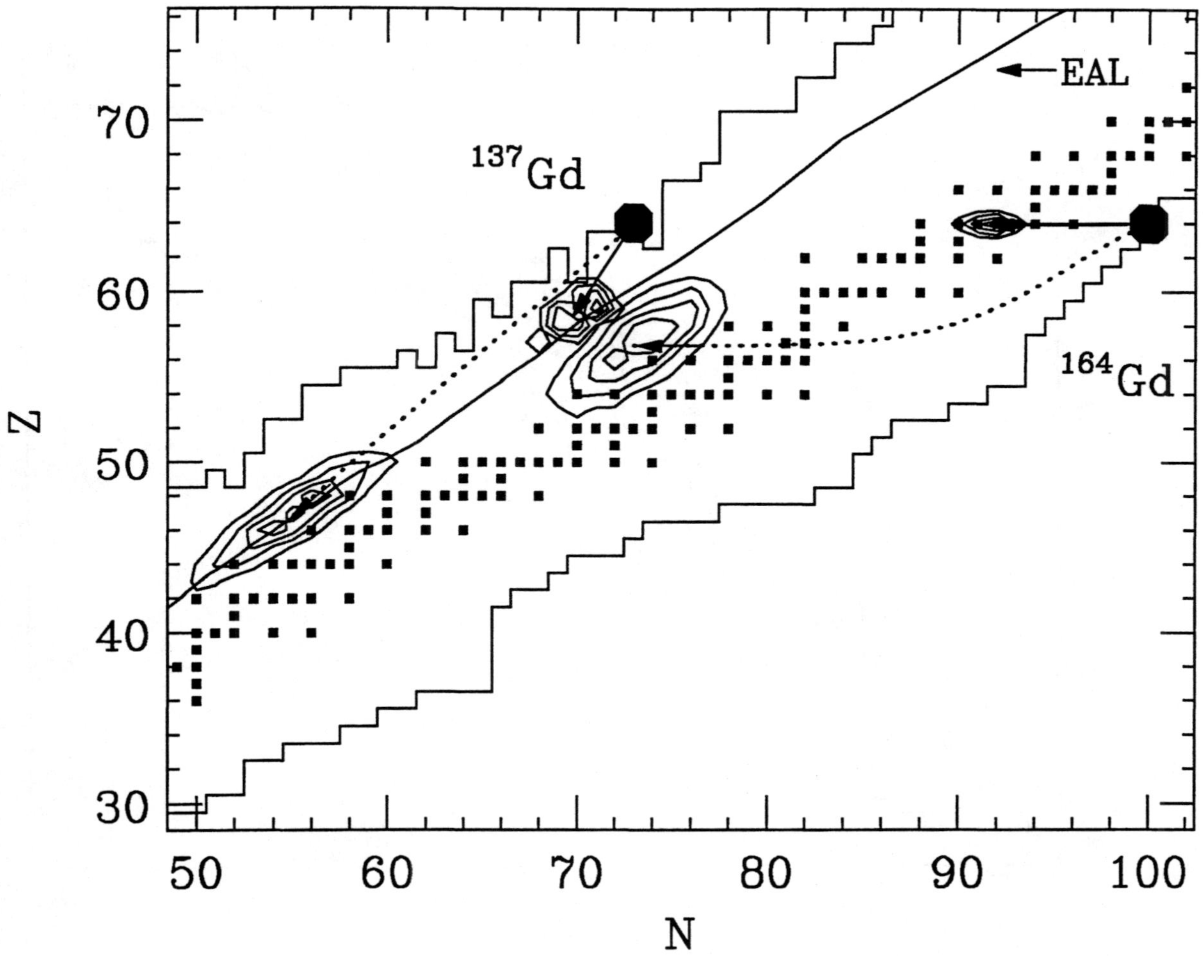

Figure 3: Contours showing the distribution of secondary fragments in the chart of nuclides predicted for ^{137}Gd and ^{164}Gd compound nuclei with 100 and 400 MeV of excitation energy. The contour interval is 20% of the peak yield. The location of the compound nuclei are indicated by the solid points and the average trajectories followed by the decaying systems are indicated by the solid and dotted curves for the E^*=100 and 400 MeV simulations, respectively. The thick curve shows the predicted location of the evaporation attractor line (EAL).

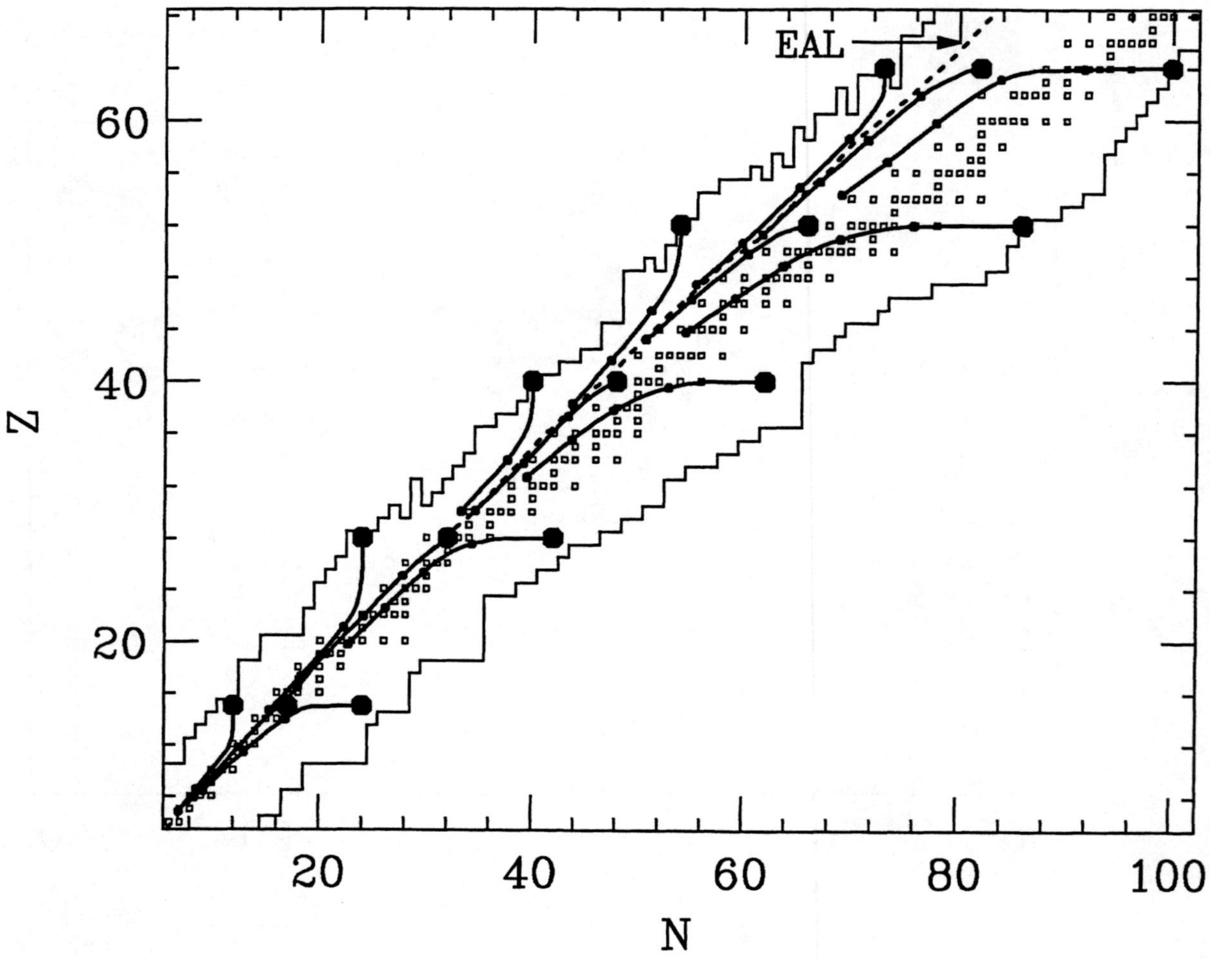

Figure 4: The curves show the loci of the predicted, average location of secondary fragments in the chart of nuclides for the compound nuclei indicated by the large solid points. The small points on the curves correspond to the average location of the secondary fragments as the excitation energy is increased in 100 MeV increments. The evaporation attractor line (EAL) is indicted.

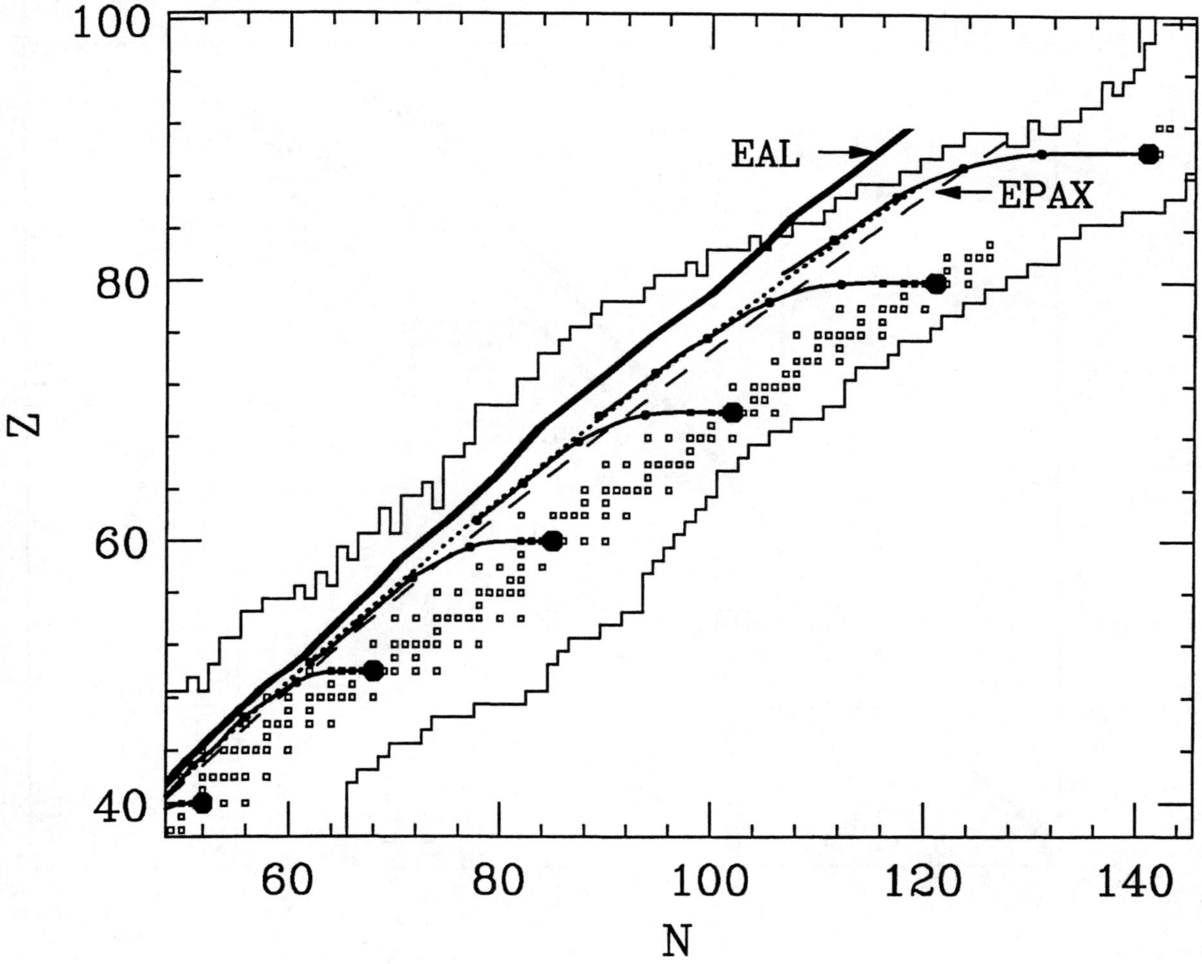

Figure 5: As in Fig. 4, but now the loci of secondary fragments positions are for primary fragments located on the β–line of stability. At high excitation energies the final fragment positions follow the dotted curve which is parameterized in the text. For comparison, the EPAX formula is indicated by the dashed curve.

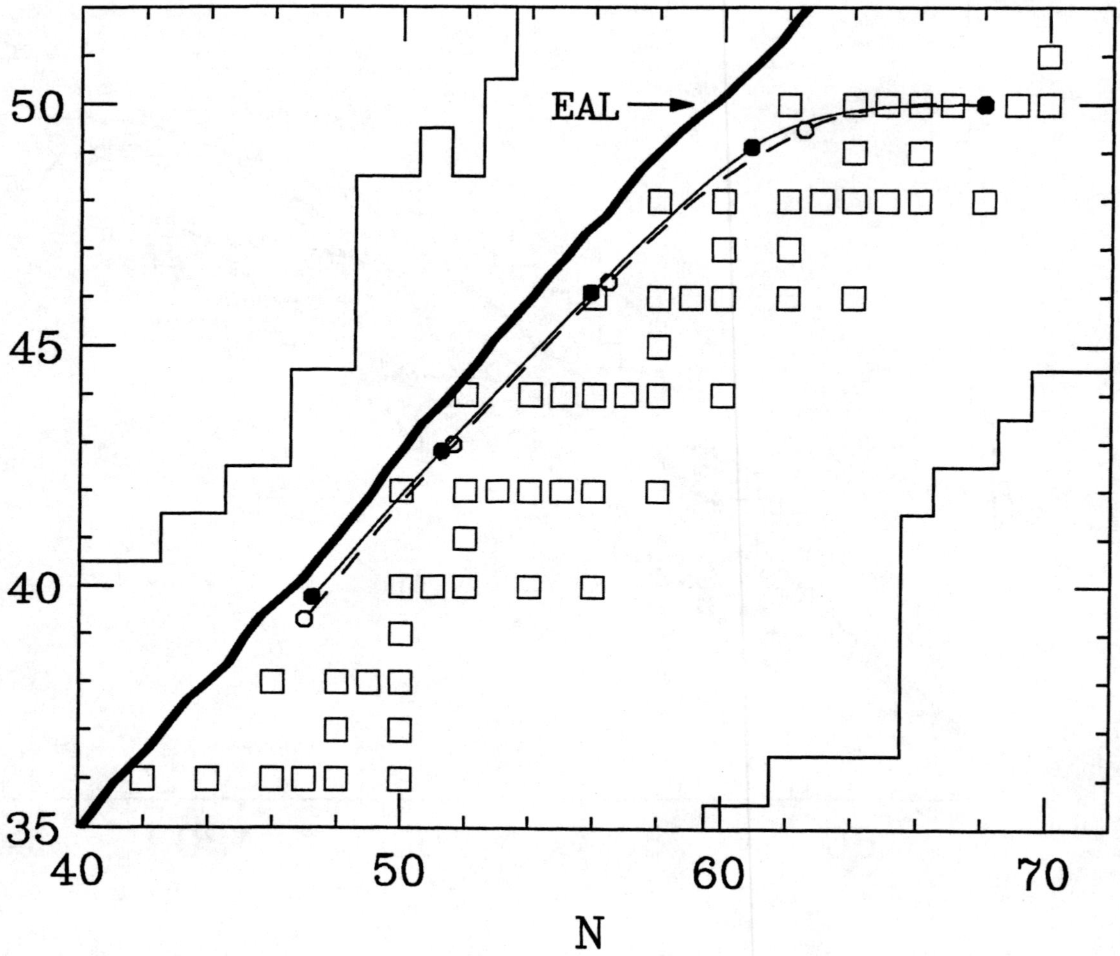

Figure 6: As in Fig. 5, but now the calculations are for ^{118}Sn primary fragments with spins of zero (solid) and 60 $\hbar$ (dashed).

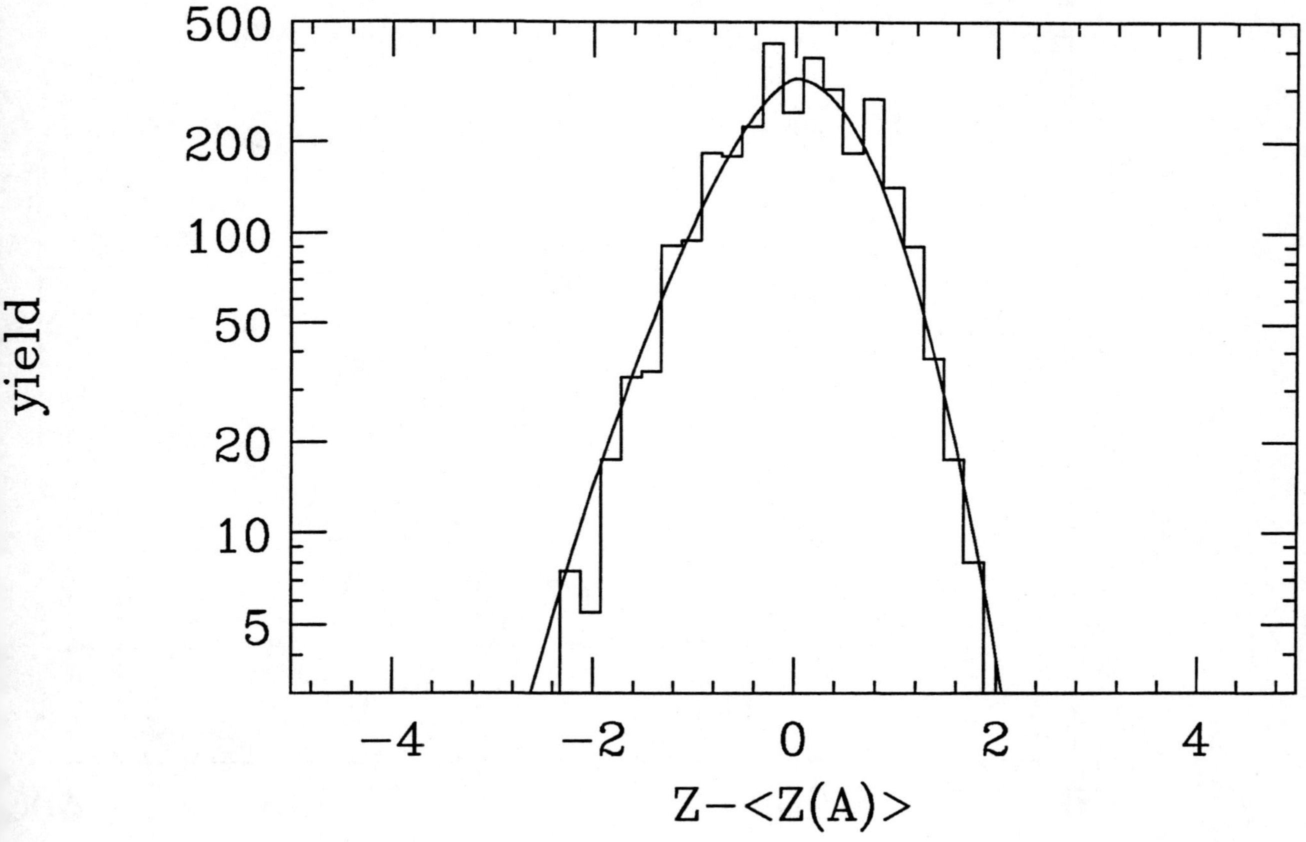

Figure 7: Distribution of Z-$< Z(A) >$ for secondary fragments predicted after evaporation from E^*=200-400 MeV ^{118}Sn primary fragments.

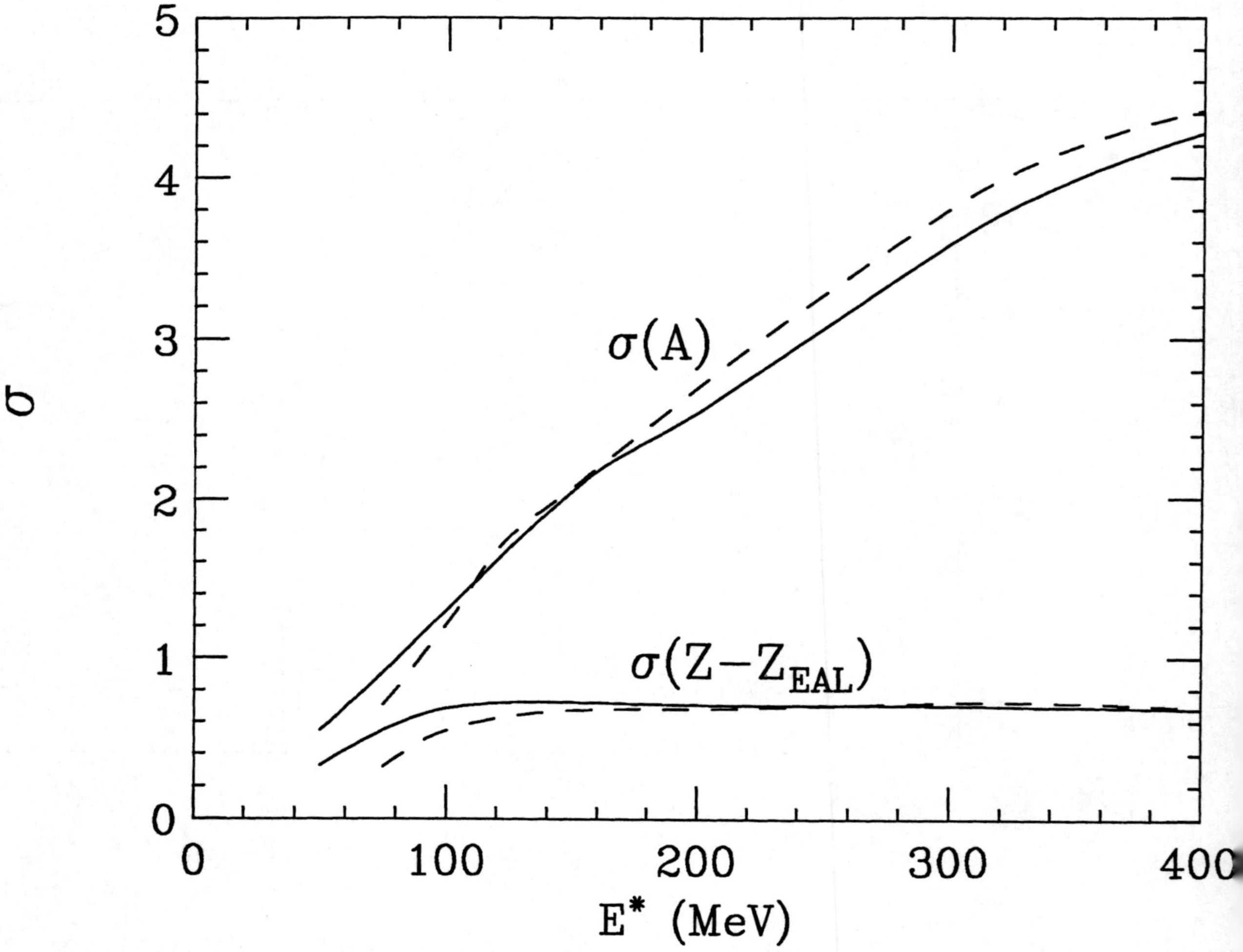

Figure 8: Excitation energy variation of the length $\sigma(A)$ and width $\sigma(Z - Z_{EAL})$ of the distributions of secondary fragments predicted for ^{118}Sn compound nuclei with spins of zero (solid) and 60 $\hbar$ (dashed).

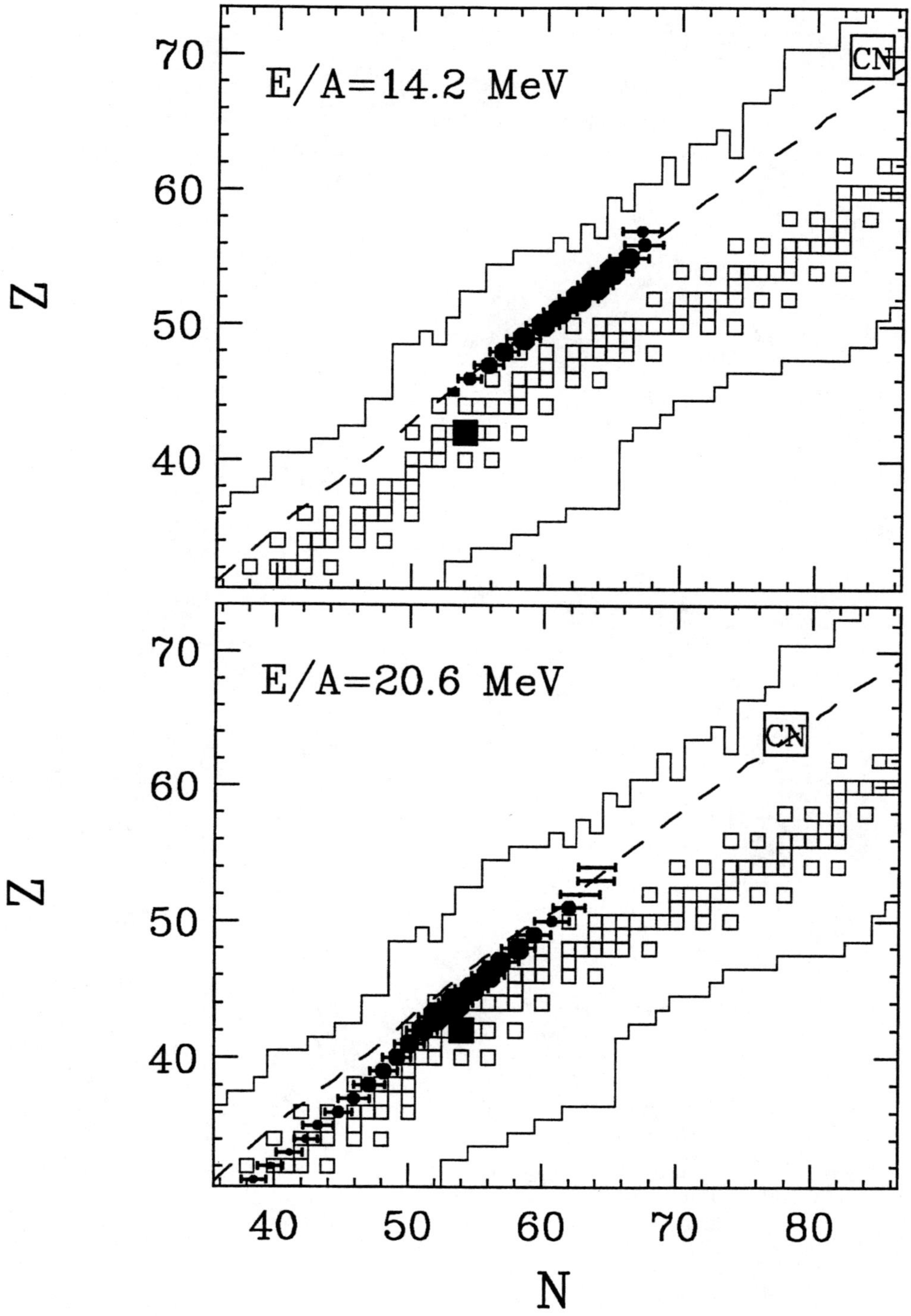

Figure 9: Experimental N–Z distributions of fragments detected at zero degrees in the $^{96}Mo + ^{58}Ni$ reactions at E/A=14.2 and 20.6 MeV. The error bars represent the 1σ widths of the measured distributions. For comparison the evaporation attractor line is indicated by the dashed curves. The average estimated compound nucleus position are indicated as the "CN" labeled squares and the projectile by the filled squares.

Isospin Effects in Fragment Production

M. B. Tsang[1], W.A. Friedman[2], W.G. Lynch[3]

Abstract

The dependence of isotope distribution on the overall isospin of the reaction system is investigated. General scaling curves are observed. The meaning of the observed isospoin effects is discussed for each of the different reaction.

In the past decade, substantial knowledge about the decay of excited systems formed in heavy ion collisions has been obtained from the measurement of the charge and energy of the emitted particles. Ideally information about these particles should include the identification of the isotopic composition, i.e., mass as well as charge. Reaction models, ranging in emphasis from statistical to dynamical, have been developed. These can account for many salient features, such as the charge and energy distributions of the elements produced in multifragmentation. However, precise experimental isotope distributions have been difficult to reproduce with theoretical models due to complications arising from the role of sequential decays [1-3]. Inclusions of nuclear spectral information into the calculations to simulate the effects of secondary decay has not been fully successful because the task is not only computationally difficult, but is also hampered by the lack of complete information about the resonances in many nuclei [3].

The importance of the isotopic degree of freedom, a) to obtain information about charge equilibration, b) to understand the behavior of the charge asymmetry features of the nuclear equation-of-state, and c) to provide strenuous tests for reaction models, has prompted the design of experiments with the specific capability of measuring isotope distributions [4,5]. Figure 1 shows the measured differential multiplicities of Li, Be, B, C, N and O isotopes at $\theta_{CM}=90°$ for three systems, $^{112}Sn+^{112}Sn$ (open circles), $^{112}Sn+^{124}Sn$ (open diamonds [6]) and $^{124}Sn+^{124}Sn$ (closed circles) measured by the LASSA collaboration at E/A=50 MeV [5]. These data are selected for central collisions by a gate on charged particle multiplicity and are subjected to a gate on rapidity of $0.46 \leq y/y_{beam} \leq 0.65$. More detailed information about the experiment can be found in Ref. [5].

The x-axis, N-Z, corresponds to the neutron excess of the isotope. For $x=0$, these are symmetric isotopes with $N=Z$. The peaks of the distributions are

[1] National Superconducting Cyclotron Laboratory and Department of Physics and Astronomy
[2] Michigan State University, East Lansing, MI 48824, USA
3 Department of Physics, University of Wisconsin, Madison, WI 53706

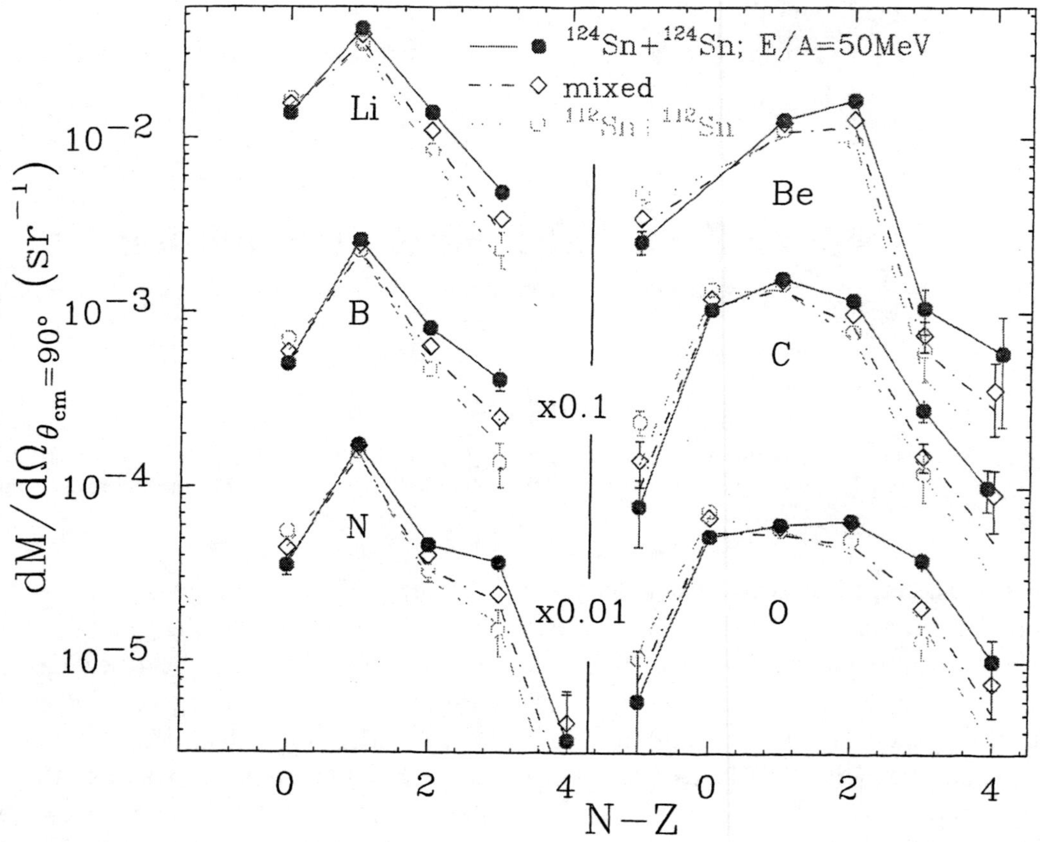

Figure 1: Differential multiplicities at $\theta_{CM}=90°$ for Li, Be, B, C, N and O isotopes as a function of neutron excess (N-Z) of the isotope. The solid points drawn by solid lines to guide the eyes are data for the $^{124}Sn+^{124}Sn$ system with $N/Z=1.48$. The open circles are data for the lightest system $^{112}Sn+^{112}Sn$ with $N/Z=1.24$. The open diamonds are the statistically averaged values from the two mixed systems, $^{124}Sn+^{112}Sn$ and $^{112}Sn+^{124}Sn$. The dashed and dotdashed lines are predictions from Eq. (4).

located around $N=Z+1$ isotopes. The yields of the boron, carbon isotopes are offset by a factor of 10 and the yields of the nitrogen and oxygen isotopes are offset by a factor of 100 in the figure. As expected, more neutron rich isotopes ($N-Z>0$) are produced by the neutron rich system, $^{124}Sn+^{124}Sn$, while the opposite is true for the proton-rich isotope yields. In general, the drop from the peak toward more proton-rich isotopes is steeper than the corresponding drop towards the neutron rich isotopes. The difference in isotope yields between the three different systems is not very large, less than a factor of two for the more abundant isotopes. The difference increases at the extremes of the isotope distributions. Thus larger differences may occur at more extreme values of the asymmetry but such data have not been obtained in this experiment.

Multifragmentation processes have been successfully described by statistical theory suggesting their cross sections are dominated by phase space. In this limit, one may obtain some guidance from the simple dependence of the population of isotopes predicted by the equilibrium limit of the Grand-Canonical Ensemble [7,8]. For this special case the yield of isotopes is governed by two nucleon chemical potentials, μ_n, μ_p and the temperature T plus the individual binding energies of the various isotopes $B(N,Z)$.

$$Y(N,Z)=F(N,Z,T)\, e^{B(N,Z)/T}\, e^{(N\mu_n+Z\mu_p)/T} \qquad (1)$$

In the Grand-Canonical limit, the chemical potentials are related to the nucleon density, $\rho_n \propto e^{\mu_n/T}$ and $\rho_p \propto e^{\mu_p/T}$. The factor $F(N,Z,T)$ includes information about the change of the yield arising from secondary feeding and the temperature of the isotope considered. It reflects the sequential feeding contribution from the particle stable states of the same nucleus or from the particle decay of heavier unbound nuclei.

From charge symmetry, mirror nuclei with $|N\text{-}Z|=1$ have similar internal partition functions. Therefore secondary feeding from the particle stable states in Eq. (1) are similar for mirror nuclei. If the sequential decay effects from unbound particle states and the influence of the Coulomb interaction with remaining system are small, the dependence of isobaric yield ratios on the binding energy difference should be exponential, i.e.

$$Y(N,Z)/Y(Z,N) \propto exp[(\mu_p\text{-}\mu_n)/T] \propto (\rho_n / \rho_p)\, e^{\Delta B/T} \qquad (2)$$

Figure 2 shows the isobaric yield ratios of three pairs of mirror nuclei, (t, ^{3}He), (^{7}Li, ^{7}Be) and (^{11}B, ^{11}C), as a function of the binding energy difference, ΔB for two reactions, ^{112}Sn+^{112}Sn (open points) and ^{124}Sn+^{124}Sn (solid points) at E/A=50 MeV. Equation (2) suggests that the dependence of the isobaric yield ratios on the binding energy difference should be exponential. In fact, the experimental data fluctuate around such a relationship. Extrapolation to $\Delta B=0$ using the best fit lines (dashed and solid lines) suggests values for ρ_n / ρ_p of 2.5 for the $^{112}Sn+^{112}Sn$ system (bottom line) and 5.5 for the $^{124}Sn+^{124}Sn$ system (top line). Both of these numbers are larger than the initial N/Z values of the two systems, 1.24 and 1.48, (marked by the two arrows in the figure) for $^{112}Sn+^{112}Sn$ and $^{124}Sn+^{124}Sn$, respectively. Statistical interpretations of the mirror nuclei yield ratios suggest a change of 2.5 in the N/Z ratio, which is clearly larger than 20% change in the total N/Z.

The mirror nuclei ratios shown in Figure 2 do not follow Eq. (2) strictly, partly because the difference in the Coulomb interaction is neglected but also because, the secondary feedings from particle unstable states are large. Statistical calculations suggest that these ratios can be reduced by as much as a factor of two due to sequential decays [9].

To search for experimental observables that are minimally affected by secondary decays, we note that the corrections to the primary yields due to secondary decays appear to be similar from reaction to reaction over a wide range of bombarding energies [10]. Taking this similarity into account, the isotope yields of two different systems with similar incident (excitation) energies but different isospins can be combined to construct ratios of the form,

$$R_{21}(N,Z) = \frac{Y_2(N,Z)}{Y_1(N,Z)}. \qquad (3)$$

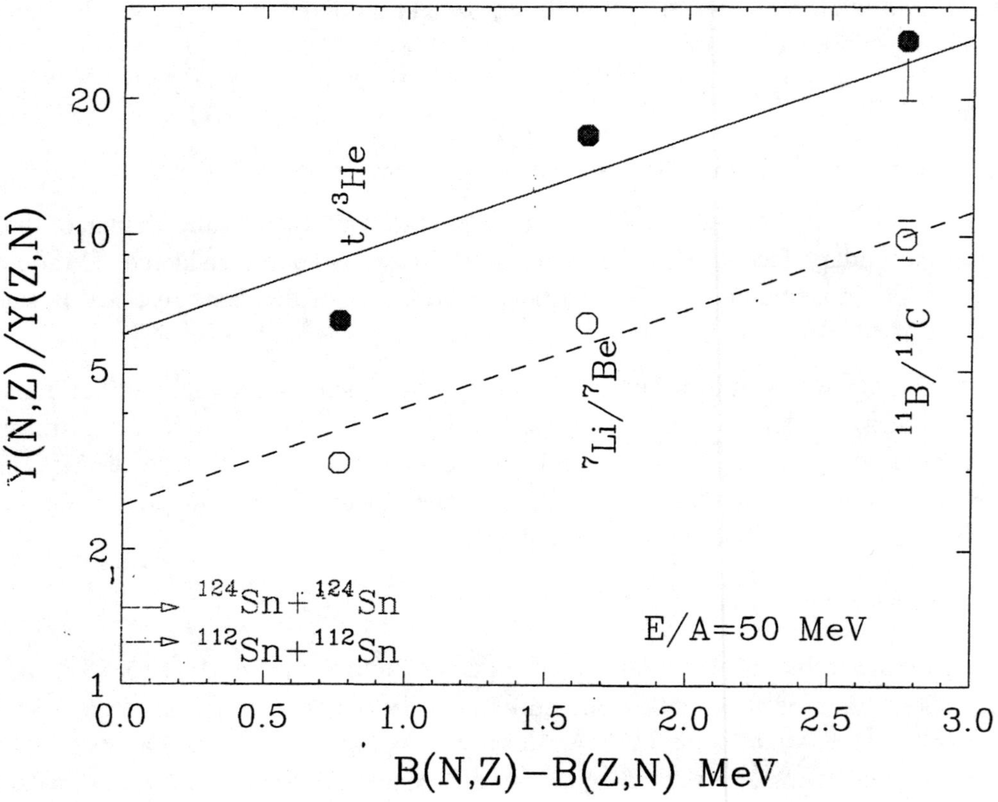

Figure 2: Isobar ratios for three pairs of mirror nuclei obtained from the $^{112}Sn+^{112}Sn$ (open circles) and $^{124}Sn+^{124}Sn$ (solid points) reactions. The lines are best-fit parameterizations of Eq. (2)

(Throughout this article, we have chosen the system in the numerator to be more neutron-rich than the one in the denominator.) Statistical calculations predict R_{21} to be insensitive to sequential decay effects. R_{21} constructed from primary yields appears to be quite close to the values of R_{21} extracted from the secondary yields [5, 9].

If the system is in equilibrium and $F_1(N,Z,T)= F_2(N,Z,T)$, one might anticipate that $R_{21}(N,Z)$ is governed by only three parameters:

$$R_{21}(N,Z) \sim Ce^{N\alpha+Z\beta} \tag{4}$$

where $\alpha=\Delta\mu_n/T$ and $\beta=\Delta\mu_p/T$ reflect the differences between the neutron and proton chemical potentials for the two reactions and C is an overall normalization constant. The N and Z dependence becomes most apparent if, for each element Z, R_{21} is plotted versus N for all isotopes on a semi-log plot. The resulting slopes would then be α for each Z. Similarly, plotting R_{21} against Z for all isotones would provide a common slope β for each N. This is demonstrated in Figure 3 where the isotope yield ratios from the central collisions of $^{124}Sn+^{124}Sn$

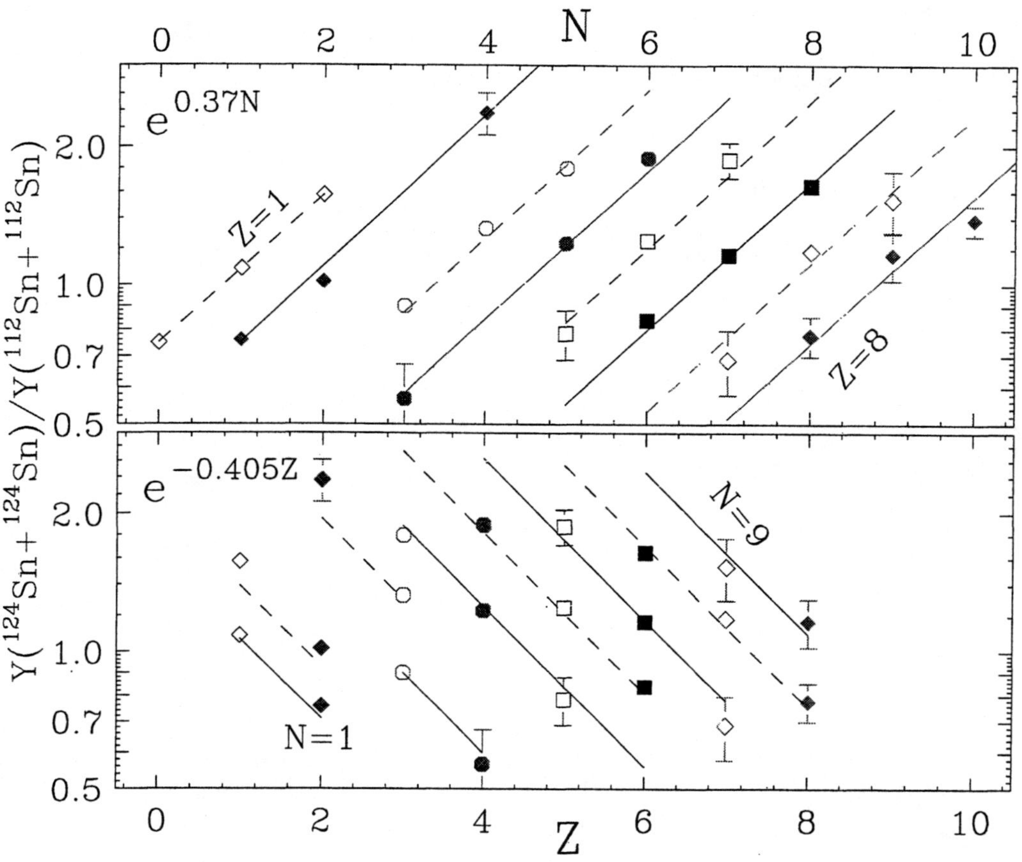

Figure 3: R_{21} for $^{124}Sn+^{124}Sn$ and $^{112}Sn+^{112}Sn$ systems, as a function of *N* (isotope data, upper panel) and as a function of *Z* (isotone data, lower panel). Solid and dashed lines are best fits to the data.

and $^{112}Sn+^{112}Sn$ are plotted as a function of *N* (upper panel) and as a function of *Z* (lower panel) for 24 isotopes spanning from *Z* =1 to *Z*=8 elements. The excellent agreement between data and Eq. (4) can be seen more clearly by comparing the experimental ratios with the best fits of straight lines with $\alpha=0.36$ and $\beta=-0.41$.

In the Grand Canonical Ensemble limit of dilute non-interacting gas, α and β are related to the relative nucleon density, $\hat{\rho}_n = \rho_n^2 / \rho_n^1 = e^{\alpha}$ and $\hat{\rho}_p = \rho_p^2 / \rho_p^1 = e^{\beta}$ [5]. The resulting values, of $\hat{\rho}_n$ and $\hat{\rho}_p$, extracted from isotope ratios for the three systems: $^{112}Sn+^{112}Sn$, $^{124}Sn+^{112}Sn$, $^{124}Sn+^{124}Sn$ are shown in Fig. 4 as a function of the *N/Z* ratio of the composite system, $(N/Z)_O$.

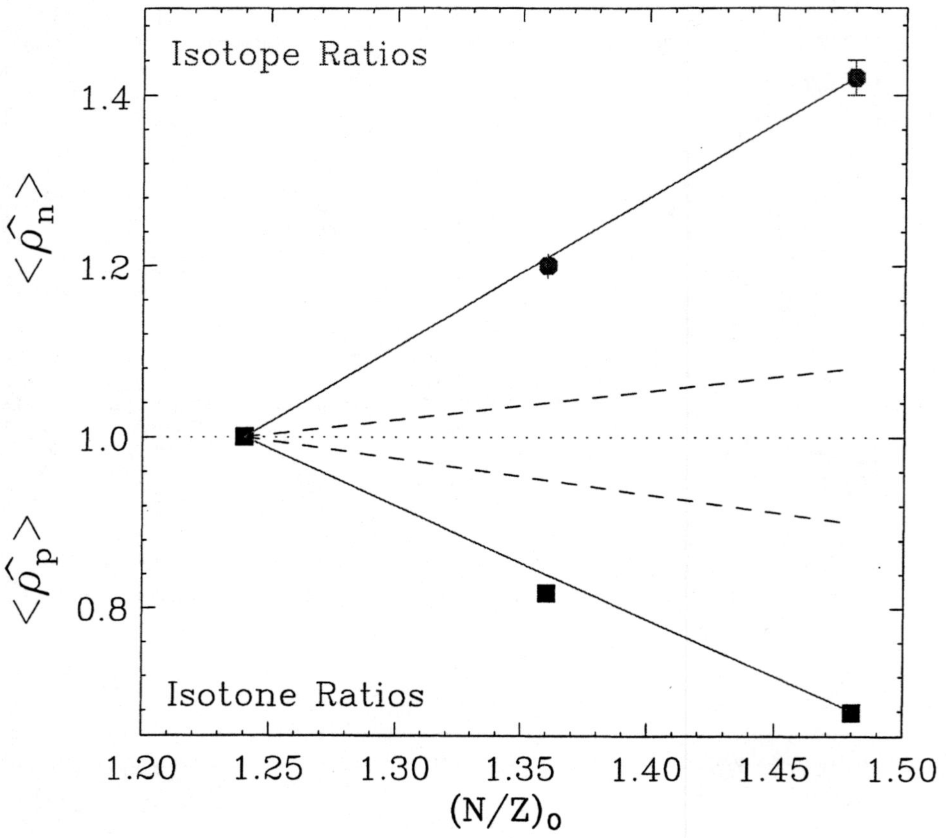

Figure 4: The relative free neutron $\hat{\rho}_n = e^{\alpha}$ and free proton density $\hat{\rho}_p = e^{\beta}$ as a function of *(N/Z)*$_O$. The solid lines are best fit through the data. The dashed lines are the expected n-enrichment and p-depletion with the increase of isospin of the initial systems.

The trend of the total neutron and proton density, assuming the same total matter density for the two systems, is given by the dashed lines in Fig. 4. The experimental data suggest that as *(N/Z)*$_O$ increases, the system responds by making the asymmetry of the gas (given by the solid lines) much greater than the asymmetry of the total system (given by the dashed lines). The extracted $\hat{\rho}_n / \hat{\rho}_p \approx 2$ at *(N/Z)*$_O$=1.48 is roughly consistent with the mirror nuclei ratios obtained in Figure 2 for the $^{124}Sn+^{124}Sn$ and $^{112}Sn+^{112}Sn$ systems.

If the equilibrium interpretation of these data is correct, the nucleon density extracted from isotope, isotone and isobar ratios is more enriched in neutrons than the liquid phase represented by bound nuclei, consistent with the predicted partial fractionation of nucleon components in the liquid gas phase transition [11, 12]. The neutron enrichment is much more enhanced in the collisions of neutron-rich systems as compared to collisions of neutron-deficient system.

A more compact way of plotting the data in Figure 3 is to construct the scaling factor $S(\beta)=R_{21}(-\beta Z)$. With the best fit value of β, all isotopes would lie on a single line if $S(\beta)$ is plotted versus N on a semi-log plot. This trend is referred as "scaling". The scaling behavior of the data shown in Figure 3 is illustrated in Figure 5. The data are plotted next to the label "multifragmentation". Isotopes from Lithium to Oxygen lie on a line. The hydrogen and helium isotopes lie slightly above the line even though each element still has the same slope. This small deviation may result if light particles such as hydrogen and helium are produced by different mechanisms, for example pre-equilibrium emission.

This systematics, suggested by the grand canonical expression, appears to be more generally respected than one might expect. For example, evaporation process displays similar scaling behavior. Figure 5 shows the scaling factor $S(\beta)$ determined for isotopes detected at backward angles ($\theta_{lab}=154°$) in the reactions $^{4}He + ^{116}Sn$ and $^{4}He + ^{124}Sn$ at an incident energy of 200 MeV [13]. These data are plotted next to the label "evaporation". The dashed line drawn through the data is a best fit to this data using Eq. 4.

For the low energy collisions of ^{16}O with two targets, ^{232}Th and ^{197}Au, at E/A=8.6 MeV [14], we have plotted the scaling factor $S(\beta)$ obtained for strongly damped reactions. At such energies, just above the Coulomb barrier, the isotopes originate from the breakup of di-nuclear systems formed by the transfer of energy and mass between projectile and target. These data are plotted next to the label "deep inelastic" in Figure 5 and the best fit line is also drawn through the data.

Figure 5 suggests that the scaling behavior appears to be a surprisingly robust feature over a range of data from deep-inelastic heavy ion reactions at lowest energies, through evaporation reactions induced by light and heavy ion projectiles, to high energy heavy ion reactions characterized by the copious production of intermediate mass fragments (IMFs) and multifragmentation. For evaporation and deeply inelastic processes, the grand-canonical ensemble approach of Eq. (1) has little relevance. Why the scaling phenomenon occurs in these cases is discussed below.

At large scattering angles of deeply inelastic reactions, it is frequently assumed that equilibrium is established in the binary system and the two fragments come to a common temperature [14]. Empirically it is observed that the isotope yields are correlated with the Q-value for the mass transfer, known as the "Q_{gg} -systematics". The observation of Q_{gg}-systematics in the $^{16}O + ^{232}Th$ and $^{16}O + ^{197}Au$ reactions, implies that the yield ratios can be expressed in terms of the difference between the binding energies of the two binary reaction partners.

$$R_{21} \propto exp\,[(BE(N_2-N,Z_2-Z)-BE(N_1-N,Z_1-Z))/T],$$

$$\approx exp\,[(-\Delta S_n \cdot N - \Delta S_p \cdot Z)/T] \qquad (5)$$

where Z_i and N_i are the total proton and neutron number of reaction i. $BE(N,Z)$ is the binding energy of a nucleus with neutron number N and proton number Z. Expanding the binding energies in Eq. (5) in Taylor series, one can obtain an approximation similar to Eq. (4) that is first order in N and Z. In this expression, the role of the chemical potential difference in Eq. (4), is played by the difference of the average neutron and proton separation energies, ΔS_n and ΔS_p.

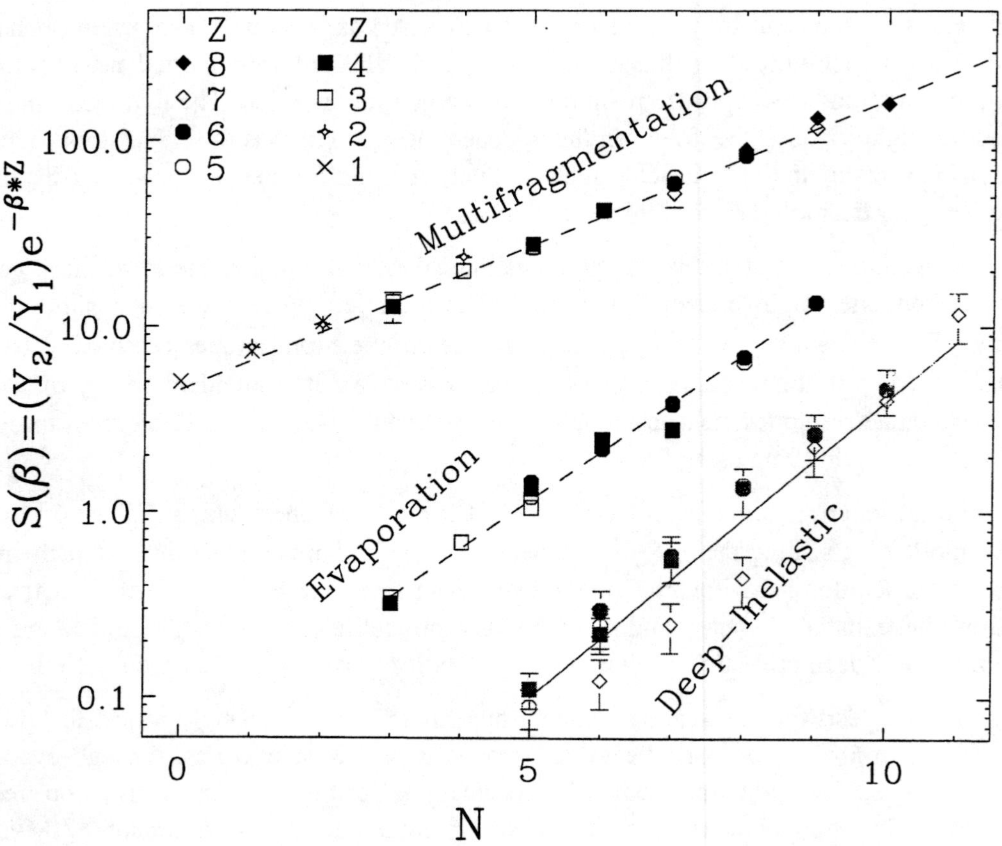

Figure 5: The scaled isotopic ratio, *S(β)* is plotted as a function of *N* using the best fit value of *β* obtained from fitting isotopes with *Z≥3*. The data points plotted next to the label "multifragmentation" denote *S(β)* extracted from multifragmentation events in proton density, assuming the same total matter density for the two systems, iscentral $^{112}Sn+^{112}Sn$ and $^{124}Sn+^{124}Sn$ collisions [5]. The scaling behavior for evaporation process is illustrated by the reactions $^{4}He+^{116}Sn$ and $^{4}He+^{124}Sn$ [13] plotted next to the label "evaporation". Systematics of the strongly damped binary collisions is represented by the data of ^{16}O induced reactions on two targets ^{232}Th and ^{197}Au [14] plotted next to the label "deep-inelastic".

The underlying origin of the scaling behavior observed in evaporation processes turns out to be similar to that found in the case of deep inelastic scattering. Using the formalism proposed by Friedman and Lynch [15] and performing another Taylor expansion, the relative isotope ratio can be written as

$$R_{21} \propto exp\ [\{(-\Delta S_n + \Delta f_n^*)\cdot N + (-\Delta S_p + \Delta f_p^* + e\Delta\Phi(Z_i - Z))\cdot Z\}/T] \qquad (6)$$

where the terms f_n*(f_p*) represent the excitation energy contribution to the free energy per neutron (proton) and Φ is the electrostatic potential at the surface of the nuclei. More detailed discussions of Eq. (5) and (6) can be found in Ref. [16]. While first order terms in the Taylor expansion dominate for small A<12, it can be expected that the scaling behavior will breakdown at high Z and N values.

Equation (6) is also consistent with multifragmentation description that follows the Expanding Emitting Source (EES) model [17]. In that model the evaporation considerations discussed above are coupled to an expansion process which reduces the density of the emitting system. The density dependence of the various terms in the exponent in Eq. (6) becomes important. If the expansion were to be isentropic, then the temperature would vary as $\rho^{2/3}$. Likewise, the Fermi energy would also be anticipated to vary with the density $\rho^{2/3}$. Thus, for a Fermi gas, $\Delta f/T$ would remain constant even with reducing density. As seen in the evaporation process, the value of ΔS_n and ΔS_p are strongly governed by the symmetry term contributions to the masses. If the system is expanding, dependence of the symmetry term on density will be important. Model calculations suggest that the low energy density dependence of the symmetry energy may vary with the density to the 2/3 power. In that case the ratio of symmetry energy to temperature for an isentropic expansion may remain constant with reducing density. Thus for isentropic expansion and the specific assumptions for the density dependences we have outlined, the EES rates will provide a constant ratio over the expansion process, so that R_{21} can be determined by the initial conditions. On the other hand, EES calculations, predict that the emission of IMFs occurs strongly at a stage of large expansion, with densities on the order of 1/3 of normal density and temperatures on the order of 5 MeV. Thus R_{21} can, more generally, be determined by the particular values of the symmetry energy under the conditions of reduced density without any assumption for a specific density dependence. Consequently, a measurement of the observable R_{21} can be used to probe the density dependence of the symmetry energy. This is one of the anticipated benefits of future studies of the R_{21} parameter.

The EES predictions also suggest that the N/Z of the emitting system may be modified by early emission of light fragments, so detailed comparison between data and freeze-out models requires an estimate of this change. At present, an analogous procedure is attempted to adjust both energy loss and mass loss in the early stages of the reaction leading to the freeze-out. Additional care will be required in estimating the isospin of the freeze-out source.

Figure 5 suggests that the scaling behavior is a general property of heavy ion reactions where some form of equilibrium is established. If two systems are employed in constructing values of R_{21}, the observation of the systematic trends does not imply that both reacting systems proceed with the same reaction mechanism. Additional information is needed to elucidate details of the underlying reaction mechanism [16].

Since most data obtained in the past do not have a large range of well measured isotope yields, one can investigate the criterions for scaling behavior using the double ratio which is related to the slope of the logarithm of the scaling function,

$$\mathrm{R4} = \frac{Y_2(N_i+1,Z_i)}{Y_2(N_i,Z_i)} \Big/ \frac{Y_1(N_i+1,Z_i)}{Y_1(N_i,Z_i)} = e^{\alpha} \qquad (7)$$

This formula relates α to specific isotope double ratios; a similar formula has been obtained by ref. [5], which relates corresponding isotone ratios to β. As shown in ref. [5], use of Eq. (7) is equivalent to fitting the systematics shown in Fig. 3. R4 has the advantage that values for α or β can be obtained very easily from a sparse data set. To illustrate its functionality, we show experimental values for *R4* in the lower panel of Fig. 6 which were obtained from ref. [13,18,19] for four systems $^{4}He+^{116}Sn$ at 200 MeV, $^{4}He+^{124}Sn$ at 200 MeV, *p+Ag* at 480 MeV and *p+U* at 5.5 GeV at backward angles, $\theta_{lab} > 150°$. While the incident energies of these systems are different, values for the temperature extracted from isotopic thermometry measurements are statistically consistent with an average value of about 2.7 MeV [20]. (At such low temperatures, the corrections to the isotopic thermometer for secondary decay should be small [2].) Thus, one may be justified with comparing the isotopic yields via an R_{21} or *R4* ratio and analyzing the result via an evaporation expression such as that given in Eq. (6). A linear dependence of the resulting *R4* values upon *N/Z* is expected (see Fig. 4), and its confirmation (solid line drawn to guide the eye) in the lower panel of Fig. 6 supports the expectation that all four systems are at a similar temperature. The slope, which reflects the lower temperature *(T~2.7 MeV)* of the evaporation systems is steeper than that shown in Figure 4 for the *Sn+Sn* systems, which have a higher temperature *(T~4 MeV)*.

The upper panel shows a similar comparison involving the isotopic yields emitted to forward angles where preequilibrium emission is important [13]. Here, *R4* values for $^{4}He+^{116}Sn$ and $^{4}He+^{124}Sn$ reactions at 200 MeV and $\theta_{lab} = 12°$ [13] as well as $^{3}He+Ag$ and $^{3}He+Au$ at 200 MeV and $\theta_{lab} = 12°$ [21] are plotted. As in the case with the backward angle measurements, we note that the isotopic temperatures in these reactions have roughly the same value [20], about 4.3 MeV. Contrary to the data plotted in the lower panel, no clear systematic trends emerge. Very likely such comparisons are complicated by the fact that the yields used in *R4* are strongly angle dependent reflecting the strong forward focusing of the preequilibrium component of the spectrum. Whether or not systematic trends would emerge if the isotopic yields for preequilibrium emission were extracted instead by a moving source decomposition of equilibrium and preequilibrium sources is unclear. In any case, the observation of systematic trends in *R4* is not guaranteed; some care to make sure that the emission mechanism is thermal and that one knows how to correct the yields for the kinematic focusing of the reaction mechanism appears to be required. Similar problems were encountered in extracting < *R4*> for 660 MeV proton induced reactions on $^{10,11}B$, $^{58,64}Ni$, $^{112,124}Sn$, $^{150,154}Sm$ at $\theta_{lab} = 90°$ [22] and 1 GeV proton induced reactions on *Ti*, $^{58,64}Ni$, $^{112,124}Sn$ at $\theta_{lab} = 60°$ [23].

In summary, we have examined the isotope yields produced in heavy ions collisions over a wide range of energy and projectile, target combinations. In situations where equilibrium is established to the extent that a temperature may be defined, a general scaling is observed. This allows a transparent characterization of the dependence of isotope distributions on the overall isospin of the system. This scaling applies to a broad range of statistical production mechanisms including evaporation, strongly damped binary collision, and multifragmentation. We have shown how this systematics can be explained by theories frequently applied to such processes. In the case of multifragmentation, within the grand canonical ensemble limit, the scaling behavior can be interpreted as isospin fractionation. The nucleon density extracted from the gas phase

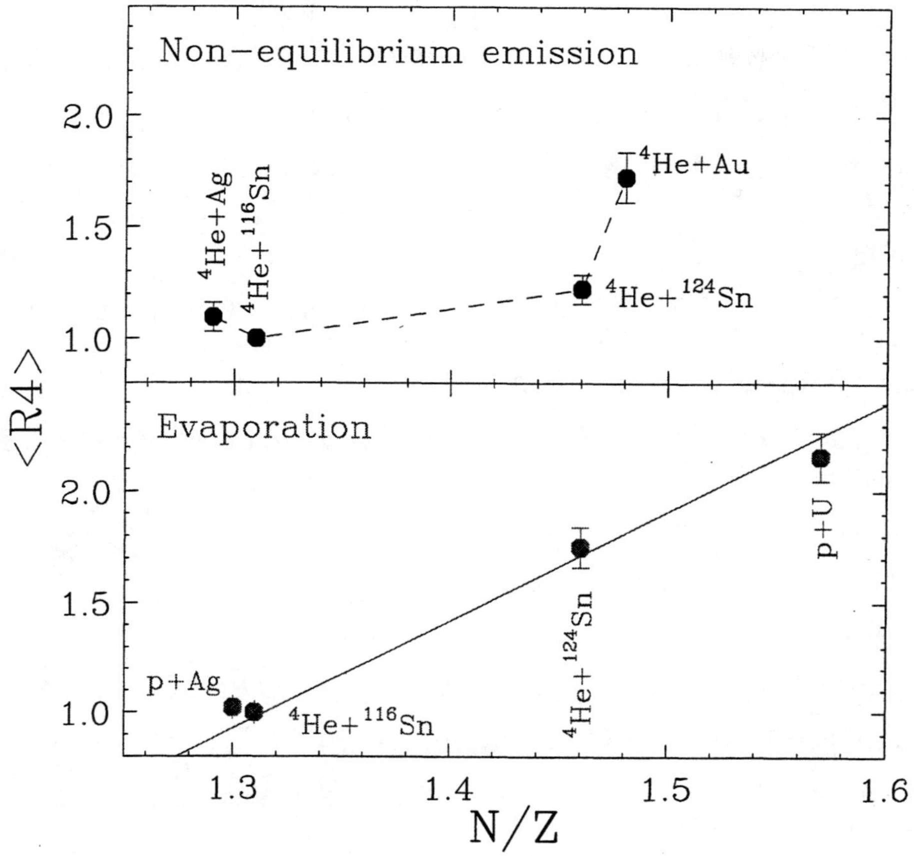

Figure 6: The *N/Z* dependence of *R4* for isotopes produced in evaporation reactions at backward angles (bottom panel) [13,18,19] and for isotopes measured in non-equilibirium emissions at forward angles (top panel) [13, 21].

using isotope, isotone and isobar ratios of four Sn+Sn systems suggest that the gas phase is more enriched in neutrons than the liquid phase represented by bound nuclei, consistent with the predicted partial fractionation of nucleon components in the liquid gas phase transition. The observed effects are significantly enhanced in neutron-rich systems relative to neutron-deficient systems.

This work was supported by the National Science Foundation under Grant Nos. PHY-95-28844 and PHY-96-05140.

References:

[1] T.K. Nayak, T. Murakami, W.G. Lynch, K. Swartz, D.J. Fields, C.K. Gelbke, Y.D. Kim, J. Pochodzalla, M.B. Tsang, H.M. Xu, F. Zhu, K. Kwiatkowski; Phys. Rev. C**45**, 132 (1992).

[2] H. Xi, W.G. Lynch, M.B. Tsang, W.A. Friedman, D. Durand; Phys. Rev C**59**, 1567 (1999).

[3] S.R. Souza, W.P. Tan, R. Donangelo, C.K. Gelbke, W.G. Lynch, M.B. Tsang; Phys. Rev. C (in press).

[4] S.J. Yennello; see article in the same Volume.

[5] H.S. Xu, M.B. Tsang, T.X. Liu, X.D. Liu, W.G. Lynch, W.P. Tan and G. Verde, L. Beaulieu, B. Davin, Y. Larochelle, T. Lefort, R.T. de Souza, R. Yanez, V.E. Viola, R.J. Charity, L.G. Sobotka; Phys. Rev. Lett. **85**, 716 (2000)

[6] The mixed system data shown in Fig. 1 and 4 are the statistical average of two systems $^{124}Sn+^{112}Sn$ and $^{112}Sn+^{124}Sn$.

[7] J. Randrup, S.E. Koonin; Nucl. Phys. **A356**, 223(1981).

[8] S. Albergo et al.; Nuovo Cimento **A89**, 1 (1985).

[9] W.P. Tan, private communication.

[10] M.B. Tsang, W. G. Lynch, H. Xi, W.A. Friedman; Phys. Rev. Lett. **78**, 3836 (1997).

[1] Horst Muller, Brian D. Serot; Phys. Rev. **C52**, 2072 (1995).

[12] F. Gulminelli, Ph. Chomaz, Phys. Rev. Lett. **82**,1402 (1999). Also see article by Chomaz in the present volume.

[13] J. Brzychczyk, D. S. Bracken, K. Kwiatkowski, K. B. Morley, E. Renshaw Foxford, V. E.Viola, Phys. Rev. **C47**, 1553 (1993).

[14] V.V. Volkov, Phys. Rep. **44**, 93, (1978).

[15] W. Friedman and W. Lynch, Phys. Rev. C **28**, 950 (1983).

[16] M. B. Tsang, W.A. Friedman, C.K. Gelbke, W.G. Lynch, G. Verde, H. Xu, NSCL_1165 (July, 2000) and to be published.

[17] W.A. Friedman, Phys. Rev. Lett. **60,** 2125 (1988); and Phys. Rev. **C42,** 667 (1990).

[18] Ray E.L. Green, Ralph G. Korteling, K. Peter Jackson, Phys. Rev. C **29**, 1806 (1984).

[19] A.M. Poskanzer, Gilbert W. Butler, and Earl K. Hyde PRC 3, 882 (1973)

[20] V. E. Viola, K. Kwiatkowski, W. A. Friedman, Phys. Rev. C **59**, 2660 (1999)

[21] J. Zhang, K. Kwiatkowski, D. Bonser, M. Fatyga, S.D. Coon, K. Stith, V.E. Viola, L.W. Woo and S.J. Yennello, Phys. Rev. C **56**, 1918 (1997).

[22] V.I. Bogatin, V.F. Litvin, O.V. Lozhkin, N.Z. Perfilov and Yu. P. Yakovlev, Nucl. Phys. **A260**, 446 (1976).

[23] E.N. Volnin, G.M. Amalsky, D.M. Seleverstov, N.N. Smirnov, A.A. Vorobyov and Yu. P. Yakovlev, Phys. Lett. **55B** 409 (1975)

Isospin Dependence of Total Reaction Cross Sections and Nuclear Radii

W. Q. Shen *

Abstract

The total reaction cross section provides direct information about the size of nuclei involved in the nuclear reaction and has been one of the standard techniques for studying the size and structure of exotic nuclei far from β-stability line. Theoretical analysis of the measured total reaction cross section can be made by using the Glauber model, Boltzmann-Uehling-Uhenbeck (BUU) equation or Quantum Molecular Dynamics (QMD) and empirical parameterized formulas. The isospin dependences of total reaction cross sections and nuclear radii have shown very interesting phenomena, especially such as the neutron halo and skin in light exotic nuclei and also possibly the proton halo and skin in light exotic nuclei. Relativistic Mean Field (RMF), Relativistic Hartree-Bogoliubov (RHB) calculations and the empirical formulas try to reproduce the isospin dependences of the nucleon density distribution in nuclei and nuclear radii and to explain the reasons of the existence of the neutron (proton) halo and skin. Normally, there are several experimental techniques for measurement of the total reaction cross section, but mostly used techniques are the transmission method and 4π-γ coincidence method. Up to now, the extraction of the nucleon density distribution in nuclei and nuclear radii from experimental total reaction cross section has been done almost exclusively by using Glauber model. But the comparison of available data at relativistic energies and data at intermediate energies has shown important difference, which make it difficult to conclude whether ^{8}B is proton halo nuclei or not. Further researches not only theoretically but also experimentally on total reaction cross sections are important, especially on the excitation function of total reaction cross sections at intermediate energies. New phenomena such as giant halo formed by up to six neutrons have been predicted by RHB. Whether the giant halo can be observed experimentally and how will the giant halo influence the total reaction cross section are interesting questions which people wait to solve.

1 Introduction

The total reaction cross section σ_R is one of most fundamental quantities characterizing the nuclei and nuclear reaction. It has been extensively studied both theoretically and experimentally for nearly fifty years[1-16]. It can give us information about the size and shape of the nuclei, the interaction potential between nuclei and the nucleon-nucleon cross section in nuclear medium etc. Such data are also very useful for application of nuclear technology. Recently the development of radioactive ion beam (RIB) technique stimulates the research of nuclear reaction induced by exotic nuclei experimentally and theoretically. The studies by

*Shanghai Institute of Nuclear Research, Chinese Academy of Science, 201800 P.O.Box 200804 Shanghai, China and Heavy Ion Research Facility (HIRFL), Chinese Academy of Science, 730000 Lanzhoou, P.O.Box 31. China

using RIB demonstrated that large enhancement of σ_R induced by neutron rich nuclei was found in some experiments. Together with the enhancement of electromagnetic dissociation cross section, the narrow longitudinal momentum distribution of fragment produced in the fragmentation reaction and the measurement of the quadruple moment etc. One considers that some neutron rich nuclei like ^{6}He and ^{8}He have thick neutron skin (but without long tail for nucleon density distribution) and some nuclei which have small separation energy for last one or two neutrons like ^{11}Li, ^{11}Be and ^{14}Be etc. have neutron halo (with very long tail in neutron density distribution)[17-31]. The discovery of these phenomena stimulates further theoretical and experimental researches on proton halo and proton skin, e.g. whether ^{8}B has proton halo is in strong discussing[32-38]. The discovery of these phenomena also stimulates further theoretical and experimental researches on the isospin dependences of the total reaction cross section and nuclear radii.

For stable nuclei the elastic scattering angular distributions cross sections of a large number of heavy ion systems have been measured at several energies and the optical model has been found to be successful in extracting the total reaction cross section and the interaction radii. In addition there are some direct measurements of the total reaction cross section. In the past, Huizenga, Bass and Alexander et al. have been given the empirical description to calculate σ_R for energies below 10 MeV/A. Gupta et al. have overcome some shortcomings of the earlier methods and have parameterized σ_R by using energy dependent interaction radii in order to produce a better fit to σ_R than the earlier methods. Later, intermediate (10 to 100 MeV/A) and high-energy heavy ion beams have become available to experimentalists. Some experimental results already established a decrease of about 30% in σ_R for a wide range of projectile-target systems when the beam energy increased from 30 to 80 MeV/A. Then σ_R saturates at high energy (several hundreds MeV/A to GeV/A). This clearly demonstrated the important role of the transparency effect in heavy ion collisions in the intermediate energy range. Such trends of the data were quantitatively reproduced by a microscopic calculation originally developed by Karol[3] and based on the Glauber theory[4]. This agreement strongly suggested that the transparency effect for heavy ion collisions is related with the behavior of the nucleon-nucleon cross section in nuclear medium (N-N cross section) σ_{NN}. Actually, DeVries et al.[6] succeeded in reproducing σ_R for nucleon-nucleus collision in terms of individual nucleon-nucleon collision. Such high energy microscopic Glauber multiple-scattering theory is based on the individual nucleon-nucleon (N-N) collisions in the overlap volume of the colliding nuclei. The optical limit of the Glauber theory has been extensively used and the excellent agreement with the experimental data has been achieved by the simplified Glauber-type model calculations for the reactions induced by stable and exotic nuclei. However, σ_R is always underestimated by a large amount at the lower energies for the exotic nuclei induced reactions in usual Glauber-type calculations[27]. In addition, most of the Glauber-type models have difficulties in investigating the dynamics and quantum effects on the total reaction cross section, and the sensitivities of some ingredients to σ_R are also difficult to study. Nuclear transport theories such as the Boltzmann-Uehling-Uhlenbeck (BUU)[39-41] model and quantum molecular dynamics (QMD)[42] have been extensively applied to the heavy ion collision in the intermediate energy range. These models incorporate the Fermi motion, mean field, individual N-N interactions and Pauli blocking effect in calculations. They become main tools to obtain the information about the nuclear equation of state (EOS) and in-medium N-N cross section σ_{NN} from the analysis of some observables in intermediate energy heavy ion reactions. These observables include the energetic particle production, flow angle, collective flow, etc. Ma and Shen et al.[43,44] have attempted to calculate the total reaction cross section and extract the interesting information about the EOS and σ_{NN} by using these dynamical models. On the other hand, parameterized formulas like Gupta's failed when the energy was higher than 10 MeV/A. They predict that σ_R increases with energy and will reach the geometrical limit. Considering transparency effects, Kox et al.[7] suggested a new parameterized formula of the total reaction cross sections which succeeded in the intermediate energy range, but failed for the energy below 10 MeV/A. Shen et al. have suggested a unified parameterized formula which can reproduce the experimental data well from low energy to the intermediate energy range[8].

The determination of nuclear size is one of the most important problems in nuclear physics. Various experimental methods have been employed to determine the nuclear size, the measurement of the total reaction cross section for heavy ion collision is one of them. The radii of the proton and neutron distributions in nuclei have been determined with various experimental methods and compared with theoretical calculations. However, due to experimental restrictions such measurements have been limited mostly to the case of stable nuclei. Some measurements using different isotopes as target[45] already demonstrate an isospin effect on the total reaction cross section, i.e. the dependence of the nuclear size on the neutron excess degree of freedom. Recently the development of RIB technique at intermediate and relativistic energy heavy ion accelerators provides a possibility for the measurement of the total reaction cross section of neutron-rich and proton-rich projectiles. New experimental results show that the square of the radius parameter r_0^2 of the interacting nuclei as extracted from the measured total reaction cross section has strong isospin dependence. The neutron halo and neutron skin and, possibly, the proton halo and skin has been discovered in light nuclei near the drip line. These nuclei have large rms radii, which could be related to the small last one or two nucleon separation energies in such nuclei. As the non-relativistic mean field calculations failed to describe quantitatively the separation energies, Bertsch et al.[21] introduced an arbitrary renormalization factor in the mean field potential based on the Hartree-Fock (HF) mean field calculations with Skyrme interaction to reproduce the extracted experimental separation energies of ^{11}Li. Further, Sagawa[46] improved the calculated separation energies and studied the effect of the centrifugal force on the density profile at large radius by using the same method. They found that the separation energy of loosely bound nucleons plays an important role in the study of halo nuclei. Relativistic mean field (RMF) theories have been applied with considerable success for quantitative description of nuclear properties over the whole periodic in recent years[47-50]. The ground state properties of ^{11}Li isotopes have been investigated in the non-linear relativistic mean field theory[51]. The results showed that the relativistic effects are important and the trend of binding energies and nuclear radii can be satisfactory obtained in the RMF theory. The RMF approximation, however, cannot reproduce the experimental data quantitatively and neutron halo in ^{11}Li. In its spirit and in its results this model is very similar to the conventional density dependent HF theory with Skyrme forces[52-53]. But the relativistic description has the advantage of treating the spin degree of freedom on a more fundamental basis in the Dirac equation and is technically somewhat simple. Zhu and Shen et al.[54] improved the results of the RMF calculation by applying a similar modification to the potential as reference[21] in order to adjust the proper size of the halo. The extracted experimental nuclear density distributions and separation energies of loosely bound neutrons of halo nuclei ^{11}Li and ^{11}Be were well described by such modified RMF calculation. Meng and Ring[55] have used the relativistic Hartree-Bogoliubov (RHB) to describe the chain of lithium isotope from ^{6}Li to ^{11}Li. Paring correlations are taken into account by a density dependent force of zero range. In contrast to RMF and

HF theory where the halo in ^{11}Li could only be reproduced by an artificial renormalization of the potential, the halo is now reproduced in a self-consistent way. The development of a proton skin as well as neutron skin has been described by RHB theory where the RHB equations are solved self-consistently in coordinate space so that the continuum and paring have been better treated[56]. A Glauber model calculation has been carried out with the density distribution obtained from RHB. A good agreement has been obtained with the measured cross section for long chain Na isotopes from ^{20}Na to ^{32}Na on ^{12}C target. Based on the above success, Meng and Ring[57] have researched the chain of Zirconium isotopes from ^{116}Zr to the drip line nucleus ^{140}Zr via RHB. For neutron numbers larger than the magic number N=82 a giant neutron halo with up to six neutrons outside the core of ^{122}Zr has been predicted. This is a new phenomena which has not been observed experimentally so far. It would allow the study of collective phenomena in neutron matter of low density.

In this article, the calculation of the total reaction cross section including Glauber model, BUU (or QMD) method and parameterized formulas will be described in section II. Section III will introduce the calculation of the nucleon density distribution and the nuclear radii including Relativistic Mean Field (RMF) calculation, Relativistic Hartree-Bogoliubov (RHB) calculation and empirical formulas. Section IV will review the measurement of the total reaction cross section and the isospin dependence of the total reaction cross section and the nuclear radii. The method of projectile fragmentation to produce RIB and experimental techniques like transmission method and 4π-γ coincidence method for measurement of the total reaction cross section will be described in detail. The extraction of the nucleon density distribution in nuclei and the nuclear radii from experimental total reaction cross section will be discussed in detail also. The neutron halo and skin in light nuclei and possible proton halo in light nuclei will be reviewed also. An overview and outlook will be given at the end.

2 Calculation of the total reaction cross section

2.1 Glauber model

Several microscopic models have been developed to account for σ_R data. In these models a nuclear reaction is described in terms of a sum of individual nucleon-nucleon scattering process. Therefore, the nuclear matter density distribution in the projectile and the target and the average nucleon-nucleon cross section $\bar{\sigma}$ are very important. Here, we briefly review the basic concept of Karol's microscopic model[3]. It is based on the Glauber theory[4] and the semiclassical optical model.

In a nucleus-nucleus collision with a given distance r between the centers of the colliding nuclei the local mean free path $\Lambda(r)$ of the projectile in the target can be given by

$$\Lambda(r) = [\bar{\sigma} \int d\tilde{r} \int_V \rho_P(r)\rho_T(r-s)ds]^{-1} \tag{1}$$

where $\rho_i(r)$ (i=P or T) is the nuclear density distribution of the projectile or target, V is the nuclear volume overlap at distance r, s is the position variable and $d\tilde{r}$ is the angular part of the variable r. One obtains he probability for no nucleon-nucleon interaction T(b) at a given impact parameter b as follows:

$$T(b) = exp \int_{-\infty}^{+\infty} -\frac{dZ}{\Lambda(r)} \tag{2}$$

where the distance to the scattering center measured along the beam axis is Z, and $r^2 = b^2 + Z^2$. Then σ_R is derived by assuming that any nucleon-nucleon scattering process contributes to the reaction:

$$\sigma_R = \int_0^\infty 2\pi[1 - T(b)]db \tag{3}$$

Some authors also take into account the Pauli blocking due to the Fermi motion[15] and also the deflection of the projectile in the Coulomb plus nuclear potential[6]. Neglecting the Pauli blocking, the deflection effect of projectile in the Coulomb plus nuclear potential and the effect of the nuclear skin, Karol[3] has derived an analytic expression for σ_R by assuming a simple "surface-normalized" gaussian shape for the nuclear density distribution:

$$T(b) = exp[-\frac{\pi^2 \bar{\sigma} \rho_T(0) \rho_P(0) a_T^3 a_P^3}{a_T^2 + a_P^2} exp(-\frac{b^2}{a_T^2 + a_P^2})] \tag{4}$$

where $a_{P,T}$, $\rho_{P,T}(0)$ represent the parameters of the gaussian shape density distribution of projectile or target and

$$\bar{\sigma}(E) = [(\frac{Z_T}{A_T})(\frac{Z_P}{A_P}) + (\frac{N_T}{A_T})(\frac{N_P}{A_P})]\sigma_{ii}(E) + [(\frac{Z_T}{A_T})(\frac{N_P}{A_P}) + (\frac{Z_P}{A_P})(\frac{N_T}{A_T})]\sigma_{ij}(E) \tag{5}$$

where σ_{ii} is proton-proton (or neutron-neutron) cross section and σ_{ij} is the proton-neutron cross section. It is well established that the high energy σ_R data can be reproduced within a few percent accuracy by using the Glauber model together with experimental N-N cross sections and the known nuclear density distribution. The results of such Glauber model calculation performed with a realistic matter distribution reproduce the behavior of σ_R for nuclei near the β-stability valley quite well also at intermediate energy. The experimental square of the radius parameter r_0^2 of the interacting nuclei in ref.[19] increases strongly with increasing N-Z of the projectile. We have converted also the experimental σ_R values from ref.[7,45] into r_0^2 via an unified parameterized formula (see section II.C) for H+$^{58-64}$Ni, ^{12}C+$^{62-68}$Zn etc. The strong increase of r_0^2 with the neutron excess of the target exhibits also, although the slope is smaller than that of neutron-rich projectile induced reaction in ref.[19]. The above Glauber model cannot explain these experimental results, because it assumes the same density distribution for neutrons and protons. Such Glauber type calculation underpredicts the experimental σ_R at intermediate energy for nuclei far from the β-stability line also. The Glauber model like Karol's should be modified by taking into account the difference between the neutron and proton density distributions.

For this purpose Shen et al.[8] have developed a modified microscopic model of Glauber calculation by introducing different neutron and proton distributions in a nucleus as given by the droplet model[58]. The neutron skin, which is predicted by the droplet model for neutron-rich nuclei, may produce an enhancement of σ_R for neutron-rich projectiles or targets. Because σ_{np} is about three times larger than σ_{nn} (or σ_{pp})at intermediate energies, the interaction N_T-N_P, P_T-P_P, N_T-P_P and P_T-N_P (where N, P denote neutron and proton; the subscripts T and P denote the target and projectile, respectively) are treated separately. We can still use the formalism of eqs.(2.1-2.3) for the calculation. For simplicity, by using the simple "surface-normalized" gaussian distributions, Karol's analytic expression of Glauber model for σ_R will be changed into

$$T(b) = exp[-\pi^2 \sum_{i=N,Z} \sum_{j=N,Z} \frac{\sigma_{ij} \rho_{Ti}(0) \rho_{Pj}(0) a_{Ti}^3 a_{Pj}^3}{a_{Ti}^2 + a_{Pj}^2} exp(-\frac{b^2}{a_{Ti}^2 + a_{Pj}^2})] \tag{6}$$

where i=N, Z; j=N, Z represent the summation over neutrons and protons in the target and projectile, respectively; $\rho_{T_i}(0)$, $\rho_{P_i}(0)$, a_{T_i} and a_{P_i} are parameters of "surface-normalized" gaussian distribution of protons or neutrons in the target and projectile, respectively. For heavier projectiles it is better to use a Feimi-type distribution:

$$\rho_i(r) = \rho_i(0)\frac{1}{1+exp[(r-C_i)/(t_i/4.4)]}, \quad i = N, Z \tag{7}$$

where

$$\rho_N(0) = \frac{3N}{4\pi C_N^3[1+\pi^2 t_N^2(19.36C_N^2)]}$$

$$\rho_Z(0) = \frac{3Z}{4\pi C_Z^3[1+\pi^2 t_Z^2(19.36C_Z^2)]}$$

C_N, C_Z, t_N and t_Z are the half density radii and the surface diffuseness of the neutron and proton distributions, respectively. C_N, C_Z are determined by the droplet model:

$$C_i = R_i[1-(b_i/R_i)^2], \; i = N, Z \tag{8}$$

where b_i=0.43t_i. The equivalent sharp surface radius of the neutron and proton density distributions, R_N, R_Z, are determined by the effective nuclear matter radius R and the neutron skin thickness d:

$$R_N = R + (Z/A)d, \tag{9}$$

$$R_Z = R - (N/A)d, \tag{10}$$

where

$$R = r_0 A^{1/3}(1+\bar{\varepsilon}), \; d = \frac{2}{3}[(N-Z)/A - \bar{\delta}]R \tag{11}$$

For the nuclei around the β-stability line some experiments have proved that the surface diffuseness of neutron and proton distribution is a constant: $t_Z = t_N = 2.4fm$. The values of $\bar{\varepsilon}$, $\bar{\delta}$, r_0 etc. are taken from refs.[58]. In this case the numerical integral is carried out according to the equations (2.1-2.3). For deformed nuclei the deformation can be taken into account in the above formalism, where the shape of the surface is treated in terms of Legendre polynomials[59]. The calculated results shown in fig 1 indicate that the modified microscopic model reproduces the experimental data taken from the literature well for σ_R induced by nuclei around the β-stability line. Imaging that the long-range tail of the neutron distribution in a neutron-rich nucleus will be a reasonable origin of the increase of r_0^2 with N-Z. Keeping other properties of the nucleus as predicted by the droplet model, we assumed that the surface diffuseness of the neutron distribution has a following dependence on N-Z:

$$t_N = t_0 + k_N(\frac{N-Z}{A} - I_\beta)(fm), \tag{12}$$

where $t_0 = 2.4fm$ is the normal surface diffuseness for a stable nucleus, I_β is the value of (N-Z)/A along the β-stability line and k_N is a free parameter used for searching a best fit to the experimental data. The value k_N=10 is finally chosen. T(b) is quite sensitive to t_N in the case of a neutron-rich nucleus at intermediate energy. For a neutron-rich nucleus the modified microscopic model in which the surface diffuseness of the neutron distribution

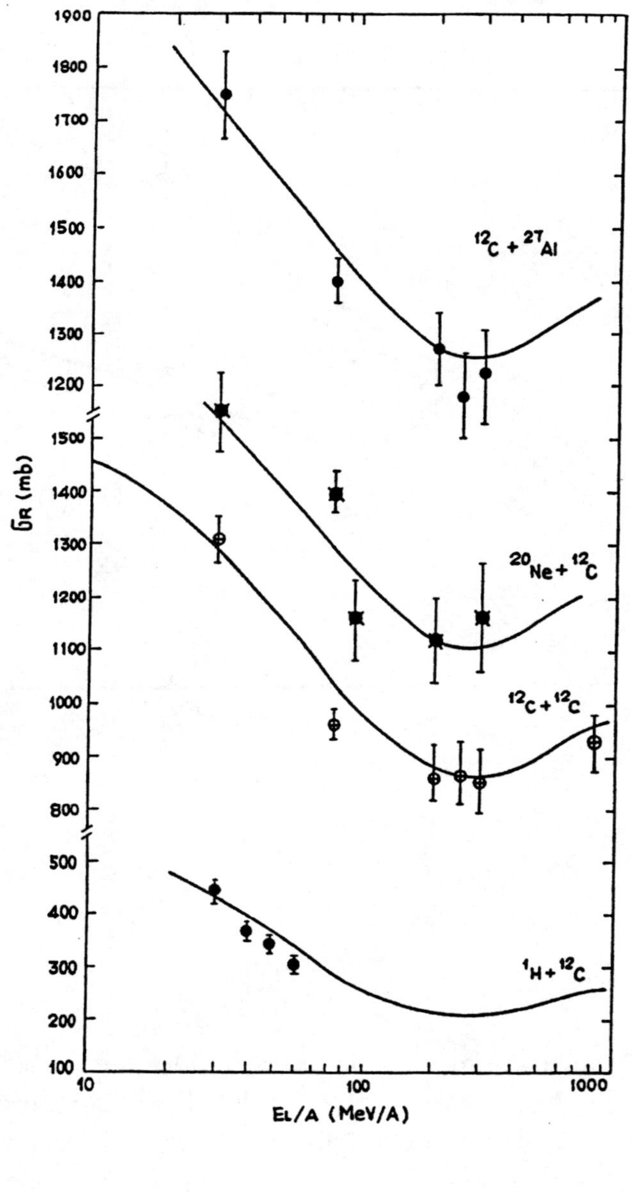

Figure 1: Comparison of the total reaction cross section σ_R in the intermediate energy range with the calculated results of the modified microscopic model based on individual nucleon-nucleon collision cross sections.

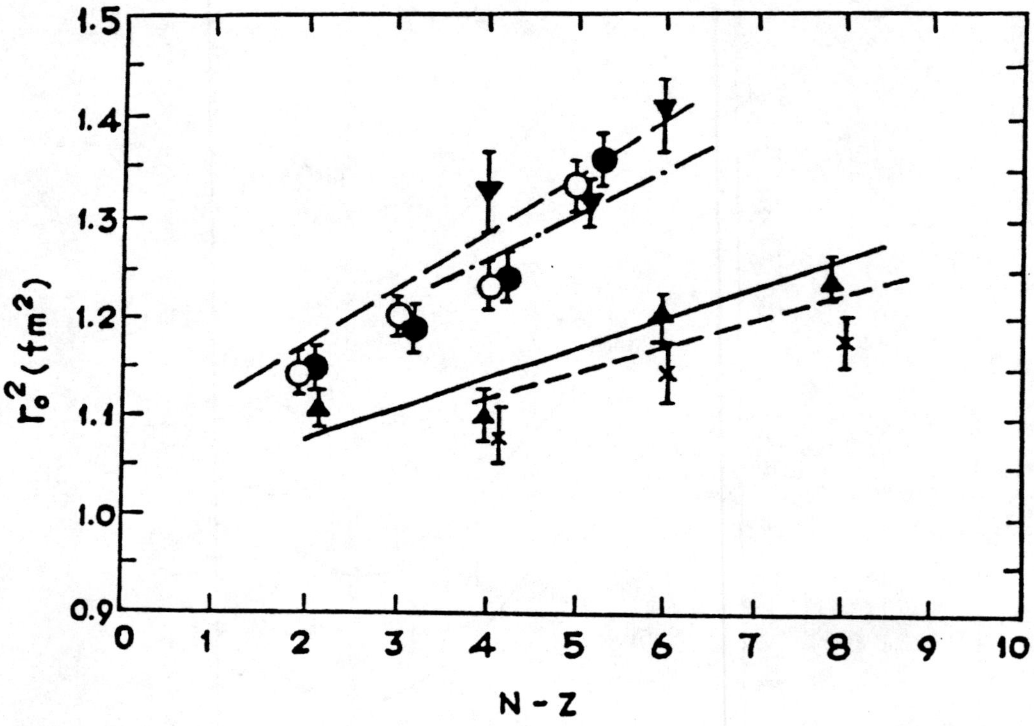

Figure 2: Square of the radius parameter of the interacting nuclei r_0^2 as a function of the neutron excess. The lines represent the calculated results according to the modified microscopic model in which the difference between neutron and proton distributions in the nuclei is considered via the formulas 2.6-2.11. The empty circles are for the $^{16-18}N+^{28}Si$ system; the full uptriangles are for $H+^{58-64}Ni$; the full downtriangles are for $^{26-28}Na+^{28}Si$; the crosses are for $^{12}C+^{62-68}Zn$ and the full circles are for $^{18-21}O+^{28}Si$. For reference see texts.

has a dependence on N-Z like in eq.(2.12) shows a good agreement with the experimental results (see fig 2). In fig 2, the calculated r_0^2 values are extracted from the calculated σ_R through formula in section II.C for nuclei far from the β-stability line. It strongly supports the idea that the increase of surface diffuseness of the neutron distribution on N-Z is an important origin for experimental increase of r_0^2 with the neutron excess degree of freedom.

For stable nuclei the diffuseness is chosen to $2.4fm$ which is predicted by the droplet model due to a nearly constant nuclear separation energy ($\sim$ 8MeV/n). For exotic nuclei the loosely bound nucleons with small separation energy will cause a long tail in the nucleon distribution. The relativistic Hartree theory calculation[60] predicts that the diffuseness of the neutron distribution for neutron-rich nuclei increases with the neutron excess degree of freedom which is related to the neutron separation energy. Considering these results, Feng and Shen et al.[61] assumed later that the diffuseness parameter t_N for neutron distribution as a function of neutron separation energy ε follows a simple relation:

$$t_N = t_0 + \frac{\hbar}{\sqrt{2N\varepsilon m}} \tag{13}$$

where N is the number of neutron in the nucleus and m is nucleon mass. The one neutron separation energy and the two neutron separation energy are compared and the smaller value among them is used. The experimental data of σ_R for a large number of stable nuclear systems can be reproduced in such way. The large increase of σ_R for exotic nuclei is reasonably explained. Fig 3 shows the large increase of σ_R as a function of isotope mass of projectiles (such as ^{6}He, ^{8}He, ^{11}Li, ^{11}Be and ^{14}Be). Although in this framework the excitation function of σ_R induced by exotic nuclei can be fitted quite well in high energy range, the calculated σ_R is still smaller than experimental value in 20-50 MeV/A energy range where the reaction dynamics changes strongly from mostly mean field effect to nucleon-nucleon collision.

2.2 BUU or QMD method

New calculation for σ_R is suggested by Ma and Shen et al. within the framework of nuclear transport theory like BUU or QMD[43,44]. In the BUU model the average N-N collision number can be obtained by assuming a reasonable parameterization of the N-N cross section σ_{NN} according to Poisson statistics, the fraction of nucleons which have experienced n times two-body collisions during the course of nucleus-nucleus reaction is given by

$$T_n(b) = \frac{N^n e^{-N}}{n!} \tag{14}$$

where N is the average N-N collision number, i.e. half of the total collision number. The sum of $T_n(b)$ over n ($n \geq 1$) represents the total probability of N-N collisions, $Q_0(b)$, and relates closely to the absorption probability of nuclear reaction. This quantity $Q_0(b)$ is expressed by

$$Q_0(b) = \sum_{n=1}^{\infty} T_n(b) = 1 - T_0(b) = 1 - exp(-N) \tag{15}$$

Therefore, we can easily obtain the total reaction cross section σ_R by

$$\sigma_R = 2\pi \int bdb[1 - exp(-N)] \tag{16}$$

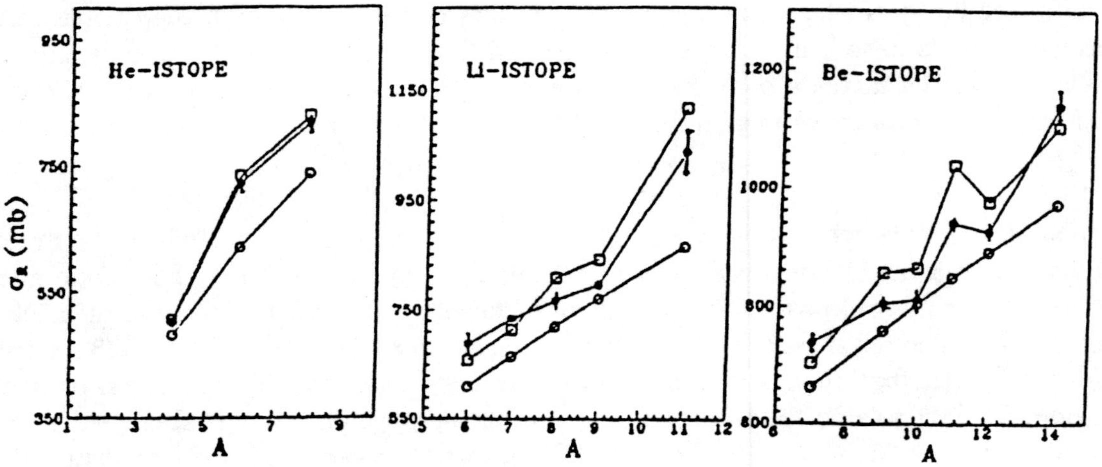

Figure 3: The reaction cross section of He, Li and Be isotopes on ^{12}C target. The solid lines connecting empty circles show the calculated results with $t_N = 2.4fm$. The solid lines connecting the boxes show calculated results obtained with t_N formula 2.13. The full circles are experimental data at high energies[61].

The details of the BUU model were described in ref.[39]. Here the test particle method has been used to solve the BUU equation and the well-known parameterization of Cugnon et al. for elastic and inelastic N-N cross sections as a function of the available center-of-mass energy has been used. The average N-N collision number N is the sum of elastic and inelastic collision numbers. The mean-field potential is parameterized as a density-dependent function and two parameter sets are taken for the nuclear potential which correspond to the values of nuclear compressibility K = 200 MeV (soft EOS) and K = 380 MeV (stiff EOS), respectively. In the initialization of the coordinate space, we distribute all the nucleon randomly in a sphere. For a ^{12}C nucleus, the nucleons are sampled randomly in the sphere with a radius of $r_0A^{1/3}$. The r_0 is chosen as 1.33 fm, corresponding to the rms radius of 2.36 fm which agrees with the empirical value 2.35±0.02 fm of ^{12}C[18]. As an example, we calculated the N-N collision number distribution at different impact parameters b for $^{12}C+^{12}C$ at 800 MeV/nucleon. By integrating these distributions, the dependence of the average collision number on b is obtained. The fraction of nucleons which have experienced n times collisions can be obtained also. In order to show the reasonableness of the extracted values of $Q_0(b)$, we researched the time evolution of $Q_0(b)$. Clearly $Q_0(b)$ reaches asymptotic values after a long enough interaction time which assures the correct extraction of $Q_0(b)$.

The energy dependences of the total reaction cross section for $^{12}C+^{12}C$, $^{12}C+^{27}Al$ and $^{14}N+^{64}Cu$ systems have been calculated. In these calculations the parameter r_0 of nuclei are chosen as 1.25 fm except for ^{12}C. These calculated results can reproduce well most of the experimental data. The total reaction cross section is more sensitive to the nuclear equation of state when the energy range is below 100 MeV/nucleon than in the higher energy range.

It is noticeable that a similar conclusion is drawn by the analysis of collective flow below 100 MeV/nucleon. This sensitivity of σ_R in this energy range may reflect the change of the dynamical mechanism with the beam energy. It changes from the dominant role of the mean field at lower energies to that of individual N-N collisions at higher energies. At lower energies the individual N-N collisions is not dominant, so the choice of a different form of the nuclear potential is more important. The insensitivity of σ_R to the EOS at higher energies may in principle provide us a direct observable, for example, σ_R, and then we can extract the value of the σ_{NN} by comparing the model calculation with the experimental data. So through the investigation of σ_R ,it is possible to give rise to a convenient approach to get some useful information about EOS and σ_{NN}. In order to investigate the medium effect of N-N collisions on σ_R, we also use $0.8\sigma_{Cug}$ and $0.6\sigma_{Cug}$ to calculate σ_R for ^{12}C+^{12}C system. The result σ_R decrease systematically with decreasing σ_{NN}. This indicates that σ_R is sensitive to σ_{NN}. In addition, a different mechanism of reaction dynamics may give a different σ_R. The results calculated by the old cascade version of Cugnon et al. overestimate strongly σ_R at energies less than GeV/n because in this model the attractive nuclear mean field which binds nucleon tightly in the nucleus is not taken into account and the Pauli-blocking effect is neglected. This indicates that the role of the mean field is always important for the total reaction cross section. With the increasing energy, the σ_R calculated by the cascade model decreases and approaches the results of BUU calculation in the relativistic energy range.

Tanihata et al. reported that the total reaction cross sections of ^{11}Li, ^{11}Be,^{14}Be, etc., induced reactions were much larger than that expected from the stable nuclei induced reactions and they call them neutron halo nuclei. Tanihata et al. deduced the R_{rms} of light nuclei using the optical limit calculation of the Glauber model. For the ^{11}Li nucleus, the extracted R_{rms} is 3.11±0.22 fm. The obtained R_{rms} is almost independent of the assumed form of the nucleon density distribution. When we took the radius of a sphere $1.6A^{1/3}$ for ^{11}Li, to which the R_{rms} agrees with the R_{rms} of Tanihata et al., a good agreement between the calculated excitation function and the experimental results for ^{11}Li+^{12}C system is obtained. Fig. 4(a) shows our calculation with stiff EOS and σ_{Cug} and the comparison with the experimental data[25]. The total reaction cross section is sensitive to the parameter r_0, i.e. the effective nucleon density distributions. By simulating the empirical nucleon density distribution given by Tanihata et al.[25] the similar fitting to the excitation function of σ_R can be obtained. We also show the agreement of σ_R between the calculation and the experimental data for exotic nuclei ^{11}Be induced reactions[23] in the bottom of fig. 4, where the open and solid circles with errors bars denote the experimental σ_R on ^{27}Al and ^{12}C targets, respectively, while the solid and dashed lines represent the results calculated with the stiff and soft EOS, respectively. In the calculation we adopt $r_0 = 1.5fm$ for the ^{11}Be nucleus[18]. Our calculations are in good agreement with the experimental data and the sensitivity of σ_R to EOS are also seen.

In the usual Glauber-type calculations at the lower energies, the total reaction cross section for exotic-nuclei-induced reactions is always underestimated[26-28]. In the present calculations at lower energies, the σ_R calculated with stiff EOS and σ_{Cug} is generally overestimated for the reactions induced by the stable nuclei. However, they adequately describe the reactions induced by exotic nuclei (see fig. 4). The σ_R is mainly dependent on EOS, σ_{NN}, Pauli blocking, etc., of which the in-medium of N-N cross sections is important. In the stiff EOS case, a good agreement between the theoretical value and the experimental data may be expected with small σ_{NN} at lower energies for stable nuclei induced reactions. Due to the weakly bound neutrons of the exotic nuclei, σ_{NN}] for the neutron-rich nuclei

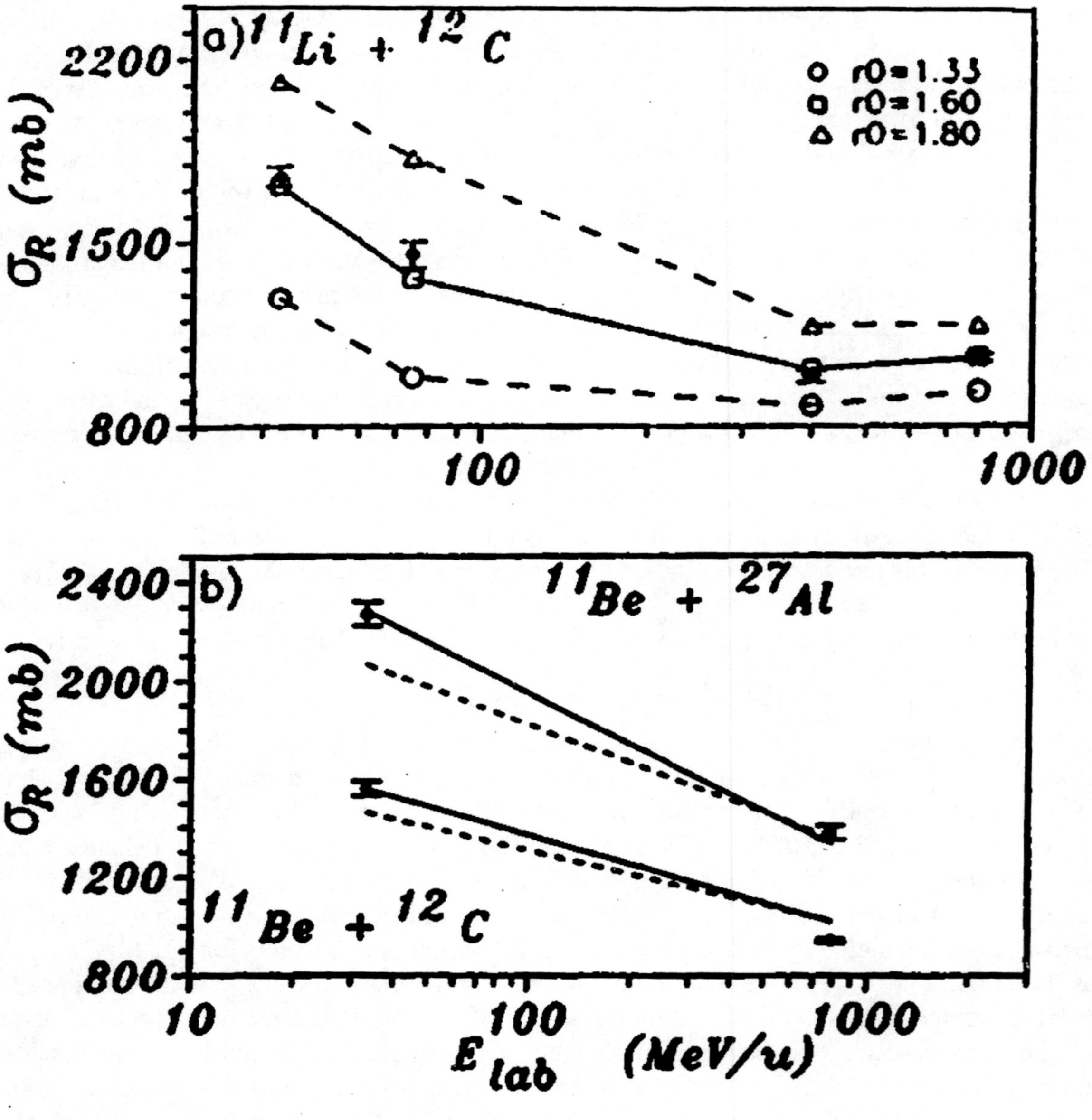

Figure 4: The calculated excitation function of σ_R for ^{11}Li+^{12}C, ^{11}Be+^{12}C, ^{27}Al reactions. The dots with error bars denote the experimental data. The r_0 for ^{12}C is always 1.33 fm. In (a) the circles, squares, and triangles on lines represent the results calculated with r_0=1.33, 1.60, and 1.80 fm for ^{11}Li, respectively. In (b) r_0=1.5 fm for ^{11}Be and r_0=1.25 fm for ^{27}Al are adopted. The solid and empty circles represent the experimental data for ^{11}Be on ^{12}C and ^{27}Al targets, respectively. The solid and dashed lines denote the results calculated with the stiff and soft EOS, respectively[43].

could be larger at this energy range than that for stable nuclei.

2.3 Parameterized formulas for the total reaction cross section

According to the strong absorption model, the total reaction cross section σ_R can be represented in terms of the interaction radius R and nuclear-nuclear interaction barrier B, as

$$\sigma_R = 10\pi R^2(1 - B/E_{c.m.})(mb), (\text{R in fm; B},E_{c.m.}\text{in MeV}) \tag{17}$$

Different parameterized formulas will have different parameterized forms for R and B. Considering the transparency and isospin effects observed at intermediate energies, Kox et al.[7] developed a new parameterization for σ_R in which they used the Coulomb barrier with r_c=1.3 fm as the nucleus-nucleus interaction barrier: $B=Z_T Z_P e^2/r_c(A_T^{1/3}+A_P^{1/3})$ and parameterized the interaction radius R as follows:

$$R = R_{vol} + R_{surf} \tag{18}$$

where

$$R_{vol} = r_0(A_T^{1/3} + A_P^{1/3}),$$

$$R_{surf} = r_0[1.85\frac{A_T^{1/3}A_P^{1/3}}{A_T^{1/3} + A_P^{1/3}} - C(E)] + \frac{\alpha(A_T - 2Z_T)Z_P}{A_T A_P}$$

The energy dependent parameter C(E) is called the transparency coefficient. It is independent of the systems. With r_0=1.1 fm and α=5, Kox's formula can fit a large number of experimental values well at intermediate energy. At low energies this parameterization cannot reproduce the experimental data and Kox et al. did not give the values of C(E) below 30 MeV/A

For the whole energy range from low energy to intermediate energy Shen et al.[8] proposed a unified parameterized formula for σ_R using:

$$B = \frac{1.44Z_T Z_P}{r} - b\frac{R_T R_P}{R_T + R_P}(MeV), \tag{19}$$

$$R_{surf} = r_0[A_T^{1/3} + A_P^{1/3} + 1.85\frac{A_T^{1/3}A_P^{1/3}}{A_T^{1/3} + A_P^{1/3}} - C'(E)] + \frac{\alpha(A_T - 2Z_T)Z_P}{A_T A_P} + \beta E_{c.m.}^{-1/3}\frac{A_T^{1/3}A_P^{1/3}}{A_T^{1/3} + A_P^{1/3}} \tag{20}$$

where

$$r = R_T + R_P + 3.2fm,\ b = 1MeV \cdot fm^{-1},$$

$$R_i = 1.12A_i^{1/3} - 0.94A_i^{-1/3},\ (i = T, P)$$

According to the trends of C(E) and by fitting procedure to the experimental data one obtains the value of C′(E). For energies higher than 30 MeV/A, C′(E) is the same as C(E) in ref.[7]. C′(E) is independent of the system. It increases dramatically at intermediate energies and reaches saturation around 200-300 MeV/A. The behavior of C′(E) for energies higher than 30 MeV/A is representative of the behavior of the free nucleon-nucleon cross section. It supports the idea that at low energies the surface transparency is small and the mean-field interaction plays the most important role, while at intermediate energies the surface transparency increases dramatically and nucleon-nucleon collisions become the main interaction mechanism. By using this formula, the data about 89 systems more than 270 energy points have been fitted. $\chi^2 = \frac{\sum[(\sigma_R^{exp}-\sigma_R^{cal})/\sigma_R \times 0.05]^2}{N}$ where N = the number of data points and σ_R^{exp}=experimental data assumed to have 5% uncertainty is 4.57. It is better than other parameterization forms for σ_R. Fig. 5 shows the comparison of experimental σ_R values taken from the literature with the prediction from this unified parameterized formula. The parameters are fixed once and for all. The best values are α=1 fm, β=0.176 MeV$^{1/3}$fm, r$_0$=1.1 fm.

Another point of interest is the so-called isospin effect, which strongly influences the σ_R data. As an example σ_R for the p+$^{58-64}$Ni reaction increases about 20% with increasing neutron excess of the target. In above unified parameterized formula for σ_R the term $\alpha(A_T\text{-}2Z_T)Z_P/A_TA_P$ describes this effect. The calculated results explain the isospin effect rather well. For neutron-rich projectiles far from the β-stability line, σ_R actually increases rather strong with the neutron excess. In such case the unified parameterized formula underestimates σ_R.

2.4 In-medium nucleon-nucleon cross section and its effect on σ_R

In Glauber model and BUU method, the free-space nucleon-nucleon cross section σ_{NN}^{free} obtained by experimental measurement is used. Sometime, fσ_{NN}^{free} where f is arbitrary parameter has been used. But the real in-medium nucleon-nucleon cross section $\sigma_{NN}^{in-medium}$ is different from the free-space nucleon-nucleon cross section because of the effects of Pauli blocking and finite nuclear matter density in heavy-ion reactions. The medium effect of nucleon-nucleon cross section is an important topic for study because, up to now it couldn't be gained directly by experiment. Theoretically, in-medium nucleon-nucleon cross section can be obtained by solving the many-body Bethe-Goldstone (BG) equation, namely G-matrix cross section. However, up to now, most of the calculations was performed only at several values of nuclear matter density and incident energy. So it is valuable if one can present some new formula to calculate the in-medium nucleon-nucleon cross section over a wide energy range. Cai et al.[62] and Kohno et al.[63] present a phenomenological formula for in-medium nucleon-nucleon cross section based on several recent theoretical studies, especially of G. Q. Li and R. Machleidt's works [64,65]. Such phenomenological formula can be used to calculate σ_R in the framework of Glauber model and BUU method.

In-medium nucleon-nucleon cross sections for the incident energies from 50MeV to 300MeV in laboratory frame and the matter densities up to 2ρ_0 (where ρ_0 is the saturation density of normal nuclear matter, ρ_0 =0.17) were calculated by G. Q. Li and R. Machleidt[64,65]. Results of their numerical calculation were fitted by a semi-empirical formula (LM formula) including the dependence on incident energy and nuclear matter density. D. Klakow et al.[66] calculated the balance energy by BUU with a soft EOS. They found that the calculated balance energy can fit experimental data best when the in-medium nucleon-nucleon cross section equals to eighty per cent of the free-space nucleon-nucleon cross section. So another empirical formula was expressed as $\sigma_{NN}^{in-medium}$=0.8σ_{NN}^{free}. This conclusion is often used now in the transport model such as BUU. But in fact, the medium effect is different in various incident energy ranges and matter densities. It seems that the conclusion of D. Klakow et al. underestimates the medium effect and describes the density and energy dependence too simply. While for the LM formula, the in-medium nucleon-nucleon cross section at ρ=ρ_0 was about fifty percent of the free-space nucleon-

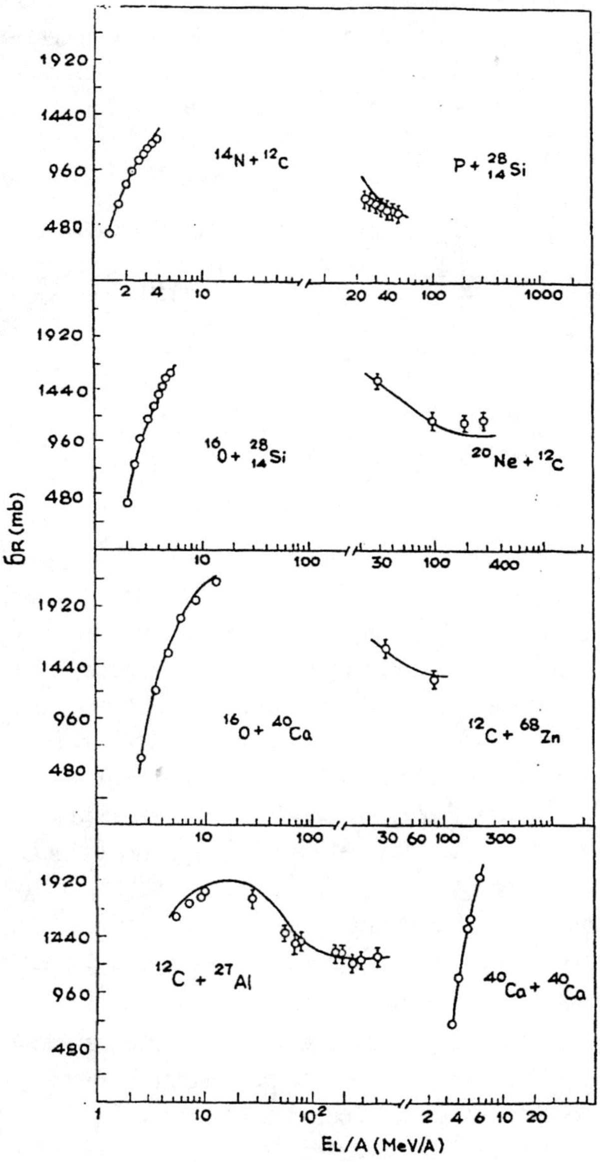

Figure 5: Comparison of the total reaction cross section σ_R with the prediction of the unified parameterized formula.

nucleon cross section around above balance energy, it is too small. LM formula overlooks the medium

effect, especially in low energy.

By considering the above results and LM formula, Cai et al. add the energy dependence and density dependence in the free-space nucleon-nucleon cross section by a least squares fit to some experimental total reaction cross section data and proposed a new phenomenological formula for in-medium nucleon-nucleon cross section. The formula is as the following expression:

$$\begin{cases} \sigma_{nn} = (13.73 - 15.04\beta^{-1} + 8.76\beta^{-2} + 68.67\beta^{4})\frac{1.0+7.772E_{lab}^{0.06}\rho^{1.48}}{1.0+18.01\rho^{1.46}} \\ \sigma_{np} = (-70.67 - 18.18\beta^{-1} + 25.26\beta^{-2} + 113.85\beta)\frac{1.0+20.88E_{lab}^{0.04}\rho^{2.02}}{1.0+35.86\rho^{1.90}} \end{cases} \tag{21}$$

$$\beta = \sqrt{1.0 - \frac{1.0}{\gamma^2}},\ \gamma = \frac{E_{lab}}{931.5} + 1.0 \tag{22}$$

where β is the ratio of projectile velocity to light velocity. The first part in Eq.(2.21) describes the free-space nucleon-nucleon cross section and the second part describes the incident energy and nuclear density dependence. From Eq.(2.21), we find that the nucleon-nucleon cross section in nuclear medium decreases with increasing density. Medium effect is important at intermediate energies and becomes smaller at higher energies but doesn't vanish. Fig. 6 shows the energy dependence of nucleon-nucleon cross section. The solid circles describe the experimental values of free-space nucleon-nucleon cross section, which are taken from Ref.[9]. The solid lines describe results of Eq.(2.21) at ρ=0. The dotted lines describe the results of LM formula at ρ=0. The solid lines can fit the experimental values well over the energy range from 10MeV to 1GeV. The dotted line deviate the true nucleon-nucleon cross section at lower and higher energy ranges because their model could not be applicable at both lower and higher interactive energy. The dashed lines show the calculated results of Eq.(2.21) at $\rho=\rho_0$.

The analytic formula proposed to calculate the in-medium nucleon-nucleon cross section over a wide energy range describes the effect of nuclear density and incident energy on nucleon-nucleon cross section. The nucleon-nucleon cross section in nuclear medium is predicted to decrease with increasing density. The energy dependence is sensitive at low energies and becomes smaller at higher energies gradually. Systematic calculations of σ_R for both stable and exotic nuclei have been performed by introducing this formula at $\rho=\rho_0$ into the Coulomb-modified Glauber model. An improved agreement has been obtained with experimental values by using new in-medium nucleon-nucleon cross section formula. But the calculations underestimate the σ_R at intermediate energies for projectiles far from the β-stability line. Actually, lots of important physical quantities, e.g. transverse collective flow, the rotational flow, their disappearance energy (balance energy), multifragmentation cross section and π-meson production cross section etc.[67,68] are sensitive to the in-medium nucleon-nucleon cross section. The work to involve the nuclear matter density dependence and interactive energy dependence of in-medium nucleon-nucleon cross section into the transport model like BUU or QMD will give the best choice of $\sigma_{NN}^{in-medium}$ which will improve the σ_R calculation in framework of Glauber model and BUU method.

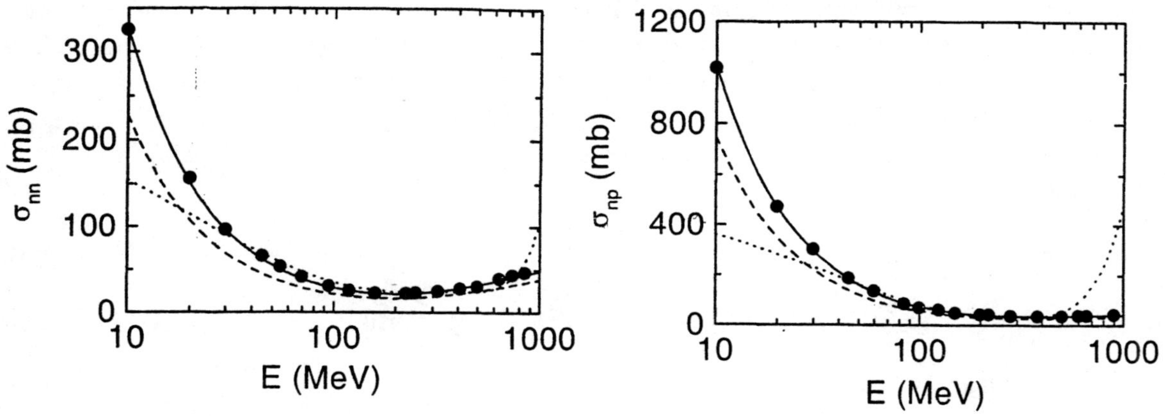

Figure 6: Nucleon-nucleon cross section as a function of incident energy. (a) neutron-neutron (proton-proton) interaction cross section. (b) neutron-proton interaction cross section. The solid circles indicate experimental free-space nucleon-nucleon cross section. The solid and dashed lines indicate the result of the formula 21 at ρ=0 and ρ_0, respectively. The dotted lines indicate results of LM formula at ρ=0.

3 Calculation of the nucleon density distribution and the nuclear radii

3.1 Relativistic Mean Field (RMF) calculation

As discussed in section II, the nuclear size and nuclear radii can be obtained from the measurement of the total reaction cross section via Glauber model and BUU method. It is called as the nuclear matter radii. From several other experimental techniques such as electron scattering and isotope shift etc., the charge radii are measured with a high accuracy. From the matter radii and charge radii, the neutron density distribution can be obtained. The data concerning the nuclear size and shape are already known for many nuclei, especially for the nuclei near the β-stability line. At present, more exotic nuclei are explored also as discussed in section II. With such experimental bases, the nuclear radii become one of the crucial quantities good for testing of every theoretical model of the nucleus. Here we introduce the RMF calculation for the nuclear density distribution, the nuclear radii, the binding energy and the separation energy of last one or two nucleon etc. which is very often used for exotic nuclei[54].

Starting from the Lagrangian

$$\begin{aligned}\mathcal{L} &= \bar{\psi}(i\gamma^\mu\partial_\mu - M)\psi + \frac{1}{2}(\partial^\mu\sigma\partial_\mu\sigma - m_\sigma^2\sigma^2) - \frac{1}{3}b\sigma^3 - \frac{1}{4}c\sigma^4 - \frac{1}{4}\Omega^{\mu\nu}\Omega_{\mu\nu} + \frac{1}{2}m_\omega^2\omega^\mu\omega_\mu \\ &-\frac{1}{4}R^{\mu\nu}R_{\mu\nu} + \frac{1}{2}m_\rho^2\rho^\mu\rho_\mu - \frac{1}{4}F^{\mu\nu}F_{\mu\nu} - g_\sigma\bar{\psi}\sigma\psi - g_\omega\bar{\psi}\gamma^\mu\omega_\mu\psi \qquad (23)\\ &-g_\rho\bar{\psi}\gamma^\mu\rho_\mu\tau\psi - e\frac{1}{2}(1+\tau_3)\bar{\psi}\gamma^\mu A_\mu\psi\end{aligned}$$

where the Dirac spinors ψ represent the nucleons with mass M. A scalar meson σ and three vector particles, the isoscalar ω-meson, the isovector ρ-meson and the photon A, are also included. In addition we take into account a non-linear coupling of the σ-mesons amongst themselves, which has turned out to be crucial for a quantitative understanding of many nuclear quantities, and the usual phenomenological constant gap approximations as discussed in Ref[50] in order to consider the pairing correlations in open shell nuclei.

The equations for the baryon wave functions in the static case contain two potentials: the repulsive potential

$$V(r) = V_{\omega-\rho}(r) + V_{Coulomb}(r) \tag{24}$$

with $V_{\omega-\rho}(r) = g_\omega\omega_0(r) + g_\rho\tau_3\rho_0(r)$ and the attractive potential:

$$S(r) = g_\sigma\sigma(r) \tag{25}$$

The coupled differential equation derived from the Lagrangian are solved self-consistently in the spherical or deformed cases with several different parameters sets[49]. The good fit to the nuclear radii and binding energy etc. can be obtained for many nuclei near the β-stability line. The neutron skin was predicted. But RMF can not produce the neutron halo density distribution in the nuclei such as ^{11}Li and ^{11}Be. Zhu and Shen et al. multiply the potential $V_{\omega-\rho}(r)$ and S(r) in Eqs(3.2), (3.3) by the same constant normalization factor f for the last loosely-bound neutron configurations, i.e., for the neutron orbit outside the core $f \neq 1$ and otherwise $f = 1$. In the numerical procedure, the factor f is regarded as input data. The coupled differential equations derived from the Lagrangian are solved self-consistently in the spherical cases with the parameter set NL1[49]. The factor f has been adjusted until a calculated mass rms radii is obtained that is very close to the extracted experimental data. For ^{11}Li and ^{11}Be, the calculated density distributions of halo nuclei show a nice agreement with the extracted experimental data. The halo tail of the density distribution is dominated by the halo neutron orbits which is shown very clearly in Fig 7. The experimental separation energy are well reproduced in this approach also. Nevertheless, a satisfactory explanation of the renormalization factor is lack. Otherwise, there could be other ways to handle the neutron-rich nuclei in a relativistic model, for example, a refit of the parameter set and a inclusion of a tensor coupling of the vector mesons.

Sharma et al.[69,70] proposed an interaction called NL-SH in RMF which can fit the properties of the nuclei far from β-stability line[69,72]. Sugahara and Toki[73]included a non-linear ω self-coupling term in addition to the non-linear σ self-coupling term and found two parameter sets, TM_1 and TM_2. Hirata et al[74] used such RMF with TM_1 set to study successfully the ground state properties for some nuclei. Chen et al.[75] investigated the properties of neutron-rich nuclei in 2s1d shell via TM_2 and NL-SH interactions and predicted that there are one proton halo in ^{26,27,28}P and two proton halos in ^{27,28,29}S where clearly ^{27}S and ^{26,28}P are the best candidate for experimental investigations.

Warda et al.[76] have researched the nuclear radii of even-even β-stable nuclei with A $\geq$ 40 and a few chains of isotopes with Z = 50, 56, 82, 94 and isotopes with N = 50, 82, 126 neutrons with RMF. Then they proposed a simple phenomenological formula for isospin dependence of the charge radii and the neutron radii for A $\geq$ 40:

$$R_{ch} = 1.241(1 - 0.154\frac{N-Z}{A} + 0.580\frac{1}{A})A^{1/3} fm \tag{26}$$

$$R_n = \frac{(R_{ch}^2 - 0.64)^{1/2}}{1.048(1 - 0.364(N-Z)/A - 3.148/A)} fm \tag{27}$$

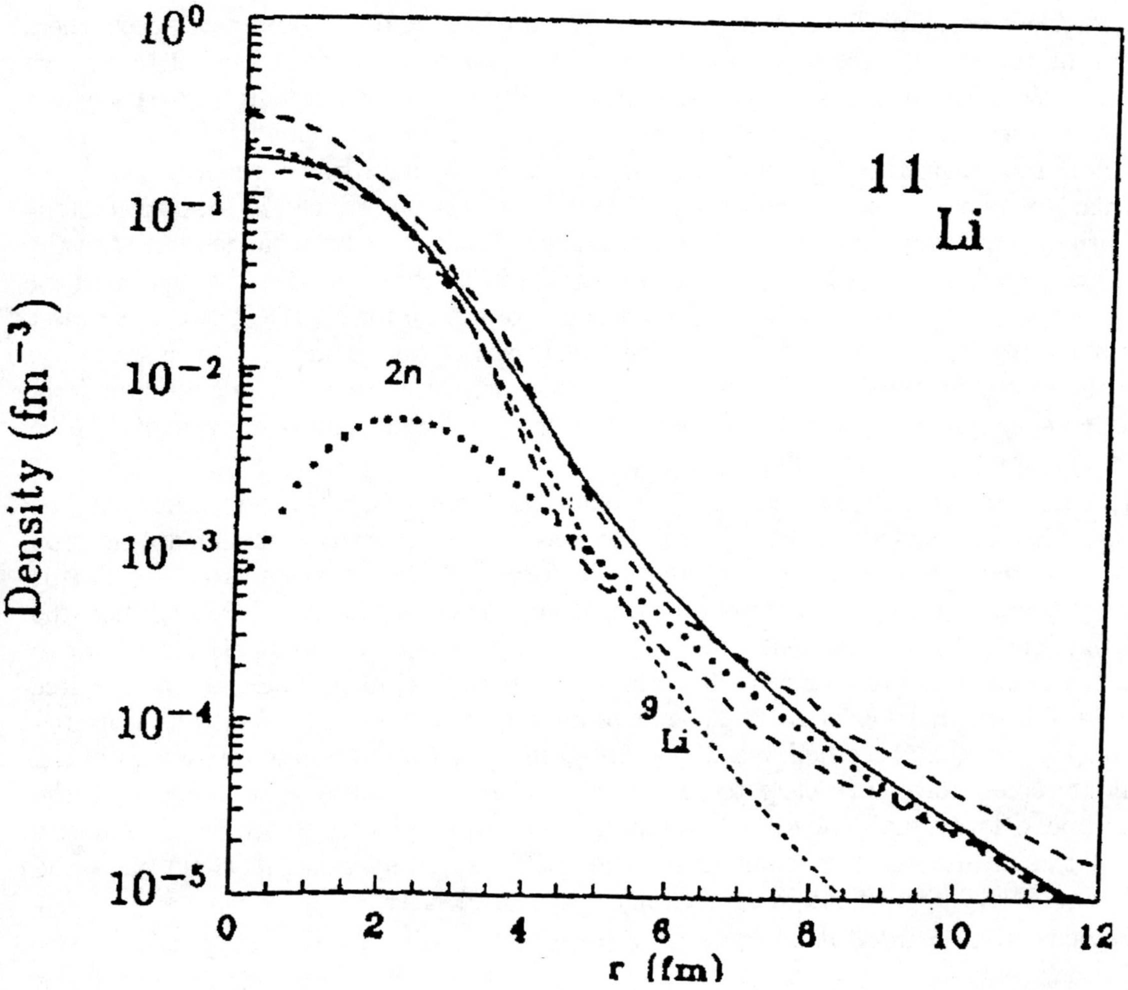

Figure 7: Calculated and extracted experimental density distribution in ^{11}Li. The solid curve shows the calculated result. The long-dashed curves give the extracted experimental data. The density of core ^{9}Li is plotted with short-dashed curve. The dotted curve presents the density of halo neutrons[54].

3.2 Relativistic Hartree-Bogoliubov (RHB) calculation

With either the non-relativistic Hartree-Fock or RMF, the pairing is always neglected or simply treated. For the pairing correlation, the contribution from the continuum is essential for the nucleus near drip lines. Meng and Ring[55] used the RHB in coordinate space where the pairing correlations are taken into account by a density dependent force of zero range and extended RMF to a continuum RHB theory. For details of formulas and parameters see the references [55-57]. In contrast to earlier investigations within RMF and density dependent HB where the halo could only be reproduced by an artificial renormalization of potential for halo nucleons. The halo such as ^{11}Li is now reproduced in a self-consistent way without further modification[55]. A Glauber calculation has been carried out with the density obtained from RHB. A good agreement has been obtained with the measured σ_R of Na isotopes on ^{12}C target[56]. The measurement of the σ_R for Na isotopes was a widest range of neutron number among all the experiments carried out so far[77]. Fig. 8 shows the neutron and proton density distribution for neutron-rich Na isotopes. The development of a neutron skin can be clearly seen. The proton density distribution is not sensitive to he neutron number, it is almost unchanged.

RHB is used to describe the chain of Zirconium isotopes from ^{116}Zr to neutron drip line nucleus ^{140}Zr. For neutron number larger than the magic number N = 82, a giant neutron halo outside the core of ^{122}Zr is predicted. It is formed by up to six neutron[57]. This is a new phenomena which has not been observed experimentally so far. In calculation the strength of the density dependent force of zero range is adjusted for known isotope ^{116}Zr to a similar calculation with Gogny's DIS force in the pairing channel. There is no other free parameters in this RHB calculation. The halos are formed by two to six neutrons scattered as Cooper pairs mainly to the levels $3P_{3/2}$, $2f_{7/2}$, $3P_{1/2}$ and $2f_{1/2}$. These resonances in the continuum come down very close to the Fermi level in these nuclei and couple with the loosely bound levels just below the continuum limit. We will wait to see whether future experiment can discover such giant halo nucleus. If they could exist, it would allow the study of collective phenomena in neutron matter of low density.

Vretenar et al.[78] used RHB to describe the ground state properties of spherical even-even isotone chain for N = 18, 20, 22 within atomic charge number from 14 to 28. This is the first application of RHB to proton-rich nuclei. They used NL3 parameter set for the effective mean field and described the pairing correlation by the finite-range Gogny interaction DIS. It is shown that the NL3 + Gogny DIS effective interaction provides a very good fit to experimental ground state properties like two proton separation energy and proton rms radii. Fig. 9 shows the proton density distribution for the N = 20 isotones. The proton radii increase gradually with the charge number Z. Inserts in fig. 9 are the values of the surface thickness S and the diffuseness parameter α. The surface thickness S is defined to be the change in radius when $\rho(r)/\rho_0$ reduces from 0.9 to 0.1. The diffuseness parameter α is determined by fitting the density profile to Fermi distribution: $\rho(r) = \rho_0[1 + exp(\frac{r-R_0}{\alpha})]^{-1}$. In going away from β-stability line, the proton surface thickness increase, but the diffuseness parameter α has a maximum at Z = 20. This should be due to the strong influence of the Coulomb barrier.

There are some other theoretical models for calculation of the nuclear radii such as the microscopic three cluster model[79]. We will not introduce here for more. From the experimentalist's point of view, simple phenomenological formulas are very useful sometime. In section III.A, we have introduced the Warda's nuclear radii formulas for A $\geq$ 40 based on RMF. Next subsection will introduce simple nuclear radii formulas for A $\leq$ 40.

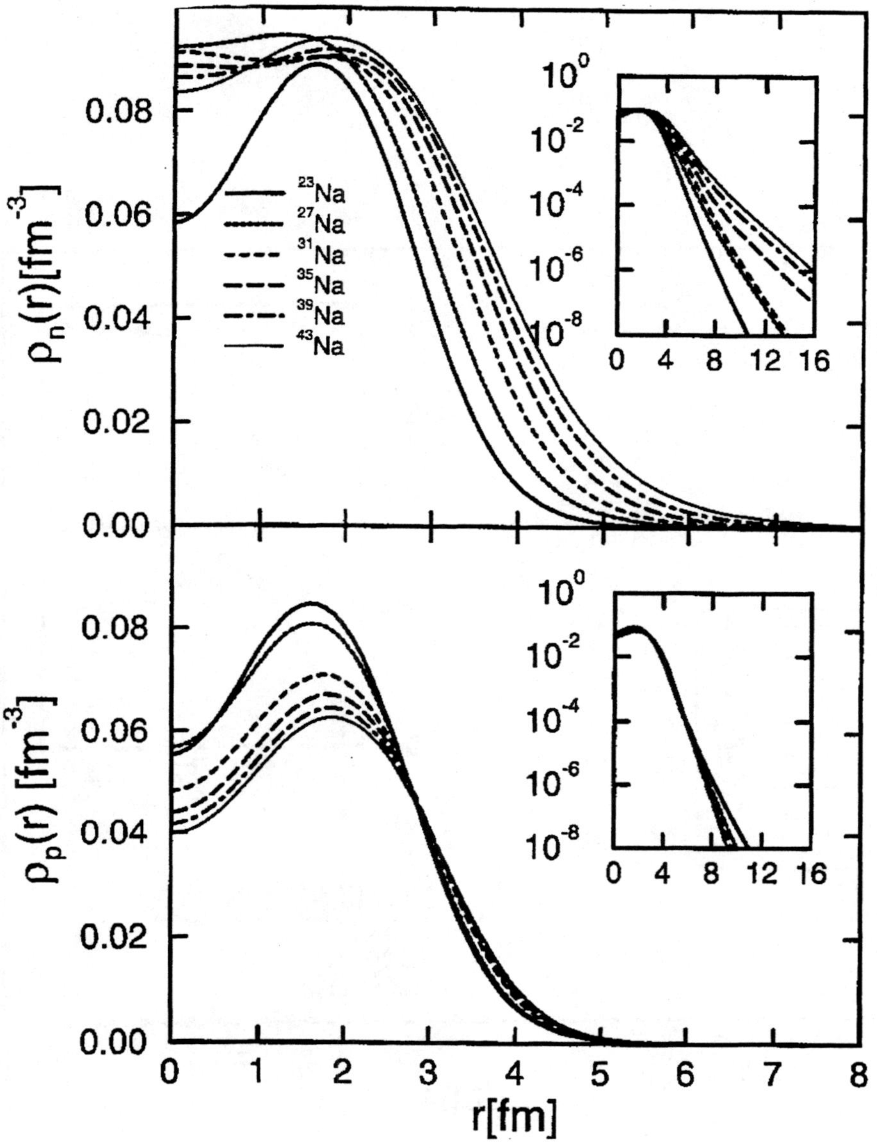

Figure 8: The neutron (upper) and proton (lower) density distributions in Na isotopes. The same figures but in logarithm scale are given as inserts to show the tail part of the density distribution[56].

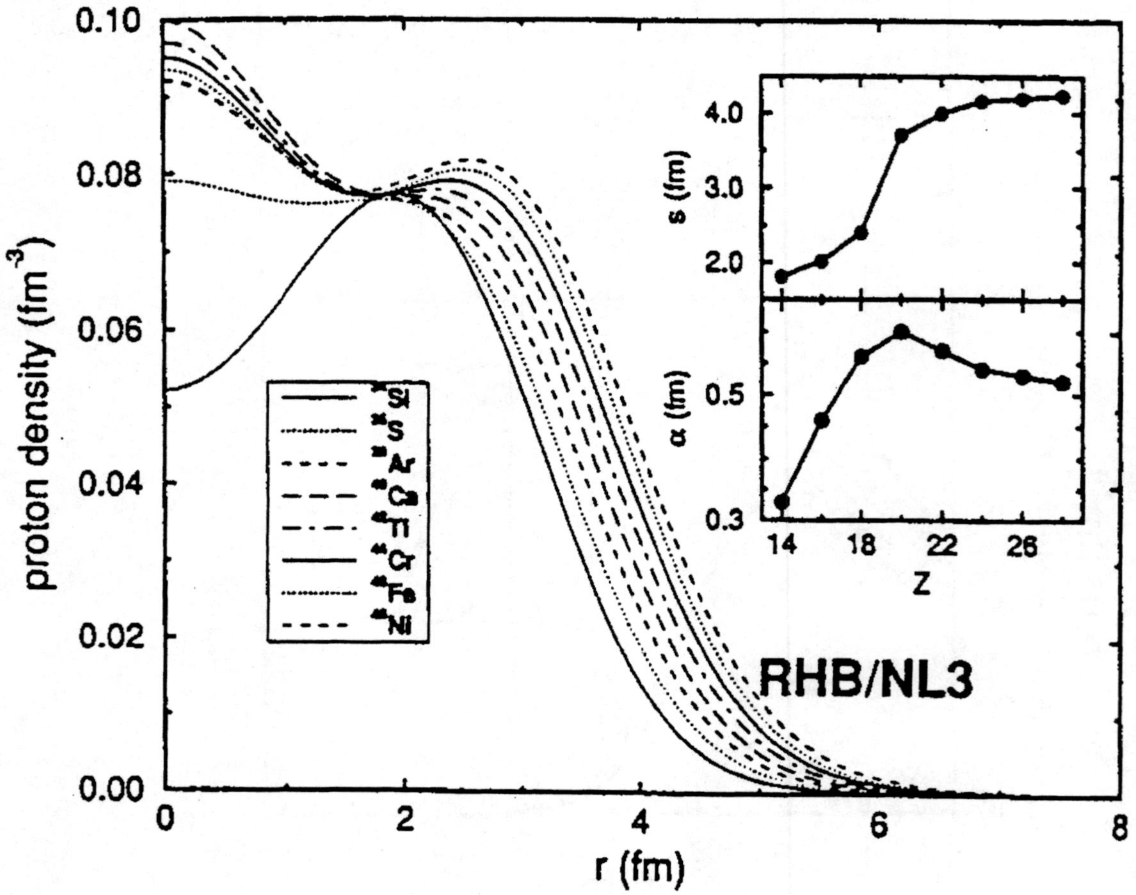

Figure 9: Self-consistent RHB proton density distributions for the N = 20 isotones, calculated with NL3+Gogny DIS effective interaction[78].

3.3 Empirical formulas for the nuclear radii

As discussed before, the experimental nuclear radii can be reproduced well by RMF model. But RMF failed to reproduce the experimental results for light neutron-rich nuclei near the drip line. Renormalization of the halo neutron potential can solve this problem. Recently the RHB model removes this artificial renormalization and reproduce the halo distribution for light neutron-rich nuclei. Based on RMF, Warda et al. introduced simple phenomenological formulas of the nuclear radii for A $\geq$ 40 already [76]. For light nuclei Lassaut et al.[80] suggested a simple empirical formula which uses different coefficients for different element. Base on experimental results and RMF calculations, Wang and Shen[81] suggested following empirical proton and neutron radii formulas for light nuclei:

$$< r^2 >_p^{1/2} = \begin{cases} (a_1\frac{N-Z}{A} + a_2 A^{\frac{1}{3}} + a_3\frac{1}{\sqrt{B}} + a_4\frac{1}{\sqrt{A \cdot SE_n}}) & if\, \frac{B}{SE_p} \geq 6.0 \\ (a_1\frac{N-Z}{A} + a_2 A^{\frac{1}{3}} + a_3\frac{1}{\sqrt{B}}) & if\, \frac{B}{SE_p} < 6.0 \end{cases} \tag{28}$$

$$< r^2 >_n^{1/2} = \begin{cases} (b_1\frac{N-Z}{A} + b_2 A^{\frac{1}{3}} + b_3\frac{1}{\sqrt{B}} + b_4\frac{1}{\sqrt{A \cdot SE_p}}) & if\, \frac{B}{SE_n} \geq 3.7 \\ (b_1\frac{N-Z}{A} + b_2 A^{\frac{1}{3}} + b_3\frac{1}{\sqrt{B}}) & if\, \frac{B}{SE_n} < 3.7 \end{cases} \tag{29}$$

where $SE_p = min(S_p, S_{2p})$ and $SE_n = min(S_n, S_{2n})$ are separation energy for proton and neutron and B is the average binding energy for the nucleon. The nuclear matter rms radii can be deduced from the neutron and proton rms radii as:

$$< r^2 >_m = \frac{N}{A} < r^2 >_n + \frac{Z}{A} < r^2 >_p \tag{30}$$

Via fitting the available experimental data by the lease square method, the coefficients have been obtained as $a_1 = -0.542 fm$, $a_2 = 0.795 fm$, $a_3 = 1.621 fm \cdot MeV^{-1/2}$, $a_4 = 0.180 fm \cdot MeV^{-1/2}$. b_1=0.420fm, b_2=0.792fm, $b_3 = 1.757 fm \cdot MeV^{-1/2}$, $b_4 = 0.308 fm \cdot MeV^{-1/2}$,

In the formulas, the first term describe the isospin dependence of nuclear neutron (proton) radii, while the second term comes from the droplet model and the third term considers the effect of binding energy. As discussed before, a small separation energy is the main reason causing the formation of the nuclear halo structure. It will have strong effect on nuclear radii when it is small enough. That is considered in the fourth term and the conditions. These formulas describe the existing data for A $\leq$ 40 quite well and follow the trend of RMF calculation. Fig. 10 demonstrates the comparison of the results from RMF calculations and simple empirical formulas.

4 Measurement of the total reaction cross section

4.1 Production of the radioactive ion beam (RIB)

There are two methods for production of RIB, i.e. projectile fragmentation (PF) and isotope separation on line (ISOL). Normally the RIB for σ_R measurement in high energy and intermediate energy is produced by PF method. For σ_R measurement of exotic nuclei table 1 shows the technical parameters of several RIB line used in the world now. As an example we introduce here the RIB line called RIBLL on Heavy Ion Research Facility in

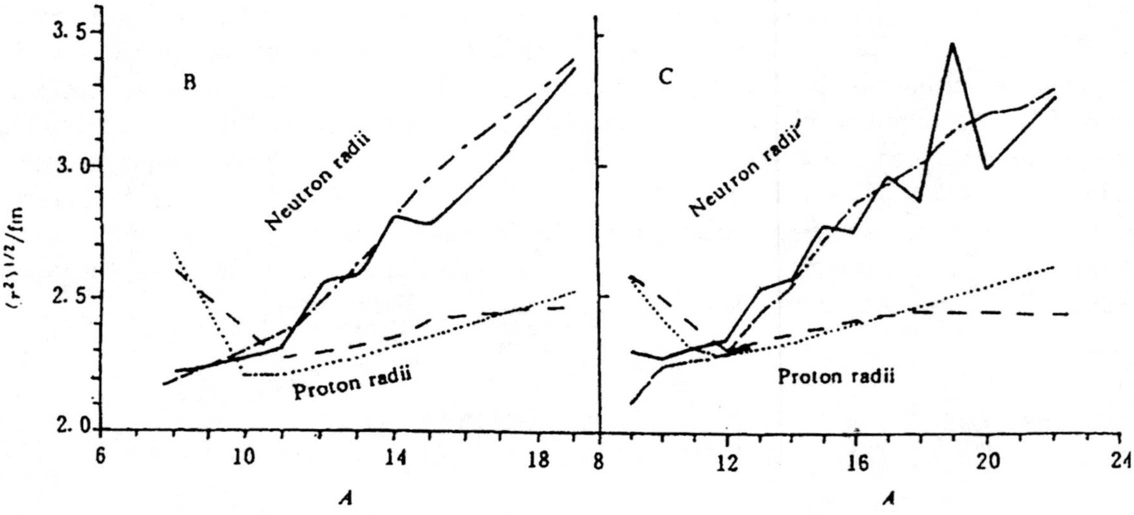

Figure 10: The comparison of the calculated results from RMF theory (short-dashed line for neutron radii and long-dashed line for proton radii) and simple empirical formulas (solid line for neutron and dotted line for proton)[81].

Lanzhuo, China (HIRFL). RIBLL is a double achromatic (momentum and position) radioaction ion beam line. Its lattice structure is $Q_{01}Q_{02}D_0$-T_0- $Q_1Q_2D_1Q_3Q_4$-C_1-$Q_5Q_6D_2Q_7Q_8$ -T_1-$Q_9Q_{10}D_3Q_{11}Q_{12}$-C_2- $Q_{13}Q_{14}D_4Q_{15}Q_{16}$-$T_2$, where Q_{ii} is quadrupole, D_i is dipole, T_i represents the focal plane and C_i represents the dispersion point. The section T_0-T_1 is same with the section T_1-T_2 which has a symmetric double achromatic structure. Against focal plane T_1 they are asymmetric each other. For production of RIB one section is enough. But two asymmetric sections T_0-T_1 and T_1-T_2 will help to obtain purer RIB. The primary beam bombards the target at T_0 and produces secondary radioactive beam with velocity V_{RIB} via projectile fragmentation reaction. The separation and identification of RIB will be realized in section T_0-T_1 by using $(B\rho_1)$-ΔE-$(B\rho_2)$-TOF mode where the dipole D_1 selects RIB first according A/Z ($B\rho_1 \propto \frac{A}{Z} V_{RIB}$), then the wedge-shaped degrader at dispersion point C_1 selects RIB according energy loss in degrader ($\frac{dE}{dx} \propto Z^2$) and finally the dipole D_2 selects RIB again. The time of flight (TOF) between T_0 and T_1 will be used for the identification of RIB. The selection T_1-T_2 will repeat above procedure in order to obtain purer RIB. The secondary target will be put in T_2. The reaction products will be measured by detectors around the target at T_2. For RIBLL there are other two operation modes: the RIB+ magnetic spectrometer mode where T_1-T_2 section used as magnetic spectrometer and secondary target will be put at T_1 and the high intensity mode where T_1-T_2 is used just as RIB beam transportation line. Normally such RIB line can produce rather pure RIB, e.g. by using 60 MeV/u ^{18}O on 550 mg/cm^2 Be, the purity of ^{17}N can reach 99%[87].

Table 1. Several Radioactive Ion Beam Lines

line's name /lab	Energy of primary beam MeV/u / intensity ions/s	mass range of projectile	angular acceptance (msr)	momentum acceptance $\Delta P/P(\%)$	momentum resolution	maximum rigidity (Tm)	reference
LISE/GANIL	$25\text{-}95/10^{12}$	≤ 136	1.0	5.0	800	3.2	82
A1200/MSU	$30\text{-}180/10^{11}$	≤ 136	0.8-4.3	3.0	700-1500	5.4	83
FRS/GSI	$100\text{-}2000/10^{8}$	≤ 238	0.7-2.5	2.0	240-1500	18	84
RIPS/RIKEN	$30\text{-}135/10^{11}$	≤ 136	5.0	6.0	1500	5.76	85
COMBA/JINR	$< 60/10^{12}\text{-}10^{14}$	< 50	6.4	20	4360	4.5	86
RIBLL/HIRFL	$< 100/10^{10}$	< 86	6.5	10	1700	4.2	87-89

The projectile fragmentation at intermediate and high energies has been extensively investigated both theoretically and experimentally. There are some parameterized formulas for the calculation of the isotope production cross section in PF reaction. For example, the Summerer's empirical analytic formulas[90] can fit the experimental isotopic distribution at a fixed atomic number Z and predict the production cross section of exotic nuclei for certain reaction system within a factor of two. That is very important for choosing the reaction system to produce the exotic nuclei which will be investigated. For very heavy projectile the formulas fit the experimental data not so well. Much effort has been devoted to the understanding of the fragment production mechanism. However the detailed explanation of some aspects is still in an unsatisfactory state. At high energy the fragmentation process is a quick clean cut between two colliding hard spheres and there is almost no dissipation. The production cross sections of the fragments can be well described by the participant-spectator model[91]. In intermediate energy regime, the interplay of nucleon-nucleon collision and mean field interaction makes the reaction mechanism more complicated. The abrasion-ablation model[92] can reproduce the experimental isotopic distribution of ^{40}Ar induced reaction but failed to explain the experimental results produced by heavy projectile such as ^{86}Kr. Brohm and Schmidt have developed the statistical abrasion-ablation model[93] which can reproduced the production cross section produced by ^{40}Ar, ^{86}Kr and ^{129}Xe etc. quite well except for neutron rich side of the isotopic distribution in reaction induced by ^{129}Xe and heavier. Considering the effect of neutron skin and one body dissipation the abrasion-ablation model was developed, then it can well reproduce the experimental results produced by ^{40}Ar and ^{86}Kr. Considering the medium effect of nucleon-nucleon cross section and the separation energy dependence of the neutron diffuseness, the statistical abrasion-ablation model was developed and it can reproduce the experimental data better[94,95].

In framework of above modified statistical abrasion-ablation model the isospin effect of projectile fragmentation reaction was investigated[95]. The isotopic distribution with a fixed atomic number Z shifts toward the neutron-rich side for neutron-rich projectile, but the shift decrease with the increase of the atomic number difference Z_{proj}-Z. In order to understand the isospin effect of the fragmentation reaction on the isotopic distribution with a fixed atomic number z more clearly, the peak positions and widths were extracted by a Gaussian fit to the isotopic distributions. The difference $\Delta A = A_{peak}\text{-}A_{\beta}$ and width of the isotopic distribution as a function of $(Z_{proj}\text{-}Z)/Z_{proj}$ were given in fig 11, where A_{peak} is peak position of the isotopic distribution, A_{β} is the mass of most β-stable nucleus determined by $Z = A_{\beta}/(1.98 + 0.0155 * A^{2/3})$. Roughly speaking, the parameter $(Z_{proj}\text{-}Z)/Z_{proj}$ can be viewed as the violence of the nuclear reaction which is related to the impact parame-

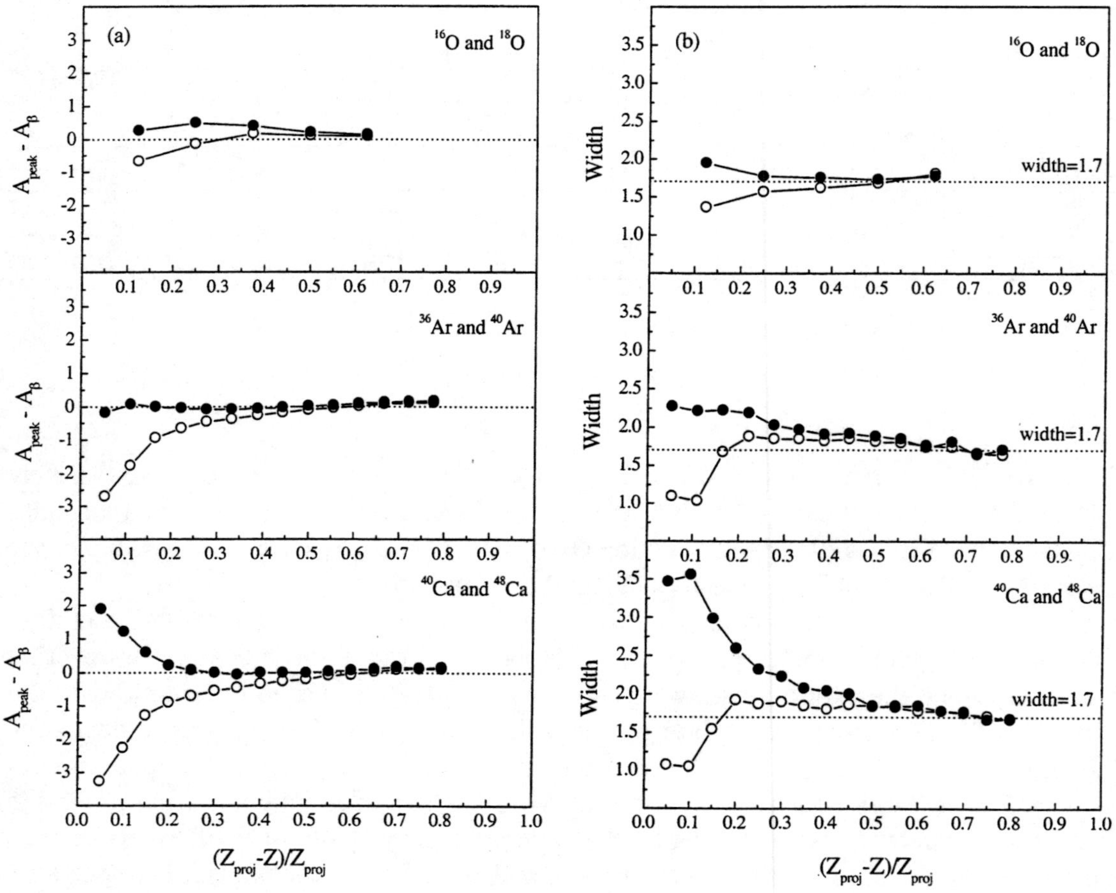

Figure 11: The difference A_{peak}-A_β (a) and width of the isotopic distribution (b) as a function of $(Z_{proj}-Z)/Z_{proj}$. The solid circles are the results of the neutron rich projectiles ^{18}O, ^{40}Ar and ^{48}Ca. The open circles are the results of ^{16}O, ^{36}Ar and ^{40}Ca.

ter. From the figures we can see that ΔA becomes larger for more neutron-rich projectile, but it approaches zero when $(Z_{proj}-Z)/Z_{proj}$ comes close to 0.5. For the width of the isotopic distribution, it also becomes larger for more neutron-rich projectile and approaches a constant value around 1.7 for $(Z_{proj}-Z)/Z_{proj}$ larger than 0.5. This demonstrates that the isospin effect of the fragmentation reaction becomes smaller with the increase of $(Z_{proj}-Z)/Z_{proj}$ or the violence of the nuclear reaction. Maybe, the evaporation effect mentioned above can account for these results. One point should be noted here. When $(Z_{proj}-Z)/Z_{proj}$ increases, other reactions such as multifragmentation will compete strongly with fragmentation reaction. The experimental isotopic distributions may be difficult to be fitted just in above model. Other reactions such as multifragmentation should be considered also. There are several empirical formulas which can predict isotope production cross section well[90]. Now before real experiment the RIB intensity can be estimated within a factor, say two. Such calculation will help the experimentalists to choose suitable reaction system before experiment.

4.2 Experimental techniques for the total reaction cross section

There are several experimental techniques for σ_R measurement. Among them three are often used, i.e. the angular distribution of elastic scattering, the beam transmission method and the 4π-γ coincidence method. The angular distribution of elastic scattering can be fit by optical model. From such fit the σ_R can be extracted. Of course this method is model dependent (optical model) and we should know optical model potential. This method has been used widely for stable nuclei. For exotic nuclei it will be very difficult to distinguish the elastic scattering from inelastic scattering because of the bad energy resolution of RIB. That is the reason why it is seldom to obtain σ_R for exotic nuclei from the measurement of angular distribution of elastic scattering. Here we will introduce other two methods for σ_R measurement, especially for σ_R measurement of exotic nuclei.

The beam transmission method or the beam attenuation method is a model independent method which directly measures the transmissive number N of projectiles after N_0 number of projectiles pass the target with thickness N_t (target atom number per unit area). The total reaction cross section should be as following:

$$\sigma_R = \frac{1}{N_t} \ln \frac{N_0}{N} \tag{31}$$

In the experiment one detector system before target identifies the projectiles and measures their number N_0 and another detector system after target identifies the particles which pass the target without reaction and measures their number N. In order to correct the loss of the projectiles on the way of RIB transportation, the experimental measurement with no target under same condition will be taken also. Then we will obtain σ_R more precisely:

$$\sigma_R = \frac{1}{N_t} \ln \frac{\gamma_0}{\gamma} \tag{32}$$

where $\gamma = \frac{N_0}{N}$ with target and γ_0 is the same ratio without target. Typical experimental setup for σ_R measurement was used by Tanihata et al.[17]. Six multiwire proportional counters were used to monitor the flight trace of the particles, two (PBT1 and PBT2) located before target and other four PAT1-4 behind target. Two scintilater detectors SBT before target and SAT behind target measure the energy loss of particles. After target a big magnet HISS deflects the particle. Above detectors together with time of flight measurement the particle before and after target can be identified well. Because the elastic scattering and the beam divergence the big angular acceptance is required for the detection system after target.

Mittig and Saint-laurent[19,20] developed another way called 4π-γ coincidence method for σ_R measurement. They think if a reaction happens, one γ-ray will be emitted at least. If the efficiency ε_γ of γ-detector is 100%, then the number of events in which one γ-ray was at least measured together with the number of the projectiles and the target thickness will give the σ_R. Of course there is no γ-detector systems which have 100% efficiency. But the γ-multiplicity in the reaction M_γ is quite high normally. The probability P which is efficiency to detect one γ-ray at least is $P = 1 - (1 - \varepsilon_\gamma)^{M_\gamma}$. If M_γ is large enough, P will close to 1, although ε_γ is less than 1. That is the reason why the principal of 4π-γ coincidence method works. Some σ_R measurement data were obtained by using 4π-γ coincidence method.

4.3 Extraction of the nucleon density distribution in nuclei and the nuclear radii from experimental total reaction cross section

Tanihata et al.[17,18] discovered the neutron halo of ^{11}Li,^{11}Be, ^{14}Be etc. by using high energy (~800 MeV/u) RIB on light target (Be, C, Al). They measured the cross section σ_I called the interaction cross section which is the probability with no change of projectile composition after the bombardment of projectile on target. At such high energy, σ_I nearly equals to σ_R. Via experiments they proved that

$$\sigma_I = \pi[R_I(proj) + R_I(target)]^2 \tag{33}$$

If we know R_I(target), the interaction radii of projectiles which is nearly same with the nuclear matter radii can be easily obtained from the measured σ_I. From the nuclear interaction radii for some light exotic nuclei extracted via this way, we see clearly that ^{11}Li,^{11}Be, ^{14}Be etc. did not follow R_I=$r_0A^{1/3}$ (r_0=1.21) and have large interaction radii which is first prove for the discovery of the neutron halo (see fig 12).

It is well established that the high energy σ_I data can be reproduced within a few percent accuracy by using the Glauber model together with N-N cross sections and the known nuclear density distribution between stable nuclei (see section II). Suzuki et al.[96] discussed the way to deduce the rms matter radius from σ_I by the Glauber model employing the Fermi-type density distribution:

$$\rho(r) = \rho_0/[1 + exp((r - r_0A^{1/3})/\alpha)] \tag{34}$$

Two parameters, a size parameter r_0 and a diffuseness parameter α, can not be determined unambiguously from the σ_R value obtained at one beam energy. By using two extreme conditions which are (1) $r_0(^ANa) = r_0(^{23}Na) = 1.095fm$ or (2) $\alpha(^ANa) = \alpha(^{23}Na) = 0.401fm$, the rms matter radii for $^{20-23,25-32}$Na were obtained. For light exotic nuclei by assuming harmonic-oscillator (HO) type density distribution or Gaussian density distribution, the rms matter radii were deduced unambiguously in the Glauber model fit to the experimental σ_R data. The width of the HO density distribution which is the only parameter in the calculation was set to reproduce the σ_I at high energy. Ozawa et al.[97] compared the σ_I experimental data at high energy and intermediate energies with the calculation results of Glauber model. In the calculations the parameter was obtained by fit the σ_I data at high energy. Fig 13 shows such comparison of the experimental data with the Glauber model calculations (solid lines in figure).

From fig 13 we see that the Glauber model calculations always underestimate the σ_R at intermediate energies. The differences between the data at intermediate energies and the calculated curves vary from nucleus to nucleus. For quantitative discussion Ozawa et al. defined a difference factor d as:

$$d = \frac{\sigma_R(exp) - \sigma_R(Gl)}{\sigma_R(Gl)} \tag{35}$$

where σ_R(exp) is the experimental σ_R at intermediate energies and σ_R(Gl) is the σ_R obtained by Glauber calculations at same energies where the density distribution was obtained by fitting the high energy data via Glauber calculation. Fig 14 shows the neutron excess (N-Z) dependence of the difference factor (d) for Be, B and C isotopes. The enhancement in the difference is considered to reflect an enhancement of the tail component in the nuclear density distribution. In Be isotopes, the d for ^{11}Be is larger than those of other isotopes. ^{11}Be was suggested as a neutron halo nucleus[23]. In B isotopes the d for ^{14}B is larger than

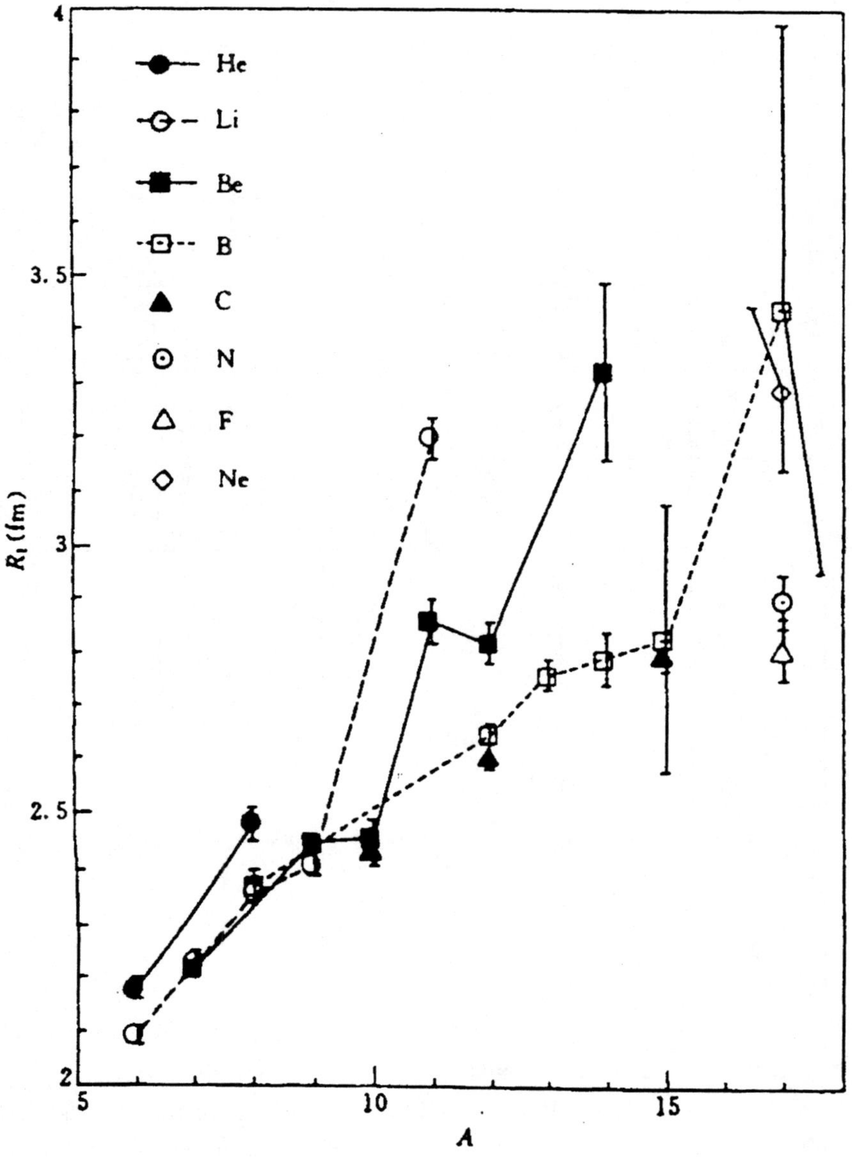

Figure 12: Nuclear interaction radii for some light nuclei[17,18].

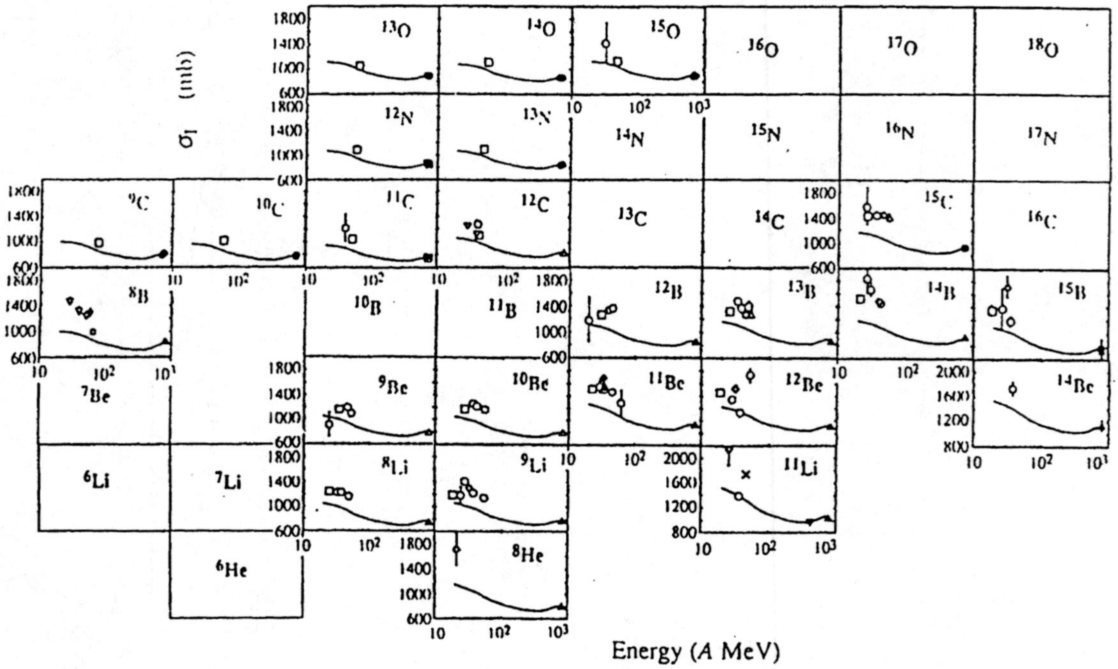

Figure 13: Excitation function of σ_R with C target. The solid curves are calculated by the Glauber model with HO-type density distribution. The figure is taken from[97], the references for experimental points, see figure 4 caption in reference[97].

those of other isotopes. Because ^{14}B has small neutron separation energy of 0.97 MeV, it is suggested that a large tail component exist in the ^{14}B density distribution. Same for the C isotopes, ^{15}C has a small neutron separation energy of 1.22 MeV and its d is also larger than other C isotopes shown in figures 14 and 15. The difference factor is about 10-20% for the nuclei near the β-stability line. D increases with increasing N-Z. For the nuclei which have halo or skin structure d increases strongly and reaches 40-50% for some nuclei near the drip line. This problem is still not solved now. The measurements of the σ_R by more reliable method at intermediate energies are necessary for conclusion statements. Also, more reliable method to extract the nuclear density distribution from σ_R measurement needs to be developed. Maybe, the BUU or QMD calculation can solve this problem. (see section II).

4.4 Isospin dependence of the total reaction cross section and the nuclear radii

B. A. Li et al. discussed the isospin dependence of total reaction cross section and nuclear radii in reference[98]. Here we will introduce first a table for summary of some σ_R measurement for exotic nuclei in table 2. Actually there are many of such measurements which are not listed in table 2.

From the isospin dependence of the total reaction cross section and the nuclear radii, the neutron halo and neutron skin in light nuclei were discovered and the possible proton halo and proton skin were discussed extensively. In the following we will describe some experimental results as example.

For light neutron rich nuclei, the rms matter radii extracted from σ_R measurements have been researched extensively as a function of the isospin (for fixed mass A) and the atomic mass (for fixed atomic charge number Z and for isotones with fixed neutron number N). The neutron halo (^{11}Li, ^{11}Be, ^{14}Be) and the neutron skin (^{6}He, ^{8}He) have been discovered for more than ten years already. Recent measurement of σ_R with $^{12-20}$C around 950 MeV/u suggested that the ^{19}C is one neutron halo nucleus[99]. Fig 16 shows the σ_I and R^m_{rms} as function of atomic mass A for C isotopes. The observations of both a large σ_I and a large one-neutron removal cross section indicate a core plus one neutron halo structure for ^{19}C. Ozawa et al.[97] measured the σ_R with 730 MeV/u ^{15}B, ^{9,10,15}C, ^{13}N and $^{13-15}$O. The nuclear rms radii have been deduced using a Glauber model calculation. The global isotope and isospin dependences of the radii were shown together with data from other experiments. The σ_I are also compared with those obtained at intermediate energies. An anomalous nuclear structure (a halo or skin) is suggested for ^{14}B and ^{15}C. Licot et al. measured the σ_R of 50-70 MeV/u $^{41-46}$Ar, $^{44-49}$Kr, $^{46-51}$Ca and $^{49-53}$Sc on Si target to extent the σ_R measurements to heavier exotic nuclei[100]. They declared the possible neutron skin for such heavier exotic nuclei. While Suzuki et al.[96] suggested neutron skin for neutron rich Na and Mg isotopes from σ_R measurement up to mass A=32. Bochkarev et al.[101] have measured the σ_I and charge-changing cross section of 950 MeV/u ^{20}N, ^{20}O, ^{20}F, ^{20}Ne, ^{20}Na, and ^{20}Mg on C target and suggested that ^{20}N has neutron skin.

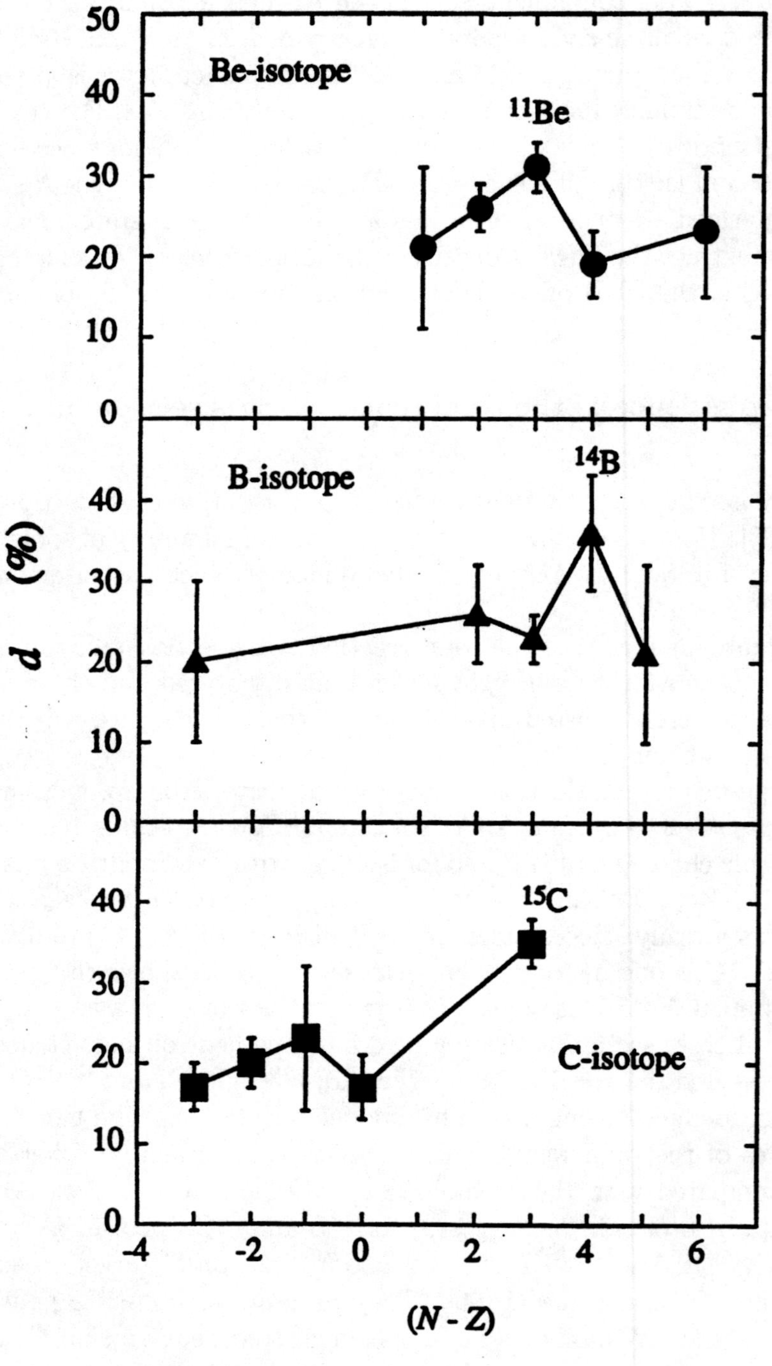

Figure 14: Neutron excess (N-Z) dependence of the difference factor (d) for Be (closed circles), B (closed triangles) and C (closed squares) isotopes. Taken from reference[97].

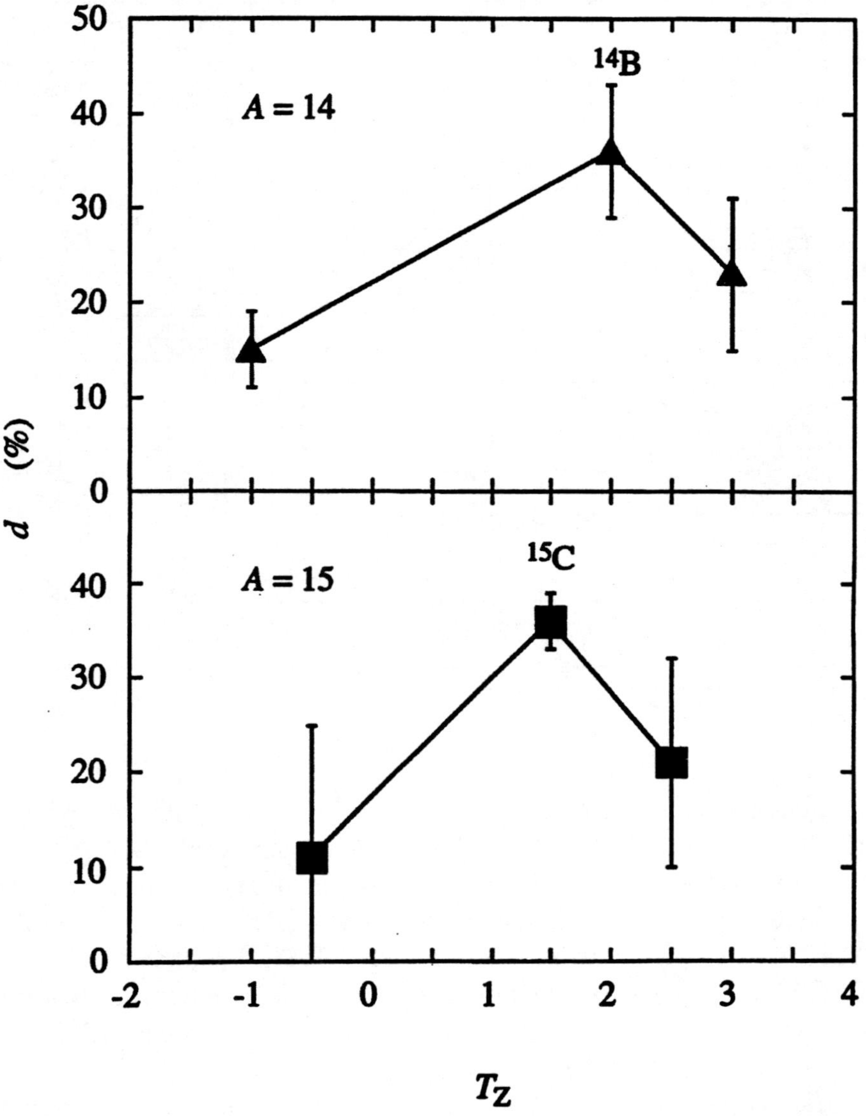

Figure 13: Isospin dependence of the difference factor (d) for A=14 (closed triangles) and A=15 (closed squares) isobars. Taken from reference[97].

Table 2. Summary of some σ_R measurement for exotic nuclei

Energy (MeV/u)	projectile	target	first author of the paper	year	reference
790	$^{3-8}$He	Be, C, Al	Tanihata	1985	17
790	$^{6-11}$Li, $^{7-10}$Be	Be, C, Al	Tanihata	1985	17
0-50	Z=5-12	Si	Mittig	1987	19
0-40	Z=2-10	Cu	Saiint-laurent	1987	20
790	^{11}Li, $^{11-14}$Be, $^{8-15}$B	Be, C, Al	Tanihata	1988	18
790	^{6,8}He, ^{11}Li	C, Pb	Kobayashi	1988	102
0-40	Z=2-13	Si, Cu	Villari	1989	103
800	^{11}Li	Be, C, Al, Cu, Pb	Kobayashi	1989	104
81	^{8,9,11}Li	C, Pb	Blank	1990	107
43, 75	^{11}Li	C	Shimoura	1991	105
Energy (MeV/u)	projectile	target	first author of the paper	year	reference
0-40	Z=2-15	Si	Villari	1991	24
33	^{11}Be	C, Al	Fukuda	1991	23
30	^{8}He, ^{11}Li, ^{14}Be	Be, Ni, Au	Riisager	1992	106
400, 800	^{11}Li	Be, C, Pb	Tanihata	1992	25
1000	^{50,56}Mn, ^{51,58}Fe, ^{54,60}Co, ^{55,62}Ni	CH2	Brohm	1992	108
80	^{11}Li	C, Al, Cu, Sn, Pb	Blank	1993	109
700	^{17}N, ^{17}F, ^{17}Ne	Be, C, Al	Ozawa	1994	110
950	$^{20-23,25-32}$Na	C	Suzuki	1995	77
20-60	^{8}B, ^{12}C, ^{14}N	Si	Warner	1995	111
20-60	^{4,6,8}He, $^{6-11}$Li, ^{10}Be	Si	Warner	1996	112
730	^{15}B, ^{9,10,15}C, ^{13}N, $^{13-15}$O	Be, C, Al	Ozawa	1996	97
790	^{8}B	C	Obuti	1996	114
40	^{12}N, ^{17}Ne	Si	Warner	1997	113
50-70	$^{41-46}$Ar, $^{44-49}$K, $^{46-51}$Ca, $^{49-53}$Sc	Si	Licot	1997	100
142,285	^{7}Be ^{8}B, ^{9}C	C, Pb	Blank	1997	37
950	^{20}N, ^{20}O, ^{20}F, ^{20}Ne, ^{20}Na, ^{20}Mg	C	Bochkarev	1998	101
950	$^{12-20}$C	C	Ozawa	1998	99
950	$^{20-23,25-27}$Na, $^{20,22,23,27,29-32}$Mg	C	Suzuki	1998	96
43	^{11}Be	Si	Negoita	1999	115

The best candidate for proton halo nucleus is ^{8}B because its last proton separation energy is only 0.137 MeV. Warner et al.[111] measured the total reaction cross section for ^{8}B on Si at 20-60 MeV/u and found the enhancement of the total reaction cross section. Blank et al.[37] measured the interaction cross sections and one proton removal cross section at energies 285 MeV/u and 142 MeV/u for ^{7}Be, ^{8}B, ^{9}C. The Glauber-type calculations need a proton halo for ^{8}B although less extended than for the known neutron halo cases like ^{11}Li to obtain good agreement between the experimental results and calculations. On the other hand ^{7}Be and ^{9}C can be described by standard density distributions. Minamisono et al.[33] measured the quadruple moment of ^{8}B and Schwab et al.[36] measured a narrow momentum distribution in ^{8}B induced fragmentation reaction. These experimental results indicate also that the matter density distribution extends to large radii due to a loosely bound proton halo in ^{8}B. But the experiment of interaction cross section induced by 790 MeV/u ^{8}B[18] has shown no enhancement of interaction cross section. Another experiment at 40 MeV/u[35] has also shown normal behavior of σ_R. Obuti et al.[114] remeasured the σ_I induced by 790 MeV/u ^{8}B on C target and can not remove the discrepancy. They extracted a much smaller radius for ^{8}B without proton halo from σ_I if the Glauber-type calculations were used to fit σ_I. More experimental and theoretical studies are desirable to fully understand the σ_R or σ_I of exotic nuclei. In particular a model that reasonably treats the reaction at lower energy is desirable as discussed in section II. Ozawa et al.[110] measured the interaction cross section and radii of the A=17 isobars. They found that the radius of ^{17}Ne is larger than that of its mirror nucleus ^{17}N. It shows that there is a possible proton halo or proton skin in ^{17}Ne. The experiment by Chulkov et al.[116] indicated a possible proton skin or

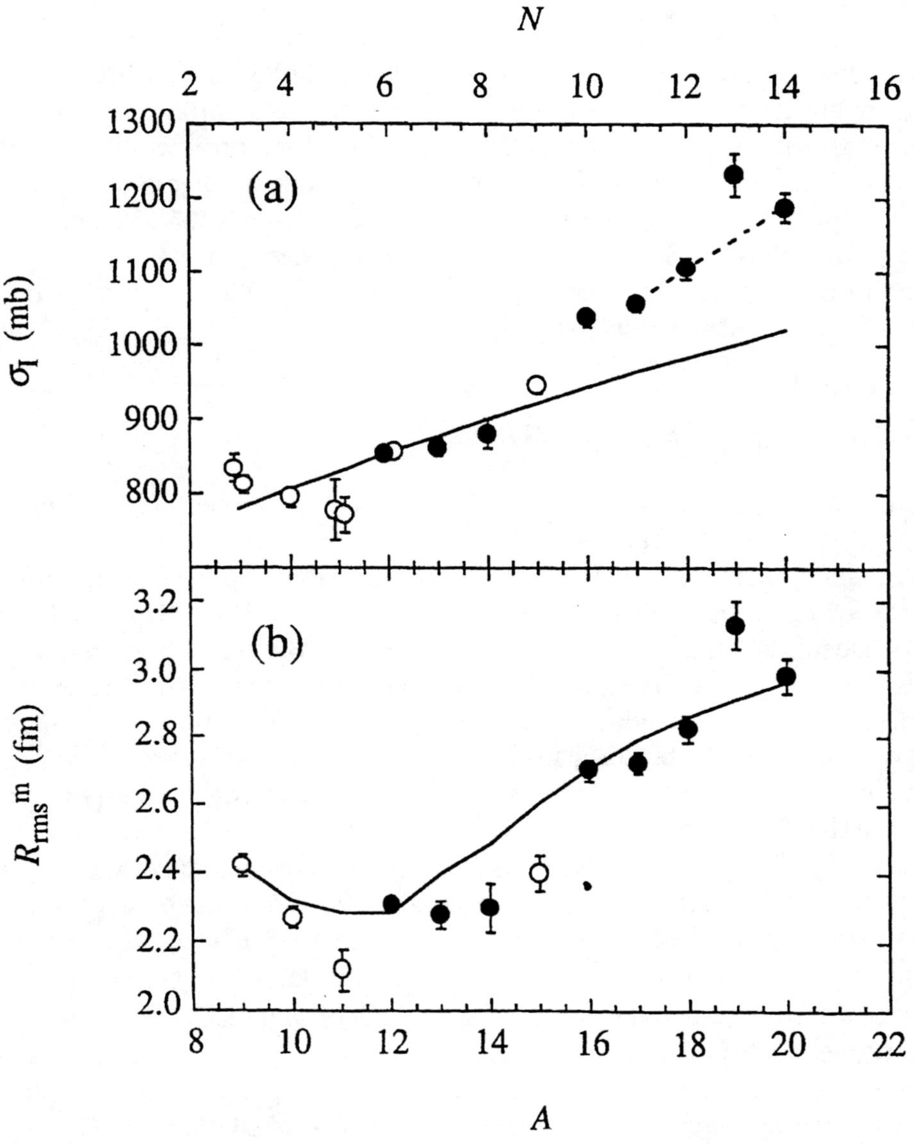

Figure 16: σ_I (a) and R^m_{rms} (b) for carbon isotopes on carbon target. The closed circles are data obtained in[99] and the open circles data from[97]. Solid line in (a) is calculated by $\sigma_I = \pi[R_I(C) + r_0 A^{1/3}]^2$ where R_I(C) is the interaction radii of ^{12}C (2.61fm). The dashed line in (a) shows the systematic increase obtained by simple fitting of ^{17}C, ^{18}C and ^{20}C values. The solid line in (b) shows the results of normal RMF calculation.

proton halo in ^{20}Mg.

The isospin dependence of the total reaction cross section are important for research of the structure of the exotic nuclei. But for the definite conclusion we need more experimental evidences such as the quadrupole moment measurements, the width of the longitudinal momentum distribution of fragment produced by the break-up reaction and the neutron angular distribution produced by the break-up reaction etc. Other experimental method like elastic scattering of halo nuclei[117], charged pion photo production for probing proton halo[118] etc. will be used also. There are some new measurements and new ideas about the isospin dependence of σ_R, nuclear radii and related topics[115,119,120]. Certainly we will see more experimental and theoretical researches on exotic nuclei especially near proton and neutron drip line including the experimental and theoretical researches on the isospin dependence of the total reaction cross section.

5 Overview and Perspectives

The theoretical calculations have predicted many interesting new physical phenomena for the nuclei far from the β-stability line, especially for the nuclei near the neutron and proton drip lines. Among them one neutron halo and two neutron halo have been confirmed by the experiments for light neutron-rich nuclei such as ^{11}Li, ^{11}Be, ^{14}Be, ^{19}C etc. The experiments show also some nuclei have neutron skin such as ^{6}He and ^{8}He. The experiment about long Na isotope chain demonstrates the development of the neutron skin with increasing the neutron number in nuclei. Even though proton halo and proton skin should be strongly hindered by the Coulomb force, the proton halo has also been suggested in some nuclei such as ^{8}B, $^{26-28}$P and ^{27,29}S and the proton skin in some nuclei such as ^{20}Mg. Several experiments have discussed whether ^{8}B is the one proton halo nuclei, but the definite conclusion needs further experiments. For Zirconium isotopes such calculations have predicted a giant neutron halo formed by up to six neutrons outside the core ^{122}Zr. We anticipate that there will be substantial experimental and theoretical activities in study of nuclear structure and new phenomena for the nuclei near proton and neutron drip lines in the coming years.

Theoretical studies also indicate that the isospin dependent EOS has significant influence on the properties of the nuclei far from the β-stability line. The radii, binding energy and deformation of exotic nuclei are shown to dependent critically on the isospin dependent nuclear EOS. The nuclear many-body theories have predicted that the N-N cross section might be significantly altered by the nuclear medium. The N-N cross section in nuclear medium dependents on the energy of nucleon, the density of nuclear medium and, may be, the isospin of nuclear medium. Because of this additional degree of freedom, isospin, new phenomena of the nuclear reactions like new structure properties for exotic nuclei have been predicted to occur in isospin asymmetric nuclear matter. In particular, the chemical instability is predicted to occur in a large configuration space. This may have significant effects on the mechanism of multifragmentation induced by isospin asymmetric nuclei. The liquid-gas phase transition such as its critical temperature etc. will dependent on isospin also. The balance energy where the effect of mean field and nucleon-nucleon collision on nuclear reaction happens to be in balance and the isotope distribution produced in fragmentation reaction etc. have isospin dependence also. There will be also substantial experimental and theoretical activities in study of the isospin dependence of the nuclear reaction in the coming years.

The isospin dependence of the properties of nuclear matter has significant influence also on the properties of neutron star. The cooling rate and neutrino emissions of neutron stars are affected significantly by the symmetry energy, especially its isospin and density dependence.

The energetic collisions of RIB with stable or neutron (proton) rich (deficit) target can create transiently nuclear matter with isospin asymmetry and provide a testing ground for studying the isospin physics. Dedicated RIB facilities in the world now we have opportunities to observe many interesting isospin effects already in exotic heavy ion collision. With the improvement of exist RIB facilities and fulfillment of new RIB facilities in the coming years, we will have excellent opportunities to achieve the goal of isospin physics. Because small RIB intensity and decreasing of RIB intensity with increasing isospin asymmetry, the measurement of the total reaction cross section for exotic nuclei is chosen very often for studying the isospin physics. Actually first proves for discovery of neutron halo and skin were based on the σ_R measurement. σ_R measurements will give us the chance to research the nuclei size and structure, the nucleon density distribution, EOS and N-N cross section in nuclear medium etc. Although there is a standard way already for studying the size and structure of exotic nuclei via σ_R measurement, there are still open problems. Among them the most puzzle problem is how to properly extract the nucleon density distribution from σ_R. The standard and almost exclusively way is based on Glauber theory. But Glauber theory underestimates σ_R in intermediate energy around 20% for exotic nuclei and even 50% for halo nuclei, if we fit σ_R well in high energy. Whether BUU or QMD model can solve this problem needs more researches. How to extract properly the N-N cross section in nuclear medium and EOS from σ_R needs more researches too. The normal isospin independent BUU model was used successfully already in calculating σ_R. The isospin dependent BUU model should be even better for calculating σ_R. The neutrons and protons can be distributed randomly in the density profile given by the RMF or RHB calculation, the isospin dependent BUU might be used to extract the EOS and N-N cross section in nuclear medium via fitting σ_R. Note that three ingredients, i.e. the nucleon density distribution, EOS and N-N cross section in nuclear medium from one total reaction cross section data can not be obtained at the same time. With the help of calculations to set two ingredients, one can extract the other ingredients by fitting σ_R via such BUU calculation. Such technique needs to be developed. But the measurements of the excitation function of total reaction cross sections are important to add more constraints on the research of the properties of isospin asymmetry nuclear matter. Especially the excitation function of σ_R of exotic nuclei in intermediate energy range where σ_R was underestimated by Glauber theory and more sensitive with N-N cross section in nuclear medium than EOS should be measured systematically for different isotope chains, different isobaric chains and different isotone chains. Comparison of their σ_R will help us to make conclusions whether the nuclei have halo or skin structure.

We anticipate that there will be substantial experimental and theoretical activities in this direction in near future. In order to extract more quantitative and definite information about exotic nuclei from the measurement of total reaction cross section, it is necessary to have systematic measurements and to combine theoretical calculations with experiments. With collaboration in world wide, the big progress in this direction will be achieved.

I would like to thank B.A.Li, W.L.Zhan, J.Feng, Y. G. Ma, Z. Y. Zhu, J. S. Wang, D. Q. Fang, X. Z. Cai, I. Tanihata, J. B. Natowitz, M. B. Tsang, J. Peter for collaboration in the study of several topics discussed in this review and helpful conversations on the subject. I am also grateful to D. Q. Fang for help in preparing the manuscript. The work was supported in part by the National Science Foundation of China under Grant No. 19675059 and Shanghai Science and Technique Development Foundation under Grant No.

96XD14011.

References

[1] H. A. Bethe, Phys. Rev. 57 (1940) 1125.

[2] H. W. Bertini, Phys. Rev. C 6 (1972) 2118.

[3] P. J. Karol, Phys. Rev. C 11 (1975) 1203.

[4] R. J. Glauber, Lectures on Theoretical Physics (Interscience, New York, (1959),Vol. I.

[5] R. Bass, Nuclear Reaction with Heavy Ion: Texts and Monographs in Physics (Springer-Verlag, Berlin, 1980).

[6] R. M. Devris and J. C. Peng, Phys. Rev. C 22 (1980) 1055.

[7] S. Kox, A. Gamp, C. Perrin, J. Arvieux, R. Bertholet, J. F. Brundet, M. Buenerd, R. Cherkaoui, A. J. Cole, Y. EI-Masri, N. Longequeue, J. Menet, F. Merchez and J. B. Viano, Phys. Rev. C 35 (1987) 1678.

[8] W. Q. Shen, B. Wang, J. Feng, W. L. Zhan, Y. T. Zhu and E. P. Feng, Nucl. Phys. A 491 (1989) 130.

[9] S. K. Charagi and S. K. Gupta, Phys. Rev.C 41 (1990) 1610.

[10] R. E. Warner and G. N. Felder, Phys. Rev. C 42 (1990) 2252.

[11] J. Chauvin, D. Lebrun, A. Lounis and M. Buenerd, Phys. Rev. C 28 (1983) 1970.

[12] C. Perrin, S. Kox, N. Longequeue, B. J. Viano, M. Buenerd, R. Cherkaoui, A. J. Cole, A. Gamp, J. Menet, R. Ost, R. Bertholt, C. Guet and J. Pinston, Phys. Rev. Lett. 49 (1982) 1905.

[13] M. Buenerd, A. Lounis, J. Chauvin, D. Lebrun, P. Martin, G. Duhamel, J. C. Gondrand and P. de Saintgnon, Nucl. Phys. A 424 (1984) 313.

[14] J. Jaros, A. Wagner, L. Anderson, O. Chamberlain, R. Z. Fuzesy, J. Gallup, W. Gorn, L. Schroeder, S. Shannon, G. Schapiro and H. Stiner, Phys. Rev. C 18 (1978) 2273.

[15] N. DiGiacomo, J. C. Peng and R. M. Devries, Phys. Lett. B 101 (1981) 383.

[16] M. S. Hussein, R. A. Rego and C. A. Bertulani, Phys. Rep. 201 (1991) 279.

[17] I. Tanihata, H. Hamagaki, O. Hashimoto, Y. Shida, N. Yoshikawa, K. Sugimoto, O. Yamakawa, T. Kobayashi and T. Takahashi, Phys. Rev. Lett. 55 (1985) 2676; Phys. Lett. B 160 (1985) 380.

[18] I. Tanihata, T. Kobayashi, O. Yamakawa, S. Shimoura, K. Ekuni, K. Sugimoto, T. Takahashi, T. Shimoda and H. Sato, Phys. Lett. B 206 (1988) 592.

[19] W. Mittig, J. M. Chouvel, W. L. Zhan, L. Bianchi, A. Cunsolo, B. Fernandez, J. Gastebois, A. Gillebert, C. Greogire, Y. Schutz and C. Stephan, Phys. Rev. Lett. 59 (1987) 889.

[20] M. G. Saint-Laurent, R. Anne, D. Bazin, D. Guillemaud-Mueller, U. Jahnke, G. M. Jin, A. C. Mueller, J. F. Bruandet, F. Glasser, S. Kox, E. Liatard, T. U. Chan, G. J. Costa, C. Heitz, Y. EI-Masri, F. Hanappe, R. Bimbot, E. Arnold and R. Neugart, Z. Phys. A 322 (1989) 457.

[21] G. F. Bertsch, B. A. Brown and H. Sagawa, Phys. Rev. C 39 (1989) 1154.

[22] B. Blank, J. J. Gaimard, H. Geissel, K. H. Schmidt, H. Stelzer, K. Summerer, D. Bazin, R. Del Moral, J. P. Dufour, A. Fleury, F. Hubert, H. G. Clerc and M. Steiner, Z. Phys. A 340 (1991) 41.

[23] M. Fukuda, T. Ichihara, N. Inabe, T. Kubo, H. Kumagai, T. Nakagawa, Y. Yano, I. Tanihata, M. Adachi, K. Asahi, M. Kouguchi, M. Ishihara, H. Sagawa and H. Shimoura, Phys. Lett. B 268 (1991) 339.

[24] A. C. C. Villari, W. Mittig, E. Plagnol, Y. Scutz, M. Lewitowicz, L. Bianchi, B. Fernandez, J. Gastebois, A. Gillbert, C. Stephan, L. Tassan-Got, G. Audi, W. L. Zhan, A. Cunsolo, A. Foti, A. Belazyorov,S. Lukyanov and Y. Penionzhkevich, Phys. Lett. B 268 (1991) 345.

[25] I. Tanihata, T. Kobayashi, T. Suzuki, Y. Yoshida, S. Shimoura, K. Sugimoto, K. Matsuta, T. Minamisono, W. Christie, D. Olson and H. Wieman, Phys. Lett. B 287 (1992) 307.

[26] K. Yabana, Y.Ogawa and Y. Suzuki, Nucl. Phys. A 539 (1992) 295.

[27] Y. Ogawa, K. Yabana and Y. Suzuki, Nucl. Phys. A 543 (1992) 722.

[28] M. Ishihara, Nucl. Phys. A 538 (1992) 309c.

[29] I. Tanihata, in Treatise on Heavy Ion Science, edited by D. A. Bromley (Plenum, New York, 1989), Vol. 8., p. 433.

[30] R. N. Boyd and I. Tanihata, Phys. Today 45(6) (1992) 45

[31] E. Roeckl, Rep. Prog. Phys. 55 (1992) 1661.

[32] J. S. Al-Khalili and J. A. Tostevin, Phys. Rev. Lett. 76 (1996) 3903.

[33] T. Minamisono, T. Ohtsubo, I. Minami, S. Fukuda, A. Kitagawa, M. Fukuda, K. Matsuta, Y. Norji, S. Takeda, H. Sagawa and H. Kitagawa, Phys. Rev. Lett. 69 (1992) 2058.

[34] H. Nakada and T. Otsuka, Phys. Rev. C 49 (1994) 886.

[35] I. Pecina, R. Anne, D. Bazin, C. Borcea, V. Borrel, F. Castoiu, J. M. Corre, Z. Dlouhy, A. Fomiche, D. Guillemaud-Mueller, H. Keller, A. Kordyasz, M. Lewitowicz, S. Lukyanov, A. C. Muller, Y. Penionzhkevich, P. Roussel-Chomaz, M. G. Saint-Laurent, N. Skobelev, O. Sorlin and O. Tarasov, Phys. Rev. C 52 (1995) 191.

[36] W. Schwab, H. Geissel, H. Lenske, K. H. Behr, A. Brunle, K. Burkard, K. Irmich,T. Kobayashi, O. Kraus, A. Magel, G. Munzenberg, F. Nickel, K. Riisager, C. Sheidenberger, B. M. Sherrill, T. Suzuki and B. Voss, Z. Phys. A 350 (1995) 283.

[37] B. Blank, C. Marchand, M. S. Pravikoff, T. Baumann, F. Boue, H.Geissel, M. Hellstrom, N. Iwasa, W. Schwab, K. Summerer and M. Gai, Nucl. Phys. A 624 (1997) 242.

[38] K. Hencken, G. F. Bertsch and H. Esbensen, Phys. Rev. C 54 (1996) 3043.

[39] G. F. Bertsch and S. Das Gupta, Phys. Rep. 160 (1988) 189.

[40] C. Gregoire, B. Remaud, F. Sebbile, F. Vient and Y. Raffrary, Nucl. Phys. A 465 (1987) 317.

[41] W. Cassing, V. Metag, U. Mosel and K. Nitta, Phys. Rep. 188 (1991) 363.

[42] J. Aichelin, Phys. Rep. 202 (1991) 233.

[43] Y. G. Ma, W. Q. Shen, J. Feng and Y. Q. Ma, Phys. Rev. C 48 (1993) 850.

[44] Y. G. Ma, W. Q. Shen, J. Feng and Y. Q. Ma, Phys. Lett. B 302 (1993) 386.

[45] J. H. Menet, E. E. Gross et al., Phys. Rev, C 4 (1971) 1114.

[46] H. Sagawa, Phys. Lett. B 286 (1992) 7.

[47] B. D. Serot and J. D. Walecka, Adv. Nucl. Phys. 16 (1986) 1.

[48] P. G. Reinhard, Rep. Prog. Phys. 52 (1989) 439.

[49] P. G. Reinhard, M. Rufa, J. Maruhn, W. Greiner and J. Friedrich, Z. Phys. A 323 (1986) 13.

[50] Y. K. Gambhir, P. Ring and A. Thimet, Ann. Phys.(NY) 511 (1990) 129.

[51] W. Koepf, Y. K. Gambhir, P. Ring and M. M. Sharma, Z. Phys. A 340 (1991) 119.

[52] D. Vautherin and D. M. Brink, Phys. Rev. C 5 (1972) 626.

[53] D. Vautherin, Phys. Rev. C 7 (1973) 296.

[54] Z. Y. Zhu, W. Q. Shen, Y. H. Cai and Y. G. Ma, Phys. Lett. B 328 (1994) 1.

[55] J. Meng and P. Ring, Phys. Rev. Lett. 77 (1996) 3963.

[56] J. Meng, I. Tanihata and S. Yamaji, Phys. Lett. B 419 (1998) 1.

[57] J. Meng and P. Ring, Phys. Rev. Lett. 80 (1998) 460.

[58] W. D. Myers et al., Ann. of Phys. 55 (1969) 395 and 84 (1974) 186.

[59] W. D. Myers and K. H. Schimdt, Nucl. Phys. A 410 (1983) 61.

[60] D. Hirata et al., Phys. Rev. C 44 (1991) 1467.

[61] J. Feng, W. Q. Shen, Y. G. Ma and Z. Y. Zhu, Phys. Lett. B 305 (1993) 9.

[62] X. Z. Cai, J. Feng, W. Q. Shen, Y. G. Ma, J. S. Wang and Y. We, Phys. Rev. C 58 (1998) 572.

[63] M. Kohno, M. Higashi, Y. Watanabe and M.Kawai, Phys. Rev. C 57 (1998) 3495.

[64] G. Q. Li and R. Machleidt, Phys. Rev. C 48 (1993) 1702.

[65] G. Q. Li and R. Machleidt, Phys. Rev. C 49 (1994) 566.

[66] D. Klakow, G. Welke and W. Bauer, Phys. Rev. C 48 (1993) 1982.

[67] Suneel Kumar, Manor K. Sharma and Rajeev K. Puri, Phys. Rev. C 58 (1998) 3494.

[68] Suneel Kumar and Rajeev K. Puri, Phys. Rev. C 58 (1998) 1618.

[69] M. M. Sharma, M. A. Nagarajan and P. Ring, Phys. Lett. B 312 (1993) 377.

[70] M. M. Sharma, G. A. Lalazissis, W. Hillebrandt and P. Ring, Phys. Rev. Lett. 72 (1994) 1431.

[71] G. A. Lalazissis and M. M. Sharma, Nucl. Phys. A 586 (1995) 201.

[72] J. P. Maharana, L. S. Warrier and K. K. Gambhir, Ann. Phys. 250 (1996) 237.

[73] Y. Sugahara and H. Toki, Nucl. Phys. A 579 (1994) 557.

[74] D. Hirata, H. Toki and I. Tanihata, Nucl. Phys. A 589 (1995) 239.

[75] B. Q. Chen, Z. Y. Ma, F. Grummer and S. Krewald, J. Phys. G 24 (1998) 97.

[76] M. Warda, B. Nerlo-Pomorska and K. Pomorski, Nucl. Phys. A 635 (1998) 484.

[77] T. Suzuki et al., Phy.s Rev. Lett. 75 (1995) 3241.

[78] D. Vretenar, G. A. Lalazissis and P. Ring, Phys. Rev. C 57 (1998) 3071.

[79] P. Descouvemout, Phys. Lett. B. 347 (1998) 7.

[80] M. Lassaut, P. J. Lombard, Z. Phys. A 431 (1992) 125.

[81] J. S. Wang, W. Q. Shen, J. Feng and Y. We, Chinese Journal of Nucl. Phys. V19 (1997) 23.

[82] A. C. Mueller et al., Nucl. Instrum. Methods in Phys. Res. B 56/57 (1990) 559.

[83] B. M. Sherrill et al.,Nucl. Instrum. Methods in Phys. Res. B 70 (1992) 298.

[84] H. Geissel et al.,Nucl. Instrum. Methods in Phys. Res. B 70 (1992) 286.

[85] T. Kubo et al.,Nucl. Instrum. Methods in Phys. Res. B 70 (1992) 309.

[86] A. G. Artukh et al., Proc. 3rd. Int. Conf. on Radioactive Nuclear Beams, East Lansing, Michigan, USA (1993) 45.

[87] J. X. Li, W. L. Zhan, Z. Y. Guo et al., High Energy Physics and Nuclear Physics V23 (1999) 231.

[88] Z. Y. Guo, W. L. Zhan et al., High Energy Physics and Nuclear Physics V23 (1999) 237.

[89] Z. Y. Sun, W. L. Zhan, Z. Y. Guo et al., High Energy Physics and Nuclear Physics V23 (1999) 298.

[90] K. Summerer, W. Bruche, D. J. Morrissey, M. Schadel, B. Szweryn and W. F. Yang, Phys. Rev. C 42 (1990) 2546.

[91] J. Hufner, K. Shafer and B. Schurmann, Phys. Rev. C 12 (1975) 1888.

[92] D. Gurreau, J. Phys. C 4 (1986) 207.

[93] T. Brohm and K. H. Schmidt, Nucl. Phys. A 569 (1994) 821.

[94] J. Feng, W. Q. Shen, Y. G. Ma, W. L. Zhan, Y. T. Zhu, Z. Y. Guo, L. Tassun-Got, C. Stephan, A. Gillibert, Y. Schutz, P. Jusun and W. Mittig, Phys. Rev. C 50 (1994) 2420.

[95] D. Q. Fang, W. Q. Shen et al., Submitted to Phys. Rev.

[96] T. Suzuki, H. Geissel, O. Bochkarev, L. Chulkov, M. Golovkov, N. Fukunishi, D. Hirata, H. Irnich, Z. Janas, H. Keller, T. Kobayashi, G. Kraus, G. Munzenberg, S. Neumaier, F. Nickel, A. Ozawa, A. Piechaczeck, E. Roeckl, W. Schwab, K. Summerer, K. Yoshida and I. Tanihata, Nucl. Phys. A 630 (1998) 661.

[97] A. Ozawa, I. Tanihata, T. Kobayashi, Y. Sugahara, O. Yamakawa, K. Omata, K. Sugimoto, D. Olson, W. Christie and H. Wieman, Nucl. Phys. A 608 (1996) 63.

[98] B. A. Li, C.M. Ko, and W. Baur, International Journal of Modern Physics E 7 (1998) 147.

[99] A. Ozawa, O. Bochkarev, L. Chulkov et al., RIKEN-AF-NP-294 (1998).

[100] I. Licot, N. Added, N. Carlin et al., Phys. Rev. C 56 (1997) 250.

[101] D. V. Bochkarev et al., Eur. Phys. J. A 1 (1998) 15.

[102] T. Kobayashi, O. Yamakawa, K. Omata, K. Sugimoto, T. Shimoda, N. Takahashi and I. Tanihata, Phys. Rev. Lett. 60 (1988) 2599.

[103] T. Kobayashi, I. Tanihata, K. Katori, K. Matsuta, T. Minamisomo, S. Shimoura, K. Suqimoto, W. Muller, D. L. Olsen, T. J. M. Symous and H. Wiemann, Phys. Lett. B 232 (1989) 51.

[104] A. C. C. Villari, W. L. Zhan, W. Mittig et al., XXVII intern. Winter Meeting on Nucl. Phys., Bormio, Italy 1989, p74.

[105] S. Shimoura in proceeding of the international symposium on structure and reaction of unstable nuclei, Niigata, Japan, 1991, p132.

[106] K. Riisager, A. S. Jensen, P. Moller, Nucl. Phys. A 540 (1992) 365.

[107] B. Blank et al., Z. Phys. A 343 (1992) 375.

[108] T. Brohm et al., Nucl. Phys. A 550 (1992) 540.

[109] B. Blank, J. J. Gaimard, H. Geissel et al., Nucl. Phys. A 555 (1993) 408.

[110] A. Ozawa et al., Phys. Lett. B 334 (1994) 18.

[111] R. E. Warner, J. H. Kelley, P. Zecher et al., Phys. Rev. C 52 (1995) R1166.

[112] R. E. Warner, R. A. Patty, P. M. Voles et al., Phys. Rev. C 54 (1996) 1700.

[113] R. E. Warner et al., Phys. Rev., C 56 (1997) 2694.

[114] M. M. Obuti et al., Nucl. Phys. A 609 (1996) 74.

[115] F. Negoita et al., Phys. Rev. C 59 (1999) 2082.

[116] L. Chulkov et al., Z. Phys. A 353 (1996) 351, and Nucl. Phys. A 603 (1996) 219.

[117] R. C. Johnson, J. S. Al-Khalili and J. A. Tostevin, Phys. Rev. Lett. 79 (1997) 2771.

[118] S. Karataglidis and C. Bennhold, Phys. Rev. Lett. 80 (1998) 1414.

[119] M. Lassaut and R. J. Lombard, Eur. Phys. J. A 4 (1999) 111.

[120] S. N. Ershov, B. V. Danilin, T. Rogde and J. S. Vaagen, Phys. Rev. Lett. 82 (1999) 908.

Effects of Isospin Asymmetry and In-Medium Corrections on Directed Transverse Flow and the Balance Energy

Wolfgang Bauer*, Frank Daffin†, and Gary D. Westfall*‡

Abstract

The effects of isospin asymmetry and in-medium corrections to the nucleon collision cross section on directed transverse flow and the balance energy are explored experimentally and theoretically. Flow and the balance energy have been measured for ^{58}Ni+^{58}Ni, ^{58}Fe+^{58}Fe, and ^{58}Mn+^{58}Fe. Flow and the resulting balance energies depend on the isospin of the system. The BUU model is used with isospin-dependent mean fields to calculate the balance energies of ^{58}Fe + ^{58}Fe and ^{58}Ni + ^{58}Ni for a range of impact parameters. We find that we are able to reproduce the measured impact parameter dependence of the balance energy, and the sign (but not the magnitude) of the shift in balance balance energy as a function of isospin asymmetry.

Introduction

The bulk properties of nuclear matter under extreme conditions are of interest to both nuclear physics and astrophysics. To generate these conditions, we collide nuclei in search of hints concerning the nature of the nucleus-nucleus interaction. Theoretical studies of these events allow us to make the connection between models and experiment though observables such as the balance energy [1, 2, 3]. The balance energy, or the beam energy at which transverse, in-plane directed flow [4] changes sign [5] as a function of beam energy, is sensitive to the forces underlying the dynamics of these collisions. Thus its study, in particular as a function of the impact parameter [6], can yield valuable information about the complex interplay of the repulsive nature of the hard scattering nucleon cross section, the repulsion and attraction of the nuclear mean field, and the repulsive contribution of the Coulomb force.

The BUU (Boltzmann-Uehling-Uhlenbeck) [7, 8, 9, 10, 11, 12] transport model is successful in reproducing and predicting experimental magnitudes and trends in collective single-particle observables [13, 14, 15], such as the balance energy, and we use it here. BUU is a theory for the single-particle phase space distribution of baryons (plus resonances and mesons). In its most popular numerical realization, the model propagates test particles

*National Superconducting Cyclotron Laboratory and Department of Physics and Astronomy Michigan State University, East Lansing, MI 48824-1321, USA, email address: bauer@pa.msu.edu

†current address: University of Alabama at Birmingham, School of Medicine, Department of Radiology, 619 19th Street South, Birmingham, AL 35249-6830

‡email address: westfall@nscl.msu.edu

with Hamilton's equations through the influence of experimentally observed nucleon collision cross sections [16], the Coulomb field, and a nuclear mean field based on a Skyrme-type interaction [17] which typically depends upon local density and momentum.

However, the inter-nucleon potential is not symmetric with respect to the isospin of the nucleon. Protons and neutrons experience different forces due to their constituents. This means that an asymmetry term exists in the nuclear mean-field. In addition, the nucleon cross section is modified from its vacuum values in nuclear matter [18, 19, 20].

Work by Müller and Serot [21] has shown that the liquid-gas phase transition may in fact be of second rather than first order. Their work, based upon a thermodynamic approach, demonstrated that local chemical instabilities are responsible for the transition when an asymmetric potential is used. This contrasts with the common view that mechanical instabilities are responsible for the onset of fragmentation in excited, diffuse systems. Studies using the percolation model [22] also have been used to explore the isospin degree of freedom in the nuclear fragmentation transition [23, 24]. The new experimental facilities coming on-line which will probe heavy-ion dynamics near the drip lines affords an opportunity to explore these results.

In stellar evolution the softening of the compressibility of nuclear matter as neutron ratios deviate from 1/2 is critical for generating a supernova explosion for massive stars ($12\,M_\odot$ to $15\,M_\odot$) [25]. Core collapse cannot sustain a shock if the nuclear Equation of State (EoS) retains its equilibrium stiffness; the result is at least a delay of such important process as nucleosynthesis.

Neutron stars cool by emitting neutrinos. After an initial fast-cooling phase, they enter a more sedate cooling epoch producing neutrinos via the modified URCA process: $(n,p)+p+e^- \rightarrow (n,p)+n+\nu_e$, $(n,p)+n \rightarrow (n,p)+p+e^-+\bar{\nu}_e$. However, if the proton density exceeds a critical value between 11% and 15%, which is dependent upon the symmetry energy, the direct URCA process can occur and become the dominant cooling mechanism[26]: $n \rightarrow p+e^-+\bar{\nu}_e$, $p+e^- \rightarrow n+\nu_e$.

The exploration of the isospin degree of freedom in heavy ion collision thus carries the promise that it will help determine the answers to these questions of astrophysical relevance [27].

The isospin dependence of various reaction observables has been studied experimentally including the isospin dependence of multifragmentation,[28] N/Z equilibration,[29] and the isospin dependence of directed transverse flow and the balance energy.[30, 31] Recently, it was experimentally shown [30, 31] that collective flow and the resulting balance energy for symmetric systems of equal mass has different values for different isospin asymmetry of the colliding nuclei. In these studies, three systems, ^{58}Ni+^{58}Ni, ^{58}Fe+^{58}Fe, and ^{58}+Mn^{58}Fe, were studied from 45 to 105 MeV/nucleon. These three systems maintained the mass symmetric kinematics while varying the isospin asymmetry.

In this chapter we discuss the isospin-dependence of the balance energy, attempt to isolate its origin, and delineate ways to find sensitivity to the isospin-dependence of the nuclear mean field.

EXPERIMENTAL RESULTS

Isospin dependence of collective transverse flow

Fig. 1 shows the extracted values of the directed transverse flow for three different fragment types from all three entrance channels plotted as a function of the isotopic ratio of the

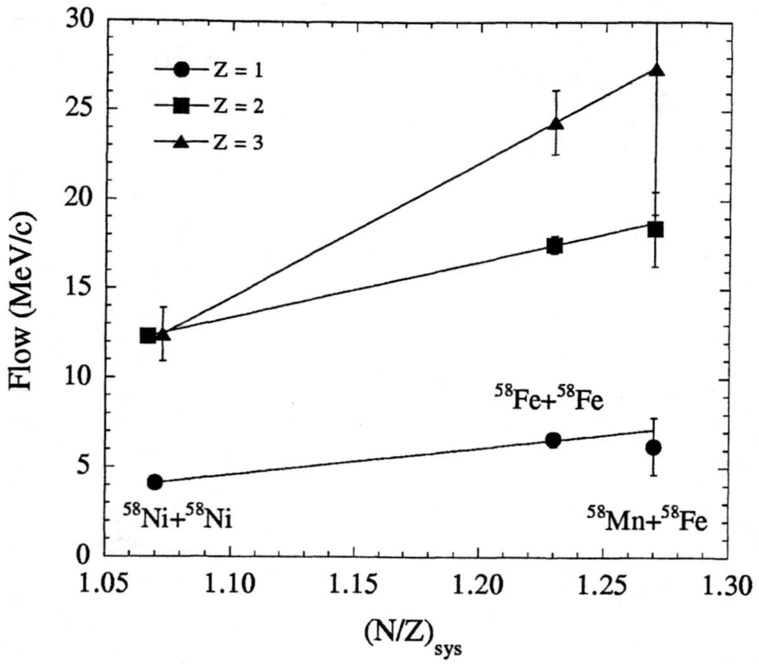

Figure 1: Measured isospin dependence of transverse flow at 55 MeV/nucleon integrated over impact parameter for three systems with different N/Z.

composite projectile plus target system where

$$(N/Z)_{sys} = (N_{proj} + N_{targ})/(Z_{proj} + Z_{targ}). \quad (1)$$

Using this definition we have: $(N/Z)_{sys}$ =1.07 for ^{58}Ni+^{58}Ni; $(N/Z)_{sys}$ =1.23 for ^{58}Fe+^{58}Fe; and $(N/Z)_{sys}$ =1.27 for ^{58}Mn+^{58}Fe.[30] The points for $Z = 2$ from ^{58}Ni+^{58}Ni and ^{58}Mn+^{58}Fe have been offset in value of $(N/Z)_{sys}$ for clarity. The errors shown are the statistical errors on the slopes of the linear fits (the systematic error associated with the range of the fitting region is +0.75 AMeV/c and -0.25 AMeV/c). The results shown in Fig. 1 demonstrate clearly there is an isospin dependence for directed transverse flow even for the impact-parameter-inclusive data. The neutron-rich system ^{58}Fe+^{58}Fe exhibits larger flow values than ^{58}Ni+^{58}Ni for all three particle types. Although the difference between the flow values extracted for ^{58}Mn+^{58}Fe and ^{58}Fe+^{58}Fe is not statistically significant, the trends are consistent with the reactions involving the two stable beams.

The extracted values of collective transverse flow in the reaction plane are displayed in Fig. 2 as a function of the reduced impact parameter for three different fragment types from ^{58}Fe+^{58}Fe and ^{58}Ni+^{58}Ni collisions at 55 MeV/nucleon.[30] The extracted values of the flow are plotted at the upper limit of each $\hat{b}$ bin. The neutron-rich system ^{58}Fe+^{58}Fe systematically exhibits larger flow values than ^{58}Ni+^{58}Ni for all three particle types at all reduced impact parameter bins displayed (except for $Z = 3$ in the most peripheral bin). The largest difference in the magnitude of the flow between the isotopic entrance channels occurs for heavier mass fragments in semi-central collisions. The impact parameter dependence of the directed transverse flow shown is in qualitative agreement with previous work [32, 33, 34], because the flow is maximal for semi-central events. The mass dependence of the directed transverse flow shown in Fig. 2 also demonstrates the well known increase in magnitude for heavier fragments [3, 32, 35].

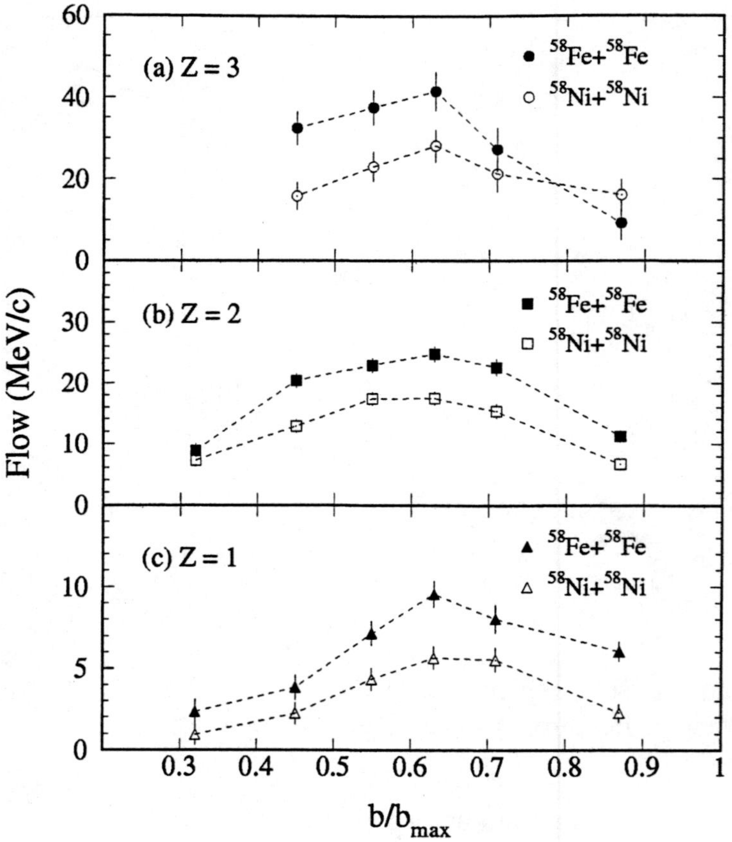

Figure 2: Measured isospin dependence of transverse flow at 55 MeV/nucleon as a function of impact parameter.

Isospin dependence of the balance energy

The measured values of the directed transverse flow plotted versus the incident beam energy are shown in Fig. 3.[31] The solid (open) squares are for fragments with $Z = 2$ from semi-central ^{58}Fe+^{58}Fe (^{58}Ni+^{58}Ni) collisions. The errors shown are the statistical errors on the slopes of the linear fits (the systematic error associated with the range of the fitting region is +3 MeV/c and -1 MeV/c). The curves are included to guide the eye. To extract the balance energy E_{bal}, the data were fit with a second-order polynomial allowing the fitting range to vary until χ^2 per degree of freedom was a minimum [35]. The second-order fits pass through minima for which the value of the abscissa corresponds to the balance energy for that particular entrance channel and $\hat{b}$ bin. The horizontal displacement of the minima of the curves in Fig. 3 clearly indicates that E_{bal} is higher for ^{58}Fe+^{58}Fe than ^{58}Ni+^{58}Ni at this $\hat{b}$ bin.

Additional experimental evidence for the isospin dependence of the balance energy is presented in Fig. 4. The solid (open) squares are the measured values of the balance energies for ^{58}Fe+^{58}Fe (^{58}Ni+^{58}Ni) extracted for four reduced impact parameter bins. These experimental values of $E_{bal}(\hat{b})$ are plotted at the upper limit of each $\hat{b}$ bin, and the values for ^{58}Ni+^{58}Ni have been slightly offset in the horizontal direction to show the error bars more clearly. The errors shown on the measured values of the balance energies are statistical (the systematic error is estimated to be +5% and -0%). The balance energy increases as

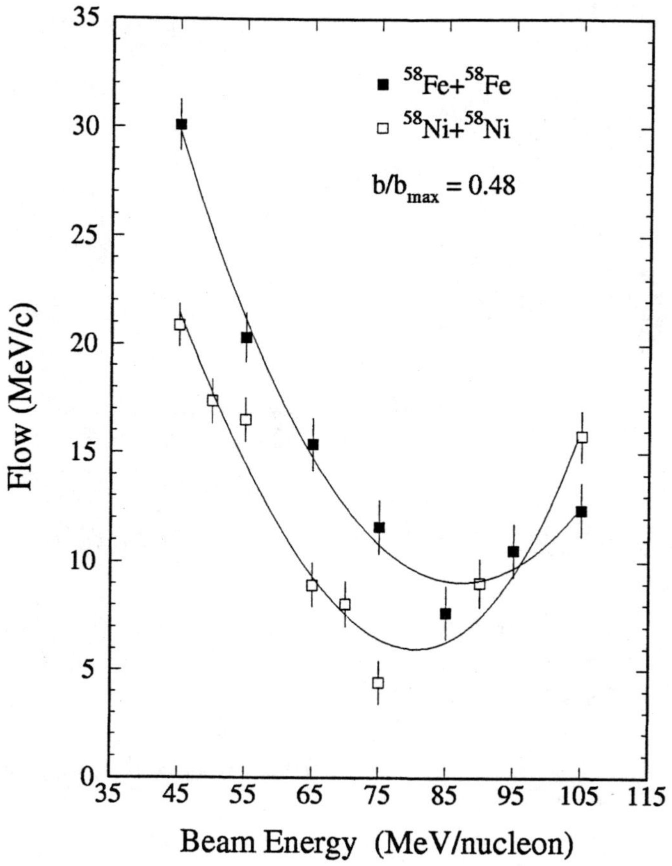

Figure 3: Excitation functions of the measured transverse flow for Z = 2 fragments from semicentral collisions (b/b_{max} = 0.48). The solid (open) squares are for ^{58}Fe + ^{58}Fe (^{58}Ni + ^{58}Ni).

a function of impact parameter for both isotopic systems in agreement with previous work [33, 34, 36], and $E_{bal}(b)$ is systematically higher for the more neutron-rich system at all measured $\hat{b}$ bins.

Model Input

The BUU transport equation evolves in time a single nucleon phase-space distribution under the influences of nucleon-nucleon collisions, the Coulomb field, and a nuclear mean field. In this study, we choose two mean fields: One recently used by Bao-An Li *et al.* [37, 38] (also see Ref. [39, 40, 41, 42]):

$$U = A\left(\frac{\rho}{\rho_0}\right) + B\left(\frac{\rho}{\rho_0}\right)^{\sigma} + C\tau_z\left(\frac{\rho_n - \rho_p}{\rho_0}\right), \tag{2}$$

where ρ_0 is the normal nuclear density, ρ_n is the neutron density, ρ_p is the proton density, and τ_z is the isospin factor which is 1 for neutrons and -1 for protons. Coefficients A, B and σ are typically chosen to match the ground state properties of symmetric nuclear matter such as the saturation density and saturation binding energy. The compressibility is a free parameter, and in this study we chose it to be 200 MeV. This gives $A = -109$ MeV, $B = 82$ MeV, and $\sigma = \frac{7}{6}$. In keeping with H.M. Xu and Bao-An Li's work, $C = 32$ MeV.

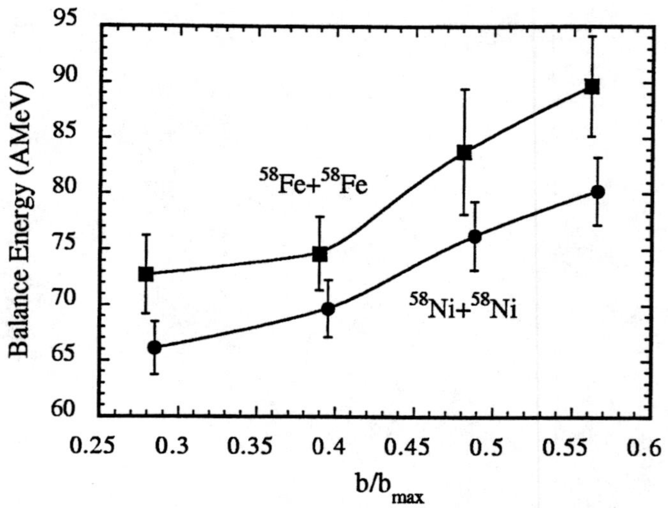

Figure 4: Measured isospin dependence of the balance energy for ^{58}Ni+^{58}Ni and ^{58}Fe+^{58}Fe.

The other mean field is derived from a Hamiltonian due to Sobotka [43]:

$$\mathcal{H} = \mathcal{KE} + \frac{4a}{\rho_0}({\rho_n}^2 + b\rho_n\rho_p + {\rho_p}^2) + \frac{4c}{\rho_0}({\rho_0}^2\rho_p + \rho_n{\rho_p}^2). \tag{3}$$

Here the coefficients $a = -3.66$ MeV, $b = 15.0$ and $c = 23.4$ MeV and KE is the kinetic energy term. Neutrons are acted upon by:

$$U_n = \frac{\partial\mathcal{H}}{\partial\rho_n} = 8a\frac{\rho_n}{\rho_0} + 4ab\frac{\rho_p}{\rho_0} + 8c\frac{\rho_n\rho_p}{{\rho_0}^2} + 4c\frac{{\rho_p}^2}{{\rho_0}^2}, \tag{4}$$

whereas protons are affected by:

$$U_p = \frac{\partial\mathcal{H}}{\partial\rho_p} = 8a\frac{\rho_p}{\rho_0} + 4ab\frac{\rho_n}{\rho_0} + 8c\frac{\rho_n\rho_p}{{\rho_o}^2} + 4c\frac{{\rho_n}^2}{{\rho_0}^2}. \tag{5}$$

In symmetric matter, this mean field reduces to the often-called "Stiff" equation of state (compressibility 380 MeV). It should be noted that there is no transformation of the mean field of Sobotka that will yield an equivalent "Soft" equation of state. Thus direct comparisons between the two formulations are impossible. One may simply alter A, B and σ in Equation 2 to give the appropriate compressibility in symmetric matter.

With the isospin asymmetry

$$\delta = \frac{\rho_n - \rho_p}{\rho_0} \tag{6}$$

and the reduced density

$$\overline{\rho} = \frac{\rho}{\rho_0} \tag{7}$$

Equations 2 and 4 can be written as:

$$U = A\overline{\rho} + B\overline{\rho}^{\sigma} + C\tau_z\delta \tag{8}$$

and

$$U_n = \overline{\rho}(4a + 2ab) + \delta(4a - 2ab) - c\delta^2 + 3c\overline{\rho}^2 - 2c\delta\overline{\rho}, \tag{9}$$

for the neutron potential, and

$$U_p = \overline{\rho}(4a + 2ab) + \delta(2ab - 4a) - c\delta^2 + 3c\overline{\rho}^2 + 2c\delta\overline{\rho}, \tag{10}$$

where a, b and c are as defined above. Fig. 5 (left side) shows Equation 8 as neutron excess and normalized nuclear density are varied. The right side of the same figure shows Equation 9 as neutron excess and normalized nuclear density are varied. One can see that Equation 9 is more attractive to neutrons in the midst of proton-rich matter for all densities than Equation 8.

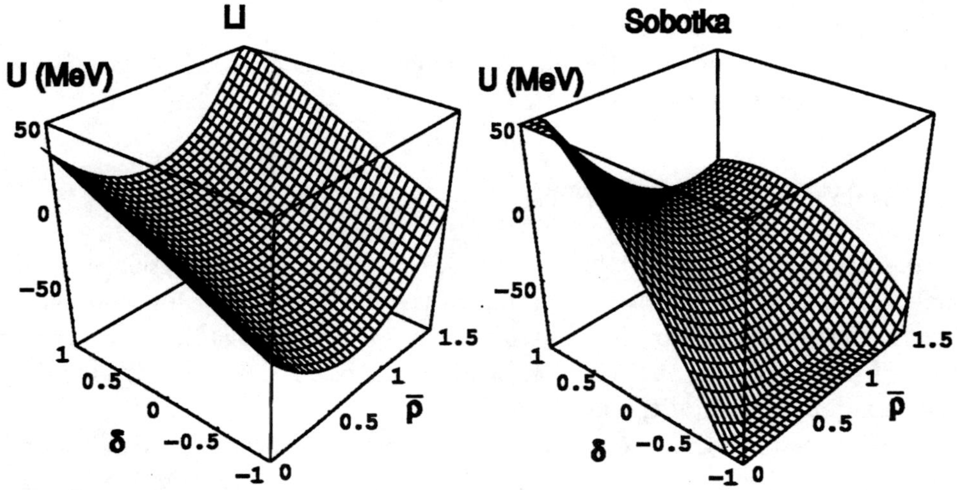

Figure 5: Mean field potential as a function of reduced density, $\hat{\rho} = \rho/\rho_0$, and isospin asymmetry, δ, as used by Li et al. (left) and Sobotka (right).

Before we proceed, we should briefly discuss the density profiles implied by the single-particle potentials used here. These are of particular importance at peripheral impact parameters, where the existence of a neutron skin can have considerable importance on the resulting reaction dynamics. Sobotka has shown [43] that the density surface profiles for Au-nuclei, for example, have neutron skins of roughly 1/3 – 1/2 fm thickness, in good agreement with what the droplet model predicts. We find similar results for heavy nuclei for the other isospin potential used here. The neutron skins for the lighter nuclei with less isospin asymmetry that we will use for the numerical studies below show a much smaller neutron skin of less than 1/6 fm thickness, a value that is small compared to the numerical surface thickness fluctuations.

Since these isospin potentials are constructed to reproduce the asymmetry term in the Bethe-Weizaecker formula, they are roughly able to reproduce the dependence of the binding energy as a function of isospin asymmetry, for nuclei not too far from the $N = Z$ limit, as shown in [43]. A global comparison of the resulting numerical binding energies to experimental values would be desirable, but has not been performed yet.

Other parameterization of nuclear mean field potential also exist [44, 45], but were not considered here.

The nucleon-nucleon cross sections are parameterization from the Particle Data Group[16] with medium modification implemented according to the density dependent prescription:

$$\sigma_{\rm NN} = \sigma_{\rm NN}^{\rm free}\,(1 + \alpha\,\overline{\rho}) \tag{11}$$

where α is the logarithmic derivative of the in-medium cross section with respect to the density, taken at $\rho = 0$,

$$\alpha = \rho_0 \frac{\partial}{\partial \rho} (\ln \sigma_{\mathrm{NN}})|_{\rho=0} \tag{12}$$

It is the lowest coefficient of a Taylor-expansion of the cross-section in powers of the density [19]. This is a parameterization of the Pauli-blocking of intermediate states is motivated by Brückner G-matrix theory [18]. One can show [20, 46, 47] that values of α in the range between -0.4 and -0.2 yield the best agreement of the simple parameterization used here and the more involved G-matrix calculation using a realistic nucleon-nucleon interaction. In addition, it has been shown that an in-medium reduction of $\alpha \approx -0.2$ provides the best reproduction of the flow signals of symmetric collisions for a broad range of energies and projectile-target combinations [48, 19, 3].

Elementary considerations

Before we proceed to show our numerical results, it is appropriate to attempt to understand the reasons why we would expect a difference in the balance energy for symmetric systems of equal mass, but different isospin. There are three isospin effects that we may consider: the average effective nucleon-nucleon cross section, the difference in the Coulomb potential, and the isospin dependence of the mean field.

In order to understand the sign of the different effects, we first remind ourselves that the balance energy marks the beam energy at which the repulsive and attractive interactions are roughly equal. At beam energies lower than the balance energy, the attractive interaction dominates, and at energies higher than the balance energy, repulsive interactions determine the flow.

The difference in the Coulomb interaction is very simple. Fe has 26 protons and Ni 28. This means that the ratio of the Coulomb forces is $(26/28)^2 = 0.862$. Since the Coulomb force adds repulsion, we thus expect that the balance energy for nickel should be lower than for iron.

The difference in the isospin dependent mean field has the opposite effect. The isospin asymmetry, δ, is three times bigger for iron than it is for nickel:

$$\delta(^{58}Fe) = 0.103, \qquad \delta(^{58}Ni) = 0.034 \tag{13}$$

A larger isospin asymmetry results in more repulsion, compare Fig. 5. Thus we expect that the isospin dependent mean field alone would favor a lower balance energy for the ^{58}Fe system than for the ^{58}Ni system.

The numbers of protons and neutrons in the colliding system also have an influence on the average nucleon-nucleon cross section, because the neutron-proton cross section is approximately a factor of 3 bigger than the proton-proton cross section in the energy regime of interest here. We can then calculate the average nucleon-nucleon cross section from the number of collisions between nucleons of equal and opposite isospin,

$$\begin{aligned} \overline{\sigma} &= \frac{N_{np}\,\sigma_{np} + (N_{nn} + N_{pp})\,\sigma_{pp}}{N_{np} + N_{nn} + N_{pp}} \\ &= \frac{3\,N_{np} + N_{nn} + N_{pp}}{N_{np} + N_{nn} + N_{pp}}\,\sigma_{pp} \end{aligned} \tag{14}$$

where N_{np} is the number of neutron-proton collisions.

We make the *Ansatz* that the number of collisions of two types of nucleons is proportional to the number of nucleons of each species, taken to some power, ζ,

$$\begin{aligned} N_{np} &\propto (N_{n,T}\,N_{p,P})^\zeta + (N_{n,P}\,N_{p,T})^\zeta \\ N_{nn} &\propto (N_{n,T}\,N_{n,P})^\zeta \\ N_{pp} &\propto (N_{p,T}\,N_{p.P})^\zeta \end{aligned} \tag{15}$$

where $N_{n,T}$, $N_{n,P}$ are the numbers of neutrons in target and projectile, and $N_{p,T}$, $N_{p,P}$ are the numbers of protons in target and projectile. In intermediate energy heavy ion collisions, we find numerically that ζ has values roughly in the range between 2/3 and 1.

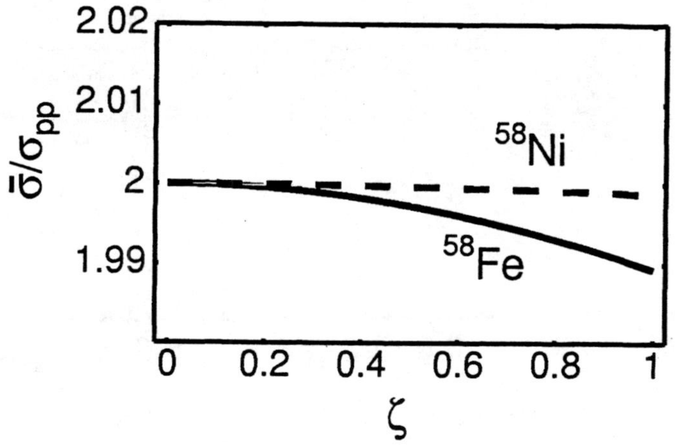

Figure 6: Dependence of the average nucleon-nucleon cross section as a function of the parameter ζ for the two different isospin systems.

Inserting this *Ansatz* into our equation for the average cross section yields the result displayed in Fig. 6. The difference in the systems with different isospin is small (only approximately 0.5% for ζ=1), but is clearly visible. Since the isospin-averaged cross section is slightly bigger for the ^{58}Ni system, this results in a higher kinetic pressure, therefore more repulsion, and thus a lower balance energy.

The isospin composition of the projectile and target also has an effect on the Pauli exclusion principle for the final scattering states of the nucleons. If there are more neutrons than protons, then these neutrons necessarily occupy are larger fraction of the available phase space for the scattered nucleons, resulting in a smaller number of Pauli-allowed *nn* collisions than *pp* ones. However, for the considerations of the isospin dependence of the average in-medium nucleon-nucleon cross section, the interplay between Pauli-principle and isospin content of the nuclei has only a small effect. To convince ourselves of this, we note that the Boltzmann collision integral contains the factor

$$\Pi_{ab} = [1 - f_a(\vec{r},\vec{p})] \cdot [1 - f_b(\vec{r},\vec{p})] \tag{16}$$

where a and b are the isospin indices of the two nucleons that take part in the scattering event. The factor Π_{ab} represents the correction of the free nucleon-nucleon cross section due to the Pauli exclusion principle for the final scattering states. If we average over all possible two-body scattering events, we find for beam energies larger than the Fermi energy:

$$\langle \Pi_{pn} \rangle \approx \tfrac{1}{2}(\langle \Pi_{pp} \rangle + \langle \Pi_{nn} \rangle) \tag{17}$$

The leading correction term to this approximation enters as δ^2. Thus the differential effect of the Pauli-exclusion principle on the isospin dependence of the average in-medium cross section is only a small correction to the determination of $\overline{\sigma}$.

Comparisons with Measured Data

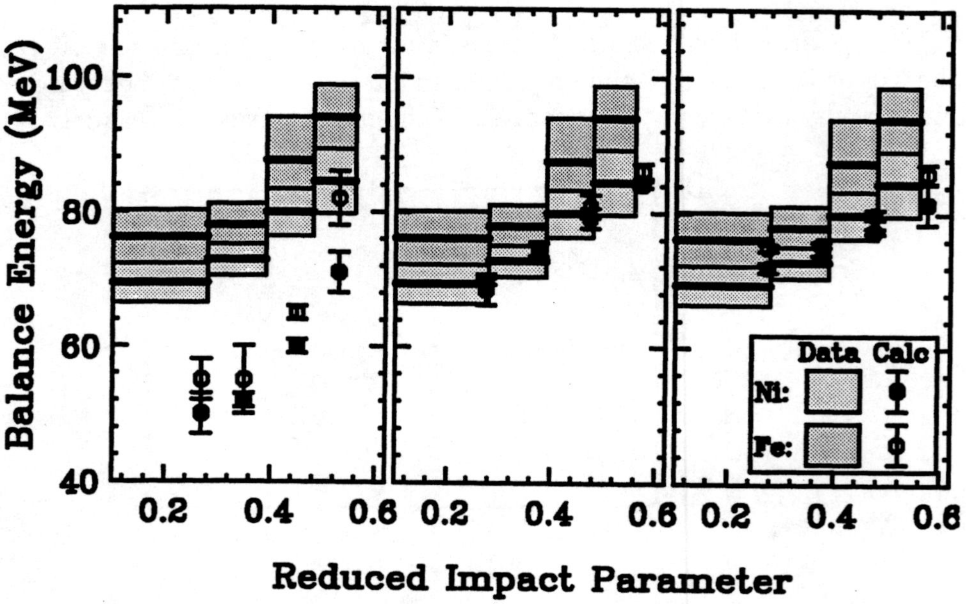

Figure 7: Impact parameter dependence of the balance energy for the systems $^{58}Ni + {}^{58}Ni$ (light shaded rectangles: data; open circles: calculations) and $^{58}Fe + {}^{58}Fe$ (dark shaded rectangles: data; filled circles: calculations). Left panel: results of B.A. Li with free nucleon-nucleon cross sections; middle panel: mean field of B.A. Li and $\alpha = -0.3$; right panel: mean field of Sobotka and $\alpha = -0.2$.

We preface this section by briefly outlining our procedure to determine the balance energies: We begin by calculating the proton freeze-out phase-space distribution at reduced impact parameters $\hat{b} = b/b_{max} = 0.275, 0.375, 0.475$ and 0.575, beam energies $50, 60, 70, 80, 90$ and 100 MeV/A, using Equations 2 and 4, and various in-medium corrections α, Eq. 11 This is followed by calculating the flow for each permutation, and then the balance energy for each $\hat{b}$, using a χ^2 minimization procedure.

In Fig. 7, we show the experimental results of Pak *et al.* [31] as the shaded rectangles. The width of these rectangles represent the width of the impact parameter bins used for the integrations of the experimental data. The height of the rectangles represent the error bars in the balance energy (standard deviation of the mean). The darker shaded rectangles represent the experimental results for the $^{58}Fe + {}^{58}Fe$ system, and the lighter gray areas those for the $^{58}Ni + {}^{58}Ni$. The thick horizontal lines through the middle of each rectangle represent the quoted experimental values. This experimental information is the same in all three panels of Fig. 7, each time compared to a different calculation.

Before we discuss the ingredients of each calculation, we should point out that the experimentally found balance energies in $^{58}Fe + {}^{58}Fe$ system are always higher than those for the $^{58}Ni + {}^{58}Ni$ system. This is what we would expect from our elementary considerations of the effects of the Coulomb interaction and the two-body collisions, but contrary to what we expect from the mean field alone. We will return to this point later.

In order to perform our calculations, we initialize the nucleons (test particles) randomly within a spherical volume of radius $r = 1.1 \cdot A^{1/3}$. We then allow the nuclei to find their most favorable density profiles by iterating for about 20-30 time steps before the surface of the nuclei come into contact. We stop the collisions after the "asymptotic" values of the flow value is reached. This corresponds approximately to the point in time when the number of test particle collisions becomes negligibly small, and when the two nuclear remnants have separated from each other. (This procedure fails for low-energy orbiting reactions, in which the reaction partners stick to each other. However, for determining the balance energy, these reactions play no role.)

The theoretical results for the ^{58}Fe + ^{58}Fe system are indicated by the open plot symbols, and those for ^{58}Ni + ^{58}Ni by the filled ones.

In the left panel of Fig. 7, we show the result of a calculation with the mean field interaction of B.A. Li [31], and with $\alpha = 0$. Pak *et al.* [31] found that BUU under-predicted the balance energies of ^{58}Fe + ^{58}Fe and ^{58}Ni + ^{58}Ni collisions. This is consistent with previous work [48, 19, 3] that has shown the BUU model utilizing free-space scattering cross sections consistently under-predicting the balance energies of various systems. However, the positive results of Pak *et al.* was the correct reproduction of the differential effect in the balance energies – the difference between the balance energies for the two systems has the right sign and approximately the right magnitude. This can be verified by examining the left panel of Fig. 7.

In the central and right panels, we use the isospin dependent mean fields of Li (center, [37, 38], Eq. 2) and of Sobotka (right, [43], Eq. 4), combined with the in-medium corrections to the elementary two-nucleon scattering cross sections. For the central panel, we use $\alpha = -0.3$, and for the right panel, we use $\alpha = -0.2$, respectively.

Let us first compare the results of the central and the left panel. For both calculations, we use identical isospin-dependent mean fields. The only difference is the change in the scattering in-medium correction. We can make two rather obvious observations:

1. The theoretical calculations are much closer to the data when using the in-medium reduction of the scattering cross section – for all impact parameter intervals. This observation is consistent with previous results that did not look at isospin-dependent effects [48, 19, 3].

2. Even though the balance energies for the iron system are still slightly higher than for the nickel system, the magnitude of the splitting has been reduced and is now at least a factor of 4 smaller than what is observed in experiment.

The same effects can be observed when using a different isospin dependent mean field, compare the right panel of Fig. 7. Here, the differences between the theoretical balance energies are somewhat larger, but still a factor of at least 2 smaller than those found in the data.

We mentioned above that the observed sign of the difference in balance energies as a function of isospin is opposite to what one would expect from the behavior of the isospin-dependence of the mean field. This assertion can be verified in our numerical calculations by switching the isospin-dependent term in the mean field on and off. This comparison has been performed in Fig. 8. Here we use the same results as in the right panel of Fig. 7 (Sobotka mean field, $\alpha = -0.3$, represented by the diamonds in Fig. 8) and compare them to a mean field without isospin dependence, but with the same value of the nuclear

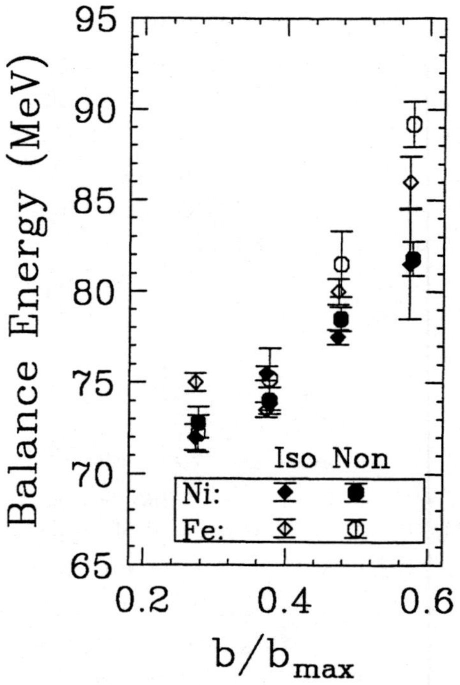

Figure 8: Comparison of the effects of the isospin-dependent mean field (diamonds) relative to the corresponding mean field without isospin dependence (circles).

compressibility (circles). In particular for the larger impact parameters, we can clearly see the expected effect: The isospin-dependent part of the mean field actually *reduces* the difference in the balance energies for the two systems. This is consistent with the results of our elementary considerations. The three-times larger isospin asymmetry in the iron system results in a stronger repulsion due to the isospin dependent mean field and is thus pushing the balance energy in the iron system more down than in the nickel system.

Concluding Remarks

We see an improvement in the performance of the BUU model's prediction of balance energies as a function of impact parameter for collisions of $^{58}Fe + {}^{58}Fe$ and $^{58}Ni + {}^{58}Ni$ by including both an asymmetry energy term in the mean field and in-medium reduction of the nucleon cross section. We observe similar performance among the different formulations used for the mean field: One formulation recently used by Bao-An Li [37, 38] and one due to Sobotka [43]. However, the mean field of Bao-An Li, Equation 2, requires $\alpha = -0.3$ in Equation 11, whereas the Sobotka formulation, Equation 4, requires $\alpha = -0.2$ for substantial improvement. Efforts to distinguish between these two mean fields will likely come from heavy ion calculations in conjunction with experiments near the drip lines. In the systems studied in the experiments by Pak *et al*, the range in isospin asymmetry was not sufficiently large to constrain our parameter space any further.

We have shown the different contributions of the Coulomb force, the NN-scattering, and the isospin-dependent mean field on the isospin dependence of the balance energy. Both our elementary considerations and our numerical results support the conclusion that the sign and the bulk of the isospin-difference in the balance energies is caused by the difference in the Coulomb interaction and the isospin dependence in the effective nucleon-nucleon two-body scattering.

While we can understand the absolute magnitude of the balance energies and the sign of the isospin-dependent difference between the two systems, our theoretical calculations do not yield the correct absolute magnitude of the difference. One can speculate on the origin of this important disagreement. A lack in our understanding of the isospin dependence of the nuclear mean field or the isospin-dependent in-medium modification of the two-body scattering cross section are the leading candidates. However, there is strong reason to suggest that the higher beam intensities of the radioactive beam facilities currently in planning or under construction will allow us to make progress in our understanding of this problem by enabling us to explore a larger isospin asymmetry in heavy ion reactions.

ACKNOWLEDGMENT

This work was supported in part by the National Science Foundation under grants PHY-9605207 and PHY-95-28844.

References

[1] D. Krofcheck, W. Bauer, G.M. Crawley, C. Djalali, S. Howden, C.A. Ogilvie, A. Vander Molen, G.D. Westfall, W.K. Wilson, R.S. Trickle, C. Gale, Phys. Rev. Lett. **63**, 2028 (1989).

[2] C.A. Ogilvie, D.A. Cebra, J. Clayton, P. Danielewicz, S. Howden, J. Karn, A. Nadasen, A. Vander Molen, G.D. Westfall, W.K. Wilson, and J.S. Winfield, Phys. Rev. C **40**, 2592 (1989).

[3] G.D. Westfall, W. Bauer, D. Craig, M. Cronqvist, E. Gualtieri, S. Hannuschke, D. Klakow, T. Li, T. Reposeur, A.M. Vander Molen, W.K. Wilson, J.S. Winfield, J. Yee, S.J. Yennello, R. Lacey, A. Elmaani, J. Lauret, A. Nadasen, E. Norbeck, Phys. Rev. Lett. **71**, 1986 (1993).

[4] P. Danielewicz and G. Odyniec, Phys. Lett. **157B**, 146 (1985).

[5] J.J. Molitoris and H. Stöcker, Phys. Lett. **162B**, 47 (1985).

[6] S. Soff, S.A. Bass, Ch. Hartnack, H. Stöcker, and W. Greiner, Phys. Rev. C **51**, 3320 (1995).

[7] G.F. Bertsch, H. Kruse and S. Das Gupta, Phys. Rev. C **29**, 673 (1984).

[8] H. Kruse, B. Jacak, J.J. Molitoris, G.D. Westfall, and H. Stöcker, Phys. Rev. C **31**, 1770 (1985).

[9] J. Aichelin and G.F. Bertsch Phys. Rev. C **31**, 1730 (1985).

[10] W. Bauer, G.F. Bertsch, W. Cassing, and U. Mosel, Phys. Rev. C **34**, 2127 (1986).

[11] H. Stöcker and W. Greiner, Phys. Rep. **137**, 277 (1986).

[12] G.F. Bertsch and S. Das Gupta, Phys. Rep. **160**, 189 (1988).

[13] G.F. Bertsch, W.G. Lynch and M.B. Tsang, Phys. Lett. **189B**, 384 (1987).

[14] C. Gale, G.M. Welke, M. Prakash, S.J. Lee, and S. Das Gupta, Phys. Rev. C **41**, 1545 (1990).

[15] G.D. Westfall, C.A. Ogilvie, D.A. Cebra, W.K. Wilson, A.M. Vander Molen, W. Bauer, J.S. Winfield, D. Krofcheck, J. Karn, S. Howden, T. Li, R. Lacey, K. Tyson and M. Cronqvist, Nucl. Phys. **A519**, 141 (1990).

[16] Particle Data Group, Phys. Lett. **B 126**, 1 (1988).

[17] T.H.R. Skyrme, Nucl. Phys. **9**, 615 (1959).

[18] K. A. Brückner, Phys. Rev. **97**, 1353 (1955).

[19] D. Klakow, G. Welke, and W. Bauer, Phys. Rev. C **48**, 1982 (1993).

[20] T. Alm, G. Röpke, W. Bauer, F. Daffin and M. Schmidt, Nucl. Phys. **A587**, 815 (1995).

[21] H. Müller and B.D. Serot, Phys. Rev. C, **52**, 2072 (1995).

[22] W. Bauer, U. Post, D. Dean, and U. Mosel, in *Proceedings of the 7th High Energy Heavy Ion Study, Darmstadt, Germany*, R. Bock, H.H. Gutbrod, R. Stock, eds., Report GSI-85-10, ISSN: 0171-4546, p. 701 (1984); W. Bauer, U. Post, D. Dean, and U. Mosel, Phys. Lett. **B150**, 53 (1985); W. Bauer, U. Post, D. Dean, and U. Mosel, Nucl. Phys. **A452**, 699 (1986); W. Bauer, Phys. Rev. C **38**, 1297 (1988).

[23] G. Kortemeyer, W. Bauer, and G.J. Kunde, Phys. Rev. C **55**, 2730 (1997).

[24] W. Bauer, in *Proceedings of the Workshop on Nuclear Matter in Different Phases and Transitions, Les Houches, France, March 1998*, ed.: J.P. Blaizot, X. Campi, and M. Ploszajczak (Kluwer Academic Publishers, Dordrecht, 1999), p. 203.

[25] E. Baron, J. Cooperstein, and S. Kahana, Phys. Rev. Lett. **55**, 126 (1985).

[26] James M. Lattimer, C. J. Pethick, Madappa Prakash, and PawełHaensel, Phys. Rev. Lett. **66**, 2701 (1991).

[27] B.-A. Li, C.M. Ko, and W. Bauer, Int. J. Mod. Phys. E **7**(2), 147-229 (1998).

[28] M. L. Miller, O. Bjarki, D. J. Magestro, R. Pak, N. T. B. Stone, M. B. Tonjes, A. M. Vander Molen, G. D. Westfall, and W. A. Friedman, Phys. Rev. Lett. **82**, 1399 (1999).

[29] S.J. Yennello, B. Young, J. Yee, J.A. Winger, J.S. Winfield, G.D. Westfall, A. Vander Molen, B.M. Sherrill, J. Shea, E. Norbeck, D.J. Morrissey, T. Li, E. Gualtieri, D. Craig, W. Benenson, and D. Bazin, Phys. Lett. **321**, 15 (1994).

[30] R. Pak, W. Benenson, O. Bjarki, J.A. Brown, S.A. Hannuschke, R.A. Lacey, Bao-An Li, A. Nadasen, E. Norbeck, P. Pogodin, D.E. Russ, M. Steiner, N.T.B. Stone, A.M. Vander Molen, G.D. Westfall, L.B. Yang, and S.J. Yennello, Phys. Rev. Lett. **78**, 1022 (1997).

[31] R. Pak, Bao-An Li, W. Benenson, O. Bjarki, J.A. Brown, S.A. Hannuschke, R.A. Lacey, D.J. Magestro, A. Nadasen, E. Norbeck, D.E. Russ, M. Steiner, N.T.B. Stone, A.M. Vander Molen, G.D. Westfall, L.B. Yang, and S.J. Yennello, Phys. Rev. Lett. **78**, 1026 (1997).

[32] H.H. Gutbrod, A.M. Poskanzer, and H.G. Ritter, Rep. Prog. Phys. **52**, 1267 (1989).

[33] R. Pak, W.J. Llope, D. Craig, E.E. Gualtieri, S.A. Hannuschke, R. A. Lacey, J. Lauret, A. C. Mignerey, D. E. Russ, N.T.B. Stone, A. M. Vander Molen, G. D. Westfall, and J. Yee, Phys. Rev. C **53**, R1469 (1996).

[34] J.P. Sullivan J. Peter, D. Cussol, G. Bizard, R. Brou, M. Louvel, J.P. Patry, R. Regimbart, J.C. Steckmeyer, B. Tamain, E. Crema, H. Doubre, K. Hagel, G.M. Jin, A. Peghaire, F. Saint-Laurent, Y. Cassagnou, R. Legrain, C. Lebrun, E. Rosato, R. MacGarth, S.C. Jeong, S.M. Lee, Y. Nagashima, T. Nakagawa, M. Ogihara, J. Kasagi, and T. Motobayashi Phys. Lett. B **249**, 8 (1990).

[35] R. Pak, O. Bjarki, S.A. Hannuschke, R.A. Lacey, J. Lauret, W.J. Llope, A. Nadasen, N.T.B. Stone, A.M. Vander Molen, and G.D. Westfall, Phys. Rev. C **54**, 2457 (1996).

[36] A. Buta, J.C. Angelique, G. Auger, G. Bizard, R. Brou, C. Cabot, Y. Cassagnou, E. Crema, D. Cussol, Y. El Masri, Ph. Eudes, M. Gonin, K. Hagel, Z.Y. He, A. Kerambrun, C. Lebrun, R. Legrain, J.P. Patry, A. Peghaire, J. Péter, R. Popescu, R. Regimbart, E. Rosato, F. Saint-Laurent, J.C. Steckmeyer, B. Tamain, E. Vient, and R. Wada, Nucl. Phys. **A584**, 397 (1995).

[37] B.-A. Li and S.J. Yennello, Phys. Rev. C **52**, R1746 (1995).

[38] B.-A. Li, Zhongzhou Ren, C.M. Ko, and S.J. Yennello, Phys. Rev. Lett. **76**, 4492 (1996).

[39] M.B. Tsang, G.F. Bertsch, W.G. Lynch and M. Tohyama, Phys. Rev. C **40**, 1685 (1989).

[40] H.M. Xu, Phys. Rev. Lett. **67**, 2769 (1991).

[41] P. Danielewicz and Q. Pan, Phys. Rev. C **46**, 2002 (1992).

[42] B.-A. Li, Phys. Rev. C **48**, 2415 (1993).

[43] L.G. Sobotka, Phys. Rev. C **50**, R1272 (1994).

[44] C. Hartnack, R.K. Puri, J. Aichelin, J. Konopka, S.A. Bass, H. Stoecker, and W. Greiner, Eur. Phys. J. **A1**, 151-169 (1998).

[45] W.D. Myers and W.J. Swiatecki, Phys. Rev. C **57**, 3020 (1998).

[46] H.-J. Schulze, A. Schnell, G. Röpke, and U. Lombardo, Phys. Rev. C **55**, 3006 (1997).

[47] A. Schnell, G. Röpke, U. Lombardo, and H.J. Schulze, Phys. Rev. C **57**, 806 (1998).

[48] C.A. Ogilvie, W. Bauer, D.A. Cebra, J. Clayton, S. Howden, J. Karn, A. Nadasen, A. Vander Molen. G.D. Westfall, W.K. Wilson, and J.S. Winfield, Phys. Rev. C **42**, R10 (1990).

Isospin Equilibration as a Probe of Nuclear Stopping Power

Bao-An Li *and Sherry J. Yennello †

Abstract

Isospin equilibration in heavy-ion collisions can be used to study the nature and degree of nuclear stopping in heavy-ion collisions. Using isospin tracer observables it has been found that only at beam energies below the Fermi energy, complete golbal isospin equilibrium can be obtained. Moreover, the isospin relaxation time is found to be shorter than the momentum relaxation time at these energies. A transition to isospin nonequilibration occurs at intermediate energies, indicating the onset of nuclear translucency. At intermediate and higher energies the dynamical instability sets in before either global isospin or thermal equilibrium is achieved. Transport model analyses indicate that the degree of isospin equilibration and thus the stopping power depends on both the in-medium nucleon-nucleon cross section and the isospin-dependence of the nuclear equation of state.

1 Nuclear Stopping Power

There has been considerable interest in studying the stopping power of nuclei from low to ultra-relativistic energies. The nuclear stopping power can be viewed as a measure of the degree to which the energy of the initial relative motion of two colliding nuclei is transformed into other degrees of freedom[1, 2, 3]. The degree of nuclear stopping power determines parameters, such as, the energy density and volume of the interaction region, which governs the reaction dynamics and the possibility of reaching conditions capable of forming new phases of nuclear matter. In heavy-ion collisions at intermediate energies, nuclear stopping power is determined by both the nuclear equation of state and the in-medium nucleon-nucleon cross sections[4, 5].

Furthermore, strong stopping is a necessary condition to reach global thermal equilibrium in heavy-ion collisions. Knowledge on the stopping power is also important for developing theoretical models to understand and predict the outcome of heavy-ion collisions at various energies. If thermal equilibrium is established a macroscopic treatment of the later stage in terms of temperature, volume and chemical potential then becomes possible although the early stage of heavy ion collisions still must be described by microscopic models. Also, the interpretation of nuclear multifragmentation in heavy ion collisions either as a dynamical or as a statistical process depends on whether golbal or local thermal equilibrium can be achieved in the collisions. Based on the assumption that thermal equilibrium is established in subsystems prior to fragment emission, statistical models have been quite

*Department of Chemistry and Physics, Arkansas State University P.O. Box 419, State University, AR 72467-0419, USA

†Cyclotron Institute and Department of Chemistry, Texas A&M University College Station, TX 77843-3366. USA

successful in describing heavy-ion reaction data[6, 7]. However, a critical examination of whether a global or partial thermal equilibrium can be reached in a model independent way is necessary.

In this chapter we review the method of using isospin observables as a probe of the nuclear stopping power in heavy-ion collisions, mainly at intermediate energies. We start by first discussing the importance of understanding the mechanism and degree of nuclear stopping in nuclear reactions. After a brief review of traditional methods of measuring the nuclear stopping power, the new technique of isospin tracing, i.e., using the isospin equilibration as a probe of nuclear stopping power is introduced. Several examples of applying this method to heavy-ion collisions from low to relativistic energies are then presented. Finally, a summary is given at the end.

2 Traditional Measurements of Nuclear Stopping Power

Usually, experiments determine the stopping power by measuring 1)final state nucleon rapidity distributions; 2)the energy remaining in forward-going baryons after the reaction; 3)transverse energy distributions or 4)quadrupole moments of momentum distribution of nucleons and fragments[5, 8, 9, 10]. All of these measurements can provide information about the amount of energy that is being transferred from initial longitudinal motion to other directions and particle production, they thus reflect the stopping power from different aspects. Among these methods, final state proton rapidity distributions have been most frequently used. The incoming projectile direction is taken as the z axis, the total energy of a particle is denoted E, the magnitude of its momentum p, and the projection of the momentum on the beam axis is p_z. The rapidity is then defined as

$$y = \frac{1}{2} ln \frac{E + p_z}{E - p_Z}. \tag{1}$$

One quantitative measure of the stopping power is the average rapidity loss[1, 8]

$$< \delta y > \equiv y_{beam} - < y_b >, \tag{2}$$

where y_{beam} is the incoming projectile rapidity and $< y_b >$ is the average nucleon rapidity after the collision. A more detailed measure of the stopping power is the shape of the net baryon rapidity distribution dn/dy. Complete stopping would lead to a rapidity distribution which peaks at the mid-rapidity, which has been observed in heavy reaction systems in a wide beam energy range. This mid-rapidity peaked rapidity distribution is a necessary, but not sufficient condition for global thermalization. Although thermal models can often describe these data successfully (see, e.g., ref. [11]), there has been little model independent confirmation that a thermal equilibrium has indeed being reached in the reactions either globally or locally. Moreover, in heavy reaction systems, such as Au+Au, a simple scaling law

$$y_{beam1} \cdot \frac{dn_1}{dy} = y_{beam2} \cdot \frac{dn_2}{dy} \tag{3}$$

has been observed in both experiments[9] and theoretical model simulations[12]at beam energies from 0.25 GeV/nucleon to 160 GeV/nucleon. Shown in Fig. 1 are the scaled rapidity distributions $dn/d(y/y_{beam})$ versus y/y_{beam}. The crosses represent the experimental systematics from analyzing central Au+Au reactions at beam energies form 0.25 to

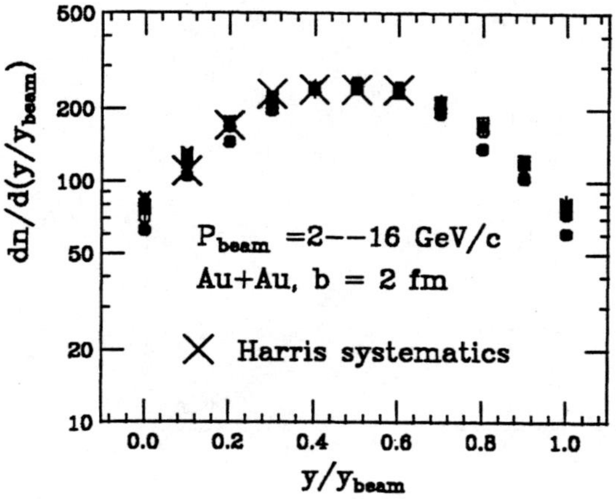

Figure 1: The scaled proton rapidity distribution from the reaction of Au+Au at an impact parameter of 2 fm and beam momentum of 2 to 16 GeV/c per nucleon. The large crosses are from Ref. [9].

160 GeV/nucleon[9]; while the calculations are based on A Relativistic Transport (ART) model[13]. It is seen that the experimentally observed scaling can be well reproduced by the model calculations. The scaling indicates a large amount of stopping in the collisions over the entire energy range. However, the physical implications of the observed scaling are not fully understood from studying the rapidity distributions along. In particulr, how strong is the stopping? Does the single-peak scaled rapidity distribution indicates a complete stopping in all of the reactions? To answer these questions one needs to explore other approachs. Results from a recent study based on a nuclear transport model[14] illustrate further the shortcomings of studying the stopping power using the rapidity distributions. Fig. 2 shows the rapidity distribution in the final state of a head-on Au+Au reaction at a beam energy of 150 MeV/nucleon. The long (short) dashed line is the contribution from the projectile (target) nucleus. The sum of these two components is given by the solid line. Although the total rapidity distribution is close to a thermal distribution which can be approximated by the squares generated by putting all nucleons in a box with periodic boundary conditions[14], the rapidity distributions of target and projectile nucleons are still clearly separated. The obvious relative collective motion of the projectile and target nucleons indicates that there is no complete stopping in the reaction. Of course, even in head-on collisions there could still be some memory effect due to nucleons close to the surfaces of the colliding nuclei. To avoid this problem, the stopping power of nucleons in a tube with a radius of 1 fm along the beam direction has been studied in the same model[14].

The ratio of target-nucleons over projectile-nucleons in the tube is shown in Fig. 3 as a function of rapidity for head-on Au+Au reactions at beam energies from 150 MeV/nucleon to 2 GeV/nucleon. Only at the mid-rapidity are there equal numbers of nucleons from the projectile and target as one expects from symmetry and geometry. At all other rapidities

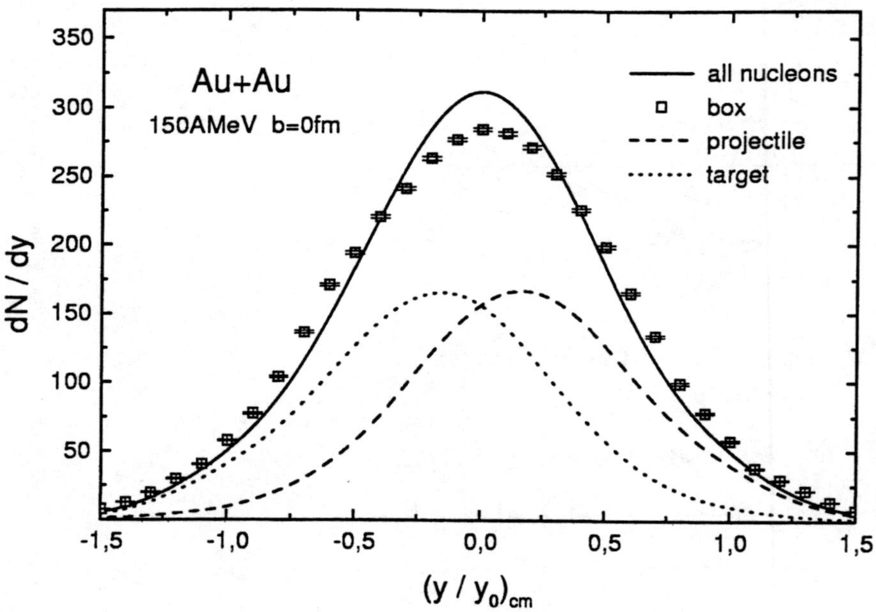

Figure 2: Baryon rapidity distribution in a central Au+Au reaction at a beam energy of 150 MeV/nucleon. The long (short) dashed line is the contribution from the projectile (target) nucleus. The squares are the results from simulating the thermolyzation by put all nucleons in a box, taken from ref. [14].)

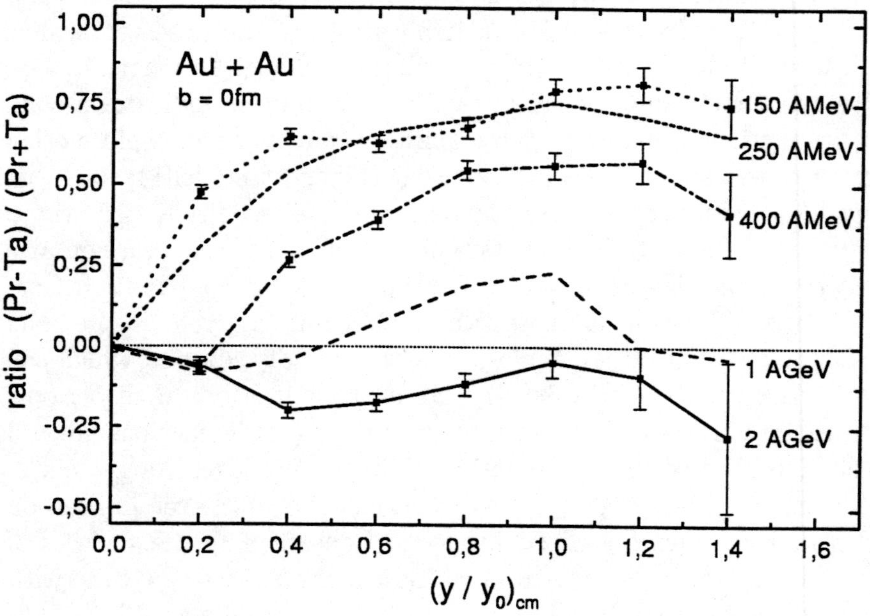

Figure 3: Ratio of target to projectile nucleons as a function of rapidity in a tube with 1 fm radius in head-on Au+Au reactions, taken from ref. [14].)

there are unequal mixing of nucleons from the target and projectile. These finding coupled with the relative motion seen in the separated rapidity distributions of projectile and target nucleons in Fig. 2 indicate that the complete stopping and thermalization are not guaranteed even if the final baryon rapidity distribution has a single peak and can be well described by thermal models.

3 The isospin tracing method

If we were able to tag the nucleons from projectile and target in experiments, the task of measuring the stopping power in a model independent way would be much easier. By using nuclei with different N/Z ratios, such a tag can be provided as final nucleons can be attributed on average to either the projectile or target. We stress that the indistinguishable nature of nucleons don't allow one to separate unambiguously nucleons originally from the target or projectile. The isospin tracing method only works on the average sense. This idea is illustrated in Fig. 4 where the neutron/proton ratio $\delta \equiv N/Z$ is examined in both coordinate and momentum space. The pictures on the left show the nucleon distributions in coordinate space, while that on the right are the rapidity distributions for the three scenarios of pile-up and bounce-back, stopping and mixing (isospin equilibrium), and translucency (partial transparency). It is seen that for a projectile and a target with very different N/Z ratios, a comparison of the rapidity distribution of N/Z before (shown in top picture on the right) and after the collision can give direct information about the degree of stopping between the target and projectile [15, 16]. This method thus allows one to study whether there is a transition from full stopping to translucency (partial transparency) as the beam energy increases. Also, if isospin equilibrium can be reached in the collisions, it is then possible to determine its time scale relative to that for thermal equilibrium. Moreover, one can study the dependence of isospin equilibrium on the nuclear equation of state and the in-medium nucleon-nucleon cross sections. We note that this method is restricted to beam energies that are below the pion production threshold; so the total nucleon charge is conserved. At higher energies corrections have to be introduced to account for the shift of the total baryonic isospin due to charged pion production[14].

4 Isospin equilibrium at low energies

The isospin degree of freedom has been found to reach equilibrium faster than all other degrees of freedom in deep inelastic heavy ion collisions[17, 18, 19, 20]. Recent experimental studies [21, 22, 23] of the isotopic composition of intermediate mass fragments (IMF) and their angular distributions in heavy ion collisions have also shown that only at low energies ($\approx$ 30 MeV/nucleon) is isospin equilibrium reached before fragment emission.

Effects of the nuclear symmetry potential on isospin equilibrium at low beam energies have been studied within a Landu-Vlasov transport model by Farine et al[20]. To include the effects of isospin on nuclear dynamics, they added to the Zamick potential energy density, which has been widely used in transport models, an asymmetric term of the form

$$w_{\rm sym} = c\rho[(\rho_n - \rho_p)/\rho_0]^2, \tag{4}$$

where the coefficient $c \approx e_{\rm sym}(\rho_0) - \frac{1}{3}E_F^0$ is the symmetry energy at normal nuclear matter density due to nuclear interactions. By varying the value of c, it has been found that a stronger symmetry potential enhances the degree of isospin equilibrium. Shown in Table

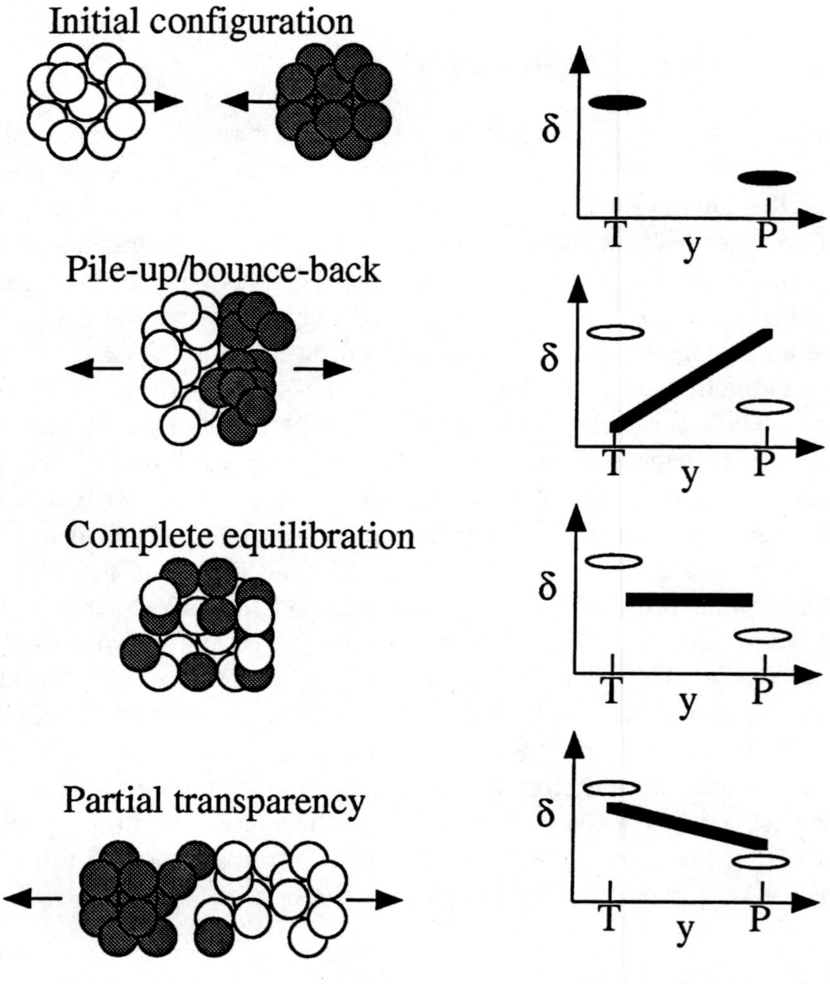

Figure 4: Illustration of using the rapidity distribution of neutron/proton ratio as a probe of nuclear stopping power. Taken from Ref. [16].

Table 1: Values of N-Z for the quasitarget(QT) and quasiprojectile (QP)

Reaction/N-Z		c=0	c=20	c=28
^{56}Ca $+^{40}$ Ca	QT	3.74	4.64	5.02
	QP	8.76	5.82	4.92
^{48}Ca $+^{48}$ Ca	QT	6.52	5.78	5.52
	QP	6.96	5.84	5.12

1 are the values of $N-Z$ in the quasitarget (QT) and quasiprojectile (QP) formed in the reaction of $^{56}\mathrm{Ca}+^{40}\mathrm{Ca}$ at an impact parameter of 7 fm and a beam energy of 15 MeV/nucleon. For comparisons, results from reactions of a symmetric system $^{48}\mathrm{Ca}+^{48}\mathrm{Ca}$ are also listed in the table. In both reactions the total mass and charge numbers are the same. The difference in the values of $N-Z$ for the symmetric system is completely due to the numerical fluctuations of the calculations. In the asymmetric reaction the initial value of $N-Z$ is 16 and 0 for the projectile and target, respectively. It is seen that significant mixing is achieved even in the case of no symmetry potential ($c=0$). However, without the symmetry potential isospin equilibrium cannot be reached, and the degree of isospin mixing increases with the strength of the symmetry potential.

5 Transition from stopping to translucency at intermediate energies

At intermediate energies the isospin tracing method has revealed a transition from stopping to translucency. In several experiments[21, 22, 23, 25] both isotopic and isobaric ratios of intermediate mass fragments from central collisions of ^{40}Cl, ^{40}Ar and ^{40}Ca with ^{58}Fe and ^{58}Ni at E_{beam}/A=25, 35, 45 and 53 MeV were studied at the National Superconducting Cyclotron Laboratory and Texas A&M University. A consistent picture appears from the different analyses of these data. For example, it was shown that at E_{beam}/A=25 and 33 MeV the isotopic ratios ^{9}Be/^{7}Be, ^{11}B/^{10}B and ^{13}C/^{12}C increase linearly with increasing $(N/Z)_{\mathrm{cs}}$ ratio of the combined target and projectile system, but are independent of the N/Z ratio of the target or projectile. Shown in the left window of Fig. 5 are typical results of reactions at 33 MeV/nucleon, which indicate that the isospin is equilibrated in the composite system formed in these reactions before the emission of fragments.

The most striking and unexpected feature was observed from the isbaric ratios in central collisions at $E_{\mathrm{beam}}/A=45$ and 53 MeV. A typical result at E_{beam}/A =45 MeV is shown in the right window of Fig. 5. It is seen that the isotopic ratios depend on the N/Z ratio of the target and projectile in reactions with target-projectile combinations having the same $(N/Z)_{CS}$ ratio, such as ^{40}Ca+^{58}Fe and ^{40}Ar+^{58}Ni. In ref. [23], similar results are also shown

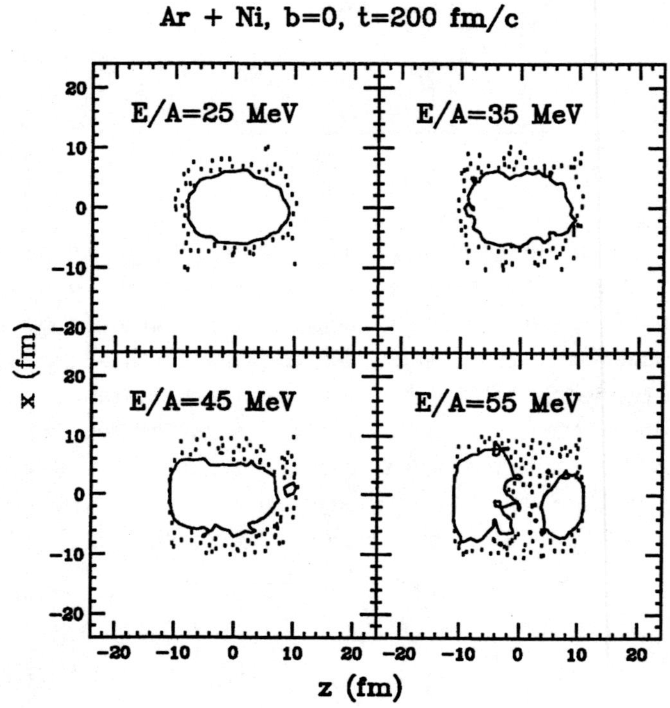

Figure 5: Isobaric ratios from central collisions plotted as a function of N/Z ratio of the combined target and projectile system at E_{beam}/A=35 and 45 MeV. Taken from Ref. [23].

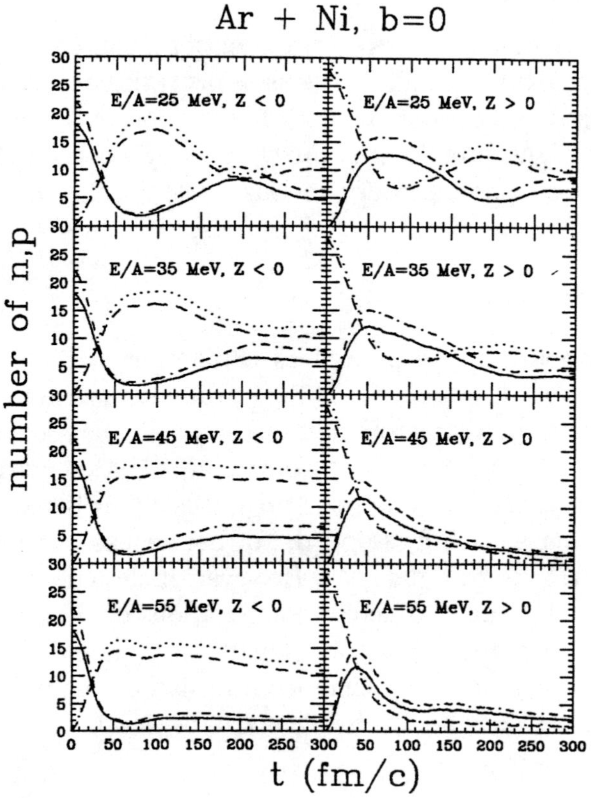

Figure 6: Isobaric ratios plotted as a function of $(N/Z)_{fitted}$ for E/A=45MeV. (a) 7Li/7Be, (b) 10Be/10B, (c) 11B/11C.

for other isotope ratios. Moreover, data at very forward and backward angles show that the isotope ratios do not simply depend on $(N/Z)_{\text{cs}}$. Instead, light fragments at backward angles are seen to have a much stronger dependence on $(N/Z)_{\text{target}}$, while at forward angles they depend more on $(N/Z)_{\text{projectile}}$. These results demonstrate that the isospin degree of freedom in reactions at $E_{\text{beam}}/A = 45$ and 53 MeV is not globally equilibrated prior to the time when fragments are emitted. Therefore, a transition from isospin equilibration to non-equilibration is observed, indicating a change from complete stopping to translucency as the beam energy increases from below to above the Fermi energy.

The above observation has profound implications on the reaction mechanism leading to multifragmentation. It not only establishes the relative time scale for multifragmentation in these reactions but also indicates that the assumption of global isospin equilibrium taken for granted in various statistical models for nuclear multifragmentation at intermediate energies is not valid. Indeed, a statistical model study was made and failed to show any entrance channel effect [21]. Calculations using an intranuclear cascade code ISABEL [26] show that the N/Z ratio of the residue is very close to that of the initial combined system [21] and thus also fails to reproduce those features observed at $E_{\text{beam}}/A = 45$ and 53 MeV. Although the exact origin of this failure is not clear, one expects that the reaction dynamics at these relatively low energies cannot be described by the nucleon-nucleon cascade alone, and should include also nuclear mean-field potential.

Although a global isospin equilibrium is not reached at beam energies above the Fermi energy, it is still possible that the observed fragments are emitted from a statistically equilibrated subsystem. Under this scenario, it should be possible to identify a source composed of the target and some fraction of the projectile that allows the isotopic ratios of the emitted fragments to be plotted on a single line. Indications of possible emission from an equilibrated subsystem have been seen in previous data[24] (see the chapter by Pochodzalla and Trautmann for detailed discussions). The isobaric ratios in refs. [21, 22, 23] have been studied as a function of $(N/Z)_{fitted}$ defined as

$$(N/Z)_{fitted} = (N_{target} + fN_{projectile})/(Z_{target} + fZ_{projectile}). \quad (5)$$

Fig. 6 shows the measured inclusive isobaric ratios plotted as a function of $(N/Z)_{fitted}$. From a global fit to all three isobaric pairs it is found that f = 0.5 produces a reasonable fit to all of the data. This would imply that approximately half of the projectile escapes the interaction zone before equilibration is obtained and the fragments are then emitted from a subsystem composed of the target and some fraction of the projectile. This is consistent with a transition point where the projectile is not fully stopped, but before a second forward source is formed as is needed to explain the 53 MeV data[21]. However, it should be noticed that the above fitting procedure is not unique since different combinations of the projectile and target can give the same fitting variable $(N/Z)_{fitted}$. Nevertheless, it is clear that fragments produced at above the Fermi energies may come from an isospin equilibriated subsystem. Thermodynamical concepts can thus be applied to the subsystem.

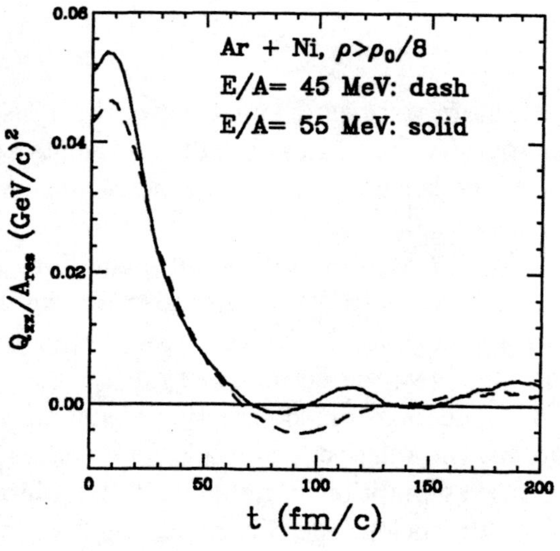

Figure 7: Density contours in the reaction plane at t= 200 fm/c in head-on collisions of Ar+Ni at $E_{beam}/A = 25$, 35, 45 and 55 MeV. Taken from Ref. [25].

The experimental observation discussed above can be explained by the Isospin-dependent Boltzmann-Uehling-Uhlenbeck (IBUU) model[25, 27, 28]. Calculations based on this model have been performed over a range of impact parameters. For peripheral collisions it shows a memory of the initial target and projectile, which is, however, gradually lost as the collisions become more central. A calculation at b=0 thus gives the most interesting test of any non-equilibrium effect. In Fig. 7, we show the energy dependence of the density contours in the reaction plane at t= 200 fm/c in head-on collisions of Ar+Ni at $E_{\text{beam}}/A =$ 25, 35, 45 and 55 MeV. The solid contours with $\rho = \rho_0/8$ correspond essentially to bound composite systems or heavy residues formed in the reactions. The dotted contours with $\rho = 0.05\ \rho_0$ are free nucleons emitted mainly before thermal equilibrium is reached. The most interesting feature in Fig. 7 is the formation of two heavy residues in the reaction at $E_{\text{beam}}/A = 55$ MeV. For reactions at $E_{\text{beam}}/A = 25$, 35 and 45 MeV, these heavy residues do not break up even after 300 fm/c. To see if the heavy residues observed above are in

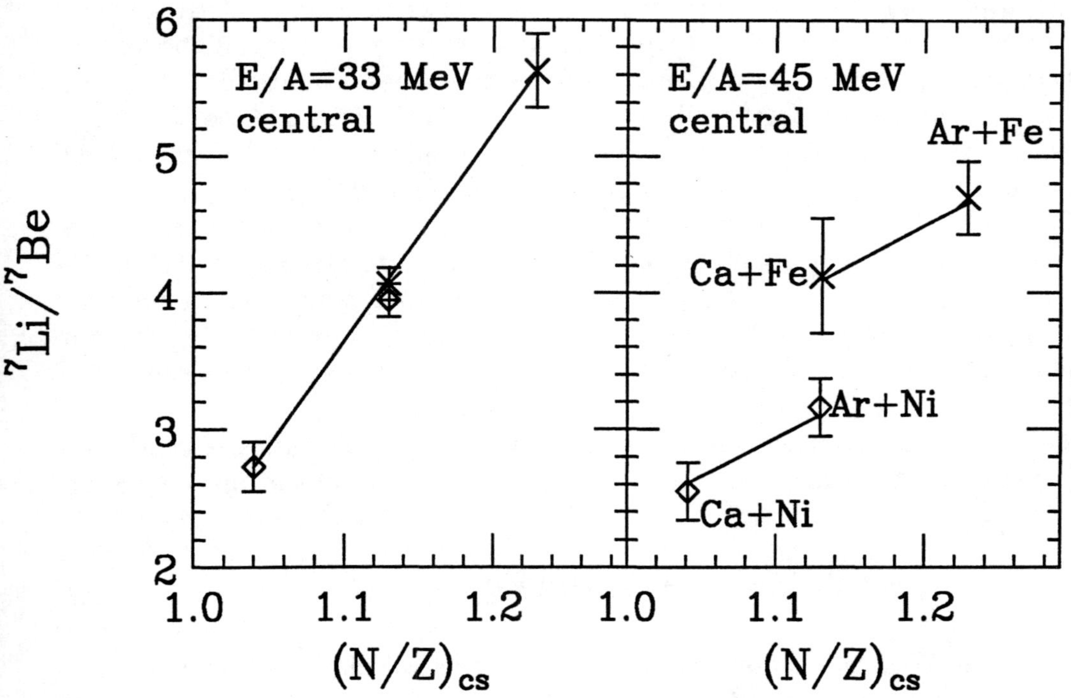

Figure 8: The neutron and proton numbers in the residues on the left ($Z < 0$) and right ($Z \geq 0$) side of the origin. Solid lines are the proton number from the projectile, while dot-dashed lines are the neutron number from the projectile which moves towards the right. Dashed lines are the proton number from the target, while dotted lines are the neutron number from the target which moves towards the left. Taken from Ref. [25].

isospin equilibrium, we show in Fig. 8 the neutron and proton numbers in the residues on the left ($Z < 0$) and right ($Z \geq 0$) side of the origin in head-on reactions of Ar+Ni at

E_{beam}/A =25, 35, 45 and 55 MeV. The solid and dot-dashed lines are, respectively, the proton and neutron numbers from the projectile, which is incident from the left. The dashed and dotted lines are, respectively, the proton and neutron numbers from the target, which moves from the right. It is seen that the neutron and proton numbers on both sides not only decrease but also fluctuate. The decreases is mainly due to nucleon-nucleon collisions and particle emissions, while the fluctuation is due to both the restoring force from the mean-field potential and nucleon-nucleon collisions. At $E_{\text{beam}}/A = 25$ MeV the neutron and proton numbers on the two sides become very close to each other and the amplitude of oscillation is rather small by the time of 300 fm/c. This indicates that the heavy residue is very close to isospin equilibrium, i.e., the proton and neutron distributions are independent of space-time. The damping of the oscillation is faster at $E_{\text{beam}}/A = 35$ MeV so the particle distribution also reaches isospin equilibrium sooner. The isotopic composition of fragments emitted from the residues in these low energy reactions after about 300 fm/c would therefore essentially reflect the $(N/Z)_{\text{cs}}$ ratio of the initial composite system, and there is little forward-backward asymmetry. These features are in good agreement with those found in the data of E_{beam}/A =25 and 35 MeV [21, 22, 23].

At higher energies, such as $E_{\text{beam}}/A = 45$ and 55 MeV, there is little oscillation in the overlapping region between the target and projectile. This is mainly because the incoming momenta of projectile-nucleons and target-nucleons are very large so the mean-field potential cannot reverse the directions of motion of many nucleons during a relatively short reaction time. As a result, there exists a large isospin asymmetry or non-equilibration at these two energies. In particular, on the left side of the origin the N/Z ratio of the residue is more affected by that of the target while on the right side it is more affected by that of the projectile. However, the N/Z ratios on both sides are not simply those of the target and projectile but a combination of the two depending on the complicated reaction dynamics. In the case of $E_{\text{beam}}/A = 55$ MeV, at the time of about 200 fm/c the heavy residue has broken up into two pieces with some longitudinal collectivity. The forward moving residue has an excitation energy of about 8.6 MeV/nucleon, while the backward moving residue has an excitation energy of about 6.8 MeV/nucleon. Both residues are found to be in approximate thermal equilibrium in their own center of mass frame but not in thermal equilibrium with each other[25].

The signature of nuclear translucency from studying the isotopic ratios is consistent with that from studying the kinematic observables, such as the scaled quadrupole moment Q_{ZZ}/A_{res} defined as

$$Q_{zz}/A_{\text{res}}(t) = 1/A_{\text{res}} \int \frac{d\vec{r}d\vec{p}}{(2\pi)^3}(2p_z^2 - p_x^2 - p_y^2) f(\vec{r},\vec{p},t), \tag{6}$$

where $f(\vec{r},\vec{p},t)$ is the Wigner function from the IBUU model calculations and A_{res} is the mass of the residue. A zero value of Q_{zz} is a necessary but not a sufficient condition for thermal equilibrium, and a positive value of Q_{ZZ} signals incomplete stopping. The quadrupole moments of heavy residues formed in the reactions at $E_{beam}/A = 25$ and 35 MeV are about zero by the time of about 300 fm/c, indicating the establishment of thermal equilibrium. Since the heavy residue formed in the reaction at $E_{\text{beam}}/A = 55$ MeV has already broken up into two pieces at about 200 fm/c, it is interesting to examine more closely the evolution of Q_{zz}/A_{res} in this reaction, which is shown in Fig. 9 for particles at local densities larger than $\rho_0/8$ for the reactions at E/A= 45 and 55 MeV. It is seen that Q_{zz} has a small positive value when the system breaks up, indicating thus some translucency in the collisions. This finding is also in good agreement with the kinematic signature of

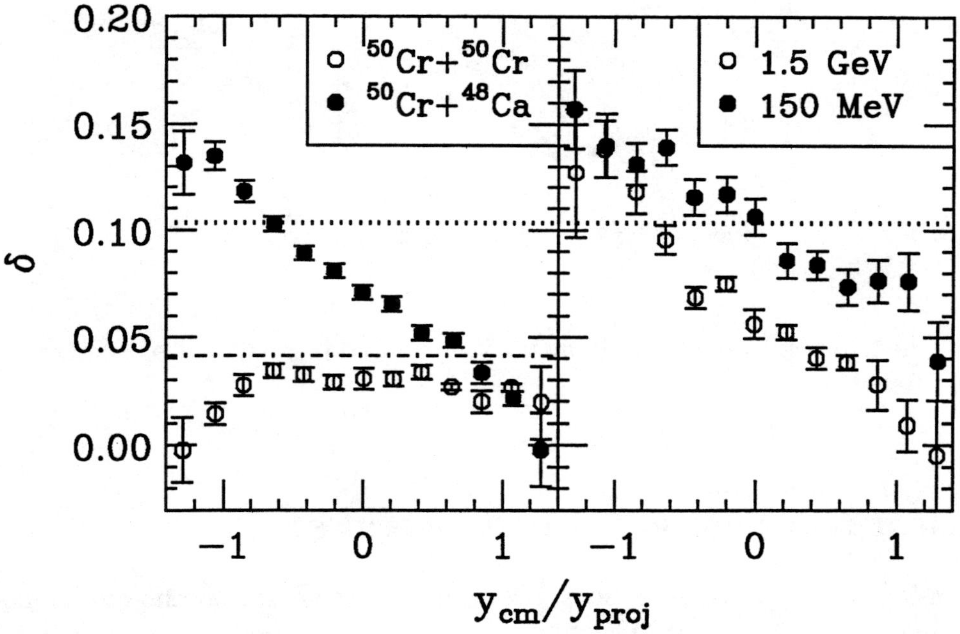

Figure 9: The scaled quadrupole moment of heavy residues formed in head-on collisions of Ar+Ni at $E_{\text{beam}}/A = 45$ and 55 MeV. Taken from Ref. [25].

translucency found recently at Ganil for the same reaction system at similar beam energies using the INDRA detector [29]. In recent calculations using the Antisymmetrized Molecular Dynamics (AMD) translucency was also found at 35 MeV/nucleon for head-on collisions of $^{40}Ca + ^{40}Ca$ [30]. The signatures of translucency from both the isotopic ratios and the quadrupole moments are also in good agreement with the systematics of nuclear stopping power [5]. Moreover, two more recent studies based on the Isospin-dependent Quantum Molecular Dynamics (IQMD) model[31, 32] and BUU model[32] have confirmed the transition from complete stopping to translucency in intermediate energy heavy-ion collisions.

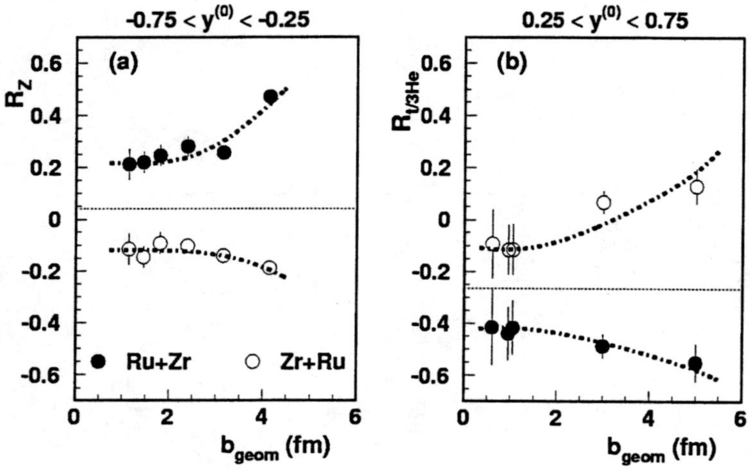

Figure 10: The neutron to proton asymmetry versus rapidity predicted by the QMD model. Taken from Ref. [15].

6 Nuclear translucency at high energies

From discussions in the previous sections, one expects nuclear translucency to be more important at higher energies. This indeed has been observed in both model calculations[14, 15, 25] and experiments[33]. In the left window of Fig. 10, the neutron to proton ratios in central collisions of $^{50}Cr + ^{48}Ca$ and $^{50}Cr + ^{50}Cr$ at $E_{beam}/A = 1.0$ GeV are compared. For the symmetric system there is a significant stopping as seen from the central-rapidity plateau. For the asymmetric system, there is clearly a strong translucency. In the right window two calculations for the asymmetric system at two different beam energies are shown, and it is seen that even at 150 MeV/nucleon the asymmetric system shows a strong translucency. Within the IQMD model it was further shown that the signature for translucency seen in the N/Z ratio is not altered by cluster formations. On the other hand, the stopping power is affected significantly by the magnitude of the in-medium nucleon-nucleon cross sections. As expected, increasing the in-medium nucleon-nucleon cross section by a factor of 5 results in a transition from translucency to full stopping in the asymmetric system[15]. More recent studies based on the BUU transport model found that a full stopping is practically never achieved even in central collisions of Au+Au in the energy range of 150 to 2000 MeV/nucleon. Further it was shown that the incompressibility of the nuclear EOS modifies the isospin equilibration signal somewhat at the low energy end, whereas in-medium modifications of the nucleon-nucleon cross sections dominantly influence the signal at higher bombarding energies. These results reflect the energy dependence of the respective roles of the nuclear mean field and nucleon-nucleon cross sections in the reaction dynamics. These findings are also in agreement with those found based on other transport model calculations[15, 20, 25, 31, 32].

Some very interesting results were recently obtained by the FOPI collaboration at GSI[33]. Four different combinations of ^{96}Ru and ^{96}Zr, both as the projectile and target, were investigated at the same bombarding energy of 400 MeV/nucleon. The degree of isospin mixing between target and projectile nucleons was mapped across a large portion of the phase space using two different isospin-tracer observables, the number of measured protons and the $^3H/^3He$ yield ratio. In the first method, the relative abundance of the projectile-target nucleons has been adopted:

$$R_Z = \frac{2Z - Z^{Zr} - Z^{Ru}}{Z^{Zr} - Z^{Ru}} \tag{7}$$

where Z is the final number of protons observed in a given cell of the momentum space; Z^{Zr} and Z^{Ru} are the values of Z in $Zr + Zr$ and $Ru + Ru$ reactions, respectively. Thus in the $Zr + Zr$ and $Ru + Ru$ reactions R_z takes the value of $+1$ and -1, respectively. In the case of a mixed reaction, $Ru + Zr$ or $Zr + Ru$, the measured proton yield Z takes values intermediate between Z^{Ru} and Z^{Zr}. If Z is close to Z^{Ru} in a $Ru + Zr$ reaction, it then indicates that the cell is mainly populated by nucleons from the Ru projectile. Similarly, an isospin tracer $R_{^3H/^3He}$ using the $^3H/^3He$ yield ratio can be defined. Results of the GSI measurements are shown as a function of centrality in Fig. 11. Here the pseudo-proton yield includes both free protons and protons still bound in deuterons. The two ratios R_Z and $R_{^3H/^3He}$ are measured in the backward and forward hemisphere, respectively. Except for an off-set, both figures display the same trend, i.e., the global isospin equilibrium is not reached even in the most central collisions.

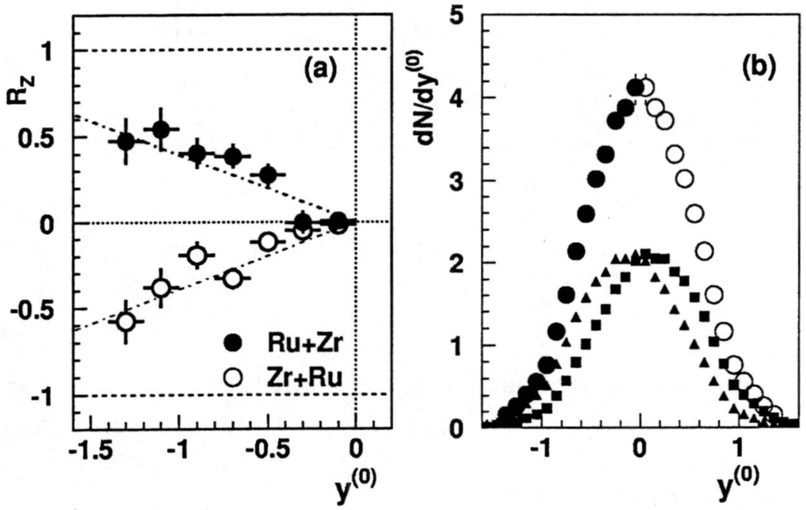

Figure 11: Two isospin tracer observables as a function of impact parameter in mixed reactions between *Ru* and *Zr* nuclei. Taken from Ref. [33].

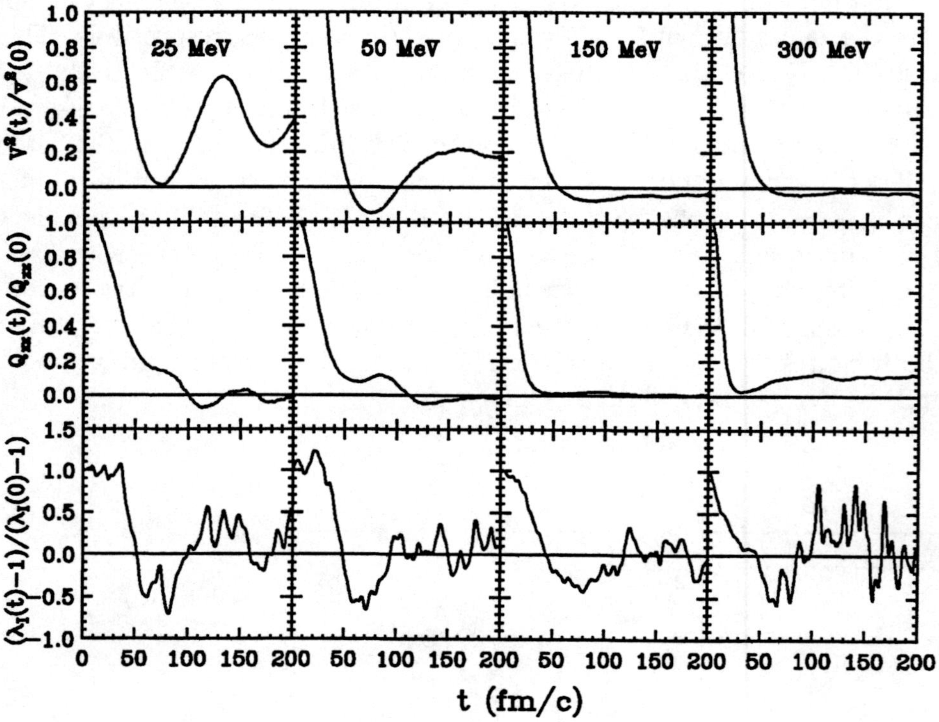

Figure 12: Panel (a) shows the isospin tracer observable R_Z as a function of rapidity in the mixed reactions between *Ru* and *Zr* nuclei. Panel (b) displays the rapidity distributions of total pseudo-protons (filled and open circles), deconvoluted 'projectile' (squares) and 'target' (triangles) components, respectively. Taken from Ref. [33].

Furthermore, the rapidity dependence of the isospin mixing has also been measured. Shown on the left of Fig. 6 are the isospin tracer R_z for the most central reactions between Ru and Zr nuclei as a function of the normalized center-of-mass rapidity $y^{(0)}$. It is seen that only at the mid-rapidity the isospin tracer R_Z is about zero. The measured variation of R_Z can be described by a linear dependence on the normalized rapidity in the form of $R_z \approx \pm 0.393 y^{(0)}$. This relation was used to deconvolute the total rapidity distribution in central $Ru+Ru$ reaction into separated distributions for the projectile- and target-nucleons. Shown on the right of Fig. 6 are the total (filled and open circles), projectile (squares) and target (triangles) components, respectively. Although the total rapidity distribution peaks at the mid-rapidity, the 'projectile' and 'target' rapidity distributions are clearly shifted relative to each other, demonstrating that a memory of the initial target/projectile relative motion survives throughout the central collision. These results are in agreement with transport model calculations as discussed earlier.

7 Isospin relaxation time in heavy-ion collisions

In past few years, extensive theoretical efforts have been devoted to investigate the rate of thermalization of nuclear matter in intermediate energy heavy ion collisions. Using the Uehling-Uhlenbeck equation, momentum relaxation both in collisions of semi-infinite nuclear matter and in infinite nuclear matter have been studied [35, 37, 38]. Collisions of finite nuclei have also been studied using the BUU equation [39, 40, 41, 42]. On the other hand, only limited studies have been carried out for understanding the isospin relaxation time in heavy ion collisions[36]. In many statistical and dynamical models it is assumed that isospin equilibrium is reached either instantaneously or as fast as momentum equilibrium. Our above discussions show that this assumption is only approximately true in deep inelastic heavy ion collisions at low energies, whereas global isospin equilibrium is never achieved at higher energies. It is thus interesting to compare the relaxation time of isospin with other characteristic times of the reaction. In the following we compare the relaxation times of isospin and momentum using the IBUU model. In this model study, a heavy residue can be identified as a collection of nucleons with densities higher than 1/10 of normal nuclear matter density. To characterize the degree of chemical equilibrium, we introduce the following quantity

$$\lambda_I(t) \equiv \frac{(n/p)_{y>0}}{(n/p)_{y<0}}, \tag{8}$$

where $(n/p)_{y>0}$ and $(n/p)_{y<0}$ are, respectively, the neutron to proton ratio for positive and negative rapidity nucleons in the rest frame of the residue. If chemical equilibrium is established this quantity then has a value of one. We further define the isospin relaxation time τ_I as the time when the quantity $(\lambda_I(t)-1)/(\lambda_I(0)-1)$ is 0.01, i.e., it is one percent from its equilibrium value. This time is an approximate measure of the rate at which the residue reaches isospin equilibrium. We note that this definition is somewhat different from that one would usually use, i.e., $\lambda(\tau_I) = 1/e$.

For describing thermal equilibrium of the heavy residue, we use the quadrupole moment $Q_{zz}(t)$ as defined earlier. Obviously $Q_{zz} = 0$ is a necessary, although not sufficient, condition for thermal equilibrium. Similar to the definition of τ_I we define the momentum relaxation time τ_p as the time when $Q_{zz}(t)/Q_{zz}(0) = 1\%$. This quantity then measures the rate at which the residue reaches thermal equilibrium.

Another important property of the heavy residue is the possible existence of dynamical instabilities (see, e.g., the chapter by Li and Ko in this volume). To study this phenomenon, we introduce the square of the adiabatic sound velocity [43, 44]

$$v_s^2 = \frac{1}{m}\left(\frac{\partial P}{\partial \rho}\right)_S = \frac{1}{m}\left[\frac{10}{9}\langle E_k\rangle + a\frac{\rho}{\rho_0} + b\sigma\left(\frac{\rho}{\rho_0}\right)^{\sigma}\right], \tag{9}$$

where $\langle E_k\rangle$ is the average kinetic energy per nucleon, a = -358.1 MeV, b = 304.8 MeV and $\sigma = 7/6$ are the parameters corresponding to a soft nuclear equation of state. For $v_s^2 < 0$, a homogeneous nuclear matter is unstable against the growth of fluctuation, leading to dynamical instability or spinodal decomposition,

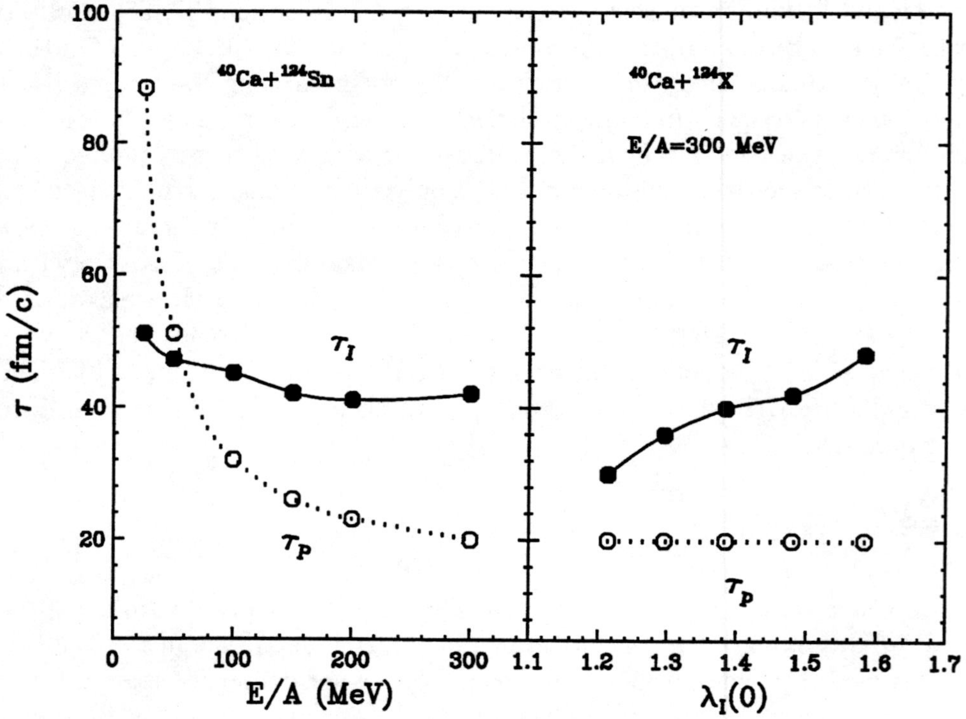

Figure 13: The development of dynamical instability (upper windows), thermal (middle windows) and chemical (lower windows) equilibrium in ^{40}Ca+^{124}Sn collisions at an impact parameter of 1 fm and beam energies of 25, 50, 150 and 300 MeV/nucleon, respectively.

It is interesting to compare the time dependence of $(\lambda_I(t)-1)/(\lambda_I(0)-1)$, $Q_{zz}(t)/Q_{zz}(0)$ and $v_s^2(t)/v_s^2(0)$. They are shown in Fig. 7 for collisions of ^{40}Ca+^{124}Sn at an impact parameter of 1 fm and at beam energies of 25, 50, 150 and 300 MeV/nucleon. The system considered here has an initial $\lambda_I(0) = 1.48$. For collisions at a beam energy of 25 MeV/nucleon the residue is found to be dynamically stable up to 300 fm/c during the collision. This time interval is long enough for both thermal and chemical equilibrium to be fully established as shown in the middle and lower windows of the first column. On the other hand, for collisions at beam energies above 50 MeV/nucleon a significant compression appears, and this is followed by expansion, leading into the adiabatic spinodal region after about 50 fm/c. We notice that at this time the heavy residue formed in the collision is still far from thermal and chemical equilibrium. Both the momentum and isospin asymmetries of the heavy residue are seen to oscillate with time.

It is seen from the middle window of the fourth column that nuclear translucency occurs at E/A=300 MeV. After spinodal decomposition the heavy residue quickly starts to break up into fragments and nucleons[44]. Although the isotopic contents of these fragments and nucleons depend on the emission angle, it is also strongly influenced by the neutron to proton ratio of the target and projectile in the entrance channel instead the average neutron to proton ratio of the combined system. This observation is consistent with recent experimental findings [21, 22, 23]. Although chemical and thermal equilibrium are not completely estab-

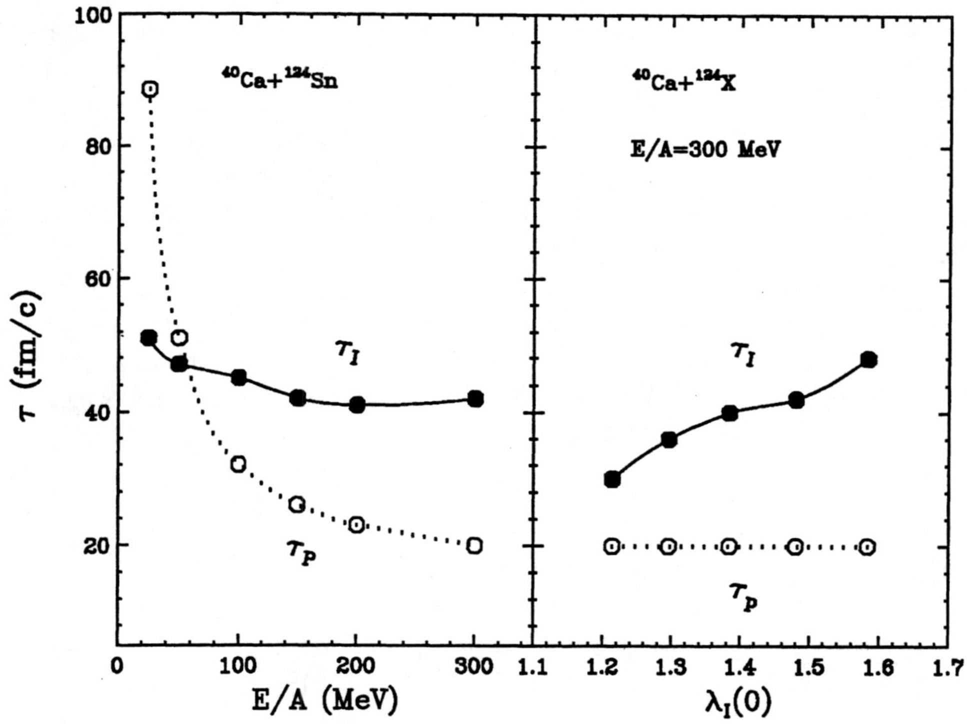

Figure 14: Left window: isospin (open circles) and momentum (filled circles) relaxation times as functions of beam energy for the reactions. Right window: relaxation times as functions of $\lambda_I(0)$ of the reactions.

lished at beam energies higher than the Fermi energy, it is still interesting to compare their relaxation times. This is shown in Fig. 7. In the left window, the comparison is made for ^{40}Ca+^{124}Sn collisions at beam energies from 25 to 300 MeV/nucleon and at an impact parameter of 1 fm. The momentum relaxation time is found to decrease with increasing beam energy. This is in qualitative agreement with that found in Refs. [35, 37, 38, 39, 40, 42]. On the other hand, the isospin relaxation time decreases slowly with the beam energy. The shorter isospin relaxation time at incident energies below about 50 MeV/nucleon is in agreement with what was found in deep inelastic heavy ion collisions [17, 18, 19]. At higher incident energies the time for reaching momentum equilibrium is found to be shorter than that for isospin equilibrium. For example, a 20 fm/c difference in the relaxation time is observed at E/A=300 MeV. In the right window, the isospin and momentum relaxation time as functions of the initial isospin asymmetry $\lambda_I(0) \equiv (n/p)_{\text{projectile}}/(n/p)_{\text{target}}$ are compared for ^{40}Ca induced reactions on several isobaric targets of mass 124 at a beam energy of 300 MeV and at an impact parameter of 1 fm. It is seen that although the momentum relaxation is almost independent of the initial isospin asymmetry, the isospin relaxation time increases with the initial isospin asymmetry.

The shorter relaxation time for isospin than momentum at low incident energies can be understood as follows. First, nucleon-nucleon collisions, which are responsible for momentum relaxation, are more likely to be suppressed due to Pauli blocking. Secondly, the repulsive symmetry potential for neutrons and the attractive symmetry potential for protons make preequilibrium emissions of neutrons more likely than protons in low energy collisions, as shown in Ref. [36], which thus enhances the isospin relaxation rate in the residue. On the other hand, in high energy collisions, Pauli blocking is less effective and the symmetry potential is also less important, leading thus to a shorter momentum relaxation time and a longer isospin relaxation time. Effects due to different forms of symmetry potential and the charge exchange reaction ($pn \rightarrow np$) on chemical and thermal equilibrium have been studied in ref. [12]. They are found to have no discernible effects on both the momentum and isospin relaxation times. Only during the later stage of the collisions do they affect slightly the momentum and isospin distributions.

Therefore, isospin and momentum relaxation times in the heavy residues formed in heavy-ion collisions at intermediate energies can be completely established only at beam energies below the Fermi energy. At higher energies the dynamical instability sets in before either chemical or thermal equilibrium is achieved. Moreover, the isospin relaxation time is shorter (longer) than that for momentum at beam energies lower (higher) than the Fermi energy.

8 Summary

In summary, isospin mixing in heavy-ion collisions can be used to study the nature and degree of stopping. Isospin asymmetry provides on average a tag to distinguish projectile- and target-nucleons in the final state. Complete isospin equilibrium can be obtained at low energies only. A transition from isospin equilibrium to non-equilibrium occurs at intermediate energies depending on the sizes of the reaction system. At higher energies global isospin equilibrium can not be reached even in the most central reactions between two gold nuclei. Theoretical studies based on nuclear transport models indicates that the time scale and degree of isospin equilibration dependens sensitively on the isospin-dependence of the nuclear equation of state at low energies. At higher energies where nuclear translucency happens, the in-medium nucleon-nucleon cross section is more important for the nuclear stopping power. Applications of the isospin tracer variables to heavy-ion collisions at relativistic energies are expected to break the scaling laws observed by using traditional methods. The isospin relaxation time is found to be shorter than the momentum relaxation time at beam energies lower than the Fermi energy and longer at beam energies higher than the Fermi energy. It has also been found that dynamical instabilities leading to multifragmentation may set in before the system reaches isospin and momentum equilibrium. With the available radioactive beams and the planned Rare Isotope Accelerator (RIA) one can envision more applications of the isospin tracer observables. These studies on the stopping power at intermediate energies can help us understand the isospin-dependence of the nuclear equation of state and in-medium nucleon-nucleon cross sections, which have important consequences extending far beyond nuclear physics (see, e.g., the contribution of U. Lombardo et al. in this book).

The work of Bao-An Li is supported in part by the National Science Foundation under Grant No. PHY-0088934 and the Arkansas Science and Technology Authority Grant No. 00-B-14. S.J. Yennello is supported in part by NSF through grant No. PHY-9457376, the Robert A. Welch Foundation through grant No. A-1266, and the Department of Energy through grant No. DE-FG03-93ER40773.

References

[1] W. Busza and R.J. Ledoux, Annual review of Nuclear and Particle Science, Vol. **38**, 119 (1988).

[2] C.Y. Wong, Introduction to High-Energy Heavy-Ion Collisions, World Scientific (Singapore), 1994.

[3] B.A. Li and C.Y. Wong, Phys. Scr. **V47**, 151 (1993).

[4] G.F. Bertsch, G.E. Brown, V. Koch and B.A. Li, Nucl. Phys. **A490**, 745 (1988).

[5] W. Bauer, Phys. Rev. Lett. **61**, 2534 (1988).

[6] J.P. Bondorf et al., Phys. Rep. **257**, 133 (1995).

[7] D.H.E. Gross, Phys. Rep. **279**, 119 (1997).

[8] F. Videbaeck and O. Hansen, Phys. Rev. C**52**, 2684 (1995).

[9] J.W. Harris, p. 401, Advances in Nuclear Dynamics 2, Eds. W. Bauer and G.D. Westfall, Plenum Press, New York, 1996.

[10] B.A. Li and C.M. Ko, Nucl. Phys. **A601**, 457 (1996).

[11] P.Braun-Munzinger et al., Phys. Lett. **B465**, 15 (1999).

[12] B.A. Li and C.M. Ko, Nucl. Phys. **A630**, 556c (1998).

[13] B.A. Li and C.M. Ko, Phys. Rev. C52, 2037 (1995); *ibid*, C53, R22 (1996).

[14] A. Hombach, W. Cassing and U. Mosel, Eur. Phys. J. **A5**, 77 (1999).

[15] S.A. Bass, J. Konopka, M. Bleicher, H. Stöcker and W. Greiner, GSI annual report, p. 66 (1994).

[16] B.A. Li, C.M. Ko and W. Bauer, Int. Jou. of Mod. Phys. E**7**, 147 (1998).

[17] B. Gatty et al., Z. Phys. **A273**, 65 (1975).

[18] F. Beck, M. Dworzecka and H. Feldmeier, Z. Phys. **A289**, 113 (1978).

[19] W. U. Schröder and J. R. Huizenga, in Treatise on Heavy-Ion Science, ed. D. A. Bromley (Plenum, New York and London, 1984) Vol. 2. P113.

[20] M. Farine, T. Sami, B. Remaud and F. Sébille, Z. Phys. **A339**, 363 (1991).

[21] S.J. Yennello, B. Young, J. Yee, J.A. Winger, J. S. Winfield, G.D. Westfall, A. Vander Molen, B.M. Sherrill, J. Shea, E. Norbeck, D.J. Morrissey, T. Li, E. Gualtieri, D. Craig, W. Benenson and D. Bazin, Phys. Lett. **B321** (1994) 15.

[22] H. Johnston, J. Winger, T. White, B. Hurst, D. O'Kelly and S. J. Yennello, Phys. Lett. **B371**, 186 (1996).

[23] H. Johnston, T. White, Bao-An Li, E. Ramakrishnan, J. Winger, D.J. Rowland, B. Hurst, F. Gimeno-Nogues, D. O'Kelly, Y-W. Lui and S. J. Yennello, Phys. Rev. **C56**, 1972 (1997).

[24] R. Wada et al., Phys. Rev. Lett. **58**, 1829 (1987).

[25] B. A. Li and S. J. Yennello, Phys. Rev. C**52**, R1746 (1995).

[26] Y. Yariv and Z. Frankel, Phys. Rev. C**26**, 2138 (1982).

[27] B.A. Li, Z.Z. Ren, C.M. Ko and S.J. Yennello, Phys. Rev. Lett. **76**, 4492 (1996).

[28] B.A. Li, C.M. Ko and Z.Z. Ren, Phys. Rev. Lett. **78**, 1644 (1997).

[29] B. Borderie *et al.*, in: Proc. of ACS Nuclear Chemistry Award Symposium, Anaheim, USA, April 2-4, 1995.

[30] A. Ono and H. Horiuchi, Phys. Rev. C**53**, 2958 (1996).

[31] L.W. Chen et al., J. phys. **G23**, 211 (1997).

[32] R. Donangelo and S. R. Souza, Phys. Lett. **B409**, 58 (1997).

[33] F. Rami et al., Phys. Rev. Lett. **84**, 1120 (2000).

[34] B.A Li and C. M. Ko, Nucl. Phys. **A618**, 498 (1997).

[35] G. F. Bertsch, Z. Phys. A **289**, 103 (1978).

[36] B.A. Li and C.M. Ko, Phys. Rev. C**57**, 2065 (1998).

[37] J. Randrup, Nucl. Phys. **A314**, 429 (1979).

[38] C. Toepffer and C. Y. Wong, Phys. Rev. C**25**, 1019 (1982).

[39] W. Cassing, Z. Phys. **A465**, 317 (1987).

[40] P. Abgrall et al., Phys. Rev. C**49**, 1040 (1994).

[41] F. Haddad et al., Z. Phys. **A354**, 321 (1996).

[42] B. Borderie et al., Z. Phys. **A357**, 7 (1997).

[43] W. Bauer, G. F. Bertsch and H. Schultz, Phys. Rev. Lett. **69**, 1888 (1992).

[44] B. A. Li and D. H. E. Gross, Nucl. Phys. **A554**, 257 (1993).

Chemical Equilibrium and Isotope Temperatures

J. Pochodzalla *and W. Trautmann †

Abstract

The measurement and interpretation of isotopic yield ratios in heavy ion reactions at intermediate and high energies are discussed and the usefulness of these observables for establishing equilibrium properties and for determining thermodynamic parameters is illustrated. The examples are mainly taken from work performed with lighter projectiles at intermediate and high energies and from studies of spectator reactions at relativistic energies. As an application, the caloric curve of nuclei, as derived for Au on Au collisions, is introduced and discussed.

1 Introduction

Isotopic yield ratios have proven to be useful observables for studying the mechanisms of heavy-ion reactions at intermediate and high energies. Production yields for isotopically resolved particles and nuclear fragments and combinations thereof can provide us with answers to the questions of mutual stopping and subsequent equilibration of the collision partners. To the extent that equilibrium is reached they permit the extraction of the corresponding thermodynamical variables. The entropy, density, and particularly the temperature of the ensemble of excited nuclear systems, formed in the course of energetic encounters, have been deduced from measured isotopic yield ratios. These parameters provide a characterization of heavy-ion reactions whose complex dynamics advocate a global description in statistical terms. Temperature observables and their correlation with the excitation energy, commonly termed caloric curve of nuclei, offer the possibility to explore the link between the liquid-gas phase transition predicted for infinite nuclear matter and the decay properties of finite nuclei.

The examples considered in this chapter are heavy-ion reactions at bombarding energies in the range of about 100 MeV per nucleon up to a few GeV per nucleon. The dynamics of these reactions, high above the Fermi energy, are predominantly governed by nucleon-nucleon collisions and by the possibility of the production (and reabsorption) of secondary particles, mostly pions. Nuclear matter properties are being probed at densities far away from the ground state density and at excitation energies up to the binding limit of nuclear matter and beyond. The corresponding part of the phase diagram of extended nuclear matter includes the spinodal and coexistence regions of the phase transition from the normal liquid phase to a gas-like phase consisting of nucleons and light complex particles. Consequences of these nuclear phase properties are expected and searched for in this class of reactions.

The isotopic degrees of freedom in the nuclear disassembly and their role in a statistical description will be the primary subject of this chapter. Isospin, therefore, will not appear

*Max-Planck-Institut für Kernphysik, D-69117 Heidelberg, Germany
†Gesellschaft für Schwerionenforschung mbH, D-64291 Darmstadt, Germany

here in its role as a symmetry principle of nuclear structure that follows from the charge independence of the nuclear forces. The third component of the isospin, $t_3 = (N - Z)/2$, is simply used to characterize isotopes according to their neutron and proton numbers N and Z. Isospin equilibration, consequently, will indicate that isotopic yield distributions correspond to the expectations for chemical equilibrium. Isospin equilibration, in this sense, is a consequence of chemical equilibration and as such a necessary condition for equilibrium that can be tested experimentally.

The experimental determination of the mass number A of a reaction product is, generally, more difficult than the determination of the element number Z alone. It involves either improved resolutions, if the $\Delta E - E$ technique is used, or measurements of additional quantities, such as the time-of-flight or the magnetic rigidity of the emitted reaction product. Yield ratios of neighboring isotopes, in return, may be assumed to be little influenced by dynamical effects such as Coulomb, recoil, or size effects, and thus should permit a fairly unbiased look at the balance equation

$$(A+1, Z) \longleftrightarrow (A, Z) + n. \tag{1}$$

In a statistical description this balance is determined by the temperature and the chemical potentials, the measurement thus gives access to these equilibrium parameters.

We will begin this chapter with a discussion of some of the earlier experiments on isotopic effects, performed with beams of fairly light ions of intermediate energy and of light particles at high energies in the GeV range. These data will allow us to make the distinction between isospin mixing and isospin equilibration. We will then more generally discuss equilibrium in spectator reactions with heavier ions of incident energies up to one GeV per nucleon. The determination of equilibrium parameters, in particular of the temperature, and the construction of a caloric curve of nuclei will be described in the final sections of this chapter.

2 Isospin Mixing and Equilibration

One of the earlier experiments on isotopic degrees of freedom has been conducted at the CERN synchrocyclotron during the eighties of the last century [1]. In this experiment, the isotopic composition of decay products was studied as a function of the neutron-to-proton (N/Z) content of nuclear systems over the range accessible with stable projectile and target nuclei. Targets of ^{58,64}Ni, ^{nat}Ag, and ^{197}Au were bombarded with ^{12}C and ^{18}O beams with incident energy 84 MeV per nucleon. The produced particles and fragments were measured inclusively with high-resolution telescopes placed at several angles between $\theta_{\text{lab}} = 40°$ and 120°. Each telescope consisted of an axial-field ionization chamber followed by three silicon detectors of increasing thickness and a bismuth germanate scintillation detector. Full isotope separation for ions up to carbon was achieved for fragment energies $E/A \geq 3$ MeV and for ions up to beryllium at somewhat lower energies.

Energy-integrated isotope yields were obtained by extrapolating from the isotopically resolved parts of the energy spectra with the help of fitting functions which were derived from the model of a Maxwellian source moving in beam direction. The measured isotope ratios were found to vary little with the detection angle. The following discussion is therefore based on the data measured at $\theta_{\text{lab}} = 40°$ which have the highest statistics and the smallest experimental uncertainties. This angle is also sufficiently backward to be outside the range of projectile fragmentation.

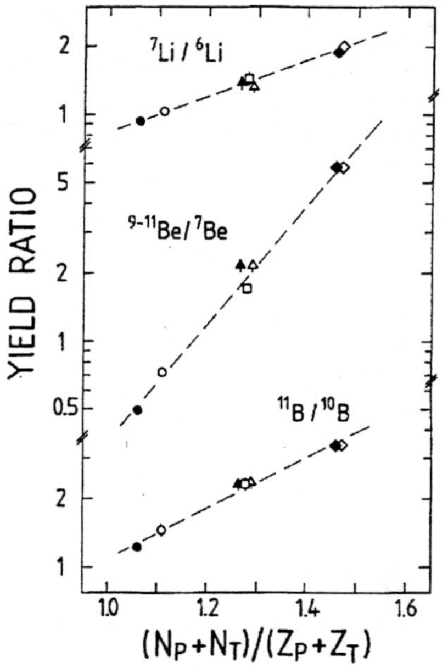

Figure 1: Ratios of energy-integrated isotope yields, measured at $\theta_{\text{lab}} = 40°$, as functions of the N/Z ratio of the combined system of projectile and target. Closed and open symbols denote ^{12}C and ^{18}O projectiles, respectively; circles, squares, triangles, and diamonds stand for ^{58,64}Ni, ^{nat}Ag, and ^{197}Au targets (from Ref. [1]).

Three yield ratios, of lithium, beryllium, and boron isotopes and for all the investigated reactions, are shown in Fig. 1, plotted as functions of the neutron-to-proton ratio N/Z of the combined system of projectile and target nuclei. This way of representing the data was motivated by the following observations: The isotope-yield ratios vary strongly with the choice of the target. The relevant property of the target, however, is neither the mass number A nor the atomic number Z but rather the ratio N/Z; the yield ratios are different for the ^{58}Ni and ^{64}Ni targets but, apparently, the same for the ^{64}Ni and ^{nat}Ag targets which have equal N/Z. Furthermore, the different N/Z of the ^{12}C and ^{18}O projectiles seem to have little influence except in the reactions with the ^{58}Ni target. The fact that in this case the ratio measured with the ^{12}C projectiles are significantly smaller than those measured with ^{18}O suggests that the N/Z of the combined system rather than that of the target nucleus alone is the relevant ordering parameter. Plotted in this way, all three isotope-yield ratios follow smooth exponential functions and increase monotonically with N/Z (Fig. 1).

We may first conclude that the combined system of the projectile and the whole target is involved in the process of emission of complex fragments in these reactions. For the moment, we will call this complete mixing, as expected for projectile nuclei that are completely stopped in the target prior to the fragment emission. The two collision partners are thought to form a sufficiently homogeneous system such that the individual contributions of the projectile and target can no longer be separately identified. Even though confirmed for a variety of reactions [2, 3] this is by no means a trivial result; there are many examples for which complete mixing is not observed: Isotope yield ratios of projectile fragments measured at small angles with ^{12}C beams at the same energy 84 MeV per nucleon are virtually constant for the same range of targets [4]. Similar observations were made at

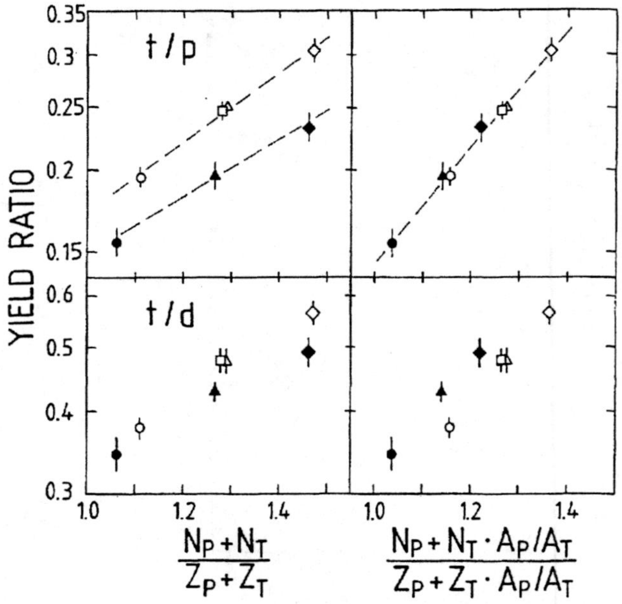

Figure 2: Ratios of preequilibrium triton-to-proton yields (top) and triton-to-deuteron yields (bottom), measured at $\theta_{\text{lab}} = 40°$ and plotted as functions of the N/Z ratio of the combined system (left-hand side) and of a source consisting of equal numbers of nucleons from the projectile and from the target (right-hand side). The symbols are chosen as in Fig. 1 (from Ref. [1]).

GANIL where the fragmentation of heavier projectiles was studied [5, 6, 7]. For the set of reactions with $A = 40$ projectiles and $A = 58$ targets, discussed in detail by Li and Yennello in Chapter 19, the stopping is found to become incomplete at incident energies of 45 MeV per nucleon and above [8, 9, 10]. For the mass symmetric reactions with ^{96}Zr and ^{96}Ru projectiles and targets, studied at the much higher energy of 400 MeV per nucleon, the mixing was found to increase with the centrality of the collision but to remain incomplete even at the smallest impact parameters (Ref. [11] and Chapter 19).

Incomplete mixing has even been observed in the same set of reactions at 84 MeV per nucleon with ^{12}C and ^{18}O projectiles, as illustrated in Fig. 2. There the ratios of hydrogen isotopes from the preequilibrium component measured at $\theta_{\text{lab}} = 40°$ are shown for different mixing assumptions. With the postulate that the observed ratios should be a unique function of the N/Z ratio of the emitting source, the light particle data are inconsistent with emission from the combined system (Fig. 2, left panels). The t/p ratios can be reconciled with this postulate if the emitting source is assumed to consist of equal numbers of nucleons from the projectile and from the target, chosen according to their respective N/Z ratios (Fig. 2, upper right panel). For the t/d ratios the same assumption overcorrects the deviations (lower right panel) but by mixing projectile and target nucleons with the ratio 1:2 a unique and monotonic increase of the measured isotope ratios can be achieved. These examples show that preequilibrium light particles originate from subsystems considerably smaller than the combined system of projectile and target.

The question of isospin equilibration will now be addressed by asking for the significance of the slopes characterizing the dependence of the isotope ratios on the source N/Z. Their trends can be most easily understood by considering the grand canonical approach [12, 13, 14]. Here the yield ratios of neighboring isotopes can be expressed as

$$\frac{Y(A+1,Z)}{Y(A,Z)} = (\frac{A+1}{A})^{3/2} \cdot \frac{\omega(A+1,Z)}{\omega(A,Z)} \cdot \exp(\frac{\mu_n + \Delta B}{T}) \tag{2}$$

where $\omega(A, Z)$ denotes the internal partition function and ΔB is the difference of the binding energies of the two nuclides. The chemical potentials μ_n of neutrons and μ_p of protons guarantee the conservation of the mean mass and charge of the disassembling nucleus within a given volume and, therefore, are functions of the N/Z ratio of the system. The assumption that μ_n depends, to first order, linearly on N/Z immediately explains the experimental finding that the logarithm of the ratio increases linearly with the N/Z ratio of the emitting source. It further explains why the slopes are about the same for the pairs of lithium and boron isotopes which both differ by one neutron (Fig. 1). The ^{7}Be isotope is, at least, two neutrons lighter than the most abundant heavier isotope, and a term $2 \cdot \mu_n/T$ (or higher) appears in the exponent on the right-hand side of Eq. 2. Consequently, the slopes should be about twice as large which is also in agreement with the observation. For this qualitative consideration, it has been tacitly assumed that the temperature T is approximately the same for the considered fragment species and reactions, a condition that is furthermore needed for ensuring constancy of the ratios of the internal partition functions. It will be shown below that this assumption is not unrealistic in the present case (Section 6).

It is an instructive exercise to estimate the N/Z dependence expected from other conceivable mechanisms [15]. In particular, the binomial distribution following from the assumption of a combinatorial mechanism of fragment formation yields a variation in proportion to $N/(N + Z)$ for ratios of neighboring isotopes which is by far too weak to describe the data. The observed dramatic variation of the isotope ratios with N/Z is a consequence of the exponential dependence on the ratio μ_n/T and thus may by itself indicate chemical equilibrium. Equilibrium in this sense does not necessarily imply complete mixing. The equilibration may be local as in the examples of incomplete mixing discussed above where strong variations as a function of the neutron-to-proton ratio of the assumed localized sources were observed. Global equilibrium, clearly, requires complete mixing.

3 The Isotopic Effect

An alternative possibility to study isospin equilibration is provided by the socalled isotopic effect [16, 17]. It is generally defined as the net effect observed when switching to another target made from a different isotope of the same element with no other change of the experimental conditions. Here we will specifically consider ratios of production cross sections of particular fragment species in otherwise identical reactions on the tin isotopes with $A = 112$ and $A = 124$. These isotopes have been frequently chosen as targets and, more recently, also as beams for isotopic studies ([18, 19, 20] and references given in [17]).

For the case of ^{112}Sn and ^{124}Sn targets and, e.g., the production of ^{7}Li fragments we may write

$$\delta(^7\text{Li}) = \frac{\sigma(^7\text{Li};^{112}Sn)}{\sigma(^7\text{Li};^{124}Sn)} = \frac{\sigma(^7\text{Li}; N/Z = 1.24)}{\sigma(^7\text{Li}; N/Z = 1.48)} = \delta(^7\text{Li}; \Delta(N/Z) = 0.24). \tag{3}$$

This notation emphasizes the role of the N/Z ratio of the source as the dominant parameter. It ignores the contribution of the projectile to the source which, however, may be justified

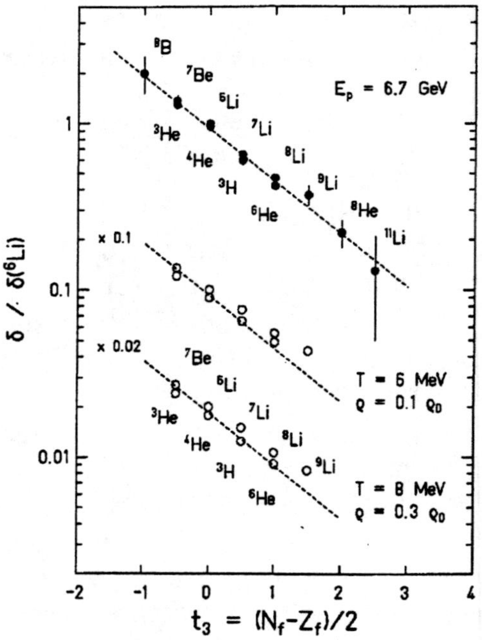

Figure 3: Reduced isotopic effect $\delta/\delta(^6\text{Li})$ versus the third component of the isospin t_3. The experimental results for protons of $E_p = 6.7$ GeV [21] are given in the upper part (full circles). The dashed line represents an exponential fit to the data. In the lower part the same fit curve is compared to the results obtained with the quantum-statistical model (open circles), calculated with two sets of parameters T and ρ, and with $\Delta(N/Z) = 0.24$ (ρ_0 denotes the saturation density of nuclei, from Ref. [17]).

for the very light projectiles (p, d, α) that were used in the examples to be discussed in the following. By defining a reduced isotopic effect $\delta(X)/\delta(^6\text{Li})$, i.e. the isotopic effect for species X in relation to that for ^{6}Li, uncertainties of the absolute normalizations of the measurements with the two targets are eliminated. The reduced isotopic effect represents a double ratio which, e.g., for X = ^{7}Li is equal to the ratio of the $^7\text{Li}/^6\text{Li}$ yield ratios measured with the two targets. It is an observable equivalent to the value of the slope of the $^7\text{Li}/^6\text{Li}$ yield ratio as a function of N/Z (Fig. 1).

Figure 3 shows the reduced isotopic effects measured at the proton synchrotron of the JINR, Dubna, for the interaction of protons of $E_p = 6.7$ GeV with the target nuclei ^{112}Sn and ^{124}Sn [21]. These experiments were performed inclusively with multi-element telescopes positioned at approximately 90° with respect to the beam. The cross section ratios differ by more than one order of magnitude for the neutron poor and the most neutron rich fragments. Plotted against the third component t_3 of the fragment isospin, the data exhibit a nearly perfect exponential dependence. This result is a generalization of what was observed for the slopes of isotope ratios (Fig. 1) and can also be understood on the basis of Eq. 2.

Quantum-statistical models (QSM) are useful for extracting more quantitatively the information contained in measured isotope yield ratios or isotopic effects. Several versions of such models have been developed for different purposes [22, 23, 24, 25]. For the following comparison the model of Hahn and Stöcker [22] has been chosen. It assumes thermal and chemical equilibrium at the breakup point where the fragmenting system is characterized by a density ρ, temperature T, and by its overall N/Z ratio. The model respects fermion

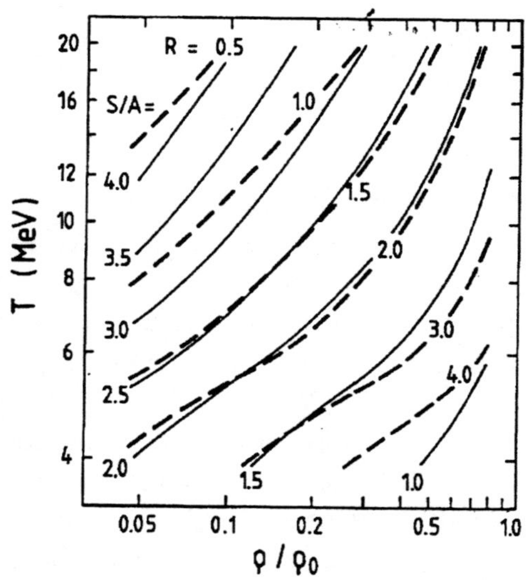

Figure 4: Results of calculations with the quantum-statistical model code [22] for the composite system ^{12}C + ^{197}Au. The thin full lines are the isentropes in the temperature-versus-density plane. The thick dashed lines are the contours of constant ratio R of the yields for ^{7}Li and ^{6}Li (from Ref. [15]).

and boson statistics which, however, is not crucial at high temperature. It does not make provisions for the finite size of nuclear systems but follows the sequential decay of excited fragments according to tabulated branching ratios.

Results of QSM calculations are given in the lower half of Fig. 3. They are compared to the experimental data represented by the dashed exponential fit curves. Pairs of temperatures T and densities ρ can apparently be found that give an accurate description of the observed t_3 dependence. The temperatures chosen in these examples are in the range of measured breakup temperatures (see below), and the matching densities correspond to large breakup volumes as assumed in the statistical multifragmentation models [26, 27]. Low densities in this range have been deduced from proton-proton correlation functions measured for reactions in which multifragmentation is dominant [28]. Their meaning in the light-ion induced reactions with small mean fragment multiplicities [29] is not fully clear at present.

The example illustrates that isotopic effects or isotopic yield ratios in this way do not permit the individual determination of either the breakup temperature or density but rather define a relation between them. This was studied in more detail in Refs. [1, 15, 30] where it was found that the loci in the temperature-versus-density plane corresponding to fixed isotopic yield ratios effectively coincide with isentropes (Fig. 4). Obviously, the N/Z ratio of the system has to be specified if this is to be used as a method to determine the entropy at breakup. Entropy values S/A determined in this way for fragmentation channels in ^{12}C and ^{18}O induced reactions at bombarding energies up to 84 MeV per nucleon are in the range of 1.5 to 2.5 [1, 30]. More recent analyses based on similar methods have shown that this range of entropies is virtually an invariant over a wide variety of fragmentation reactions [17, 31]. The interest attracted by the entropy as an observable in heavy ion reactions is related to the approximately isentropic nature of the expansion phase [32, 33]. Entropy measurements thus permit a characterization of the excited systems formed during the initial impact and before expansion [22].

4 Equilibrium in Spectator Reactions

In this section we will temporarily lift the restriction to isotopic degrees of freedom and more generally search for observables with the potential of giving further evidence for equilibrium in fragmentation reactions. We will also proceed to a new type of reaction and investigate spectator decays initiated by collisions of heavy nuclei at relativistic bombarding energies [34]. These reactions are best viewed within the participant-spectator model [35] which distinguishes between the 'hot' hadronic system, formed by fast nucleons and secondary hadrons produced in hard collisions, and the 'cold' spectators consisting of the remnants of the incident nuclei. It turns out that these spectators can also be quite highly excited, up to and beyond their total binding energies. Because of the nature of the excitation mechanism, predominantly nucleon knockout and absorption of slow nucleons recoiling from hard collisions, high degrees of equilibration may be expected.

We will focus on two recent experiments with the ALADIN spectrometer at the heavy-ion synchrotron SIS of the GSI Darmstadt in which reactions of ^{197}Au + ^{197}Au in the regime of relativistic energies up to 1 GeV per nucleon were studied. In the first experiment, the ALADIN spectrometer was used to detect and identify the products of the projectile-spectator decay [36]. The setup offers full coverage for projectile fragments with a dynamic range extending from helium isotopes to the original projectiles. Individual charge resolution and, for lighter fragments up to $A \approx 20$, also individual mass resolution is obtained. The Large-Area Neutron Detector (LAND) was used to measure coincident free neutrons emitted by the projectile source. In the second experiment, three multi-detector hodoscopes, consisting of a total of 216 Si-CsI(Tl) telescopes, and three high-resolution telescopes were positioned at backward angles to measure the yields and correlations of isotopically resolved light fragments of the target-spectator decay [37]. From these data excitation energies and masses, temperatures, and densities were deduced.

For the presentation of these data, the quantity Z_{bound} has emerged as a useful sorting variable. Z_{bound} is equal to the sum of the atomic numbers Z_i of all projectile fragments with $Z_i \geq 2$. It reflects the variation of the charge of the primary spectator system and is therefore correlated with the impact parameter of the reaction. In the second type of experiments, with high-resolution detectors looking at the target spectator, the symmetric collision system ^{197}Au + ^{197}Au was studied. Z_{bound} was determined for the projectile decay with the time-of-flight wall of the ALADIN spectrometer, and it was assumed that its mean values for the target and projectile spectators are identical for a given event class.

The universal features of the spectator decay, as apparent in the observed Z_{bound} scaling of the measured charge correlations, were the first and perhaps most striking indications for equilibrium [36]. The target invariance of the M_{IMF} versus Z_{bound} correlation was first observed for collisions of ^{197}Au projectiles with C, Al, Cu, and Pb targets at 600 MeV per nucleon [38, 39]. In Fig. 5 (top) this correlation is shown for ^{238}U projectiles at 1000 MeV per nucleon and for a set of seven targets, ranging from Be to U. The data for the lighter targets extend only over parts of the Z_{bound} range, more clearly visible in the bottom part of the figure where the differential cross sections $d\sigma/dZ_{\mathrm{bound}}$ for four out of the seven targets are displayed. From the cross sections, by assuming a monotonic relation between Z_{bound} and the impact parameter, an empirical impact parameter scale can be obtained. Central collisions lead to the smallest values of Z_{bound} that can be reached with a given target,

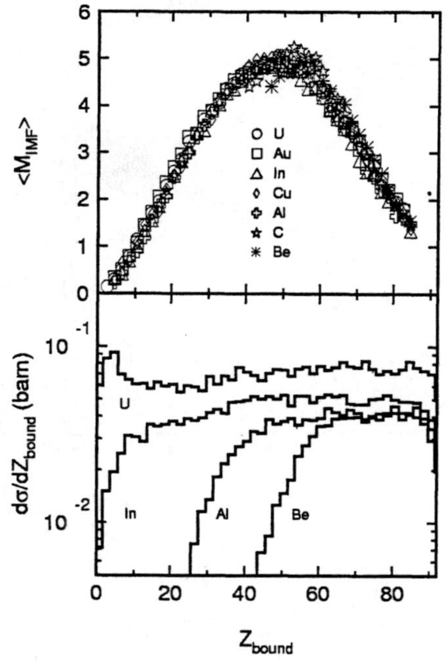

Figure 5: Top: Mean multiplicity of intermediate-mass fragments $\langle M_{\rm IMF} \rangle$ as a function of $Z_{\rm bound}$ for the reactions of ^{238}U projectiles at $E/A = 1000$ MeV with the seven targets Be, C, Al, Cu, In, Au, and U. Bottom: Measured cross sections $d\sigma/dZ_{\rm bound}$ for the reactions of ^{238}U projectiles at $E/A = 1000$ MeV with the four targets Be, Al, In, and U. Note that the experimental trigger suppresses the very peripheral collisions at $Z_{\rm bound} \geq 70$ which have much larger cross sections than indicated here (from Ref. [36]).

and given intervals of $Z_{\rm bound}$ may be reached with different targets but in collisions with different impact parameters. The partitioning, apparently, depends only on $Z_{\rm bound}$ and not on the impact parameter or target individually, as long as they populate a given range of $Z_{\rm bound}$.

The invariance of the fragmentation patterns, when plotted as a function of $Z_{\rm bound}$, suggests that the memory of the entrance channel and of the dynamics governing the primary interaction of the colliding nuclei is lost. This extends to other observables; the transverse-momentum widths of the fragments, as shown in Fig. 6, do not change with the bombarding energy, indicating that collective contributions to the transverse motion are small. The equilibration of the three kinetic degrees of freedom in the moving frame of the projectile spectator was confirmed by the analysis of the measured velocity spectra [36]. The square-root dependence on the atomic number Z implies kinetic energies nearly independent of Z and hence of the mass.

The success of the statistical multifragmentation models in describing the observed population of the partition space may be seen as a further argument for equilibration. Here the main task consists of finding an appropriate ensemble of excited nuclei to be subjected to the multi-fragment decay according to the model prescription. Starting from the entrance channel with models describing the primary phase of the collision may not necessarily provide sufficiently realistic ensembles, even though a good description of the fragment correlations was obtained with the quantum-molecular-dynamics model coupled to the statistical multifragmentation model [40]. An alternative method consists of using empirical ensembles

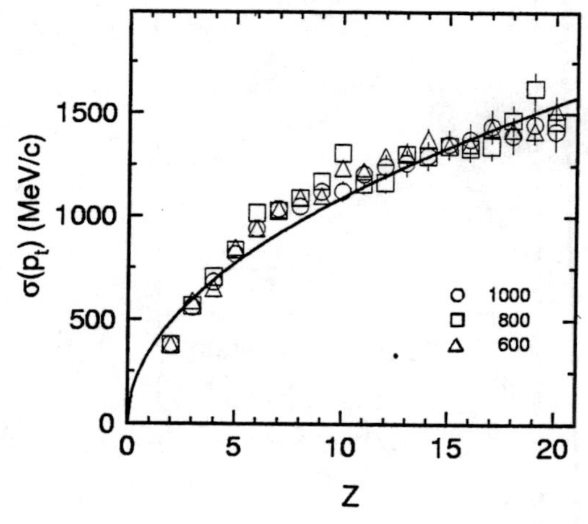

Figure 6: Widths of the transverse-momentum distributions $\sigma(p_t)$ as a function of the fragment atomic number Z for the reactions ^{197}Au + ^{197}Au at $E/A = 600$, 800, and 1000 MeV and for $20 \leq Z_{\mathrm{bound}} \leq 60$. The line is proportional to $\sqrt{Z}$ (from Ref. [36]).

derived by searching for an optimum reproduction of the observed partitioning. Near perfect descriptions of the measured correlations, including their dispersions around the mean behaviour, can be achieved [37, 41]. The mathematical procedure of backtracing allows for studying the uniqueness of the obtained solutions and their sensitivities to the observables that were used to generate them [42]. As an example, the ensemble derived empirically for the reaction ^{197}Au + ^{197}Au at 1000 MeV per nucleon is shown in Fig. 7. It extends over wide ranges of mass and excitation energy with both quantities being correlated as expected within the participant-spectator picture.

The spectator source, well localized in rapidity [36] and, apparently, exhibiting so many signs of equilibration, seems an excellent candidate for studying the nuclear phase diagram. Limitations arise from the fact that nucleons and very light particles from the early reaction stages may appear at spectator rapidities. Their contributions are difficult to suppress or to identify which makes it difficult to extract global variables such as the excitation energy of the system at the equilibrium stage. We will address this point in more detail when discussing the caloric curve of nuclei.

5 Measurement of Temperature

When measuring the temperature of excited nuclear systems one has to keep in mind that nuclei are closed systems with no external heat bath. Consequently, the temperature of the system cannot be pre-determined but has to be reconstructed from its decay products. For a microcanonical ensemble, the thermodynamic temperature of a system may be uniquely defined in terms of the total-energy state density. An experimental determination of the state density and its energy dependence is, however, hitherto impossible. Therefore, nuclear temperature determinations take recourse to 'simple' observables of specific degrees of freedom which constitute a good approximation to the true thermodynamic temperature. In

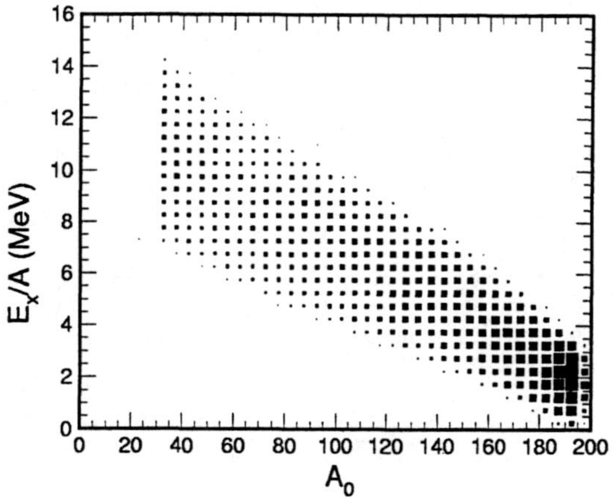

Figure 7: Ensemble of excited spectator nuclei used as input for the calculations with the statistical multifragmentation model as a function of their excitation energy E_x/A and mass A_0. The area of the squares is proportional to the intensity (from Ref. [37]).

the ideal situation, the system will act as the heat bath for the chosen probe whose response can then be treated canonically.

Several techniques have been developed for the measurement of temperatures of excited nuclear systems [43]. In the work leading to the caloric curve of nuclei the method suggested by Albergo *et al.* [13] following earlier considerations of Hoyle [44] has been used. It is based on the assumption of chemical equilibrium and requires the measurement of double ratios of isotopic yields.

In the limit of thermal and chemical equilibrium, the double ratio R built from the yields Y_i of two pairs of nuclides with the same differences in neutron and proton numbers is given by

$$R = \frac{Y_1/Y_2}{Y_3/Y_4} = a \cdot \exp(((B_1 - B_2) - (B_3 - B_4))/T) = a \cdot \exp(\Delta B/T) \tag{4}$$

where B_i denotes the binding energy of particle species i and the constant a contains the mass numbers and internal partition sums. The chemical potential appearing in the expression for single isotope ratios (Eq. 2) cancels in the appropriately chosen double ratio. The expression can therefore be solved with respect to the temperature T provided the internal partition sums are known. The approximation usually made to circumvent this difficulty is to use the corresponding expression for the population of the ground states of the four isotopes, i.e. with the spin degeneracy factors for the ground states, and to treat modifications due to decays of higher-lying states as a perturbation. The problem of sequential decay, presently one of the main limitations of temperature measurements, will be discussed in more detail below.

A meaningful temperature scale can only be derived if the ratio R is sufficiently sensitive to the temperature of the system. By differentiating Eq. 4 we see that a relative error of the double yield ratio $\Delta R/R$ results in a relative modification of the extracted temperature

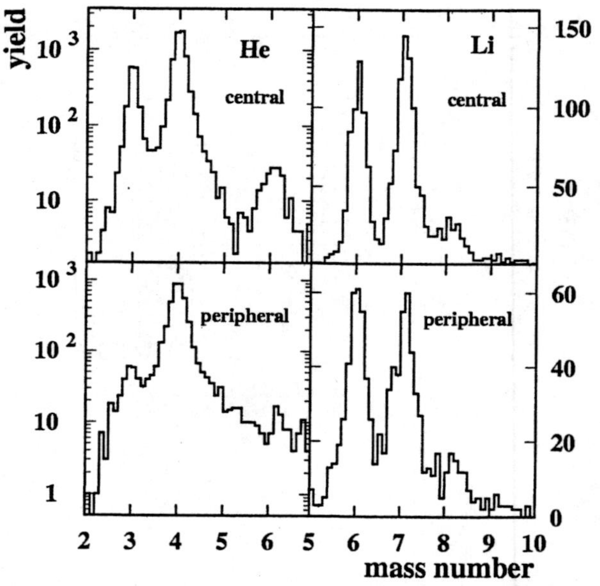

Figure 8: Mass spectra of He fragments (left panel) and Li fragments (right panel) from the projectile spectator following ^{197}Au + ^{197}Au collisions at E/A = 600 MeV. The upper and lower panels correspond to central and peripheral collisions, respectively (from Ref. [46]).

by

$$\frac{\Delta T}{T} = -\frac{T}{\Delta B} \cdot \frac{\Delta R}{R}. \tag{5}$$

Thus, a stability of this thermometer against uncertainties of $\Delta R/R$ will only result if the binding energy difference $\Delta B = (B_1 - B_2) - (B_3 - B_4)$ is larger than the typical temperature to be measured. The analysis by Tsang *et al.* [45] of more than 1000 possible isotope thermometers for the same reaction supports this rule of thumb.

Particularly large values for ΔB are obtained if a $^3He/^4He$ ratio is involved because the difference in binding energy of the two helium isotopes is 20.6 MeV. It may be combined with, e.g., the lithium yield ratio ^{6}Li/^{7}Li or with the hydrogen yield ratios p/d or d/t. Mass spectra for helium and lithium isotopes, measured for the reaction ^{197}Au + ^{197}Au at E/A = 600 MeV with the ALADIN spectrometer, are shown in Fig. 8. The strong variation of the ^{3}He yields reflects the sensitivity of this less strongly bound nuclide to the variation of the temperature with impact parameter.

Solving Eq. 4 for this case of $^3He/^4He$ and ^{6}Li/^{7}Li and in the ground-state approximation yields the following expression:

$$T_{\mathrm{HeLi},0} = 13.3MeV/\ln(2.2\frac{Y_{^6\mathrm{Li}}/Y_{^7\mathrm{Li}}}{Y_{^3\mathrm{He}}/Y_{^4\mathrm{He}}}). \tag{6}$$

The subscript 0 of $T_{\mathrm{HeLi},0}$ is meant to indicate that Eq. 6 is strictly valid only for the ground-state population of the considered isotopes at the breakup stage. The experimentally measured yields, however, contain all contributions from γ-unstable higher-lying states of the same isotope and from particle decays of other isotopes feeding the ground or γ-unstable states. In the work of the ALADIN collaboration [47], the expected magnitude of this effect

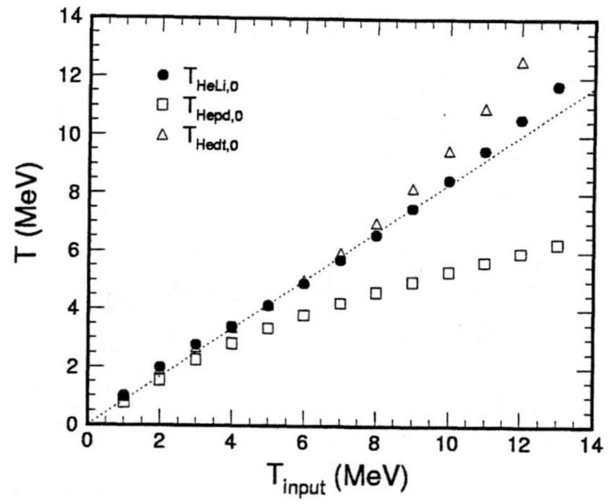

Figure 9: Temperatures $T_{\mathrm{HeLi},0}$, $T_{\mathrm{Hepd},0}$, and $T_{\mathrm{Hedt},0}$, according to the quantum statistical model, as a function of the input temperature T_{input}. A breakup density $\rho/\rho_0 = 0.3$ is assumed. The dotted line represents the linear relation $T_{\mathrm{input}}/1.2$ (from Ref. [37]).

was investigated by performing calculations with the quantum-statistical model [22, 23], a sequential-evaporation model [48], and a statistical multifragmentation model [49]. The primary fragmentation process is treated very differently in these models but they all follow explicitly the sequential decays of excited primary products.

As an example, results obtained with the quantum statistical model are shown in Fig. 9. The assumed breakup density is $\rho/\rho_0 = 0.3$ with ρ_0 denoting the saturation density. Despite the strong feeding of the light particle yields through secondary decays, the relation between the input temperature of the model and $T_{\mathrm{HeLi},0}$ as well as $T_{\mathrm{Hedt},0}$ was found to be almost linear. Other temperature probes, as illustrated for $T_{\mathrm{Hepd},0}$ in the figure, may be more strongly affected by sequential decays. Variations of the input density within reasonable limits in this model and the comparison with results obtained with the other decay models suggested that the accuracy of these estimates may lie within $\pm$ 15%. In order to account for the systematic deviation from unity, a constant calibration factor $T_{\mathrm{HeLi}} = 1.2 \cdot T_{\mathrm{HeLi},0}$ was adopted (corresponding to the dotted line in Fig. 9).

The consequences of sidefeeding from higher lying states have also been investigated by other groups with different methods (see, e.g., Refs. [24, 25, 50, 51] and references given therein). The results differ considerably in magnitude as well as in the sign of the required correction [24, 52] and, in some cases, exceed the $\pm$15% margin quoted above [25]. The treatment of the continuum part of the excitation spectrum of emitted fragments is conceptually and practically difficult [25, 50] and, furthermore, each isotope thermometer will require an individual calibration (Fig. 9 and Refs. [53, 54]). It is obvious that quantitative experimental information on the amount of sequential feeding will be needed, such as recently obtained for ^{129}Xe + Sn reactions [55], and that experimental cross comparisons with alternative thermometers are mandatory in order to possibly reduce the uncertainties of the isotopic temperature scale.

A first cross comparison of the T_{HeLi} thermometer with excited-state temperatures deduced from excited state populations of lithium, beryllium and boron isotopes for the ^{36}Ar

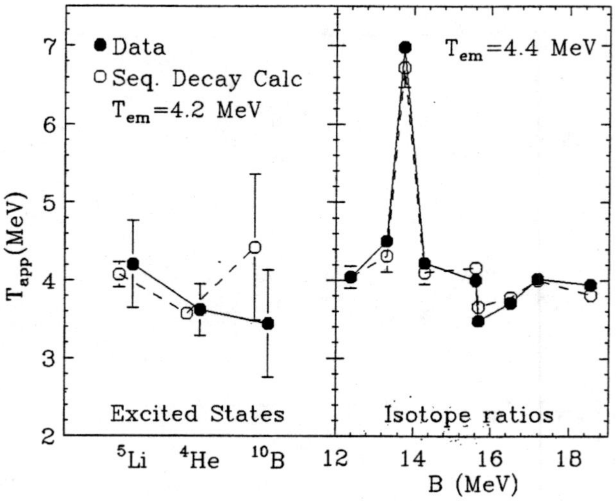

Figure 10: Apparent temperatures obtained from relative populations of excited states for ^{5}Li, ^{4}He, and ^{10}B nuclei (left panel) and from isotopic ratios (right panel, $T_{HeLi,0}$ is given by the data point at B = 13.3 MeV). The closed points are the data and the open points are the predictions of sequential decay calculations (from Ref. [56]).

+ ^{197}Au reaction at 35 MeV per nucleon gave compatible results [54]. A similar consistency was also observed for central collisions of the heavier Au + Au system at the same energy [56]. The quality of the agreement between the different thermometers investigated in the latter reaction is shown in Fig. 10. In the nomenclature used there, a distinction is made between the 'true' emission temperature at breakup T_{em} and the uncorrected apparent temperature T_{app} obtained with a particular thermometer, such as $T_{HeLi,0}$ in Eq. 6 [57]. All temperature values, deduced either from relative populations of states (left panel) or from isotopic yield ratios (right panel), are consistent with an emission temperature of 4.3 ± 0.1 MeV. The figure also demonstrates that the sequential-decay calculations used for this case reproduce the observed behaviour of the apparent temperatures rather well, including the large excursion of the apparent isotope temperature deduced from ^{3,4}He and ^{9,10}Be (ΔB = 13.8 MeV). This anomaly has been traced down to the imbalance in the number of low-lying excited states in the ^{9,10}Be pair [58]; it reflects the potentially dramatic influence of sequential decays in special cases.

A cross calibration over a wider range of excitation energies and reaction types was performed by studying central ^{197}Au + ^{197}A collision at 50 to 200 MeV per nucleon incident energy with the ALADIN spectrometer. For this purpose, the setup had been supplemented by three hodoscopes and several high-resolution telescopes positioned at angles close to mid-rapidity [59]. The angular resolution and granularity of the hodoscopes was optimized in order to permit the identification of excited particle unstable resonances in light fragments from the correlated detection of their decay products. Yields of isotopically resolved light fragments were measured with the individual telescopes which each consisted of three Si detectors of increasing thickness backed by a CsI(Tl) scintillation detector.

The obtained values for two isotope and three excited-state temperatures are given in Fig. 11. The isotope temperatures T_{HeLi} and T_{Hedt} were derived as described above, and a correction factor of 1.2 was applied in order to account for the effects of sequential feeding.

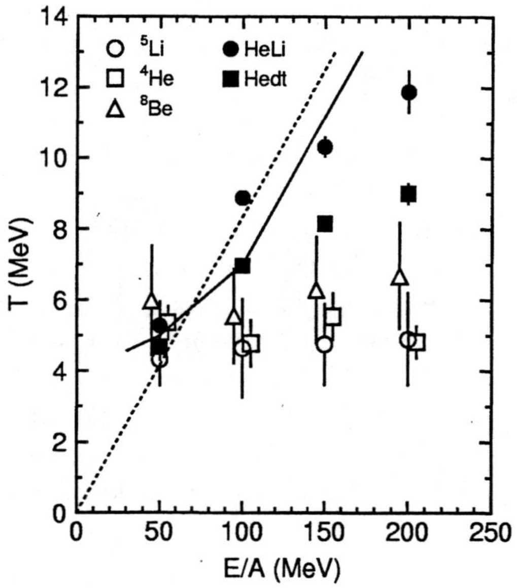

Figure 11: Measured isotope temperatures (full symbols) and excited-state temperatures (open symbols) as a function of the incident energy per nucleon. The indicated uncertainties are mainly of systematic origin. The meaning of the lines is explained in the text (data from [59]).

Excited-state temperatures were determined for ^{4}He, ^{5}Li, and ^{8}Be, fragments with two widely separated states for which modifications of the apparent emission temperatures due to feeding were expected to be small (Fig. 10 and Refs. [60, 61]).

At 50 MeV per nucleon, all temperature values coincide within the interval $T = 4$ to 6 MeV in good agreement with the result of Huang *et al.* [56] obtained at 35 MeV per nucleon (Fig. 10). With rising beam energy, however, the isotope temperatures rise linearly up to $T_{\rm HeLi} \approx 12$ MeV and $T_{\rm Hedt} \approx 9$ MeV at 200 MeV per nucleon (closed symbols). The excited-state temperatures, on the other hand, appear to be virtually independent of the bombarding energy. They scatter closely around their individual mean values of about 4.5 MeV up to 6 MeV for the three cases. The same striking divergence of the two types of thermometers has been made for central ^{86}Kr + ^{93}Nb reactions [62] and for spectator reactions in ^{197}Au + ^{197}Au at 1000 MeV per nucleon [63].

Lacking at the moment a quantitative explanation of this surprising observation, it might be instructive to recall a similar phenomenon during the cosmic big-bang. Also there different degrees of freedom freeze out at various stages of the big-bang evolution, hence signaling different temperatures. Of course, this cooling is intimately related to the existence of collective radial flow. It may, therefore, not be too surprising that the discrepancy between the two thermometers, in this case, emerges at incident energies exceeding 50 MeV per nucleon at which radial collective flow starts to represent a significant part of the available collision energy [64]. Rather crude estimates for the expected breakup temperatures if flow is taken into account are given by the lines in Fig. 11. They qualitatively indicate the tendency exhibited by the isotope temperatures. The interpretation of the excited-state temperatures should also consider that the resonances used for the temperature evaluation are very specific quantum states with widths of 1 MeV or less. They are unlikely to exist in the nuclear medium in identical forms [65, 66, 67]. In the case of volume breakup, the asymptotic states that finally will be observed can develop and survive only at very low densities that may not be experienced by the cluster before it is effectively emitted into vacuum. Excited-state temperatures may thus represent the internal fragment temperatures at the very final stages of the fragmentation process. The obtained mean value near 5 MeV corresponds well to results of dynamical calculations based on transport models [69].

Apparent temperatures deduced from purely thermal interpretations of the Maxwellian-like kinetic-energy spectra are in clear disagreement with the isotope temperatures. There are many effects that may influence the kinetic energies in these reactions, such as sequential decay, collective motion, Coulomb interaction, preequilibrium emissions and Fermi motion. For spectator decays following ^{197}Au + ^{197}Au collisions, e.g., the kinetic temperatures are of the order of 15 to 20 MeV and much larger than the breakup temperatures discussed so far. It has recently been shown, however, that taking into account the effect of the nucleonic Fermi motion within the colliding nuclei will reconcile these seemingly contradictory observations [70, 71].

6 A Caloric Curve of Nuclei

Caloric curve is the term commonly adopted for a relation between temperature T and energy E. In thermodynamics, the caloric equation of state is given by the function $E(T)$ under well defined conditions, such as fixed volume and particle number in the simplest case. Here we will speak of a caloric curve of nuclei in the looser sense in that we use our experimental means to change the energy content of a nuclear system and simultaneously measure its temperature [72]. The obvious motivation for this endeavour is provided by the hope of establishing a link to the hypothetical liquid-gas phase transition in extended nuclear matter [73].

At relativistic energies, only small fractions of the initial collision energy are converted into excitation energy of the spectator. The energy deposition there can only be reconstructed from the exit-channel configuration which, however, requires a complete knowledge of all decay products with their atomic numbers, masses, and kinetic energies and including neutrons. Since this is not easy to obtain, at the least event by event, assumptions and approximations have to be made.

A method to determine the excitation energy from the experimental data along this line was first presented by Campi *et al.* [74] and applied to the ^{197}Au + Cu data measured by the ALADIN collaboration in their first experiment [39]. The caloric curve presented in Ref. [47] was constructed from the more recent data for the ^{197}Au + ^{197}Au reaction at 600 MeV per nucleon. Here also experimental information on neutron production, collected with LAND, was used. Hydrogen isotopes were not detected and assumptions concerning the overall N/Z ratio of the spectator, the intensity ratios of protons, deuterons, and tritons, and the kinetic energies of hydrogen isotopes had to be made. The masses of the heavier fragments were not measured with sufficient precision and assumed to follow the EPAX parameterization [75]. The uncertainties resulting from the variation of these quantities within reasonable limits were included in the errors assigned to the results.

The spectator masses A_0 and the specific excitation energies E_0/A_0, obtained from this analysis, are given in Fig. 12 as functions of $Z_{\rm bound}$ and $Z_{\rm max}$, with $Z_{\rm max}$ denoting the maximum atomic number detected within an event. The mean mass $\langle A_0 \rangle$ decreases with decreasing $Z_{\rm bound}$, in good agreement with the expectations from the geometric participant-

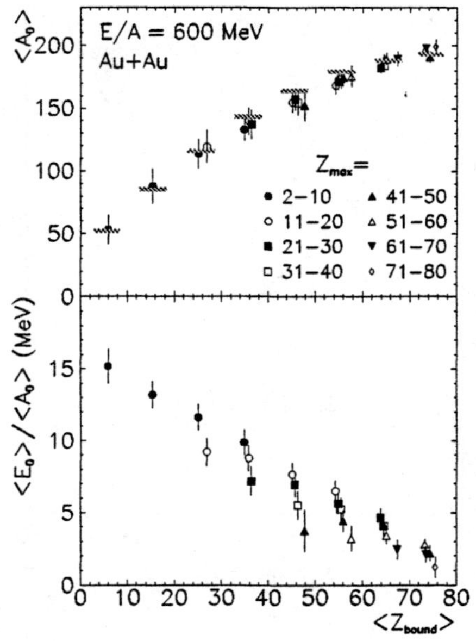

Figure 12: Reconstructed average mass $\langle A_0 \rangle$ (top) and excitation energy $\langle E_0 \rangle / \langle A_0 \rangle$ (bottom) of the decaying spectator system as functions of Z_{bound} and Z_{max}. The data symbols represent averages over 10-unit-wide bins in these two quantities as indicated. The shaded horizontal bars (top panel) represent the masses according to the participant-spectator model at the empirical impact parameter deduced from $d\sigma/dZ_{\mathrm{bound}}$ (from Ref. [47]).

spectator model [35]. Within a given bin of Z_{bound}, $\langle A_0 \rangle$ is fairly independent of Z_{max}. The smallest mean spectator mass in the bin of $Z_{\mathrm{bound}} \leq 10$ is $\langle A_0 \rangle \approx 50$. The excitation energy E_0 appears to be a function of both Z_{bound} and Z_{max}; events with smaller Z_{max}, i.e. the more complete disintegrations, correspond to the higher excitation energies. The maximum number of fragments, observed at $Z_{\mathrm{bound}} \approx 40$, is associated with initial excitation energies of $\langle E_0 \rangle / \langle A_0 \rangle \approx 8$ MeV. With decreasing Z_{bound} the deduced excitation energies reach up to $\langle E_0 \rangle / \langle A_0 \rangle \approx 16$ MeV.

The pairwise correlation of the isotope temperature, deduced as described in the last section, with the excitation energy leads to the caloric curve shown in Fig. 13. Besides the data from projectile decays following ^{197}Au + ^{197}Au collisions at 600 MeV per nucleon, results from earlier experiments with ^{197}Au targets at intermediate energies 30 to 84 MeV per nucleon (c.f. Section 2) and for compound nuclei produced in the ^{22}Ne + ^{181}Ta reaction are included. For the light-ion induced reactions at intermediate energies, the excitation energy of the target residues was obtained by subtracting the energy lost in preequilibrium emissions from the collision energy while, in the compound case, it was assumed to be equal to the collision energy.

We first like to draw attention to the fact that the whole range of reaction channels from compound evaporation to near vaporization can be covered with a single temperature observable. The required helium and lithium isotopes are still produced in sufficient quantity at these extreme ends of the range of excitation energies. We further notice the consistency of the data obtained for different types of reactions, suggesting that the smooth S-shaped curve may represent a more general property of excited nuclei. In fact, at low energies, the

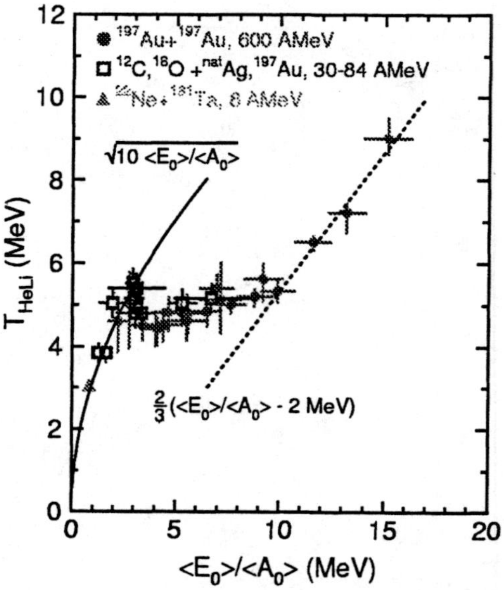

Figure 13: Caloric curve of nuclei as constituted by the temperature T_{HeLi} as a function of the excitation energy per nucleon. The lines are explained in the text (from Ref. [47]).

deduced temperatures T_{HeLi} follow the sqare-root behaviour of a Fermi-liquid, as represented by the full line calculated for a level density parameter $a = A/10\ \mathrm{MeV}^{-1}$. At high excitation energies, the rise of the temperature seems to approach a linear function with the slope 2/3 of a classical gas. In the limit of a free nucleon gas, the offset should be ≈ 8 MeV, corresponding to the mean binding energy of nuclei. A smaller offset may be caused by the finite density at freeze-out and by the finite fraction of bound clusters and fragments of intermediate mass that are present even at these high excitation energies. The offset of 2 MeV is consistent with a breakup density ρ/ρ_0 between 0.15 and 0.3 [47].

Within the range of $\langle E_0\rangle/\langle A_0\rangle$ from 3 MeV to 10 MeV, the temperature increases very little while the system passes from the 'liquid' regime, governed by the degrees of freedom of a heavy residue, to that of the 'vapor' of light fragments and particles. This transition in the dominant characteristics of the reaction channel is illustrated in Fig. 14 which shows the event distribution in the Z_{max}-versus-Z_{bound} plane. At large Z_{bound}, corresponding to small excitation energies, the ridge line starts at $Z_{\mathrm{max}} \approx Z_{\mathrm{bound}}$, i.e. with events containing only one large fragment or heavy residue. The most probable Z_{max} then drops fairly rapidly near $Z_{\mathrm{bound}} \approx 50$ and approaches very small values at the smallest Z_{bound}. The observation of $Z_{\mathrm{max}} \ll Z_{\mathrm{bound}}$ in this region of large excitation energies implies that the system has disintegrated into a larger number of smaller fragments. The energy needed for the formation of smaller constituents, correspondingly, limits the rise of the temperature.

The qualitative interpretation of the caloric curve along these lines is motivated by its reminiscence of what we expect for a first-order phase transition. It has to be refined and possibly confirmed with model studies. Statistical multifragmentation models have been used to identify the individual contributions to the total excitation energy and to quantify the origin of the second rise at $E_x/A \approx 10$ MeV [27, 77]. It was found that the internal degrees of freedom of fragments of intermediate mass are still quite important at these high excitations [27]. Transport theoretical approaches, based on antisymmetrized molecular-

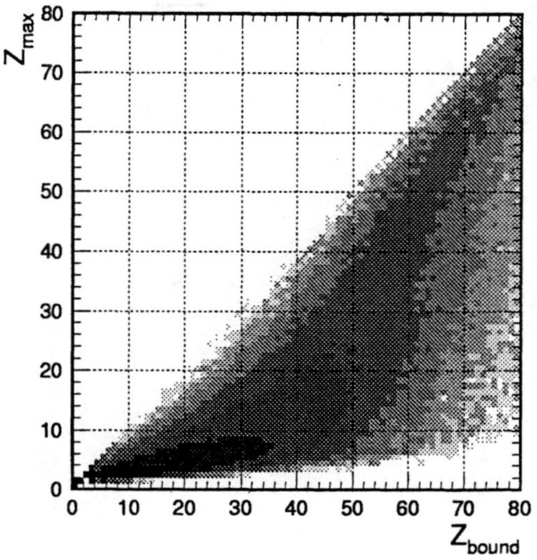

Figure 14: Z_{max}-versus-Z_{bound} distribution measured in the reaction ^{197}Au + ^{197}Au at 1000 MeV per nucleon. Conventional fission events are removed, the shadings follow a logarithmic scale (from Ref. [76]).

dynamics models, give the possibility to study nuclear systems at finite pressure [78, 79]. Caloric curves obtained in this way in model experiments include a rise at high excitations that has been shown to correspond to a nucleon gas with Van-der-Waals properties [79].

Alternative interpretations, in particular for the temperature rise at the highest energies, have been presented by several groups. In the interpretation of Natowitz *et al.* [80], the variation of the system mass (Fig. 12) is seen as the primary reason for the observed variation of the temperature and related to the mass dependence of the temperatures limiting the stability of excited nuclei [81]. In the expansion scenario modeled by Papp and Nörenberg [33], the temperatures in the plateau region are found to be consistent with a spinodal decomposition in the dynamically unstable region of the temperature-versus-density plane. The upbend at high excitation energies, in this model, indicates a concentration of the deduced breakup densities at a minimum value around $\rho/\rho_0 \approx 0.3$.

Another line of interpretation emphasizes the time dependence of the emission process. The time evolution has been studied with the expanding-emitting source model which assumes statistical emissions from an expanding and thereby cooling source [82]. According to these calculations, the less bound nuclei are more likely to be emitted at earlier reaction times, with the effect that integrated yield ratios represent a complex convolution of the temperature profile and the emission probabilities as a function of the reaction time [51]. Temperatures obtained from integrated yield ratios, consequently, have the character of weighted mean values over finite distributions. Differential analyses have been performed in order to extract cooling curves [83] or to obtain temperature values for fixed reaction times [84]. For some reactions a systematic difference of the kinetic energy spectra of the ^{3,4}He nuclei has been observed, suggesting that the two isotopes originate from different reaction stages [85, 86]. It is therefore desirable to extend the range of temperature observables to other combinations of isotopes that do not include the helium nuclei. The T_{BeLi} temperature, derived from the ^{7,9}Be and ^{6,8}Li isotope ratios appears to be a promising candidate;

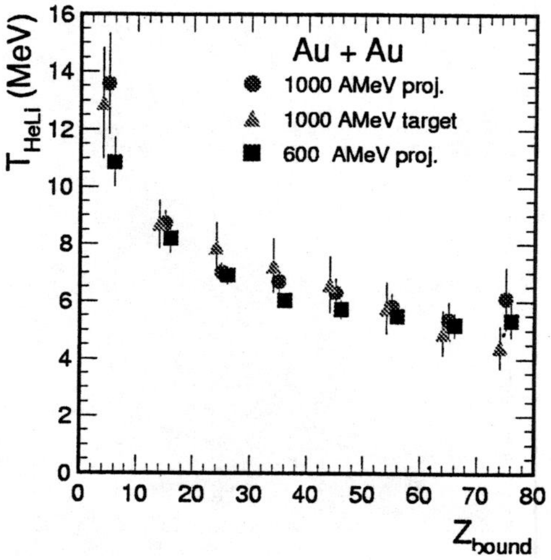

Figure 15: Temperatures T_{HeLi} for target (E/A = 1000 MeV) and projectile spectators (E/A = 600 and 1000 MeV) as a function of Z_{bound}. The data symbols represent averages integrated over Z_{max} and over 10-unit wide bins of Z_{bound} (from Ref. [76]).

the second rise at high excitation energies is also seen with this observable [87].

The discussion clearly shows that asking for a final interpretation of the caloric curve of nuclei would be premature at the present time. Furthermore, ongoing analyses and new data have also led to small modifications of the originally published curve. As a result of previously not included experimental information and corrections, the temperatures deduced for projectile fragments in the ^{197}Au + ^{197}Au reaction at 600 MeV per nucleon have increased by between 10% and 20% [37]. The new temperature values T_{HeLi} are shown in Fig. 15 as a function of Z_{bound} and, unlike Fig. 13, averaged over Z_{max}. They are compared to the results obtained at 1000 MeV per nucleon, for both the projectile and target spectators. For the target spectator, isotopic yields were measured at $\theta_{\mathrm{lab}} = 150°$ with a four-element telescope while Z_{bound} was determined for the coincident projectile decay with the ALADIN time-of-flight wall. The agreement with the temperatures obtained for the projectile decay at the same energy is expected from the symmetry of the collision and only illustrates the accuracy achieved in these experiments. The coinciding results for 600 and 1000 MeV per nucleon incident energy, on the other hand, indicate an invariance with respect to the bombarding energy that is well known from other observables (Section 4). Calculations with the statistical multifragmentation model, using the ensemble of excited spectator nuclei shown in Fig. 7, are in good agreement with these measurements [37]. The simultaneous reproduction of the observed charge partitions and of this temperature-sensitive observable represents a necessary requirement for a consistent statistical description of the spectator fragmentation.

More recent data and analyses have also shown that the excitation energies obtained from calorimetry do not exhibit a similar invariance with respect to the incident energy. It was found that the mean kinetic energies of neutrons, measured at three bombarding energies with LAND, increase considerably with the bombarding energy [88]. Together with the contribution of the protons, this leads to a difference of about 40% in the reconstructed excitation energy for the collisions at 600 and 1000 MeV per nucleon. Most likely, the experimental nucleon yields contain contributions from earlier reaction stages even if they are measured in the kinematic regimes corresponding to spectator decays. This may also provide the explanation for the excess of the experimentally determined energies over those obtained from backtracing with the statistical multifragmentation model (Figs. 7, 12, and Ref. [36]). Detailed studies of light particle emissions in these reactions, including excitation functions, will be needed in order to disentangle the various sources of light particles. At present, this uncertainty prevents the construction of a fully invariant caloric curve from spectator decays.

With the same or similar techniques caloric curves have been derived by other groups for other types of reactions [84, 89, 90, 91, 92]. The transition from the liquid branch to a plateau-like behaviour, coinciding with the onset of fragment production, seems to be a general feature of all of these curves. The second rise, however, is not seen as pronounced as for the ^{197}Au + ^{197}Au spectator reactions (Figs. 13,15). With lighter projectiles, the cross sections for reaching the high excitation energies at which this rise occurs are rather small (cf. Fig. 5). On the other hand, the vaporization of light systems into hydrogen and helium isotopes, as observed for the ^{36}Ar + ^{58}Ni reaction at 95 MeV per nucleon [93], has been described with a chemical-equilibrium model and temperatures of up to 24 MeV [25].

7 Concluding Remarks

This chapter has been meant to illustrate the usefulness of isotopic yield ratios for a statistical description of heavy-ion reactions. The main topics were the questions of how to obtain evidence for equilibration and of how to actually measure thermodynamical variables. From there a very natural path has led to phase transitions in finite nuclei, a topic of highest current interest. It is obvious that none of these subjects has been covered exhaustively, and we refer the reader to the cited literature and to the relevant articles collected in this volume.

The underlying concept of equilibrated breakup states necessarily constitutes an idealization in view of the rapid dynamic evolution of energetic reaction processes. Its limitations were evident in the discussion of multifragment emissions, most notably in the comparison of the different temperature observables. The dominant role of phase space, on the other hand, is apparent from the successful description of a variety of global characteristics of these reactions that is possible with the assumption of a statistical breakup.

The increasing availability of secondary beams for nuclear reactions will allow future studies of isospin degrees of freedom to be extended over a wider range of neutron-to-proton asymmetries. The two-fluid nature of nuclear matter promises new phenomena to be seen in experiments with asymmetric nuclei, as discussed elsewhere in this volume. Isotopic yield ratios can be expected to remain in the center of interest as important observables probing the source composition as well as its thermodynamical properties. It is also not excluded that spectator reactions at high energies will prove unique for producing globally equilibrated asymmetric systems over the desired wide range of excitation energies.

The authors would like to thank their colleagues of the ALADIN collaboration and at the GSI for support and discussions during the preparation of this manuscript.

References

[1] R. Wada *et al.*, Phys. Rev. Lett. 58, 1829 (1987)

[2] F. Deák *et al.*, Phys. Rev. C 43, 2432 (1991)

[3] L.N. Andronenko *et al.*, Z. Phys. A 350, 1 (1994)

[4] J. Mougey *et al.*, Phys. Lett. 105B, 25 (1981)

[5] D. Guerreau *et al.*, Phys. Lett. 131B, 293 (1983)

[6] R. Lucas *et al.*, Nucl. Phys. A 464, 172 (1987)

[7] For a review see, e.g., B. Borderie *et al.*, Ann. Phys. Fr. 15, 287 (1990)

[8] S.J. Yennello *et al.*, Phys. Lett. B 321, 15 (1994)

[9] H. Johnston *et al.*, Phys. Lett. B 371, 186 (1996)

[10] H. Johnston *et al.*, Phys. Rev. C 56, 1972 (1997)

[11] F. Rami *et al.*, Phys. Rev. Lett. 84, 1120 (2000)

[12] A.Z. Mekjian, Phys. Rev. C 17, 1051 (1978)

[13] S. Albergo *et al.*, Il Nuovo Cimento 89 A 1 (1985)

[14] H.W. Barz *et al.*, Phys. Lett. B 211, 10 (1988)

[15] W. Trautmann *et al.*, Nucl. Phys. A 471, 191c (1987)

[16] V.I. Bogatin *et al.*, Yad. Fiz. 19, 32 (1974) [Sov. J. Nucl. Phys. 19, 16 (1974)]

[17] O.V. Lozhkin and W. Trautmann, Phys. Rev. C 46, 1996 (1992)

[18] V. Avdeichikov *et al.*, Physica Scripta 50, 624 (1994)

[19] G.J. Kunde *et al.*, Phys. Rev. Lett. 77, 2897 (1996)

[20] H.S. Xu *et al.*, preprint MSUCL-1137 (1999)

[21] V.I. Bogatin *et al.*, Yad. Fiz. 31, 845 (1980) [Sov. J. Nucl. Phys. 31, 436 (1980)]; 34, 104 (1981) [34, 59 (1981)]

[22] D. Hahn and H. Stöcker, Nucl. Phys. A 476, 718 (1988)

[23] J. Konopka *et al.*, Phys. Rev. C 50, 2085 (1994)

[24] Z. Majka *et al.*, Phys. Rev. C 55, 2991 (1997)

[25] F. Gulminelli and D. Durand, Nucl. Phys. A 615, 117 (1997)

[26] D.H.E. Gross, Rep. Prog. Phys. 53, 605 (1990)

[27] J.P. Bondorf *et al.*, Phys. Rep. 257, 133 (1995)

[28] S. Fritz *et al.*, Phys. Lett. B 461, 315 (1999)

[29] W.-C. Hsi *et al.*, Phys. Rev. Lett. 79, 817 (1997)

[30] R. Trockel *et al.*, Phys. Rev. C 38, 576 (1988)

[31] C. Kuhn *et al.*, Phys. Rev. C 48, 1232 (1993)

[32] G.F. Bertsch, Nucl. Phys. A 400, 221c (1983)

[33] G. Papp and W. Nörenberg, APH Heavy Ion Physics 1 (1995) 241

[34] For a introductory survey see, e.g., W. Trautmann, *in:* "Correlations and Clustering Phenomena in Subatomic Physics," edited by M.N. Harakeh *et al.*, Plenum Press, New York (1997)

[35] J. Gosset *et al.*, Phys. Rev. C 16, 629 (1977)

[36] A. Schüttauf *et al.*, Nucl. Phys. A 607, 457 (1996)

[37] Hongfei Xi *et al.*, Z. Phys. A 359, 397 (1997)

[38] C.A. Ogilvie *et al.*, Phys. Rev. Lett. 67, 1214 (1991)

[39] P. Kreutz *et al.*, Nucl. Phys. A 556, 672 (1993)

[40] J. Konopka *et al.*, Prog. Part. Nucl. Phys. 30, 301 (1993)

[41] A.S. Botvina *et al.*, Nucl. Phys. A 584, 737 (1995)

[42] P. Désesquelles *et al.*, Nucl. Phys. A 604, 183 (1996)

[43] D.J. Morrissey *et al.*, Ann. Rev. Nucl. Part. Science 44, 65 (1994)

[44] F. Hoyle, The synthesis of the elements from hydrogen, Monthly Notices of the Royal Astronom. Soc. 106, 343 (1946)

[45] M.B. Tsang *et al.*, Phys. Rev. Lett. 78, 3836 (1997)

[46] T. Möhlenkamp *et al.*, *Proceedings of the XXXIII International Winter Meeting on Nuclear Physics*, Bormio, 1995, edited by I. Iori (Ricerca Scientifica ed Educazione Permanente, Milano, 1995), p. 383

[47] J. Pochodzalla *et al.*, Phys. Rev. Lett. 75, 1040 (1995)

[48] R.J. Charity *et al.*, Nucl. Phys. A 483, 371 (1988)

[49] D.H.E. Gross *et al.*, Phys. Rev. Lett. 56, 1544 (1986)

[50] Hongfei Xi *et al.*, Phys. Rev. C 54, R2163 (1996)

[51] V.E. Viola *et al.*, Phys. Rev. C 59, 2660 (1999)

[52] H.F. Xi *et al.*, Phys. Lett. B 431, 8 (1998)

[53] A. Kolomiets *et al.*, Phys. Rev. C 54, R472 (1996)

[54] M.B. Tsang *et al.*, Phys. Rev. C 53, R1057 (1996)

[55] N. Marie *et al.*, Phys. Rev. C 58, 256 (1998)

[56] M.J. Huang *et al.*, Phys. Rev. Lett. 78, 1648 (1997)

[57] The term emission temperature is sometimes exclusively used for temperatures extracted from relative populations of excited states, see e.g. Ref. [72]

[58] J. Bondorf *et al.*, Phys. Rev. C 58, R27 (1998)

[59] V. Serfling *et al.*, Phys. Rev. Lett. 80, 3928 (1998)

[60] J. Pochodzalla *et al.*, Phys. Rev. C 35, 1695 (1987)

[61] C. Schwarz *et al.*, Phys. Rev. C 48, 676 (1993)

[62] H.F. Xi *et al.*, Phys. Rev. C 58, R2636 (1998)

[63] C. Groß, PhD thesis, Universität Frankfurt, 1998

[64] W. Reisdorf and H.G. Ritter, Annu. Rev. Nucl. Part. Science 47, 663 (1997)

[65] M. Schmidt *et al.*, Ann. Phys. (N.Y.) 202, 57 (1990)

[66] T. Alm *et al.*, Phys. Rev. C 53, 2181 (1992)

[67] P. Danielewicz and Q. Pan, Phys. Rev. C 46, 2002 (1992)

[68] P.B. Gossiaux *et al.*, Nucl. Phys. **A619**, 379 (1997)

[69] C. Fuchs *et al.*, Nucl. Phys. A 626, 987 (1997)

[70] T. Gaitanos *et al.*, Phys. Lett. B 478, 79 (2000)

[71] T. Odeh *et al.*, Phys. Rev. Lett. 84, 4557 (2000)

[72] For a more comprehensive status report, see J. Pochodzalla, Prog. Part. Nucl. Phys. 39, 443 (1997)

[73] For a recent review of the field see, e.g., Proceedings of the International Workshop XXVII on Gross Properties of Nuclei and Nuclear Excitations, Hirschegg, Austria, edited by H. Feldmeier *et al.* (GSI, Darmstadt, 1999)

[74] X. Campi *et al.*, Phys. Rev. C 50, R2680 (1994)

[75] K. Sümmerer *et al.*, Phys. Rev. C 42, 2546 (1990)

[76] T. Odeh, PhD thesis, Universität Frankfurt, 1999

[77] Al.H. Raduta and Ad.R. Raduta, Nucl. Phys. A 647, 12 (1999)

[78] J. Schnack and H. Feldmeier, Phys. Lett. B 409, 6 (1997)

[79] Y. Sugawa and H. Horiuchi, Phys. Rev. C 60, 064607 (1999)

[80] J.B. Natowitz *et al.*, Phys. Rev. C 52, R2322 (1995)

[81] P. Bonche *et al.*, Nucl. Phys. A 436, 265 (1986)

[82] W.A. Friedman *et al.*, Phys. Rev. C 42, 667 (1990)

[83] H.F. Xi *et al.*, Phys. Rev. C 57, 462 (1998)

[84] J. Cibor *et al.*, Phys. Lett. B 473, 29 (2000)

[85] R. Bougault *et al.*, in Ref. [73], p. 24

[86] W. Neubert and A.S. Botvina, Eur. Phys. J. A, in print (2000)

[87] W.F.J. Müller *et al.*, in Ref. [73], p. 200

[88] C. Groß, PhD thesis, Universität Frankfurt, 1999

[89] Y.-G. Ma *et al.*, Phys. Lett. B 390, 41 (1997)

[90] J.A. Hauger *et al.*, Phys. Rev. C 57, 764 (1998)

[91] K. Kwiatkowski *et al.*, Phys. Lett. B 423, 21 (1998)

[92] M. D'Agostino *et al.*, Nucl. Phys. A 650, 329 (1999)

[93] B. Borderie *et al.*, Eur. Phys. J. A 6, 197 (1999)

Index

F

G

H

I

L

M

N

O

P

R

S

T